ACCE '88 PLANNING COMMITTEE

Dr. Kaneyoshi Ashida
University of Detroit

Dale A. Bityk
ICI Americas

Daryl R. Brace
Ashland Chemical Company

Shu-Sing Chang
National Bureau of Standards

Dr. Douglas L. Denton
Battelle

Claude A. Di Natale
General Motors Corporation

Dr. Subi Dinda
Chrysler Motors

Dr. Lawrence T. Drzal
Michigan State University

James E. Eaton
Donnelly Corporation

Dr. Robert W. Ellis Jr.
Lawrence Institute of Technology

Joseph M. Gaynor
Consultant

J. Michael Grubb
Ashland Chemical Company

William E. Haskell III
Army-Materials Technology Laboratory

Dr. Donald L. Hunston
National Bureau of Standards

Richard A. Jeryan
Ford Motor Company

Carl F. Johnson
Ford Motor Company

Thomas A. Kosakowski
Mobay Corporation

Edward J. Lesniak Jr.
Chrysler Motors

Michael E. Liedtke
General Motors Corporation

Carl H. Luther
General Dynamics

Gullmar V. Nelson
Davidson Technology

Irvin E. Poston
General Motors Corporation

Stephen C. Schwab
Sterling Engineered Products

Jerry V. Scrivo
Micromatic TEXTRON

Sharad B. Shah
University of Toledo

Roy H. Sjoberg
Chrysler Motors

Lester E. Smith
National Bureau of Standards

Dr. Richard D. Streeper
3M Corporation

William H. Todd
General Motors Corporation

PREFACE

This volume is the culmination of a year-long effort by the 1988 ASM/ESD Advanced Composites Conference and Exposition (ACCE) Planning Committee to bring together an advanced composites conference of international scope, focussing primarily on automotive applications. The committee and staff members at ESD and ASM have worked diligently in this endeavor, resulting in an expanded program for the current year.

The major theme for ACCE '88 is **automotive energy management** through use of composites. This subject includes papers covering energy absorption during material distortion and collapse, an important concern in the design of automotive structures.

Another focus area for this conference is the growing interest in use of composites for US Army applications such as ground combat vehicle hull and turret structures, as covered in the **military applications** session. Other composite design topics are covered in a more general **design** session.

Liquid molding, encompassing resin transfer molding (RTM) and structural reaction injection molding (SRIM), is an important manufacturing development area for automotive and other composite structures. A related development area is cost-effective fiber **preform** technology. Use of these two forming technologies, one for the fiber and the other for the final part, could result in significantly reduced tooling and part costs for structural composite vehicle components, compared to other composite forming methods, and formed metal parts.

There is also the question of selection of a thermoset or thermoplastic resin matrix, with many new polymer and polymer prepreg types available. Several newer thermoplastic composites have excellent elevated temperature and solvent resistance and can be final formed at attractive rates compared to thermoset composite curing times. The generally higher materials and tooling costs for thermoplastic composites may, however, be a detriment to their use in ground vehicle applications even though they are more readily recycleable. These topics, and others covering metal matrix, hybrid resin and carbon fiber composite application are included in sessions titled **emerging materials.**

In view of the growing interest in **recycling,** this conference has two sessions to address how many different composite and plastic materials might be reformed into a second product.

Other important characteristics of composite materials are included in conference sessions titled: **reliable property data, flammability and smoke** concerns, **thermoset composite cure** monitoring methods, **fastening and joining** (two sessions), **NDT and mechanical testing, manufacturing** equipment and planning and **economics** of composite fabrication.

This conference, and others like it, bring together key technologists, managers and planners in industry, academia and government who share ideas and developments. This interchange of technical, economic and marketing information is instrumental to the continued expansion of composites into a growing number of applications.

I would like to personally thank the members of the planning committee and staff members of ASM and ESD for their enthusiastic support and many hours committed to the preparation and execution of this 4th Annual Conference. I would also like to acknowledge FMC-Defense Systems Group, Santa Clara, CA for their support of my involvement in this Conference.

James E. Hill, Ph.D.
Executive Chairman

TABLE OF CONTENTS

ECONOMICS OF VARIOUS PROCESSES

ENERGY MANAGEMENT

COMPOSITE CURE DEVELOPMENT

MANUFACTURING

MILITARY

FLAMMABILITY AND SMOKE

RELIABLE PROPERTY DATA

FASTENING AND JOINING

LIQUID MOLDING

NDT & MECHANICAL TESTING

PREFORM DEVELOPMENT

RECYCLING

SELECTING CAPITAL EQUIPMENT – MYTHS, PITFALLS AND OPPORTUNITIES

Jerry V. Scrivo
Micromatic TEXTRON
Holland, Michigan   USA

SELECTING CAPITAL EQUIPMENT –
MYTHS, PITFALLS AND OPPORTUNITIES

ABSTRACT

When manufacturing technologies for production of composites are discussed, low initial cost is frequently mentioned as a primary goal. Often low mold pressure and/or moderate injection rates are projected to translate into low capital equipment cost. Sometimes prototypes have been produced using one process or another at surprisingly low capital cost, and this is projected as indicating a "low cost" process. This paper looks at the problem of selecting capital equipment, the myths, pitfalls and opportunities. What increases acquisition cost, what reduces operational cost, and how to adequately evaluate capital equipment for volume production.

INTRODUCTION

Accounting textbooks are full of methods for evaluating the desirability of capital equipment projects. In general, the following parameters are analyzed.(1)

1. Size of investment
2. Riskiness of investment
3. Size of benefits
4. Timing of benefits
5. Riskiness of benefits

(1) Accountants Cost Handbook, James Bulloch, Donald E. Kelly, Louis Valasho, John Wiley & Sons, New York 1983. Page 22-16.

Evaluation seeks to address three major concerns:(2)

1. Security - How safe?
   How soon will positive cash flow result?

2. Recompense - How profitable?
3. Predictability - How sure are the returns?

Table #1 shows a number of procedures which are presently used by financial departments to make these evaluations. Unfortunately, their is no one "right" method, and in fact different methods often produce different choices based on the same imput data.

In recent years, return on investment where (ROI) = Net Income/Average Investment has been increasingly used as a measurement of performance over time. However, for capital intensive businesses, ROI can be extremely volume sensitive.(3) It is also deficient for evaluation of capital equipment in that it doesn't account for the time value of the returns.(4) This short- coming is overcome by using either the Internal Rate of Return (IRR) or the Net

(2) Restoring our Competitive Edge, Robert H. Hayes, Steven C. Wheelwright, John Wiley & Sons, New York 1984. Page 138

(3) Using Locigal Techniques for making better decisions, Double N. Dickson, Editor John Wiley & Sons, New York, 1983. Page 290.

(4) Accountants Cost Handbook, James Bulloch, Donald E. Keller, Louis Valasho, John Wiley & Sons, New York 1983. Page 22-16.

Present Value (NPV) methods. There are two basic differences between the (IRR) and (NPV) methods that lead to differences in ranking mutually exclusive investments.(5)

1. The internal rate of return (IRR) assumes all cash flows are reinvested at the internal rate of return while Net Present Value (NPV) assumes they are reinvested at the discount rate.

2. The Net Present Value Index gives explicit consideration to investment sizes. The (IRR) method does not.

Because long term commitments of major capital expenditures effect the nature and flexability of a company; careful consideration must be given to the risks of uncertainty associated with the investment.

These include volume estimates, economic forecast, productivity projections and inflation estimates. Sensitivity analysis, decision trees and various simulations are frequently utilized to consider these factors. (6)

MYTHS

Existing methods of evaluating capital expenditures when used by knowledgeable financial personnel work reasonably well. However, the emphasis on return on investment has had an unintended result in the non-financial portion of the business community. In a simplistic effort to maximumize return on investment, we often erroneously seek an illusive "low cost" manufacturing process. This myth is based on the following view: First we look at the benefits of acquiring proposed equipment as a fixed entity. Thus,if we make the investment, a fixed and predictable stream of benefits will accrue over a known period of time. From the equation:

ROI = S/P

If the value of S is a constant, we can maximize ROI by minimizing our investment in the proposed equipment.

NOTHING COULD BE FURTHER FROM THE TRUTH!

(5) Cost Accounting - Processing, Evaluating and Using Cost Data, Wayne J. Morse, Addison - Wesley Publishing Co., Reading, 1978. Page 477.

(6) Cost Accounting - Processing, Evaluating and Using Cost Data, Wayne J. Morse, Addison - Wesley Publishing Co., Reading, 1978. Page 477.

First, S the annual net return is not known, and can at best only be estimated. The value is strongly influenced by factors totally beyond our control such as overall economic conditions. For example, the presence or absence of inflation alone can cause wild swings in ROI performance for capital intensive industries. Secondly, the value of S is strongly controlled by many business and process factors which are within our control and it is in this area that this paper will be focused.

PITFALLS

Recognizing hidden cost:

The total cost of capital equipment includes much more than the price paid to the supplier. It also includes all other costs associated with the purchased machinery. When too much emphasis is place on initial purchase price, a classical example of short term sub-optimization occurs.(7)

The so-called "hidden costs" that should be taken into account in choosing a supplier are: (8)

1. Geographic Location - Time and cost factor for products and services
2. Transportation Alternatives - Time and cost for transportation
3. Inventory Control - Turnaround time and costs for delivery delays
4. Quality - Time and cost associated with quality problems
5. Productive Capacity - Ability to meet demand economically
6. Responsiveness - Ability to meet customer's demands
7. Technology - Cost associated with early obsolescense of equipment

The existence of these "hidden cost" factors, and others not captured in quoted prices, may be widely known, but too few companies rigorously account for them.(9) Spending time to evaluate suppliers, and

(7) Hahn, Chan K., Kina, K.H. and Kim, J.S., "Cost of Competition: Implications for Purchasing Strategy, "Journal of Purchasing and Materials Management, Fall 1986, Page 6.

(8) "Contracting for Machining and Tooling: The Hidden Costs of Sourcing Abroad", A report for The National Tooling and Machining Association, Quick, Finan and Associates, September 1987, Page 9

(9) Willets, Walter E., "Supplier Evaluation, SPQR: The Total Way to Size Up Suppliers," Purchasing, February 20, 1969.

their products to determine total costs can result in major improvements in the economic benefits received from a capital investment. Remember, in the long run, a company's competitive posture will be strongly influenced - for better or worse - by its capital investment decisions.(10) The machinery you buy will save or cost the company thousands of production and maintenance dollars over its lifetime.(11)

LOOK AT THE TOTAL COST.- during its lifetime a machine costs much more than the initial purchase price.

Initial Cost + Life Expectancy + Downtime Cost + Operating Cost = Total Cost

OPPORTUNITY

Focus on the Process

Dr. W. Edward Deming has taught us much about quality control.(12) In the process, he has also taught us about process variation and predictability. He has shown that high quality equals LOW manufacturing cost and that operational results are inherent in the variability of the process. Trying to change results without changing the process is futile. The only thing that can change the output of a common-cause system is to change the process itself. His concept of continuous improvement states that no matter how good a process is, it can always get better. These concepts are fundamental to improving productivity! The purchase of new capital equipment is, therefore, a two edge word:

First it gives an unmatched opportunity to improve the process thereby enhancing both productivity and profitibility.

Secondly, once purchased, the new equipment will lock the buyer into a level of process capability that will control productivity and profitability over the machinery's useful lifetime.

(10) Mayer, G., "Making Capital Investments Simple and Sound,"Manufacturing Systems, December 1987. Page 36.

(11) Miles, Michael, "How to Evaluate the True Cost of a Valve" Engineer's Digest, November 1987, Page 16.

(12) Propst, Annabeth L., "The Deming Philosophy," Machine and Tool Blue Book, December 1987, Page 38-42.

Since high volume process equipment can have a lifetime of 10 to 40 years, selection errors have grievous consequences! This is the opportunity and challenge inherent in selecting capital equipment.

Subjective issues in supplier selection (13)

1. Flexibility - Are they known for supplying only one type of equipment or do they have a broad base of expertise and products?

2. Knowledge - Does the company have application knowledge and engineering skill to properly apply their products to your process needs. Will they take time to evaluate your needs and supply an optimum process?

3. Delivery - Will they give dependable delivery? Can production committments be made with assurance that scheduled delivery dates will be met?

4. Training - Does the manufacturer invest in training? Can you rely on the supplier for start up assistance, training and adequate documentation to optimize your start up activities?

5. Quality - Does the supplier have proper inspection equipment, materials traceability, materials certification, and a reputation for high quality standards?

6. Research and Development - Does the supplier have an active research and engineering department? Is he committed to continuous product upgrades? Are new process upgrades routinely field retrofittable? Is he selling obsolete process technology at bargain prices?

7. Service - Does the supplier have a field service department? Are they busy? Are field service personnel qualified? Can supplier be relyed on to support and maintain your production capability in the event of operational problems? How likely are such problems?

8. Safety - Does the proposed equipment pose a safety hazzard in your plant? Have such hazzards been considered in its design and manufacture? Does the supplier have product liability suites pending on this equipment? Does he carry product liability insurance? If a liability problem arises, can the supplier be relied on to assist in the defense?

(13) Miles, Michael, "How to Evaluate the True Cost of a Valve" Engineer's Diest, Nov. 1987, Page 17.

9. References - Who are the manufacturer's customers? Talk to them about the company's products, service and personnel!

Initial Cost vs Total Cost

If total cost is the best criteria; why are most capital equipment selections based on initial cost? Can't initial cost be used as an indication of expected total cost to simplify selection? Two good questions which must be considered!

First, from the above listing of supplier selection criteria, the difficulty of supplier evaluation can be seen. Also, since these criteria are subjective, they present obstacles to evaluation and ranking. However, equating initial cost with anticipated total cost is fatally flawed. To illustrate, Table II compares factors which increase initial cost with those that reduce operational costs.

It is easy to see that in many cases low initial cost is exactly opposite to lower operating cost. For example, consider the trade off between the initial cost for increased accuracy and repeatibility compared to the benefit of reduced scrap.

If a molder is to produce 250,000 bumper beams in a high volume facility on an 83 second cycle (dual cavity tools), running three shifts, five days a week; he theoretically needs 2882 hours of production time. (Table III) This is increased by the downtime for preventive maintenance, repair, change over, etc. This can be expressed as manufacturing efficiency where:

Mfg Eff - (Theoritical Prod. Time/Actual Prod. Time) X 100

Therefore, in our example:

Mfg Eff	Act. Prod. Time (Hr)
80	3603
85	3391
90	3202
95	3034
100	2882

If the cost are as follows:

Raw Material	$1.29/lb
Machine Operating Cost	$121.45/hr

Then we can calculate operating cost and theoritical profitability verses manufacturing efficiency, injection accuracy/repeatability and scrap rate. (See figures 1, 2 & 3)

These comparisons show that a 5% loss in manufacturing efficiency is equal to a .5% loss in theoritical profitability. Similarily, .5% decrease in injection accuracy is equal to a .5% loss in theoritical profitability. Finally, a 1.0% increase in scrap rate, over this production run, is equal to a 1.0% loss in theoritical profitability.

The value of equipment flexability can also be demonstrated by extending this example:

The 11.8 lb part has a production life of 3 years. The initial cost of the molding machinery plus installation was $550,000.00 and the process equipment has a potential useful life of 20 years. After 3 years a new product must be produced which weighs 15 lbs, requires 10% more platen area, 10% more mold weight capacity, 20% more clamp stroke and a different material chemistry. New capital equipment to mold this product now costs $652,000.00 If purchased three years ago it would have cost $580,000.00. (Present cost minus inflation) Installation adds $30,000.00 to the total new equipment cost. The old equipment can be sold for $400,000.00. (residual value) The production downtime to acquire and install new equipment is six months plus a two month start up on the new equipment at an average efficiency of 40%. Thus, change over cost is:

(New equipment cost - residual value) + (Downtime) + (Reduced efficiency) = Changed Over Cost

If all manufacturing cost parameters remain the same except for the equipment conversion, then the cost of not having a flexible manufacturing facility would have been $530,000.00. Thus additional cost could have been incurred for flexibility at the first purchase without adversely effecting profitability during this period.

How to Evaluate Capital Equipment (14)

First, develop a list of the cost factors that are critical including hidden cost factors. Second, assign each a subjective weight representing its importance in the selection decision. Third, each supplier is "scored" for each factor -- with the scores often subjective, but based where possible on previous data.

(14) Quick, Finan & Associates, "Contracting for Machining and Tooling: The Hidden Costs of Sourcing Abroad" a report for The National Tooling and Machining Association, Sept. 1987, Page 13-36.

Finally, develop a composite weighted score for each supplier to reflect a total cost for the process equipment. Although this procedure is time consuming and costly, the potential savings justify the effort.

Proper evaluation requires participation of various members of the organization from production, engineering, quality control and even marketing. It is usually necessary to visit each supplier's plant. Finally, to ascertain a suppliers reputation, it is recommended that trade associations, knowledgeable experts and his customers be contacted.

A secondary benefit of this procedure is that it documents the decision process producing an audit trail which can be reviewed later to reveal the logic on which the decision was based. This produces a high degree of confidence that a sound and intelligent decision has been made.

When a company sources abroad it must take under consideration additional hidden cost:

1. Customs Duties
2. Financing costs and paperwork
3. Foreign exchange rate fluctuations
4. Travel cost
5. Contracting Terms
6. Delays
7. Modifications
8. Communications

The most frequently cited communications problem is the differences between U.S. and foreign standards and differences in meaning for technical terms. Differences in completion at time of delivery can involve unanticipated cost and delays. Problems with distance, language and interpretations extend negotiating time and increase the risk of delays. Liaison costs are also likely to increase due to these problems.

A true cost worksheet has been developed to provide a frame work for identifying and organizing the most common hidden cost. (15) It is based on a cost-ratio method of supplier evaluation. If precise cost-ratio estimates cannot be pinpointed, cost ranges can be used. The key is a systematic approach which recognizes hidden cost.

(15) Quick, Finan & Associates, "Contracting for Machining and Tooling: The Hidden Costs of Sourcing Abroad" a report for the National Tooling and Machining Association, Sept. 1987, Page 29.

The worksheet contains two parts. Part A can be used to evaluate any supplier. Part B facilitates estimates of additional costs for a foreign supplier.

The left hand column of Part A contains factors important to total cost. These may have to be refined for specific jobs in particular companies.

The evaluator uses available information and knowledge of the supplier to assign an expected cost associated with his potential performance for each factor.

The expected cost is the cost that would be incurred if a particular problem occurs, multiplied by the probability that it will occur. In some cases, a supplier's performance could be superior. Then the expected cost should be negative -- that is, it should act to reduce true cost relative to the quoted price. These are expressed as a percent of the quoted price. The cost-ratios are summed and the total plus 100% is used to multiply the quoted price giving the true cost. Alternately, the expected cost can be expressed in dollars. Then the true cost will equal the sum of the quoted price plus the hidden cost in dollars.

Part B is a specialized version of the worksheet which includes estimates of additional cost associated with a foreign supplier; expressed as a percent of the U.S. supplier's quoted price. (16) These were developed by Quick, Finan and Assocites (1987) based on interviews with NTMA, their suppliers and others.(17) According to their bench marks, customs duties when not included in foreign quotes can offset 4 to 6 percent of any price advantage, while packaging, freight, insurance, financing and exchange rate can account for another 5 to 15 percent. Similarily, cost to bring foreign machines into compliance with U.S. standards can offset anywhere from 5 to 35 percent of the price difference.

While these ranges do not apply in every case, and can not be added to express a total hidden cost offset, they do reflect experiences which reveal sizable cost

(16) Quick, Finan & Associates, "Contracting for Machining and Tooling: The Hidden Costs of Sourcing Abroad" a report for the National Tooling and Machining Association, September 1987, Page 37.

(17) Quick, Finan & Associates, "Contracting for Machining and Tooling: The Hidden Costs of Sourcing Abroad" a report for The National Tooling and Machining Association, September 1987, Page 33.

offsets to foreign sources prices. (18)

Summary

Myths - The key to maximizing ROI is in minimizing initial cost. Factors which increase initial cost frequently are required to minimize operating cost, and effect profitibility more strongly than capital costs.

Pitfalls - Hidden costs make or break capital investment results. Supplier evaluation is critical.

- Long term business viability is heavily dependent on good capital investment decisions.

Opportunity - Focus on the process because operating results are inherently tied to process variability.

- Purchase of new capital equipment gives an unmatched opportunity to improve productivity and profitibility.

- Once purchased, new equipment tends to lock a company into a level of productivity and profitibility for its lifetime.

The 19th Century advice of John Ruskin is still valid:

"It's unwise to pay too much, but it's unwise to pay too little too. When you pay too much, you lose a little money....that is all. When you pay too little, you sometimes lose everything because the thing you bought was incapable of doing the thing it was bought to do. The common law of business balance prohibits paying a little and getting alot....it can't be done. If you deal with the lowest bidder, it is well to add something for the risk you run, and if you do that, you will have enough to pay

(18) Quick, Finan & Associates, "Contracting for Machining and Tooling: The hidden costs of Sourcing Abroad" A report for The National Tooling and Machining Association, September 1987, Page 34.

for something better." (19)

(19) Quick, Finan & Associates, "Contracting for Machining and Tooling: The Hidden Costs of Sourcing Abroad" a report for The National Tooling and Machining Association, September 1987, Page 13-36

BIBLIOGRAPHY

1. Accountants Cost Handbook, James Bulloch, Donald E. Keller, Louis Valasho, John Wiley & Sons, New York 1983

2. Restoring our Competitive Edge, Robert H. Hayes, Steven C. Wheelwright, John Wiley & Sons, New York 1984

3. Using Logical Techniques for Making Better Decisions, Douglas N. Dickson, Editor, John Wiley & Sons, New York, 1983

4. Cost Accounting - Processing, Evaluating and Using Cost Data, Wayne J. Morse, Addison-Wesley Publishing Company, Reading, 1978.

5. Hahn, Chan K., Kina, K.H. and Kim, J.S., "Cost of Competition: Implications for purchasing strategy, "Journal of Purchasing and Materials Management, Fall 1986.

6. "Contracting for Machining and Tooling: The Hidden Costs of Sourcing Abroad", A report for the National Tooling and Machining Association, Quick, Finan and Associates, September 1987.

7. Willets, Walter E., "Supplier Evaluation, SPQR: The Total Way to Size up Suppliers," Purchasing, February 20, 1969.

8. Mayer, G., "Making Capital Investments Simple and Sound", Manufacturing Systems, December 1987.

9. Miles, Michael, "How to evaluate the True Cost of a Valve" Engineer's Digest, November 1987.

10. Propst, Annabeth L., "The Deming Philosophy", Machine and Tool Blue Book, December 1987.

11. Miles, Michael, "How to evaluate the true cost of a Valve" Engineer's Digest, November 1987.

12. Ellerbe, Gilbert, "Economic Evaluation of a new structural composite", SPE RETEC '87, Paper 21-5.

TABLE I
COMMONLY USED EVALUATION METHODS

METHOD	FORMULA	TYPE	ADVANTAGE	DISADVANTAGE
Incremental Net Income	Net cash flows by year	Accounting Profitability	Shows Inpact on oper. profits	Ignores time value of Cash flows
Accounting Return Investment (ROI)	$= S/P$ S=Annual Net Return P=Initial Investment	Accounting Profitability	Straight forward calculations- gives a comparative % value.	Ignores time value of cash flow
Residual Income	$= S-(Pxr)$ S=Annual net return P=Avg. Initial Investment r=Cost of capital	Accounting Profitability	Positive value indicates an economic profit	Ignores time value of cash flows
Internal rate of Return (IRR)	$=r(S_1 + S_2 .. Sn)=P$ r=Discount Rate S=Annual Return P=Initial Inv. Present value - Inv.	Time value analysis	Compares to rate of interest on cash in banks	Difficult to calculate on non uniform income
Net Present Value (NPV)	$=4(S_1 + S_2 .... Sn)-P$	Time value analysis	Calculation is straight forward	Requests an established rate of return.
Present Value Index (PVI)	$=\dfrac{\text{Net Present Value NPV}}{\text{Initial Investment P}}$	Time value analysis	Useful for ranking various size investments.	
Payback Period	$=P/S$ P=Initial Investment S=Annual net return	Risk analysis	Straightforward concept and calculation	Ignores time value of cash flows
Discounted Payback Period	$=P/Sd$ P=Initial Investment Sd=Discounted annual net returns	Risk analysis	Accounts for time value of cash flow	Doesn't measure economic return
Maximum Exposure	=Sum of cumulative neg.	Risk analysis	Highlights consequences of aborting project	
Affordable risk	$=\dfrac{C}{2p}$ C=Investment Cost P=Gross Profit	Risk analysis	Balances risk with cost of capacity	Static Model - Valid over a discrete period of time
Sensitivity Analysis	-------------	Risk analysis	Ascertains impact of changes	More involved calculations - results must be summarized
Decision Tree	-------------	Risk analysis	Greater analitical capability	Cost and time to construct model
Simultion	-------------	Risk analysis	Greater analitical capability	Cost and time to construct model

TABLE II
INITIAL COST vs OPERATIONAL COST

INCREASED INITIAL COST	REDUCED OPERATIONAL COST
1. Faster Cycles/Reduced reset time	1. Faster Cycles
2. Centralized Controls Automatic control Functions Easy material change over	2. East Set Up
3. Core junctions Ejection Quick mold change Robotics	3. Reduced manual operations
4. Safety guarding Safety interlocks	4. Safe Operation
5. Automatic Lubrication On Board Diagnostics	5. Reduced Maintenance
6. Larger or variable dimensions -Platen size -Open daylight -Stroke -Clamp Tonnage -Platen Booking or shuttle -Greater mold weight capacity -Shot size -Injection Rate -Tank Size -Additional material streams -Ability to handle fillers/reinforcements/nucleation -Ability to handle multiple mix heads	6. Flexibility to handle any mold
7. Greater accuracy and repeatibility -Better parallelism -Lower deflections -Better Positional accuracy -More controlled parameters -Greater Control precision -On board SPC capability -Communication capability for CIM	7. Reduced scrap and rework

TABLE III

EFFICIENCY CALCULATIONS (13a)

Dow Chemical Company Economic Model

Part: SRIM bumper beam

Construction: SRIM with internal mold release
Preform - Spray Up

Volume: 250,000 per/year

Tooling: Two cavity/Single lid

Cycle Time: 83 seconds (41.5 sec/cavity)

Direct Cost:

Operating Cost @ $121.45/hr		$ 1.40
Capital	.27	
Tooling	.74	
Energy	.14	
Labor	.75	
Materials @ $ 1.29/lb		15.18
Resin	$ 8.25	
Preform	$ 6.93	
Total		$ 16.58

Production Time Required:	2882 hours
	120 days
	24 weeks
Molding Machinery Cost:	$550,000.00
Monthly Equipment Cost:	$ 11,250.00
Cost per piece (Theoritical)	$ .27

(13a) Ellerbe, Gilbert, "Economic evoluation of a structural composite," SPE
RETE 87 Paper 215-5 Page 352.

TRUE COST WORKSHEET
PART A
HIDDEN COST RATIOS
(Percent of Quoted Purchase Price)

Hidden Costs	Supplier 1	Supplier 2	Supplier 3
Freight and packaging costs			
Insurance costs			
Financing terms			
Travel and communications costs for up-front consultation, review, etc.			
Paperwork problems (include paperwork inaccuracies and additional paperwork required)			
Inventory costs			
Internal inspection costs			
Internal rework costs			
Lost time for rejected parts			
Lost time for delayed delivery (initial orders and to correct errors)			
Customs duties			
Exchange rate risk			
Additional features required to bring product in line with U.S. standards	_____	_____	_____
Total Hidden Costs (%)	+ 100 %	+ 100 %	+ 100 %
x Quoted Purchase Price ($)	x_____	x_____	x_____
= Adjusted Total Cost ($)			

TRUE COST WORKSHEET
PART B
ADDITIONAL HIDDEN COSTS OF FOREIGN SOURCING

Hidden Costs	U.S. Supplier	Benchmark Estimates of Added Foreign Costs (% of U.S. Quoted Price)	Foreign Supplier
Freight and packaging costs		⎫	
Insurance costs		⎬ + 4 to 12 combined	
Financing terms		⎭	
Travel and communications costs for up-front consultation, review, etc.		+ 3 to 8	
Paperwork problems (include paperwork inaccuracies and additional paperwork required)		+ 2 to 4	
Inventory costs		+ 5 to 10	
Internal inspection costs		⎫	
Internal rework costs		⎬ Too variable to set benchmarks	
Lost time for rejected parts			
Lost time for delayed delivery (initial order and to correct errors)		⎭	
Customs duties	None	+ 4 to 6	
Exchange rate risk	None	+ 1 to 3	
Additional features required to bring product in with U.S. standards	None	+ 5 to 35	
	_____		_____
Total Hidden Costs (%)			
	+ 100%		+ 100%
x Quoted Purchase Price ($)	x_____		x_____
= Adjusted Total Cost ($)			

8

COST AND PROFITABILITY EFFECTS
OF MANUFACTURING EFFICIENCY

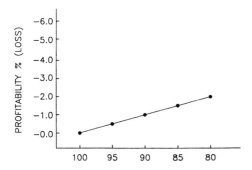

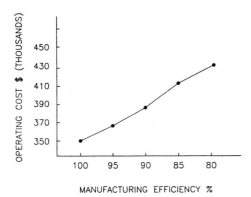

MANUFACTURING EFFICIENCY %

FIGURE 1

COST AND PROFITABILITY EFFECTS
OF INJECTION ACCURACY/REPEATABILITY

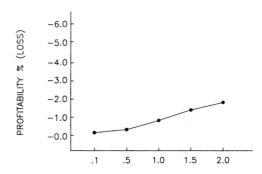

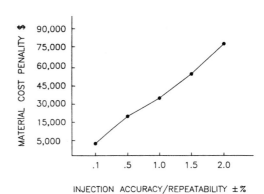

INJECTION ACCURACY/REPEATABILITY ±%

FIGURE 2

COST AND PROFITABILITY EFFECTS
OF SCRAP RATE

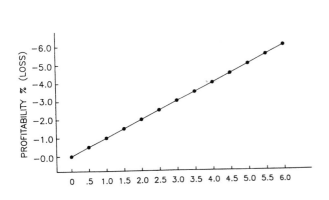

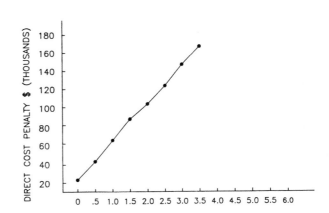

FIGURE 3

USING SPREADSHEETS TO ESTIMATE
SRIM FABRICATION COSTS

John V. Busch, Scot A. Arnold
IBIS Consulting
Cambridge, Massachusetts 02139   USA

ACCURATE PART COST ESTIMATION is fundemental not only to the OEM's, but also to the parts, equipment, and material suppliers that serve these OEM's. Part cost analysis is used to:

1. Select from alternative materials, fabrication processes, or part designs,
2. Establish component prices that will secure orders while ensuring profits,
3. Promote the use of one process or material over other alternatives, and
4. Verify that the price paid for a component is not out of line with its cost of manufacture.

In order to perform an accurate cost analysis, a company must:

1. Establish realistic design parameters for the component in question in each of the viable material alternatives.
2. Define the prodution scenario for each alternative,
3. Establish appropriate equipment, size, type, and quantity,
4. Estimate realistic processing parameters, including production rates, scrap rates, and labor requirements,
5. Establish realistic production goals, and
6. Synthesize all of this information into a cost estimate.

Dealing with these factors requires a well thought-out procedure. However, a conflicting requirement is the relatively short time period available in which to make most cost estimates.

Computer-based cost estimation programs have been found to be effective and reliable in accomplishing these objectives. In addition to speed, computer programs offer many other potential advantages over more traditional cost estimating methods. These advantages include consistency, ease of verification, flexibility, detail, availability, and controllability. In this paper, a computerized approach to SRIM part cost analysis, based upon technical, economic, and statistically derived inputs and relationships, is described.

ELEMENTAL COST ANALYSIS

Elemental cost analyses are performed by dividing total cost into individual elements, then estimating each of these elements separately. Estimates for the elements can be derived from basic engineering principles, from the physics of manufacturing processes, and from clearly defined and verifiable economic assumptions. This approach reduces the complex problem of cost analysis to a series of simpler estimating problems, and brings engineering expertise to bear on solving these problems.

When dividing cost into its contributing elements, a distinction is made between costs elements that depend upon the number of components manufactured annually, and those that do not. For example, in most instances raw materials contribute the same amount to the cost of a part, regardless of the number that are produced. On the other hand, on a per piece basis, the cost of tooling varies with changes in production volume. If more pieces are produced with a given tool, the tooling cost per piece is reduced. These two types of cost elements are known as variable and fixed costs, respectively, and they form a natural division of the elements of manufactured part cost.

An incomplete list of elements that contribute to the total cost of reaction injection molding, divided into categories of variable and fixed, is presented below. In some elements, for instance maintenance costs, the distinction between variable and fixed costs becomes blurred. Nevertheless, it is useful to place these costs into one category or the other for purposes of constructing a cost model.

Variable Cost Elements	Fixed Cost Elements
Materials	Molding Machines
Process Energy	Machine Installation
Direct Labor	Machine Maintenance
Variable Overhead Labor	Tooling
	Tooling Maintenance
	Buildings
	Fixed Overhead Labor
	Auxiliary Equipment
	Opportunity Costs

COST ESTIMATION

While it is easy to identify a long list of contributing cost elements, it is more difficult to arrive at an appropriate basis for estimating these elements. Fortunately, most of the direct cost of reaction injection molding can be accounted for with only four elements. These are the materials, labor, and capital investments in machines and tooling. The contribution of these to the total cost can be estimated as follows:

$$Material = Weight \cdot Price / (1 - Scrap) \qquad (1)$$

$$Labor = No.\ of\ Laborers \cdot Wage \cdot Cycle\ Time / Productivity \qquad (2)$$

$$Capital = \frac{No.\ of\ Machines \cdot Installed\ Price\ Per\ Machine}{Machine\ Life \cdot Annual\ Production\ Volume} \qquad (3)$$

With appropriate values for each of the parameters in equations 1 through 3, it is possible to arrive at reasonably accurate estimates of the cost of molding most components. This being the case, the reader may wonder why bother constructing a computer model, where a calculator should suffice. The difficulty lies is establishing appropriate values for the above parameters.

Consider the materials cost equation involving the part weight, material price, and scrap rate. None of these parameters are static. Material prices are subject to price-volume discounts. Scrap rates can be effected by the size of the part, the material being molded, and the type of tooling. The weight of a part is design specific, and can vary with changes to the material and the process.

The four parameters in the machine or tooling cost equation are even more difficult to estimate. Consider, for instance, the number of machines. The number of machines can be estimated from the production volume, production rate, and available time. An example of this estimation is:

$$\frac{20,000\ parts\ per\ year \cdot 533\ s\ cycle\ time}{240\ d/yr \cdot 16\ hr/d \cdot 3600\ s/hr \cdot 70\%\ productive} = 1.1\ machines$$

Rounded up to the next whole number, the example indicates that a minimum of two machines is required. However, if two is entered into the machine cost equation, the full cost of the second machine is distributed onto the parts being produced, in spite of the fact that it is only used for 10% of the year.

Alternatively, 1.1 can be entered into the machine cost equation, modeling the scenario of flexible equipment. When this is done, the investment is treated as if the equipment were rented - only the fraction of the machine cost associated with the time required for the production run is distributed unto the parts being produced.

The choice between estimating machine costs on the basis of dedicated or flexible equipment must be made on a case by case basis*. Neither approach is universally appropriate.

To further muddy the waters, consider the effect of changing the number of cavities in the tool. Increasing the number of cavities can reduce the number of machines required, but those that are required must be larger and more expensive. Additionally, it is likely that the cycle time will increase and machine productivity may decrease, resulting in further changes in the cost.

ADVANTAGES OF COMPUTER MODELS

These and similar complexities are the motivation for building computer based cost estimating models. While equations 1 - 3 appear simple, the parameters in these equations are often complex functions of interrelated parameters.

Another motivation for building a computer model is to be able to answer the frequent 'what if' questions associated with cost analyses; for example, what if the annual production volume changes. With a computerized model, this question can be easily analyzed.

Furthermore, the model can be used as a repository of cost estimating data. For instance, labor wages and material prices can be stored in the model, instead of requiring that they be input each time the model is used. This concept can be extended to the more complex relations, as discussed below.

---

* This is not true for estimating the cost of tooling. Tooling is always dedicated to a given production run, and the costs of tooling must be accounted accordingly.

Until now, the price of the equipment required for SRIM molding has been treated as an exogenous input. By collecting data from the major RIM equipment manufacturers, and regressing these data onto explanatory variables, IBIS Consulting has arrived at the following relationship for estimating the investment in SRIM clamps:

Cost = $20.70 * Platen Size (sq in) + $395 * Clamp Force (tons) + $89,800

This relationship and the necessary process data to estimate the platen size and clamp force can be built directly into the cost model. By replacing input parameters with relationships, the work load faced by the model's user is reduced.

APPLICATIONS OF COMPUTER COST ESTIMATING

Once a computer model has been built, it requires constant testing and monitoring to maintain both its validity and utility. This maintenance process includes updating cost data and adding new process derivatives to the model.

An example of a cost estimate generated by the IBIS reaction injection molding model is presented in Figure 1. This estimate corresponds to the Chrysler Louisville Slugger mid-sized pickup truck bed as described in January, 1988 Modern Plastics (1).

In addition to the cost estimate, the underlying parameters are presented below the estimate in the section titled "Additional Information". This section summarizes the assumptions upon which the cost analysis is founded. These values are either estimated from techno-economic relationships within the model, extracted from databases within the model, or, for the values labled "Linkage", transferred from other cost estimating models**.

While many of the "Additional Information" parameters are generated by the model, they can be overridden by the user interested in either exploring alternatives or performing sensitivity analyses. A sensitivity analysis involves systematically varying an input or process parameter over a range of values and calculating the cost consequences of this variation. A sample sensitivity analysis focusing on the effects of cycle time for the molded truck bed is presented in Figure 2.

The importance of sensitivity analysis lies in the fact that uncertainties surround most of the parameters in a cost analysis. In many cases, equipment costs, production volume, processing rates, tooling costs, etc. are only imprecisely known, especially before actual production begins. By knowing the extent to which part cost changes with variations in these parameters, it is possible to identify critical uncertainties and focus attention on determining actual values.

SUMMARY

Accurate cost estimating is fundamental to all sides of the business of reaction injection molding. When done well, it can be used to secure bids, sell products, select materials (and processes), and maintain profitability in a competitive market. However, accounting for all of the interrelated technical and economic factors that effect cost is complex.

Good computer models can relieve the computational drudgery associated with generating accurate cost estimates. These tools offer the additional advantage of enabling detailed scrutiny of the estimate, allowing the user not only to produce a credible value, but also to explore its rationale and underlying assumptions.

ACKNOWLEDGEMENTS - The authors wish to acknowledge the assistance of Dr. Nippani Rao, Advanced Composites Specialist at Chrysler Motor Company.

REFERENCES

1. "120-lb. Truck Beds Start SRIM Rolling", Modern Plastics, January 1988, p 15.

** Separate models are used to estimate the cost of molding foam cores and preparing glass mat preforms.

```
VARIABLE COSTS                     Per Part    Annual    Percent
                                  ---------------------------------
               Material Cost      $222.57   $4,451,460    72.2%
                  Labor Cost       $23.15     $462,949     7.5%
                 Energy Cost        $1.32      $26,415     0.4%
                                  ---------------------------------
        Total Variable Cost       $247.04   $4,940,825    80.2%


FIXED COSTS                        Per Part    Annual   Percent   Investment
                                  ------------------------------------------------
                  Press Cost        $3.14      $62,834    1.0%     $439,835
         Mix/Meter Unit Cost        $1.04      $20,892    0.3%     $146,247
                Tooling Cost       $20.24     $404,762    6.6%   $1,619,048
               Building Cost        $0.73      $14,650    0.2%     $293,007
         Overhead Labor Cost       $11.72     $234,395    3.8%
           Installation Cost        $0.84      $16,745    0.3%     $117,216
     Auxiliary Equipment Cost       $1.47      $29,304    0.5%     $205,129
            Maintenance Cost        $6.03     $120,513    2.0%
                Capital Cost       $15.86     $317,204    5.1%
                                  ------------------------------------------------
           Total Fixed Cost        $61.06   $1,221,300   19.8%

                                  ================================
        Total Cost =====>         $308.11   $6,162,125    100%
                                  ================================
```

                         ADDITIONAL INFORMATION

```
MATERIALS CALCULATIONS
       Resin System Volume    2016.00 cu in/pc
              Bulk Density       1.65 g/cc

                                   lbs        $/pc
                                 --------------------
        Resin Weight & Cost      69.60       $84.25
       Filler Weight & Cost       0.00        $0.00
    Fiber Mat Weight & Cost      48.00      $126.51  (LINKAGE)
  Plenum Preform Wgt & Cost       0.00        $0.00  (LINKAGE)
  Molded Core Weight & Cost       2.40       $11.81  (LINKAGE)
                                 --------------------
                 Total ==>      120.00      $222.57

CAPITAL COST CALCULATIONS
    Mold Load, Unload & Prep.     140 sec
              Fill & Cure         533 sec
                              ==========
        Total Cycle Time ===>     533 sec

     Run Time for One Machine      96%
   Number of Parallel Streams     1.00
         Minimum Clamp Force      403 tons
         Minimum Platen Area    9,216 sq in
     Mix/Meter Unit Capacity        8 cu in/sec
       Mold Complexity Factor     1.75
                  Mold Cost  $1,433,000
               Shuttle Cost    $186,182
        Productive Mold Life     4.00 yrs
   Electric Power Requirement    89.15 kW
     Required Building Space    3,907 sq ft
```

Figure 2:

Cost vs. Annual
Production
Volume

Chrysler
Louisville
Slugger Pickup
Truck Bed

120 lb Total
Weight
70 lb Resin
50 lb Glass
4′ x 8′x 1′

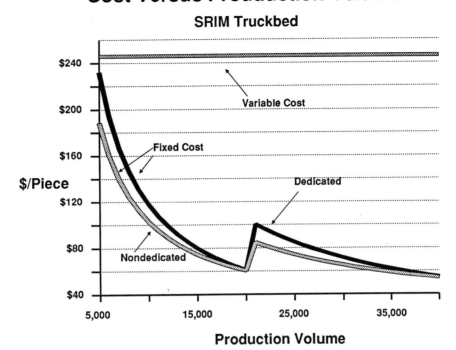

Cost Versus Prouduction Volume
SRIM Truckbed

$/Piece

Variable Cost

Fixed Cost

Dedicated

Nondedicated

$240

$200

$160

$120

$80

$40

5,000   15,000   25,000   35,000

Production Volume

COMPARATIVE COST ANALYSIS OF COMPOSITE AND STEEL AUTOMOTIVE DOOR INNER PANELS

Scot A. Arnold, Frank R. Field III, Joon C. Park
Materials System Laboratory
Massachusetts Institute of Technology
Cambridge, Massachusetts 02139   USA

ABSTRACT

A comparative cost analysis of composite and steel door inners was performed. The steel door inner part cost includes the stamping and assembly cost of the inner panel, the hinge pillar reinforcement, and the belt reinforcement. The composite versions of the door inner are speculative designs with added fiber reinforcement in the place of reinforcing beams. Using Lotus 1-2-3® spreadsheet fabrication cost models, for steel stamping and assembly, RTM, SRIM, SMC, and TPS, it is shown that the composite inners, due partially to parts consolidation of the reinforcing beams, have a lower manufacture cost at low production volumes. This result holds true under both "optimal" and "worst case" steel stamping and assembly conditions.

THE MSL AUTOMOTIVE GROUP has developed numerous case-specific cost analyses and materials comparisons for the automotive market in the past several years. Since the summer of 1987, the MSL has worked on developing more comprehensive models of automotive finishing and assembly operations, particularly as these processes have an effect on the automotive materials selection process.

The automotive materials project is focused on material alternatives for automobile doors. Automobile doors were selected for several reasons:

Steel and Polymer Alternatives - Recently both material vendors and OEM's have been exploring new materials that are available, both RP/C's, new steels, and improved steel manufacturing techniques. As a result, many potential alternative materials systems are viable, and some preliminary design work has been done. Modular door systems are examples of design innovations which could incorporate new materials.

Assembly Issues - Door assembly is directly affected by materials choices. For example conventional joining techniques such as resistance welding can no longer be used with composite doors.

Parts Consolidation - The possibility of consolidating parts using polymer materials makes them attractive alternatives.

This paper will present preliminary results from this ongoing study regarding material substitution of steel in the door inner panel. For this part comparison four structural composite fabrication methods are compared, structural reaction injection molding, (SRIM), resin transfer molding, (RTM), SMC compression molding and thermoplastic sheet stamping, (TPS), and steel stamping.

For this analysis the MSL and IBIS Consulting has developed Lotus 1-2-3® spreadsheet based cost models of the fabrication processes as well as for the subassembly of the steel door. The details of technical cost modelling have been described in an earlier article in this series.[*]

The purpose of the door study is to develop a tool for estimating the cost of a finished part. It has been demonstrated that often polymer composites do not compete well with conventional steel stampings on a piece by piece basis. This is generally because the composite part has not been designed to optimize on its material properties and does not perform well in a design for the incumbent material. Therefore the MSL has been working to develop a set of models for estimating the cost of part manufacture from the point of initial piece fabrication to final finishing and trimming. In this way the influence of design on manufacture cost can be accounted. Of course this

[*] See J.V. Busch and S.A. Arnold, "Using Spreadsheets to Estimate SRIM Fabrication Costs" ACCE IV Conference, 1988.

method will not account for non-cost related utilities of the design.

Steel parts are generally reinforced by joining smaller reinforcing beams to regions of high stress in the main part. Such is the case with the door inner panel, which requires a hinge pillar reinforcement, to give additional strength and stiffness at the hinges, and the belt rail reinforcement to give strength along the top of the door. By using structural composites however the part may be redesigned to have extra reinforcement in regions of high stress thereby eliminating the need for attaching reinforcing beams. In this way parts consolidation could ostensibly make composite parts more cost effective than steel. To make this comparison it is necessary to be able to estimate the cost of assembling the steel door inner.

PART DESCRIPTIONS

STEEL DOOR DESIGN For this analysis a generic steel door system includes the following components.

Outer Panel: 12.5 lb, one side galvanized, Cold rolled steel - commercial quality - 0.032 in, stamped

Inner Panel: 9.8 lb, Cold rolled steel, draw quality, 0.032 in, stamped

Intrusion Beam: 10 lb, Ultra high strength steel, 0.045 in, roll formed and stamped

Hinge Pillar: 3.25 lb, Cold rolled steel draw quality, 0.062 in, stamped

Belt Reinforcement: 2.45 lb, Cold rolled steel, draw quality, 0.032 in, stamped

Total weight: 38 lb, excluding any cosmetic inner trim materials and functional components.

Actual part weights will, of course, vary with different models, but these parameters were established as the baseline "generic" steel design based on information from Ford, Chrysler, and Budd. The door inner panel is generally two sided, hot dip galvanized or electrogalvanized steel (0.028 in. - 0.032 in. thick).

For the purposes of this analysis only those pieces which make up the door inner will be considered, namely the inner panel, the hinge reinforcement, and the belt reinforcement, where the total weight is 15.50 pounds.

The intrusion beam is not considered in the present analysis. The intrusion beam is required to comply with FMVSS side impact regulation, and is best placed between the front hinge and rear latch along a straight line, so that the entire structure can withstand maximum impact from both the front and sides of the car.

The original FMVSS measures performance as the ability of a door to resist the inward pressure of a rigid steel cylinder powered by a piston.[*] A recent proposal suggests, in addition to this requirement, a full-scale crash test in which the car is struck on either side by a simulated vehicle.[**] If this rule becomes effective, radical door redesign will be required. Furthermore composite panels would need an intrusion beam acting as a support for the hinges and extending laterally across the door to support the striker. This design for the intrusion beam is required to eliminate the stress concentrations which would develop by attaching the hinges or striker directly to the composite panel. With this uncertainty surrounding the design of future intrusion beams both for the steel and composite doors, this piece of the door inner is not included in the analysis.

Table I shows the part dimensions for the steel door inner panel and its reinforcements as well as important production parameters.

Table I. Door Inner Panel Dimensions	
Projected Area	1082 sq in
Total Surface Area	1477 sq in
Panel Depth	3 in
Hinge Pillar Projected Area	185 sq in
Belt Reinf. Projected Area	271 sq in
Steel Price	$0.35 /lb
Reject Rate	6%
Total Part Weight	15.50 lbs

STEEL DOOR SUBASSEMBLY Currently, inner and outer door panels and the intrusion and reinforcement beams are subassembled on a dedicated assembly line. Dedicated subassembly lines have a separate line for each door type (left, right, front, back, etc.) that has a productive lifetime limited to the lifetime of the product (except for such components of the line as toggle presses and programmable robots).

The door subassembly line generally has two entry points, one for the inner panel and one for the outer panel. The inner panel will have the hinge pillar, the belt reinforcement, the intrusion beam, and possibly some brackets spot welded to it. The outer panel will have the window channel welded to it, a bead of adhesive applied to its perimeter, and sound deadener applied laterally across its midline.

The door inner is restruck so that the edge, which is usually bent from previous handling, is straight and the two panels are then joined. A tap weld is applied to the joined panels to prevent the

[*] Federal Motor Vehicle Standard No. 214, Side Door Strength.
[**] Federal Register Vol. 53, No.17, p 2239, January 27, 1988.

Table II Composite Panel Production Data[*]	SMC	SRIM	RTM	TPS
Resin	Vinyl Ester	Polyurethane	Vinyl Ester	PC/PBT
Fiber	Chopped Glass	CSM	CSM	Swirled Glass Mat
Fiber Wt. Fraction (%)	60	40	40	35
Cycle Time (min.)	1.82	7.27	12.3	3.26
Material Cost ($/lb w/ Scrap)	0.86	1.47	1.58	1.90
Part Weight (lb)	16.95	15.66	14.3	13.2
Avg Thickness (in)	0.24	0.16	0.16	0.24

[*] Material Price Information: Dow Chemical Company and General Electric Company.

two panels from slipping apart before the hemming step. The hemming operation consists of a 45 degree edge bend followed by a 90 degree bend. Several spot welds are then applied or the perimeter is induction heated to give the hem flange sufficient handling strength.

In practice, levels of automation on assembly lines differ in a range from none, as in a fully manual line, to lines with robotic welders, adhesive appliers, and magazine-fed robotic loaders. Line flexibility apparently is not augmented by the introduction of robots, unless the line provides some sort of quick fixture changing capability. Figure 1 shows the process flow for a typical door subassembly line.

Currently, door subassembly lines can produce up 350 parts an hour, while downstream door demand is usually one per minute per plant. This situation usually leads to idle capacity. Several plants are implementing rapid fixture changing capability which will enable multiple products to be assembled on a single line. Ford Australia is using such a line capable of building six door types. It is a robotic line capable of producing 120 doors per hour, and is utilized two shifts per day (the third shift is used for preventative maintenance).

In the present study, however, only the assembly of the door inner is considered. This includes the portion of the line where the hinge pillar and the belt reinforcement are welded onto the inner panel.

COMPOSITE DOORS Other than the space frame concept and the Chevrolet Corvette, modular doors systems are prime candidates for material substitution. Modular doors are innovative designs that have been introduced largely by parts and materials suppliers, usually in partnership with customer OEMs. The term modular door system can describe anything from a "hardware cassette"-- a board equipped with pre-installed components like the window winding system, the lock, and glass -- to a door consisting of an inner and outer panel, fully trimmed.

The hardware cassette system is made of either stamped steel or RP/C (reinforced plastics/composites) and can be installed in the door in one step, to eliminate several trimming steps.

Currently, two types of processes are used for door hanging/trimming. "Doors on" assembly (the conventional process) is where the doors are fitted onto the car and all the subsequent operations, including trimming and painting, are performed with the doors on the car. The presence of the doors hinders trimming the interior and increases the required floorspace. "Doors off" assembly is a common Japanese manufacturing technique where the doors

Figure 1. Process Flow Diagram of Steel Door Subassembly

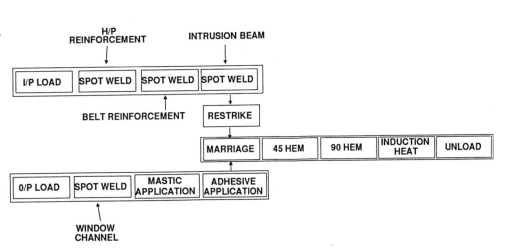

are removed after the painting operation and are trimmed offline.[*] The trimmed doors are re-hung at the end of the chassis line. "Doors off" assembly is becoming more popular in domestic companies such as Ford and GM. "Doors off" assembly provides better access to the car interior, reduces overall floor space, and increases the number of effective assembly actions per vehicle, thereby reducing WIP (work-in-progress). The door trimming cassettes are compatible with either the "doors on" or "doors off" assembly process.

The other modular door system includes an inner and outer panel, but the door is completely trimmed when it is delivered to the assembly plant. These fully trimmed modular doors could be readily adapted to "doors off" assembly.

Structural composites currently do not compete well with steel in appearance applications, but could offer advantages in hidden structural applications such as a door inner. The ultimate objective is to mold a functional inner panel from a single mold, and to eliminate the need for the reinforcements required in conventional steel doors.

The inner door panel is the subject of this preliminary analysis, since it is the non-appearance structural member most adaptable to the use of structural RP/Cs. Using RP/C should make it possible to consolidate the hinge and belt reinforcements into a single molding. RP/C panels could also include molded-in armrests, although the cost savings associated with this consolidation will only be realized further downstream when considering the door trimming operations.

[*] M. Horike, Journal of Japanese Society of Automotive Engineering Review, July 1984, 51-61.

Just as the conventional steel door is generalized, the various composite door systems that are likely to be developed will be treated in a general, prototypical way. Four RP/C panels are compared for production costs. The fabrication processes and materials used are: 1) SRIM urethane with glass thermoformable continuous strand mat and woven roving mat reinforcement at the pillar beam and belt reinforcement; 2) RTM vinyl ester matrix, reinforced with glass thermoformable continuous strand mat and woven roving mat at the pillar beam and belt reinforcement; 3) SMC vinyl ester with 60% by weight chopped glass roving; and 4) stamped thermoplastic sheet (35% Glass mat reinforced blend of PBT and PC).

COMPARATIVE PART COST ANALYSIS

COMPOSITE DOOR INNERS Table II shows some costs and other production data important to the manufacture of the inner panel from four RP/C materials. The panel thickness is an average of both thinner areas and areas where more stiffness is required, such as near the hinge. The projected and total surface area of the inner panel is assumed to be the same for all the processes. *The area of the regions of extra reinforcement is assumed to be the same as the sum of the projected areas of the hinge pillar and the belt reinforcement belt in the steel inner.* The dimensions of the parts are listed in Table III.

The general manufacturing conditions are normalized for all fabrication and assembly processes and are summarized in Table IV. Under these manufacturing conditions, Figure 2 shows the cost of the above described inner panel molded via RTM,

Figure 2.

Comparative Cost Analysis of Composite Door Inner Panels Fabricated Via SMC, RTM, SRIM, and TPS, as a Function of Production Volume.

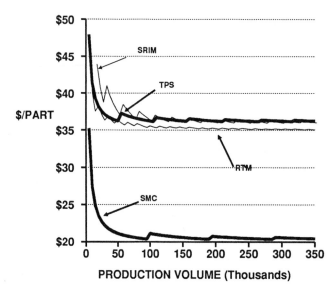

COMPOSITE DOOR INNER COST COMPARISON

SRIM, SMC, and TPS. The exact points of crossover should not be taken as absolute, since the models generate calculations based on industry cost averages. Furthermore, the actual details and feasibility of molding, in one step, a part comparable to the steel door inner-assembly is uncertain. Certainly this is true with an SMC panel not reinforced with steel beams as the Corvette door is, for this is not a very stiff material (modulus = 1.27 psi) compared to cold rolled steel (modulus = 30 psi). However, the trends are important since the models represent a consistent set of assumptions.

Table III Door Inner Panel Dimensions	
Projected Area	1082 sq in
Total Surface Area	1476 sq in
Panel Depth	3 in
Reinforcement Area	456 sq in

As expected SMC is less expensive than the other processes at most volumes. It is not clear whether it is possible to produce a single SMC molding that will be able to match the strength and modulus of the other composite parts. RTM, SRIM, and TPS are very close in cost particularly at high production volumes. TPS in this application appears to be expensive, both due to its long cycle time and the high material cost. The cycle time is high because the thickness of the part requires a long cooling time.

Table IV General Manufacturing Conditions	
Model Years	4 years
Labor Cost	$24 /hour
Electricity Cost	$0.08 /kwh
Shifts/Day	2 shift
Hours/Shift	8 hours
Days per Year	240 day
Cost of Capital	12 %
Maintenance	6.0%

STEEL MANUFACURING SCENARIOS There is presently a gap between the efficiency of "world class" and United States metal stamping operations, although there has recently been some narrowing of this gap. It is important in this analysis to consider the costs of metal stamping as a moving target which should decrease as U.S. stamping conditions become more efficient.

The processing variables that strongly influence the cost of stampings are the batch size, the efficiency or downtime, the die change time, and the die cost. The effects of these variables are described as follows:

Batch size - Lower batch sizes will lead to reduced inventory costs both in terms of operating costs and working capital costs.

Downtime - Reduction in press downtime will increase hourly production rates. It is well known

that "world class" manufacturers have far less downtime than U.S. manufacturers.

Die change time - As with reducing downtime, reductions in the die changing time will lead to increases in production rates. Die change time in "world class" manufacturing plants takes about 10 minutes, whereas it takes on average 4 hours in U.S. plants.

Die Cost - The cost of the die will vary according to the part being produced, the design of the die, and manufacturing practices -- but, generally, U.S. die costs are much higher than those of "world class" dies. The present simulation assumes that the cost of an optimally designed die is about 40% of the current U.S. die cost.[*]

Additional factors in manufacturing costs are part rejection rate and the total number of presses in use, variables that, again, are lower in "world class" plants than in the average U.S.

The disparity between Japanese and U.S. stamping efficiencies is summarized in Table V.

Table V Two Different Manufacturing Conditions		
	Typical U.S.	Efficient
Down Time	40 %	10 %
Die Change Time	4 hours	10 minutes
Batch Size	5000 parts	1000 parts
Die Cost	$1.2 million	$0.5 million
Part Rejection Rate	5 %	1 %
Total Presses	7 presses	4presses

Individual reduction of the above factors from current U.S. highs can lead to lower stamping costs; however, most of these parameters are interrelated. For instance small lot sizes will only bring cost reductions if die changes are rapid enough for the press lines to be utilized in the production of another part. Higher production rates, i.e. low press downtime, coupled with the rapid die changes will improve the economics of small lots even more. Achieving lower part rejections, higher press uptime, cheaper dies, and press lines with fewer passes requires redesign of the part for easier manufacture.

Other considerations may prevent reduction of lot size. For example, larger lot sizes may be preferable if the stamped parts are to be shipped to an assembly plant located far away. One must conclude that simply adopting the manufacturing practices of others, e.g., a JIT manufacturing system, is not readily possible if it is difficult to achieve all the related manufacturing conditions that contribute to the success of that system. However in comparing RP/C parts with stamped steel parts it is important to

[*] D. N. Smith and P. G. Heytler, "An Emerging Model for Future Automotive Plants", Advances and Trends in Automotive Sheet Steel Stamping, SAE Paper 880211.

Figure 3.

Cost Comparison of
Stamped Steel Door
Inners Under U.S. and
the "World Class"
Conditions.

STEEL DOOR INNER COST COMPARISON

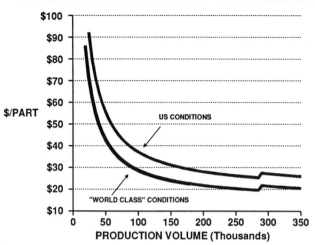

recognize that current steel manufacturing practices are improving in order to remain competitive.

In this study the analysis of part cost sensitivity has been confined to the effect of production volume. This analysis is important in order to see at which production volumes a material or process becomes more cost effective.

Three steel door manufacturing scenarios will be simulated in order to fully asses the economic position of steel relative to composites as door inners. The first scenario is the current U.S. steel stamping conditions described in Table V with fully dedicated assembly. In this case the door assembly line is treated as tooling, with a maximum productive lifetime limited to the market life of the product. The second scenario is the "efficient" manufacturing conditions described in Table V with dedicated assembly. In the third scenario, both conditions are used but with flexible or nondedicated assembly. In this case the assembly line is assumed to be "rented" only for the amount of time required to produce the required number of parts. For each of these scenarios the same general manufacturing conditions listed in Table IV apply to both the stamping and assembly operations.

Figure 3 shows the total piece cost of a stamped and assembled door inner (three parts) as a function of production volume under U.S. and Japan manufacturing scenarios. Figure 3 shows that if the annual volume is relatively high (greater than 200,000 units), the cost difference is relatively stable at about $6.00 per part. But as the annual volume becomes smaller (less than 100,000 units), the cost of the part increases. The Japanese have an obvious cost advantage in producing steel at low volumes, which may be one of the reasons they use less plastic in body applications.

Figure 4 shows the door inner cost as a function of production volume for both U.S. and Japan stamping conditions with and without flexible assembly. This plot shows a clear advantage in keeping assembly capacity fully utilized.

Figure 5 shows the costs of the door inner produced in composite and steel. Shown are the best and worst case scenarios for steel stamping and assembly, namely the worst being current U.S. practice with dedicated assembly and the best being the "efficient" practice of Japan with flexible assembly. RTM, SRIM, and TPS have been combined into a single thick band, whose width represents the cost range, since at the scale of the figure they appear to almost overlap.

It is important to remember that the steel door inners include the hinge pillar reinforcement, the belt reinforcement, and the cost associated with joining the three pieces. These parts are consolidated into a single molding with the RP/C parts by adding extra blanks in certain places, in the case of SMC and TPS, or adding stronger reinforcement to regions of high stress, in the case of RTM and SRIM. However, the cost estimate does not include the metal inserts required for various trim and structural fastenings.

The important point of Figure 5 is that it shows how composite materials save costs associated with parts consolidation, particularly at low volumes. As stated previously the actual cross-over points are not necessarily exact since the parts represent a generic door system. Furthermore the assumptions regarding exogenous manufacturing costs have been normalized for the sake of comparison but in fact will vary over each industry and firm.

However, the general trends are accurate and indicate that RP/Cs should offer significant savings at lower volumes over steel in the door inner.

Figure 4.

Cost Comparison
Between Stamped
Steel Door Inners,
Under Typical U.S. and
"World Class"
Conditions, both with
and without Flexible
Assembly .

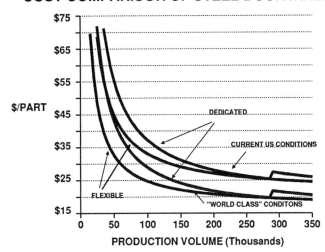

COST COMPARISON OF STEEL DOOR INNERS

However this observation is highly conditional on the feasibility of the composite door design.

FUTURE AND ONGOING WORK

With respect to composites and modular systems, the sort of analysis which led to the cost comparison in Figure 5 is only preliminary. The goal of this study is to trace the cost of material substitution through the entire fabrication and assembly process. This will require further work in the area of assembly. It is also important to get a better understanding of what would be an acceptable functional composite door. This is required for more accurate cost estimations, since the system cost will be strongly coupled to the material strength requirements and the part geometry. This is particularly true for the SMC part whose design is "speculative". The design assumes that, either by adding extra thickness or continuous fiber reinforcement to regions of high stress, the part will be functionally equivalent to the steel inner assembly.

SUMMARY

Through the use of Lotus 1-2-3® based composite and steel fabricating and assembly cost models is has been shown demonstrated that under certain conditions composite doors inners could be produced at a lower cost than the current steel parts. The composite parts are speculative single molding designs which are assumed to be functionally equivalent to a steel assembly consisting of the inner door panel, the hinge pillar reinforcement, and the belt reinforcement. The cost analysis for the steel part was done under three production scenarios: 1) current average conditions; 2) "world class" or optimal conditions; and 3) flexible versus dedicated assembly conditions. The composite panels compete favorably at low production volumes, under 100,000 parts annually, with steel parts produced under "worst case" condi-

Figure 5.

Cost Comparison
Between Stamped
Steel and Structural
Composite Door
Inners

The Composite
Fabrication Processes
are RTM, SRIM, &
TPS (All Shown in the
Same Line) and SMC

"Best and Worst"
Case Steel Stamping
Conditions are Shown

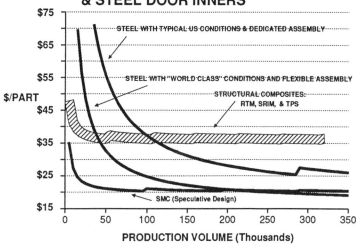

COST COMPARISON BETWEEN COMPOSITE & STEEL DOOR INNERS

tions, current average U.S. with dedicated assembly, and "best case", "world class" with flexible assembly, conditions.

ACKNOWLEDGEMENTS

Special thanks to Don Ponagajba at the Warren Stamping Plant of Chrysler Motors Corporation and Gerard Fertig and Steven Good of General Electric Company. Also special thanks to Dr. John Busch of IBIS Consulting for his assistance in preparing the Lotus 1-2-3® Cost Models.

DESIGN OF DELAMINATION RESISTANCE IN COMPOSITE PLATES SUBJECTED TO LOW-VELOCITY IMPACT

Seongho Hong, Dahsin Liu
Department of Metallurgy, Mechanics, and Materials Science
Michigan State University
East Lansing, Michigan 48824   USA

ABSTRACT

Delamination has been found to be the major damage mode in the thin composite plates subjected to low-veloxity imapct. It has also been verified from experiments that the structural degradation of the delaminated composite plate is proportional to the delamination area. Therefore, it raises an interesting question: how to design a composite material to have higher delamination resistance? The present study discusses the idea of delamination initiation in thin composite plates and the parameters involved in the formation of delamination. Based on a proposed Mismatch Theory, this study present some keys in controlling the delamination resistance.

PREMISES

Both the studies of impact and composite material involve many factors. In order to have a clear understanding about a particular phenomenon, some parameters have to be isolated. In this study, only the delamination in the thin composite plates induced by low-velocity impact will be discussed.

---

1. graduate research assistant, 2. assistant professor

1. Low-Velocity Impact

Fiber-reinforced composite materials have very high stiffness-to-weight and strength-to-weight ratios. However, they are very susceptible to impact loading. The damage area of a composite plate caused by impact loading is strongly associated with the impact velocity. Generally speaking, if the impact velocity is very high, the composite plate will most likely be perforated and the damage area will usually be localized in the impact zone [1]. However, if the impact velocity is low, the damage area of the composite plate will cover a larger region and no penetration will be observed [2]. Experimental results have revealed that the behavior of a composite plate, especially a thin plate, subjected to low-velocity impact is very similar to that under global bending [3]. It has also been concluded from studies that the damage in a composite plate impacted at low velocity is mainly attributed to bending [4]. In this study, only the damage caused by low-velocity impact would be discussed.

2. Thin Composite Plates

In addition to impact velocity, the thickness of a composite plate also plays an important role in impact. Generally speaking, if a composite plate is very thick, the damage caused by impact will be confined to the impact zone [2]. This is because that the bending stiffness of a thick plate is very high. Consequently, indentation and local damage in fiber and matrix will be observed. On the contrary, since a thin composite plate has lower bending stiffness than that of a thick plate, as a result of impact, the damage area, which is mainly affected by bending, will be relatively larger than that of a thick plate. In the present study, only thin composite plates would be investigated. However, slight changes in thickness would be allowed in studying the thickness effect on the impact-induced damage.

3. Delamination Area

During an impact event, part of the impact energy is converted into elastic deformation and vibration and then dissipates away in the form of heat. However, the remaining part of the impact energy is absorbed by the specimen and results in permanent deformation and damage in the specimen. Experimental results have shown that the damage modes of thin composite plates subjected to low-velocity impact are matrix cracking and delamination [5,6]. It has also been verified from experiments that delamination is the major damage mode in thin composite plates subjected to low-velocity impact [7]. Accordingly, in order to understand the impact resistance of a composite plate, it is important to investigate the relationship between impact energy and delamination area.

FUNDAMENTALS

There are several approaches and viewpoints leading to the understanding of the current problem. Each has its own emphasis. A short discussion regarding the thoughts and techniques are described as follows:

1. Energy Dissipation and Energy Absorption

As indicated in a previous section, the impact energy can be divided into energy of dissipation and energy of absorption. Both material property and the geometry and dimensions of a structure can influence the distribution of impact energy. In the thin composite plates subjected to low-velocity impact, the ones with lower stiffness and higher damping coefficient will experience higher energy dissipation. However, the energy absorption is essentially a function of material properties. Depending on the design philosophy and application purpose, it is possible to manipulate both geometrical and mechanical properties of a composite mateiral to have different types of damage such as perforation and subperforation. Since the damage modes in the present study of low-velocity impact are mainly delamination and matrix cracking, the energy absorption is strongly dependent on the strength and fracture toughness of matrix and the bonding strength between fiber and matrix. Consequently, in order to achieve higher impact resistance, a composite plate is required to be able to dissipate higher percentage of impact energy and suffer less damage with the energy absorbed.

2. Interlaminar Stress and Strength

Another viewpoint for the study of impact is to examine the interlaminar stress state and interlaminar strengths of the composite plate. The former is associated with specimen geometry and loading parameters, e.g. dimensions of specimen, type of boundary condition, shape of impactor, impact energy, etc. However, the latter is strongly related to the material properties such as the fracture toughness of matrix and bonding strength between fiber and matrix. If some composite plates are made of the same material and have the same dimensions, the difference in interlaminar stress state will be affected by the difference of stacking sequence. However, the final delamination area will be dependent on the interlaminar strengths. Therefore, it remains feasible for an engineer to adjust the stress state and final damage shape by tailoring the stacking sequence and controlling the interfacial property even when a certain composite material is selected. However, the measurement of the interlaminar stresses is not an easy job. Due to the complexity of the stress analysis in the three-dimensional laminated thin orthotopic plate, successful presentation for the stress state on the interface is still lack.

3. Mismatch Theory of Delamination

Insteading of studying the delamination resistance from stress viewpoint, experimental results have suggested that the bending stiffenss mismatching between adjacent laminae is responsible for the delamination on the interface. From Classical Laminate Theory [8], bending stiffness is dependent on elastic constants and stacking sequence. There are six terms of bending stiffness D_{ij}, in which i,j= 1, 2, and 6. Generally speaking, D_{11} and D_{22} have higher values than the remaining four terms. If only D_{11} and D_{22} are considered, the mismatch of bending stiffness between adjacent laminae can be defined as the difference of the bending stiffness:

$$M = D_{11}(h_b) - D_{11}(h_t) \, ,$$

where h_b and h_t represent for the angle of the bottom and top lamina respectively. Experimental results and Mismatch Theory seem to agree with each other qualitatively in low-velocity impact.

DESIGN PARAMETERS

There are some general rules in designing the delamination resistance of a thin composite plate subjected to subperforation low-velocity imapct.

1. Elastic Constant

The mismatch of bending stiffness is dependent on elastic constants. A composite material with higher difference between E_{11} and E_{22}, which are the Young's moduli in the

fiber and matrix direction respectively, will have higher value of M, as shown in Figure 1. Thus, in order to have lower delamination area, lower value of $E_{11}-E_{22}$ should be chosen.

2. Fiber Orientation

The delamination area in the interface of a h_t/h_b plate is proportional to the difference of the angle between h_t and h_b. Shown in Figure 2, both experimental results and Mismatch Theory indicate that the higher the angle between the adjacent laminae, the larger the delamination area.

3. Lamina Thickness

The thickness of each lamina also has strong influence on the bending stiffness mismatching. From Mismatch Theory, it is apparent that higher thickness in a laminae can result in higher mismatching and cause larger delamination area. Figure 3 verifies this conclusion.

4. Fiber-Matrix Bonding Strength

The bonding strength between fiber and matrix holds the important key in determining the delamination area. It is understandable that bad bonding can result in larger delamination area than good bonding. However, it has also been recognized as a rule of thumb that bad bonding will result in higher fracture toughness.

EPILOGUE

The design of the impact resistance of a composite laminate can be simply divided into two major categories: penetration and subperforation. If the residual strength and residual stiffness of an engineering structure is critical, a composite plate with high delamination resistance may be required. However, if a structure is involved in life-protection, a composite plate which is capable of converting the impact energy into large-area damage to avoid penetration is important. Based on the design philosophy and application purpose, a structure engineer can control the impact resistance of a composite plate up to some degree by using the rules described in the previous section.

ACKNOWLEDGEMENTS

The authors would like to acknowledge the State of Michigan for the financial support under the Research Excellence/Economical Development Fund.

REFERENCES

1. Zukas, J.A., Nicholas, T., Swift, H.F., Greszczuk, L.B., and Curran, D.R., Impact Dynamics, Chapter 5, John Wiley & Son, 1982.
2. Zukas, J.A., Nicholas, T., Swift, H.F., Greszczuk, L.B., and Curran, D.R., Impact Dynamics, Chapter 3, John Wiley & Son, 1982.
3. Hong, S. and Liu, D.," On the Relationship between Impact Energy and Delamination Area," (submitted to Experimental Mechanics).
4. Sierakowski, R.L., Malvern, L.E. and Ross, C.A.," Dynamic Failure Modes in Impacted Composite Plates," Failure Modes in Composite III, AIMMPE, N.Y., NY, 1976.
5. Liu, D. and Malvern, L.E.,"Matrix Cracking in Impacted Glass/Epoxy Plates," J. Composite Materials, Vol. 21, pp.594-609, 1987.
6. Joshi, S.P. and Sun, C.T.,"Impact Induced Fracture in a Laminated Composite," J. Composite Materials, Vol. 19, pp.51-66, 1985.
7. Liu, D.,"Impact-Induced Delamination - A View of Material Property Mismatching," (to appear in J. Composite Materials).
8. Jones, R.M., Mechanics of Composite Materials, Chapter 5, McGraw-Hill Co., 1976.

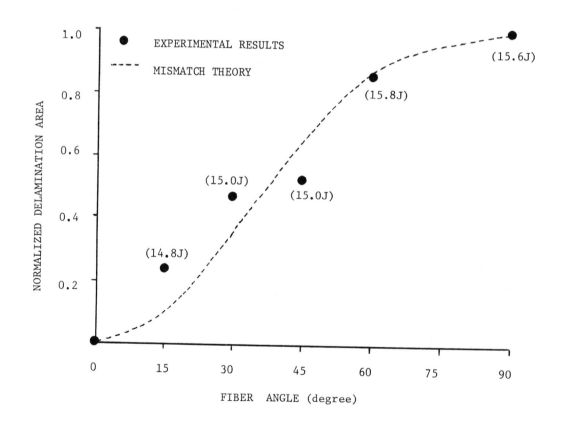

Fig. 1 - The relationship between delamination area and fiber orientation.

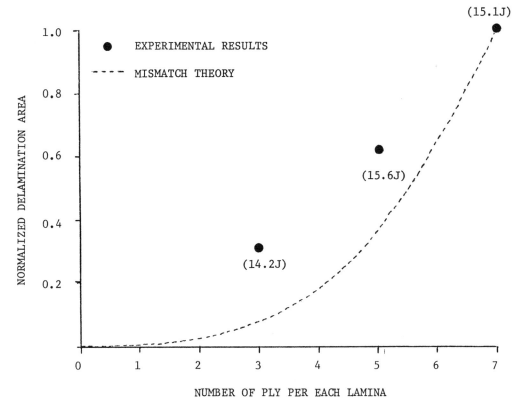

Fig. 2 – The relationship between delamination area and lamina thickness.

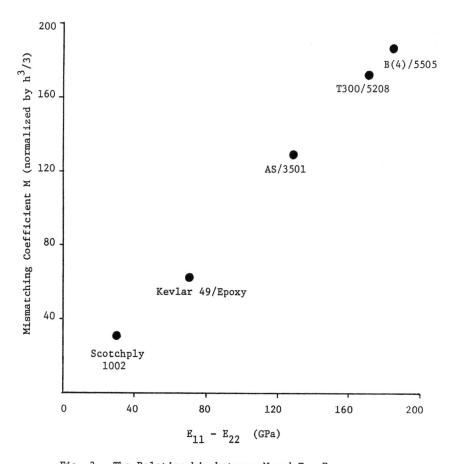

Fig. 3 – The Relationship between M and $E_{11}-E_{22}$.

TETRACORE/ULTRACORE—LOW COST QUASI-ISOTROPIC ENERGY ABSORBING STRUCTURE

I. E. Ward Figge
Atlantic Research Corporation
Gainesville, Virginia   USA

ABSTRACT

A mass producible, low cost structural concept called Tetracore/Ultracore (TC/UC) has been developed that offers potentially greater crashworthiness, in particular, the ability to withstand crash forces from all sides. The concept was originally developed to provide highly redundant, ballistically tolerant structures (23 mm HEI treat) for U.S. Army helicopter components and was successfully demonstrated in helicopter drive shafts, fuselages, rotor blades, flight control components, and crashworthy structures. The ability of these structures to absorb impact energy provided the impetus to explore TC/UC potential for automotive applications.

The concept, based on the use of tetrahedrons, can be fabricated from any formable, castable, moldable, or filament windable material using a wide variety of standard, industry-recognized techniques. All three major U.S. automotive producers and the Japanese are evaluating the concept for applications including load floors, bumpers, and chassis.

The results of analysis comparing Kevlar Tetracore beams with equal thickness steel and Kevlar rectangular tubes indicate that the Tetracore beam is significantly stiffer under uniform and point loading. Analysis of a sheet-molding compound load floor and multi-axis static tests of thermoplastic panels demonstrated Tetracore's quasi-isotropic nature. Comparison of flatwise impact tests between plastic Tetracore

and typical riveted aluminum box beam helicopter structures at 14 feet per second indicated a reduction in peak deceleration from 78 g's to 24 g's, demonstrating the energy absorbing capabilities of the TC/UC structure.

THERE IS A SIGNIFICANT NEED for improved energy absorbing structures, particularly with the ability to handle off-axis impacts. A patented, low cost structural concept called Tetracore/Ultracore (TC/UC) offers this potential. The concept, which can be fabricated utilizing a wide variety of low cost materials and fabrication techniques, was originally developed to provide highly redundant ballistically tolerant structures for U.S. Army helicopter structures. Testing of plastic TC panels demonstrated their ability to absorb significantly more energy than either aluminum honeycomb or simulated helicopter skin/stringer construction under flatwise impact loading conditions. These results, coupled with TC/UC's demonstrated quasi-isotropic in-plane behavior, provided the impetus to explore its potential for energy absorbing automotive application.

In 1985, the patent holder, Atlantic Research Corporation (ARC) and Entech, signed a contract giving Entech the sole right to market designs using the TC/UC structure to the transportation industry. To date, Entech has targeted the automotive chassis, load floors, A-pillars, and bumpers for TC/UC applications. Entech has blow-molded a plastic Tetracore A-pillar for an experimental "Consortium Vehicle" and has experimented with a sheet molding compound (SMC) Tetracore reinforcement to shore up the an automotive liftgate where it had been cracking. The company has also made an experimental load floor and is developing TC/UC for the bumper energy absorber and the lower part of the instrument panel where the riders' knees impact. An entire crashworthy chassis forward of the firewall has been designed and tooled using SMC.

STRUCTURAL ARRANGEMENT

Tetracore/Ultracore is a patented, unique, structural concept based on the use of tetrahedrons. Tetracore utilizes full tetrahedrons, while Ultracore uses truncated tetrahedrons (see Figs. 1 and 2).

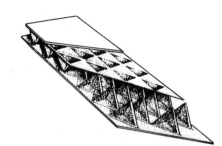

Fig. 1. Tetracore - Full Tetrahedrons.

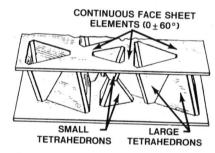

Fig. 2. Ultracore - Truncated Tetrahedrons.

It is important to emphasize that TC/UC is a basic structural concept that can be fabricated from any formable, castable, moldable, or filament windable material using a wide variety of standard, industry recognized fabrication techniques. The tetrahedron, the basic building block of TC/UC, has a higher surface area per unit volume than any other three-dimensional shape and, thus, provides the highest structural efficiency.

A unique aspect of formed TC/UC is that its areal density (lbs/sq. ft.) is independent of the height of the structure so long as the wall thickness of the tetrahedrons remains identical, e.g., a structure 1-inch-high weighs exactly the same as a structure 5-inches-high. This feature is extremely important in bending-stiffness-critical structures, since the moment of inertia is a cubic function of the height. Thus, the stiffness of a given structure can be increased significantly with absolutely no increase in weight. The in-plane properties remain unchanged regardless of height.

MATERIALS/FABRICATION TECHNIQUE

As indicated earlier, TC/UC can be fabricated from virtually any moldable, formable, castable, or windable material using a wide variety of fabrication techniques. The following indicates some of the materials and fabrication processes available:

Materials	Fabrication Processes
° Plastics (Thermoset/ Thermoform)	° Thermoforming
° Metals	° Injection Molding
° Cement	° Blow Molding
° Wood Products/Paper	° Rotational Molding
° Fibers	° Casting
° Advanced Composites	° Stamping/Forming
° Sheet Molding Compound	° Lay-up
	° Filament Winding
	° Superplastic Forming (Metals)

The selected materials/process must reflect both the structural requirements and cost targets. For example, a high performance aerospace material such as graphite/epoxy, although providing exceptional structural performance, may not be suitable for high-rate, low-cost production due to its high cost ($20 to $50 per pound) However, selective local reinforcement using the high performance materials, as discussed in the section on formed/nested/bonded Ultracore, may prove cost effective. Other processes such as injection or blow molding clearly lend themselves to cost-effective production suitable for automotive applications.

STRUCTURAL VARIATION

TETRACORE - FORMED/NESTED/BONDED - In the formed/nested/bonded arrangement, the structure is fabricated in halves and bonded together. A series of hollow tetrahedrons is formed tip-to-tip, producing a sheet with a base of nonconnected triangular elements. The triangular elements enhance the buckling stability of the tetrahedrons in the completed structure and provide additional bonding surface if face sheets are used. A second identical sheet is inverted, nested, and bonded to the first sheet in such a way that the edges of each tetrahedron on the inverted sheet nest with the edges of three mutually positioned tetrahedrons on the first sheet (see Fig. 1). In this manner, the placed and inverted tetrahedrons form a series of continuous three-axis sloped planes oriented at 0 degree, +60 degrees, -60 degrees, accounting for the quasi-isotropic nature of TC/UC. The resulting structure is self-contained and provides full structural integrity. Face sheets may be added to one or both sides of the Tetracore if additional bending stiffness is required. Furthermore, face sheets produce closed outer surfaces required for body panels and may also be advantageous in applications to prevent the accumulation of dirt or other contaminants.

ULTRACORE - FORMED/NESTED/BONDED - The basic structural arrangement of Ultracore is similar to that of Tetracore except that it employs truncated tetrahedrons instead of full tetrahedrons, and the tetrahedrons are spaced apart in each half. This arrangement provides continuous planar elements in the outstanding (upper and lower) faces oriented at 0 degree, +60 degrees, -60 degrees and, thus, enhances the bending stiffness of the structure (see Fig. 2). Additional high strength material can be added along these planar elements to further improve structural efficiency. If additional flatwise compressive strength is required, a small truncated tetrahedron can be added between the larger tetrahedrons and bonded truncated-apex-to-truncated-apex at the midplane of the structure. These small tetrahedrons also enhance the buckling stability of the planar elements and improve the shear strength of the structure.

ULTRACORE - PLANAR SHEET - In this arrangement, the truncated apexes of a single formed sheet of Ultracore are bonded directly to a planar sheet. This approach does not produce continuous in-plane elements and, thus, structural efficiency is reduced. The planar sheet lends itself to less critically loaded structures such as hoods, trunk decks, etc., where one finished surface is required.

ULTRACORE - DOUBLE THICKNESS - In this case, the truncated apexes of two sheets of Ultracore are bonded apex-to-apex producing a structure without continuous in-plane elements. However, the height of this structure is twice that of the nested approach so that even without continuous in-plane elements, its bending stiffness is substantial.

TETRACORE - SINGLE PIECE MOLDING/CASTING - Injection molding or casting techniques produce three-axis sloped planar elements similar to Tetracore with the exception that the triangular members in the outstanding faces are eliminated. This approach produces a complete structure in a single operation and eliminates the secondary bonding required in the formed techniques.

TETRACORE/ULTRACORE - SINGLE PIECE - Blow, rotational molding or twin sheet thermoforming techniques can produce a single-piece structure to which face sheets may be added if desired. Local high-strength

reinforcements can be employed in the in-plane continuous elements of the Ultracore structure.

CURVATURE - Compound curvature can be achieved in all of the above cases. In the formed concepts, curvature is achieved by varying the size/shape of the tetrahedrons in either one or both halves such that the upper half nests exactly with the lower half when each is curved to its final shape. In the cast or molded approaches, curvature is built directly into the tooling. Curvature in the formed Ultracore/planar sheet is achieved during the bonding operation; i.e., the pieces are bonded after each piece has been curved to its final shape.

TETRACORE - FILAMENT WINDING - Tetracore lends itself to filament winding utilizing any of the advanced composite materials such as glass, Kevlar, and graphite (see Fig. 3). The filament-wound approach was originally developed and demonstrated successfully for aerospace applications in which ballistic survivability was critical. Tetracore's high notch insensitivity/ballistic survivability is primarily due to its multiplicity of load paths, which also contributes to Tetracore's efficient energy-absorbing capability. Virtually any shape can be filament wound. Attachments/hard points/flanges can be mechanically captured directly during the winding process. Automated techniques have been developed to wind both flat and closed structures.

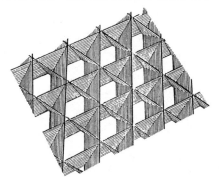

Fig. 3. Filament Wound Tetracore.

STRUCTURAL ANALYSIS

The Boeing Company has developed a detailed elasto-plastic finite element computer program to analyze TC/UC. A weight optimization program selects the optimum height, slope, and wall thickness for a given complex loading condition. The analysis examines a section within the total structure, and the local loading conditions for that section must be known. A major automotive firm has developed a finite element analysis program to evaluate crashworthy automotive chassis. Entech has also developed a finite element model to analyze TC/UC automotive components.

ANALYTICAL RESULTS

Entech conducted three analytical studies to evaluate the structural performance of TC/UC. The first evaluated the behavior of a pseudo representation of an automotive floor pan, while the second study investigated TC/UC's ability to improve the local stiffness characteristics of an automotive liftgate. In these cases, components were fabricated to demonstrate the production feasibility of the TC/UC concept. The third study compared the structural performance of a Kevlar Tetracore beam to rectangular Kevlar and steel tubes.

ULTRACORE – FLOOR PAN DESIGN - A 889 x 825.5 mm test panel design was established to provide a pseudo representation of an automotive floor pan application (see Fig. 4). Outer longitudinal deep sections (76.2 mm deep x 165.1 mm wide) represented a rocker section stepping up to a uniform platform center section thickness (38.1 mm deep). Material thickness was 1.5 mm, in keeping with a lightweight automotive design.

Fig. 4. Simulated Automotive Floor Pan.

Forty-five percent glass SMC composite was used for the analysis. Its properties were as follows:

° Modulus of Elasticity - 1.7×10^6 psi,
° Poisson's Ratio   - 0.33,
° Specific Gravity   - 1.88.

The calculated weight for the model was 18.5 lb wt (82.35 N). The results of the static linear analyses were performed on the model using NASTRAN version 63 are as follows:

Loading Condition (500 lbs)	Max. Defl. (in)	Max. Stress (ksi)
Lateral Crush Direct loading	.0009	2.0
Lateral Crush Angular (30 degree) Loading	.008	2.8
Center Plate Load Cases	.0062	1.4
Cantilever Loading	1.015	9.0
Torsional Loading	1.035	8.0

The results indicate the ability of Ultracore to effectively react a wide variety of loading conditions.

AUTOMOTIVE LIFTGATE - The intention of this study was to illustrate a possible application of the TC/UC concept as a method of improving the stiffness characteristics of an existing product.

The Ultracore design concept represented a local bonded-in reinforcement applied to the liftgate inner panel. The reinforcement provided additional bending stiffness by spanning the gap created at the lower outer corner of the backlight opening and tail light clearance dogleg kick-in corner (see Figs. 5 and 6).

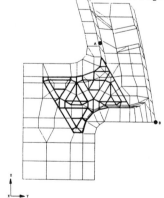

Fig. 5. Ultracore Application Study - Ford Aerostar Liftgate.

Fig. 6. Finite Element Model.

Finite element geometry models reflected both the baseline production and Ultracore inner panels. The outer panel was not used since this analysis was comparative and its outer panel's stiffness contribution would be a constant. Both models were subjected to a loading analysis simulating a rearward bending force applied at the gas strut attachment point.

NASTRAN plate elements defined the corner section of the panel. Beam elements simulated the remaining symmetrical about the centerline boundaries of the panel for loading and restraint purposes. Results of the analysis indicated that the reinforcements would add 0.9 pound to the weight of the liftgate, based on SMC with 20-percent glass content and 0.125-inch thickness.

Table I highlights analysis data comparisons between the baseline production and Ultracore liftgate inner panels. The results confirm that incorporation of the Ultracore reinforcement greatly improved the model's rigidity, especially in the lateral direction.

Table I - Analysis Summary Data.

Static Defl. (mm)	Base- line	UC	% Increase, Stiffness
Max Longitudinal	6.04	2.87	52
Max Lateral	0.89	0.12	86
Max Vertical	2.58	1.55	40
Movement (A)			
Longitudinal	4.62	2.79	40
Lateral	-0.24	0.03	87
Vertical	2.11	1.34	36
Movement (B)			
Longitudinal	6.04	2.80	53
Lateral	0.89	0.06	93
Vertical	2.36	1.26	46
Static Stress			
Avg. Max Von Mises Stress (psi)	2300	1700	(26%)

( ) Decrease

TETRACORE BEAM ANALYSIS -To determine the performance of Tetracore in a beam application, a 12-inch-long x 1.5-inch-high x 1.88-inch-wide Kevlar Tetracore beam section was compared with both a Kevlar and steel rectangular beam (see Fig. 7). The Tetracore beam and the Kevlar tube model were assumed to be fabricated from Kevlar 49 woven fabric and polyester resin. The analysis was conducted for both equal weight and equal thickness models subjected to 500-pound total pressure, 500-pound point load, and 10 ft-lb torsional moment. The results are presented as relative stiffness in Fig. 8. On an equal thickness basis, the Kevlar Tetracore beam

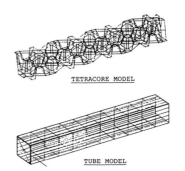

TETRACORE MODEL

TUBE MODEL

Fig. 7. Beam Models.

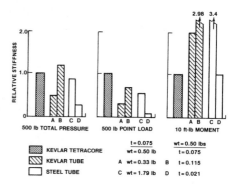

Fig. 8. Tetracore Beam Analysis Results.

was stiffer than either the Kevlar or steel tube under total pressure and point loading conditions, while for equal weight models, only the Kevlar tube under total pressure load proved stiffer than the Kevlar Tetracore beam. Not surprisingly, both the Kevlar and steel rectangular tubes were torsionally stiffer than the Tetracore beam. This can be attributed to the fact that the Tetracore beam contained only two upstanding tetrahedrons and one downward facing tetrahedron due to the width constraint of 1.88 inches; thus, the actual closed torque box structure was effectively triangular in cross-section with dimensions of 0.9 inch per side. More favorable results for torsional loading could be expected for wider sections, which would allow the torque box portion of the Tetracore element to approach those of a comparably wide rectangular section.

TETRACORE COMPONENT TESTING (STATIC AND DAMAGE)

SPECIMEN FABRICATION - The majority of the Tetracore test specimens were fabricated from cellulose butyl acetate (UVEX). The UVEX material was plasticized to provide a ductile failure mode during impact testing. The two halves of the Tetracore elements were vacuum formed on a mold consisting of a series of male tetrahedrons producing a wall thickness of one half the original thickness of the sheet material. In general, the two halves of the Tetracore elements were then nested and bonded together. Samples were also fabricated by bonding one half panel of Tetracore to a flat sheet of plastic. For some tests, aluminum face sheets of various thicknesses were bonded to either one or both sides of the completed Tetracore panel with epoxy resin.

For comparison, an aluminum riveted skin/stringer specimen (typical of helicopter fuselage construction) was also tested. This specimen is shown in Fig. 9. It should be noted that the Tetracore and aluminum specimens were not designed to the same strength/stiffness criteria and, thus, absolute comparisons were not possible.

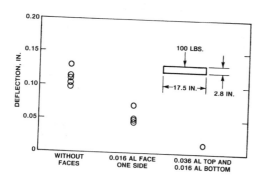

Fig. 9. Simulated Helicopter Fuselage.

STATIC TESTS - Three-point bending tests were conducted using nominally 12.5 x 23.5 x 2.8 inch long and .040 inch thick specimens with no face sheets, specimens with face sheets on one side, and specimens with face sheets on both sides. The influence of face sheets on the three-point deflection (17.5-inch total span) is shown by increases in stiffness by factors of approximately 2 and 10 for specimens with face sheets on one side and both sides, respectively, compared to specimens without face sheets (see Fig. 10).

The flatwise static compression test results of 2.8-inch- and 1/2-inch-thick panels are presented in Figs. 11 and 12, respectively. The results for 2.8-inch-thick specimens indicate yield stresses of approximately 26.8 psi and 63.5 psi for the

Fig. 10. Effect of Face Sheet on Three-Point Bending Behavior.

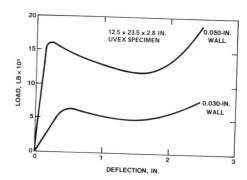

Fig. 11. Effect of Wall Thickness on Flatwise Compression Behavior.

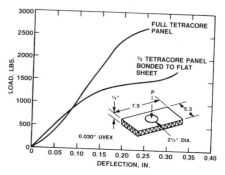

Fig. 12. Load/Deflection Comparison Between 1/2-Full Panels.

0.030- and 0.050-inch wall thickness material, respectively. After yielding, the load dropped off slightly until the specimen crushed approximately 1.5 inches, at which time a sharp rise in load occurred.

In comparison, the 1/2-inch-thick (UVEX) full Tetracore panel (0.030-inch wall thickness) yielded at approximately 460 psi (see Fig. 12). This difference can be attributed to the buckling stability of the tetrahedrons' triangular faces. Thus, it is important to size TC/UC for the required loading conditions recognizing that areal density is independent of the height of the element.

Flatwise compression tests were also conducted on a half Tetracore panel bonded to a flat sheet (Fig. 12). As expected, yielding occurred at approximately 255 psi or 55 percent of the full Tetracore panel since there are only half as many tetrahedrons to carry the load. Similar tests were also conducted on half panels of Tetracore bonded to flat sheets using three different materials UVEX, ABS, and LXB, which demonstrated yield stresses of approximately 255, 402, and 510 psi, respectively (see Fig. 13). These results indicate the need for appropriate material selection.

Figure 14 presents results of multi-axis static deflection tests on four 1/2-inch-thick UVEX full tetracore panels bonded together. Stresses for the flatwise loading condition were based on load footprint area, whereas the stresses for the end loading conditions at 0 and 30 degrees were based on net section area. End axis testing at 0 and 30 degrees indicated an approximate 10-percent reduction in yield stress for the 30-degree loading condition with respect to the 0-degree loading reflecting the relative quasi-isotropic nature of Tetracore.

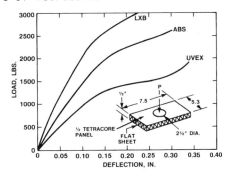

Fig. 13. Effect of Material on Load/ Deflection.

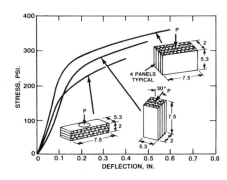

Fig. 14. Static Deflection Curves for Vacuum Formed Tetracore Panels.

IMPACT TESTS - After static testing, the 2.8- x 12.5- x 23.5-inch UVEX specimens discussed earlier were mounted in an impact machine and subjected to crash-impact testing at 10 to 14 feet per second. The weight of the falling mass was 448 pounds. Two impact footprints were used: an 8- x 10-inch flat plate and a 1-inch-wide picture frame with outside dimensions of 8 by 10 inches. Six 20-kip load cells were mounted beneath the specimen to measure load. Two piezoelectric accelerometers, one at the center of the falling mass and one 4 inches from the center, measured deceleration. The output of the accelerometers and load cells was recorded on magnetic tape and analyzed. Fig. 15 is a schematic of the impact testing equipment.

The results of the impact testing at 10 and 14 fps are presented in Figs. 16 and 17 and in Tables II and III. The failure mode was characterized by an accordian-type effect with progressively larger wavelengths emanating from the apexes of both the upstanding and inverted tetrahedrons.

Typical load and deflection time histories are presented in Fig. 18. In general, the specimens without face sheets demonstrated slightly higher values of transmitted load and deceleration at the 14 fps impact velocity. Tetracore specimens impacted by the 8- by 10-inch plate resulted in approximately a factor of 2 reduction over the aluminum skin/stringer specimen impacted with the 8- by 10-inch picture frame.

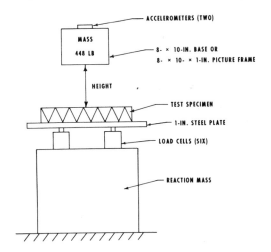

Fig. 15. Impact Test Equipment.

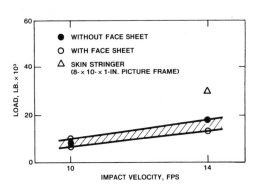

Fig. 16. Effect of Impact Velocity on Transmitted Load.

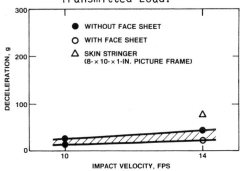

Fig. 17. Effect of Impact Velocity on Deceleration on UVEX Tetracore.

Table II - Test Results for Cellulose
Butyl-Acetate Tetracore
Specimens Without Face Sheets.

Wt. Gm	Impact Vel., fps	Total Transmitted Load, lb.	Peak Decel. g
840	14.0	18,885	44.7
1038	9.7	7,831	17.2
1015	10.1	6,689	14.3
1015	10.0	7,982	20.5

Table III - Test Results for Cellulose
Butyl-Acetate Tetracore
Specimens With Aluminum Face
Sheets.

Wt. Gm	Impact Vel., fps	Total Transmitted Load, lb.	Peak Decel. g
1333	14.0	13,375	24.3
1348	10.1	10,618	29.3
1318*	14.0	31,000	78.0

*Aluminum Skin/Stringer Specimen.

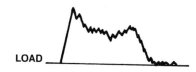
LOAD

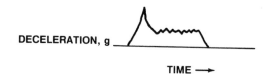
DECELERATION, g

TIME ⟶

Fig. 18. Typical Load and Deceleration
Time Histories.

CONCLUSIONS

Based on analyses and limited test results, wherein specimens compared were not all of the same material and design, the following conclusions have been made relative to the structural performance and crash-impact energy-absorbing behavior of TC/UC:

° TC/UC appears to be a quasi-isotropic structural concept and has demonstrated superior impact energy-absorbing characteristics compared to both aluminum honeycomb and typical aircraft aluminum riveted skin-stringer construction.

° The stiffness of Kevlar Tetracore beams compares favorably with Kevlar and steel rectangular tubes under both point and uniform loading conditions.

° TC/UC can be used effectively to provide local stiffening in existing structures.

° TC/UC lends itself to a wide variety of low-cost materials and fabrication techniques.

COLLAPSE TRIGGERING OF POLYMER COMPOSITE ENERGY ABSORBING STRUCTURES

Michael J. Czaplicki
The University of Michigan
Ann Arbor, Michigan   USA

Peter H. Thornton
Ford Motor Company
Scientific Research Labs
Dearborn, Michigan   USA

Richard E. Robertson
The University of Michigan
Ann Arbor, Michigan   USA

Abstract

A collapse triggering mechanism is needed in order for crash energy absorbing structures to deform in a controlled predictable manner. A trigger is made by producing some type of stress concentration from which failure initiates and then propagates through the body of a structure. Since polymer composite structures are presently being investigated for use as energy absorbing structures, the need was recognized to investigate triggers for this class of materials. Polyester and vinyl ester pultruded tubes were tested to better understand the fundamental aspects of the triggering of polymer composite structures. Triggers of different types were tested and found to produce different failure modes as well as different levels of energy absorption. Triggering was also found to be a material dependent process among polymer composite materials.

POLYMER COMPOSITE MATERIALS are being considered for replacement of sheet steel in a number of automotive structural applications. Polymer composite structures offer the potential benefits of parts integration, reduced weight, and improved performance. Before successful application, however, it must be determined if polymer composites can satisfy the requirements necessary for structural applications. In particular, satisfactory energy absorption or crashworthiness of composites is necessary for structural parts. One of the issues that must be resolved concerning crashwortiness, and the focus of this paper, is the triggering of suitable failure behavior. This paper details the behavior of several types of triggering mechanisms in order to gain a better fundamental understanding of the triggering process.

TRIGGERING

A trigger is a stress concentrator that causes failure to initiate at a specific location within a structure and propagate through the body in a controlled predictable manner. A trigger performs two primary functions:

1) It reduces the initial load peak which usually accompanies failure initiation.

2) It allows stable collapse to occur after failure initiation.

The elimination of the initial load peak is necessary to prevent deleterious high impulses at impact. A trigger accomplishes this by creating a specific mode of failure initiation, which allows collapse to propagate in a controlled manner. Without a trigger, a structure would likely fail in a catastrophic manner and absorb little energy. Since polymer composite structures do not deform plastically, a catastrophic failure would result in essentially complete loss of energy absorbing potential.

Four categories of triggers can be identified for polymer composite structures:

1) Trigger produced by modification of the end of structure to produce a stress concentration. This approach has been used frequently. Examples include the bevel, notch (1), and tulip triggers (2).

2) Triggers produced by modification of the body of the structure by either thinning or removing part of the cross-section.

3) Some structures with non-constant cross-sections are self-triggering. Hull found this to be the case for cones of constant thickness (3).

4) Structures may be triggered by using a splitting die or an extrusion die (4).

A fair amount of work has been done concerning the first category of triggers. In a number of studies concerning crash energy absorption of polymer composite structures, the bevel trigger has been used as a primary triggering mechanism (1-3,5,6-11).

Although the triggers are grouped in distinct categories, their effectiveness and functioning mechanisms may differ within a specific category. For example, Thornton (2) found differences in both the triggering initiation and subsequent failure propagation between tubes which were triggered with bevel and tulip triggers. He found that the initial load peak was lower and the mean crush load (P_m) was higher for tulip triggered specimens than bevel triggered specimens.

An exhaustive study of triggering mechanisms is difficult since the range of variables that can be tested, just in geometrical terms, is enormous. Consequently, testing in this study was limited to the analysis of representative specimens from the first, second , and fourth categories. This was done in order gain better fundamental understanding of the triggering process. Particular emphasis was placed on the study of triggering mechanisms from the second and fourth categories, as these triggers are the least well understood at this time.

To quantitatively evaluate the effectiveness of a trigger, the ratio of the mean load (P_m) to the absolute peak load (P_p) was compared.

$$\varepsilon = P_m/P_p$$

The peak load usually occurs at failure initiation. However, for triggers which are effective at eliminating the initial load peak, this value will occur in the stable failure region. The triggering effectiveness value can be skewed by a high initial load peak, lack of failure stability, or a combination of the two factors. Since a triggering mechanism affects both factors, the ratio is viewed as a useful comparison tool.

A triggering effectiveness value (ε) of 1.0 would represent a perfectly triggered structure. In reality, due to load variations, it is probably not possible to obtain values greater than 0.9. Values of 0.75 or greater generally represent a well triggered specimen and produce a near square wave curve contour.

Although this study primarily focused on tubes of square cross-section, some phenomena involved in the triggering process are generic. Square cross-section tubes were used in preference to circular cross-section tubes due to their higher flexural stiffness. In addition, square tubes are more similar to prototype vehicle structures currently being tested. Some of the results from square tubes should be easily transferrable to other structures, particularly those which possess corners.

COMPOSITE TUBE BEHAVIOR

The behavior of a polymer composite structure depends upon the individual components of the tube. Kevlar reinforced tubes, for example, fail in a manner similar to ductile metals (12). Glass and graphite tubes, however, fail in a brittle fracture mode, provided the matrix is stiff enough to accommodate this process. In this study, continuous glass reinforced polyester and vinyl ester pultruded tubes were used. All specimens used, fail in some form of progressive fracture mode. Glass reinforced composites, like those used in this study, are the composite materials of primary interest to automotive manufacturers due to their relatively low cost. Polyester and vinyl ester resins are the resins most likely to be used for automotive applications (10). Higher performance fiber reinforced materials such as graphite/epoxy are of primary interest to the aerospace industry.

The fracture of a polymer composite structure is a complex process. Factors affecting the failure include delamination, bending, kinking, matrix fracture, and fiber fracture (13). As one would expect, two tubes never collapse in exactly the same fashion. Thus it is not surprising that there is some variance in the data obtained. Because of data variance, values listed in this report are generally the average of three to five tests.

EXPERIMENTAL

All specimens were crushed axially on a 20,000 lb load capacity Instron machine at a strain rate of 0.5 in/min. All tests were also conducted at room temperature. Polyester (PE) and Vinyl ester (VE) glass reinforced pultrusions supplied by Morrison Molded Fiberglass were used. Results for 2 inch square cross-section tubes are reported. Polyester tubes with wall thicknesses of 0.125" and 0.25" were tested. Only .125" vinyl ester tubes were used. A schematic of the tube cross-section is shown in figure 1.

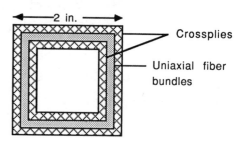

Figure 1: Cross-section of tube lay-up.

BEVEL TRIGGER

A bevel trigger is produced by grinding the end of a structure to a knife edge. A schematic of a bevel triggered tube is shown in figure 2. Because the bevel trigger is the most frequently used and best understood type of trigger, values from bevel triggered tubes are used as a baseline in this study. Average values for five tests are shown in Table 1. High and low values are also included to give an indication of the variance possible.

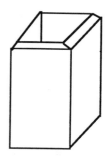

Figure 2: Schematic of a bevel triggered tube.

Table 1: Baseline data obtained from bevel triggered tubes.

Tube type	P_m(lbs)	High	Low	P_p (lbs)	High	Low	P_m/P_p
0.125" PE	5,540	6,390	5,120	10,000	12,000	8,690	0.55
0.125" VE	6,750	7,400	5,970	15,250	14,900	15,600	0.44
0.25" PE	12,480	11,200	13,500	18,460	20,000	16,700	0.68

Trigger initiation and crush propagation were different for vinyl ester and polyester tubes of identical dimensions. For tubes of each resin, a high stress concentration is created when the tube end contacts a hard crushing plate (mild steel). For polyester tubes, advancement of the crushing plate flattens the trigger knife edge and results in delamination of the beveled region. The initial load peak reaches its apex as the bevel becomes completely flattened. At this point cracks form simultaneously at the tube's four corners, resulting in a load decrease and downward crack propagation. When the cracks have propagated approximately 1 cm, a buckling of fibers occurs on each of the four tube faces, resulting in outward splaying of each wall. The crack formation permits peeling of the walls to occur. A load drop accompanies the crack formation and buckling. A curve generated for a polyester tube triggered with a bevel trigger is shown in figure 3.

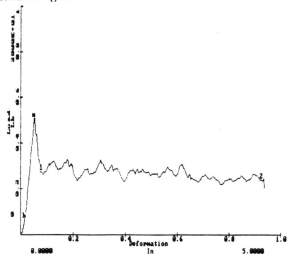

Figure 3: Curve from bevel triggered polyester tube.

Approximately 50% of the cross-section splays outward. The remainder is forced inward. The process of crack propagation and fiber buckling travels the length of the tube. At each buckling point, a whitened kink band is created. The load oscillations observed are largely the result of this phenomenon. Simultaneous observation of the curve generated and the crushing specimen showed that load drops accompany the formation of significant cracks. After buckling and formation of a kink band, the load rises until the propagation of cracks allow for the formation of another kink band.

In the crushing of vinyl ester tubes, cracks initiate at the corners almost immediately after load application. These cracks then propagate approximately 2 cm before fiber buckling. Peeling then initiates before the buckling occurs. The initial load apex occurs just prior to buckling, which is accompanied by a load drop and audible noise. The process of buckling occurs about every 2 cm and travels the length of the tube. After the initial buckling, however, simultaneous buckling does not occur again. Thus any time a length of one wall buckled, the load was buoyed by the three supporting walls. The load drops that accompany the individual wall bucklings are significant, however. Figure 4 shows the load/displcaement curve for the crushing of a vinyl ester tube.

The failure of vinyl ester beveled tubes differs from the polyester beveled tubes in five ways:

1) The apex of the initial load peak occurs at crack initiation at the four tube corners for polyester tubes. The apex of the initial load peak occurs immediately prior to the simultaneous buckling of the walls of the vinyl ester tubes.

2) The distance between kink bands is about twice as long in the vinyl ester tubes.

3) The fibers are more completely fractured at the kink bands of the vinyl ester tubes. Many fibers in the polyester tubes are not fractured in the kink bands and thus simply splay back.

4) Upon load release there is significant relaxation of fibers in the polyester tubes. No visible relaxation occurs for vinyl ester tubes.

5) There are more broken fibers among the vinyl ester debris.

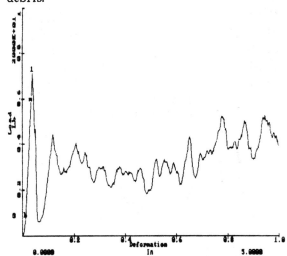

Figure 4: Curve from bevel triggered vinyl ester tube.

Examples of crushed vinyl ester and polyester tubes are shown in figures 5 and 6.

From tests performed with polyester bevel triggered specimens it was realized that the bevel triggering process in polyester tubes is controlled by two factors:

1) The ability to induce sufficient delamination to reduce the stiffness and strength of a given tube portion.

2) The formation and propagation of mode I cracks at the corners of the tube to allow peeling of the walls to occur.

For the vinyl ester tubes triggering is also controlled by two factors:

1) The initiation of mode I corner cracks.

2) The stress required to cause buckling of the tube walls.

The fact that beveled vinyl ester tubes do not trigger as well as polyester tubes (see table 1) is likely the result of the triggering mode differences. This highlights the fact that triggering is not only structurally dependent, but also material dependent.

It is possible to modify the bevel trigger to reduce the initial load peak. The modification is produced by making an angled cut on each corner to produce a knife edged v-notch. This results in failure initiation at a lower

Figure 5: Crushed bevel triggered polyester tube.

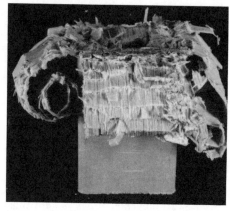

Figure 6: Crushed bevel triggered vinyl ester tube.

load since corner cracks are initiated at lower loads. By doing this the ε value is increased. However, P_m remains unchanged since an identical mode of failure as the standard beveled specimens occurs.

TULIP TRIGGER

Tulip triggered specimens were tested to determine if a mechanistic difference exists between the failure of bevel and tulip triggered tubes. A schematic of a tulip trigger is shown in figure 7. Test results supported Thornton's findings of increased energy absorption and increased triggering effectiveness. For 0.125 in. thick polyester tubes, P_m increased to about 7,250 lbs and ε increased to 0.70. These values represent more than a 20% increase over the baseline values listed in Table 1. For the vinyl ester tubes, P_m increased to 12,900 lbs and ε increased to 0.82. These figures represent a nearly 100% increase over the baseline or bevel triggered values. Representative curves for each tube type are shown in figures 8 and 9.

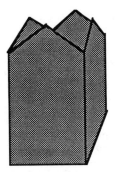

Figure 7: Schematic of the tulip trigger.

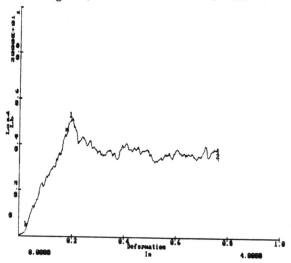

Figure 8: Curve for polyester tube with a tulip trigger.

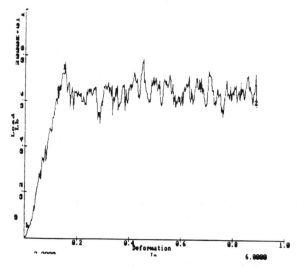

Figure 9: Curve for a vinyl ester tube with a tulip trigger.

No clear mechanistic difference was observed between the crush propagation of polyester bevel and tulip triggered specimens. During the triggering process, however, less delamination is induced by the tulip trigger. The reduced delamination may carry through after triggering. The appearance of the spent tubes, however, is macroscopically indistinguishable.

For vinyl ester tubes, there was a significant difference in the fracture of bevel and tulip triggered specimens. Unlike bevel triggered vinyl ester tubes, triggering did not result from simultaneous wall buckling. Instead, the crush zone propagated in a stable fashion throughout the entire triggering region. At the end of the triggering region failure occurred by a nonsimultaneous buckling mode. The change of failure initiation is largely a result of the elimination of the corner cracks which propagated well in front of the crush zone for bevel triggered vinyl ester tubes. The formation of those cracks results in the splaying of the walls, which reduces the structure's stiffness. Failure initiates when the applied moment to the splayed tube walls is great enough to result in fracture of the wall.

The failure behavior of tulip triggered tubes is different from bevel triggered tubes for the entire crush length. Figure 10 shows the a crushed tulip triggered vinyl ester tube. The biggest difference between the spent tubes is the distance between kink bands. Kink band distances range from 5 mm to 2 cm for bevel triggered tubes, but range from 1 mm to 5 mm for tulip triggered tubes. Accordingly, the debris obtained from the tulip triggered tubes was smaller.

Figure 10: Crushed tulip triggered vinyl ester tube.

The decrease in the distance between kink bands is the probable reason for the increase in energy absorbing ability. The distance between kink bands decreased for the tulip triggered tubes because cracks are prevented from propagating in the corners. Thus the fibers stay upright, which prevents the walls from peeling, resulting in increased fiber breakage. The decrease in crack propagation is in part supported by the fact that acoustic emission is much less for the tulip triggered tubes. No loud noises occur when the tulip triggered tubes are crushed.

A probable cause of the lower crush loads obtained for bevel triggered polyester tubes is the high degree of longitudinal delamination which is initiated. As mentioned previously, the bevel trigger is quite effective at inducing delamination in polyester tubes. Although this does aid in triggering initiation, propagation of this delamination below the crush zone would result in lowered stiffness and strength of the material at the crush zone, resulting in lower energy absorption.

HOLE TRIGGER

The hole trigger is representative of a different category of triggers. This trigger was produced by drilling four adjacent holes near the top of the tube as shown if figure 11. The hole trigger works by creating a stress concentration in the columnar pieces of material between the holes. Load application results in buckling in the middle of the columns. The buckling produces a shearing effect in which the inner and outer plies of material are ripped away from the wall. The shearing initiates the delamination. The delamination, however, is less extensive than for bevel triggered tubes.

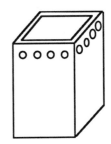

Figure 11: Hole trigger.

The crush zone propagates from the middle of the holes. When the holes have been completely closed, the material from the upper portion of the tube begins to be wedged into the delaminated wall section. This results in the formation of corner cracks due to the swelling of the wall thickness. By the time the platen contacts the bottom portion of the tube, corner cracks have been initiated and peeling of the walls and stable crush propagation have begun.

A curve generated from a specimen with a hole trigger is shown in figure 12. As with the bevel trigger, an initial load peak occurs. The load then drops dramatically at failure initiation. The load remains low until the crush has propagated past the trigger region. Unlike the bevel trigger, the load returns to a value approximately equivalent to the initial load peak.

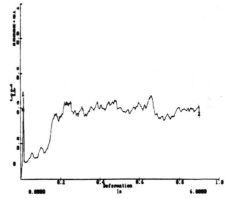

Figure 12: Curve obtained for hole triggered polyester tube.

A modified version of this hole trigger can be made (figure 13). By staggering the holes, a smoother curve results.

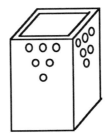

Figure 13: Modified hole trigger.

43

Results from 0.125 in. thick polyester tubes using different hole sizes for the modified hole trigger are shown in Table 2. P_i refers to the initial load peak. P_p refers to the absolute load peak. P_m divided by the baseline (β) refers to the ratio of hole triggered specimens to the baseline values listed in Table 1 (bevel triggered tubes).

Table 2: Results for hole size variation in the modified hole trigger.

Hole size	P_i (lbs)	P_m (lbs)	P_p (lbs)	P_m/P_p	P_m/β
.15 in. or less		Does not trigger			
.20 in.	10,700	8,000	10,700	.75	1.44
.25 in.	7,800	7,700	10,100	.77	1.39
.28 in.	5,000	7,200	9,400	.77	1.30

The results listed above are similar to the results for tulip triggered tubes. The results were also similar for vinyl ester tubes. P_m is about 12,000 lbs, which is quite close to the value obtained for the tulip triggered tubes. The failure mode after triggering was identical to the tulip triggered tubes. As with the tulip triggered tubes, the inhibition of corner crack running resulted in a shorter distance between kink bands.

The similar results for the hole and tulip triggered tubes suggest that a similar mechanism is operating in the two cases. Since both triggers exhibit reduced delamination (for polyester tubes) relative to the bevel triggered tubes, it supports the idea that large longitudinal cracks are responsible for the low crush loads of bevel triggered tubes. Thornton (2) has suggested that the tulip trigger produces more cracks than the bevel trigger, but the cracks do not penetrate as deeply. Thus it is quite possible that this mechanism is also operating in hole triggered tubes.

Trigger placement in relation to the top of the tube had little effect on the triggering process. Placing the trigger farther from the top of the tube changed the location of trigger activation. This altered the initiation part of the curve but left the rest of the curve unchanged.

The length of the tube was not a factor in the crush stability. Tubes up to 10 inches long were tested with no change in failure activation or propagation. As long as the tube length is less than the buckling length, the failure of the tube should remain unaffected by the tube length.

MECHANICAL TRIGGERING

A new type of mechanical trigger was used in this study (figure 14). Previous mechanical triggers worked by forcing a beveled tube over a splitting die. The extrusion trigger works by forcing an untriggered tube through a tapered die. The die opening is of the same shape as the tube cross-section. For a square tube, a truncated pyramid is used.

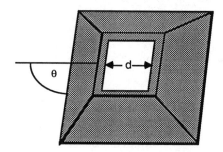

Figure 14: Extrusion trigger.

The geometry of the trigger can be described by two variables: the angle of inclination (Θ) and the diameter of the opening (d). Three different triggers with the following parameters were tested:

#1 d= 1.75 in. Θ= 45 degrees   #2 d= 1.75 in. Θ= 60 degrees
#3 d= 1.5 in. Θ= 45 degrees

When the trigger is first placed over the tube, the opening is about 0.5 in. from the top of the tube. Loading results in the combined effect of blunting the sharp tube edge and reducing the cross-section of the tube. Continued blunting of the edge results in an increased load level. By the time the end of the tube reaches the opening, considerable biaxial deformation has occurred. The frictional force from the normal force applied to the tube accounts for the load level prior to tube eruption. As the tube exits the die, material is compressed in the longitudinal direction, primarily at the corners, resulting in material being broken and then scraped off of the tube. After tube exit, the load stabilizes to a characteristic level.

Results are shown for the three tube types using this trigger in Table 3. In addition, characteristic curves and a crushed specimen are shown in figure 1. The following results were obtained:

Table 3: Average values obtained using the extrusion trigger.

Material	Wall	Trigger	P_m	P_p	P_m/P_p	P_m/β
Polyester	0.125 in.	#2	3,800	4,900	0.78	0.68
Polyester	0.125 in.	#1	4,100	5,200	0.79	0.75
Polyester	0.125 in.	#3	5,700	7,100	0.80	1.02
Polyester	0.25 in.	#2	9,100	11,200	0.81	0.73
Polyester	0.25 in.	#1	10,100	13,000	0.78	0.81
Polyester	0.25 in.	#3	13,600	17,200	0.79	1.09
Vinyl ester	0.125 in.	#1	6,400	8,600	0.74	0.95
Vinyl ester	0.125 in.	#3	7,300	9,100	0.80	1.08

Figure 15: Spent specimen with trigger still attached.

The curves obtained were close to the idealized case. Unlike the three triggering mechanisms previously described, this trigger does not rely primarily on a fracture process. Rather, this trigger relies largely on the frictional force required to push the tube through the die. A smoother curve contour results from the reduced reliance on fracture figure 16. The reduction of crack formation and inhibition of the limited number of cracks that did form are the primary reasons for the load stability. As stated previously, the formation and running of large cracks is thought to be one of the reasons for reduced load stability.

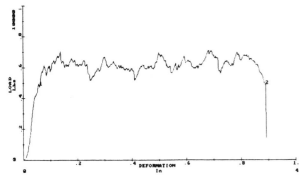

Figure 16: Curve from extrusion triggered specimen

Crack propagation well below the crush zone is inhibited for extrusion triggered tubes. In bevel triggered tubes, for example, mode I crack propagation is encouraged due to the splitting nature of the bundle wedge of material as described by Hull (8). The compressive transverse loading produced by an extrusion trigger inhibits the propagation of cracks in this manner. As indicated previously, the formation and running of significant cracks is thought to be a reason load instability occurs. Therefore, by inhibiting this behavior, it should be possible to stabilize the loading behavior. Figures 17 and 18 contrast the difference between the loading conditions of the two cases.

The crush loads obtained using the extrusion trigger were lower for two of the triggers and slightly greater for the trigger with the smallest opening. The most distinguishing factor is the consistency of the trigger effectiveness ratio. The value was quite uniform despite changes in the trigger geometry and the tube material.

This suggests that changes in the test parameters do not affect the failure mode.

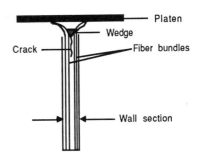

Figure 17: Tensile transverse loading (wall thickness).

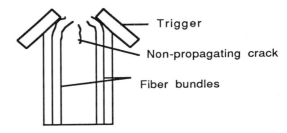

Figure 18: Compressive transverse loading (tube face).

The extrusion trigger could be classified as an alternative method of absorbing energy and not simply a triggering mechanism. The nature of the extrusion trigger suggests that it could be applied successfully to other material classes other than polymer composites.

CONCLUSION

The behavior of polymer composite energy absorbing strucures was found to depend significantly on the type of triggering mechanism used. The levels of energy absorption, particluary for the vinyl ester type of pultruded tube differed greatly among different types of triggers. Large differences of energy absorption can be primarily attributed to the mechanism of failure produced by the triggering mechanism. In addition, the mode of failure produced is not universal for all types of triggers.

A number of aspects of triggering need to investigated. These include the rate effect of loading, effect of geometrical scaling of the structure, as well as relation to other types of structure geometries. From the initial work, it appears as if triggering is largely specific to a given system application and it may be difficult to generate universal triggering rules.

REFERENCES

1) G.L. Farley, Energy absorption of composite materials, *J. Compos Mater.,* 17 (1983) pp. 267-79.

2) P.H. Thornton, Effect of trigger geometry on energy absorption of composite tubes, *ICCM-V,* Eds Harrington, Strife, and Dhingra, Metallurgical Society, Pennsylvania, (1985), pp. 1183-99.

3) J.N. Price and D. Hull, Axial crushing of glass-polyester composite cones, *Compos. Sci. and Tech.,* 28 (1987), pp 211-230.

4) P.H. Thornton, private communication, Ford Motor Co., Dearborn, Michigan, 1986.

5) D. Hull, Energy Absorbing Composites Report,

University of Liverpool, September, 1984.

6) P.H. Thornton, Energy absorption in composite structures, *J. Compos. Mater.*, 13 (1979) pp. 247-62.

7) P.H. Thornton and P.J. Edwards, Energy absorption in composite tubes, *J. Compos. Mater.* 16 (1982) pp.521-545.

8) A.H. Fairfull and D. Hull, Energy absorption of polymer matrix composite structures: Frictional effects. *International symposium on structural failure*, MIT, June 1988.

9) A.H. Fairfull and D. Hull, Effects of specimen dimensions on the specific energy absorption of fiber composite tubes. *ICCM-VI* Eds. Matthews et. al., pp. 3.36-3.45.

10) P. Beardmore and C.F. Johnson, The potential of composites in structural automotive applications, *Compos. Sci. and Tech.*, 26 (1986), pp. 251-281.

11) J.N. Price and D. Hull, The crush performance of composite structures, *Composite Structures*, Ed. Marshall, 1987, Elsevier, pp.2.32-2.44.

12) G.L. Farley, The effects of crushing speed on the energy absorbing capability of composite material, *NASA Technical Memorandum*, March, 1987.

13) P.H. Thornton, J.J. Harwood, and P. Beardmore, Fiber reinforced plastic composites for energy absorption purposes, *Compos. Sci. and Tech.*, 24 (1985) pp. 275-298.

ENERGY ABSORPTION IN CRUSHING FIBER COMPOSITE MATERIALS

Wess H. Tao
The University of Michigan
Department of Materials Science
and Engineering
Ann Arbor, Michigan USA

Richard E. Robertson
The University of Michigan
Department of Materials Science
and Engineering
Ann Arbor, Michigan USA

Peter H. Thornton
Ford Motor Company
Scientific Research Laboratory
Dearborn, Michigan USA

ABSTRACT

The nature of the crush process, factors affecting crush stability, and the contribution of various energy sinks to the total energy absorption in crushing of fiber composite structures have been examined. Crush appears to be composed of a repeated series of bending and breaking of fibers and matrix accompanied by plastic deformation in the matrix. Adventitious cracks ahead of the crush zone affected the stability of crush. Factors such as interfacial bonding, hoop fibers and a hemispherical top were found to be important in stabilizing these cracks. Increased interfacial area by using smaller fibers did not give higher energy absorption. Most of the energy of crushing was absorbed by plastic deformation in the matrix and friction. Friction involved both that between debris in the crush zone and between debris and the crushing platen.

FIBER REINFORCED PLASTIC STRUCTURES are known to be very good energy absorbers when crushed. Energy is absorbed by fracturing the materials in a progressive manner through the structures. The most efficient energy absorption occurs when the load is kept at a constant value during crush. Extensive investigations have been conducted on crush energy absorption in different composite material systems and lay-up structures [1-4]. The effects of other factors such as rate dependence, temperature dependence, trigger geometry, and section geometry have been examined [4-7]. Few attempts have been made to understand the crush process, crush stability, and the contribution of various energy sinks to the total energy absorption. Crush stability affects the efficiency of energy absorption. Through an understanding of the crush process, one may better control the stability of crush. An appreciation of the contribution of various energy sinks to the total energy absorption is important in understanding the performance of an energy absorber. Among the possible energy sinks are fracture surface energy from the matrix, fibers and interface between them, plastic deformation energy from the matrix, and friction energy expended during crush.

Hull et al. found that the friction energy between the fiber composite material and the contact surface in axial compression accounted for about one-half of the total energy expended during static compression [7]. They also analyzed the fine fracture debris of crushed Graphite/Epoxy tubes and calculated the fracture surface energy expended during axial compression [8]. The results showed that the contribution from fracture surface energy was small, only 1-2% of the total energy. As for the contribution from plastic deformation in the matrix, it is not well understood yet.

The primary object of this study was to determine the major energy sinks. Various methods were used, such as varying interfacial area, fiber diameter, fiber content and using a hemispherical to crush the specimens. Attempts also were made to understand the process and stability in crushing of continuous fiber composite materials.

SPECIMENS AND PROCEDURES

ROD SPECIMENS - Conventionally, square or round fiber composite tubes were used in energy absorption tests. To avoid the complication from specimen geometry, simple rod specimens were used in most of this study. Composite rods were fabricated in glass tubes 7 mm O.D. and 5 mm I.D. Continuous fibers were laid inside glass tubes before resin was injected. Different amounts of fiber were put into these glass tubes to vary the fiber content. The whole system was cured at an elevated temperature (90°C) for 25 minutes and post-cured at 120°C for 3 hrs. The long composite rods (5 mm in diameter) were removed from the glass tubes by smashing the glass shell of a cured system and then cut into small rod specimens. Small rod specimens with one end beveled were compressed by a flat plate or by a concave hemispherical top. The beveled ends were produced by grinding in order to initiate or trigger fracture. The materials used were continuous E-glass fibers by Owen-Corning Fiberglas Co. and vinyl ester by Dow Chemical Co. Three different fiber diameters were used, 13, 17 and 23 microns. Fibers were unidirectional along the compressive longitutinal axis. Rod specimens were semi-transparent as shown in Fig 1. They were crushed in an Instron machine at 2.54 mm/min.

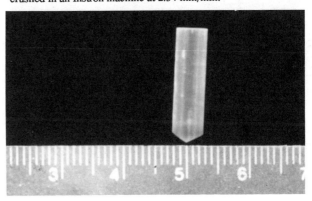

Fig. 1 - A fiber composite rod specimen

PULTRUSION TUBES - In order to compare the crush phenomena between simple fiber composite rod specimens and conventional structural members, such square or round tubes, fiber composite pultrusion tubes were tested. Square pultrusion tubes were made of polyester resin and continuous E-glass fibers using two different fiber diameters, 13 microns or 23 microns. This gave two kinds of pultrusion tubes with the same bulk geometry, lay-up structure and fiber content (42.6 vol%). Tubes with one end beveled, as shown in Fig 2, were crushed in an Instron machine at a strain rate 2.54 mm/min.

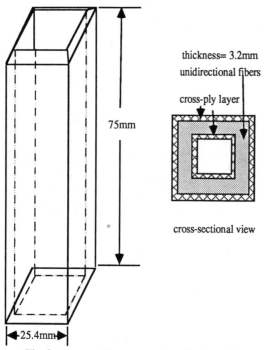

Fig. 2 - A square fiber composite pultrusion tube

RESULTS

CRUSH PHENOMENA OF ROD SPECIMENS BY A FLAT PLATE - A rod specimen with one end beveled crushed by a flat plate is shown in Fig 3. A load vs crush distance plot is shown in Fig 4. The area under load-crush distance curve like that in Fig 4 gives the energy absorbed by crushing the specimen. Since the rod specimens were semi-transparent, the following observations were made on the nature of crush.

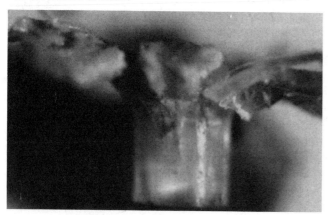

Fig. 3 - A crushed fiber composite rod

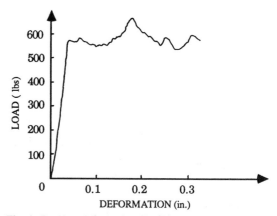

Fig. 4 - Load vs deformation plot in crushing a fiber composite rod

A narrow crush zone composed of fractured fibers and matrix developed between the unfractured part and the top platen. This crush zone propagated through the specimen in a progressive manner as the crosshead came down. For crush with a large number of extremely short cracks ahead of the crush zone, the load was stable as in Fig 4. Unstable crush often resulted from a long crack (or cracks) ahead of the crush zone. This may due to there being a longer moment arm present to open the cracks, which then destroys the structural integrity.

Cracks were seen as light-reflecting or white planes going through the semi-transparent rod specimens during crush. The cracks in beveled rod specimens tended to be parallel with the ridge line. Further investigations were conducted by using specimens with round tipped triggers as shown in Fig 5 to obtain the relationship between crack direction and the trigger geometry. The crack shape in specimens with round tipped triggers were different from those in beveled trigger by the observations of the white planes during crush. Two rod specimens, one with beveled trigger and the other with round tipped trigger, were crushed for 2 mm on the top. The crushed specimens were sectioned and polished before put under a microscope to examine the morphology of the crushed tops. Crack directions were related to the trigger geometries as shown in Fig 5.

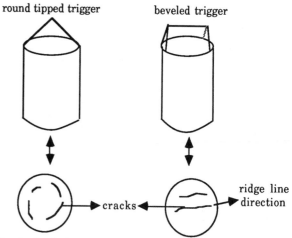

Fig. 5 - Crack directions are related to the trigger geometry

Higher magnification of the crushed top showed that the cracks went through the interface between the matrix and fibers.

Another vinyl ester resin fabricated by Hitachi Chemical Co. was used to make rod specimens similar to those in Fig 1. The same fibers were used. The crush of these rod specimens was always unstable. A large number of cracks penetrated to the bottom in the early stage of the crush. The load gradually dropped after the initial peak. In contrast to these rod specimens, similar rod specimens made from Dow Chemical Co. vinyl ester had a much more stable crush.

Since the specimen geometry, fiber content, fiber size and the mechanical properties of the matrix were similar, a possible cause for the difference in stability may have been a difference in interfacial bond strength.

EFFECTS OF INTERFACIAL AREA, FIBER SIZE AND CONTENT ON ENERGY ABSORPTION - Fracture surface energy, which may come from fibers, the matrix and interface between them, was assumed to be one of the major energy sinks in the past. The total fracture surface energy from fibers is not likely to be a major energy sink because the fracture surface energy and total fracture surface area of fibers are small. Crack propagation along interface and broken fibers with clean surface found in crushing debris indicated that most of the fracture surface may come from interface between fibers and the matrix.

The importance of the fracture surface energy from interfacial area was investigated by crushing specimens containing each of the fiber diameters, 13, 17 and 23 microns and different fiber contents. Effects of fiber diameter and content on energy absorption could also be seen here. Since all rod specimens in the crushing series had the same cross-sectional area, the average load can be used to indicate energy absorption per unit volume. For the rod specimens with 23-micron fibers, specimens with four different fiber contents were crushed. The average load vs fiber volume percentage plot is shown in Fig 6. Similar results for rod specimens with 17-micron fibers and 13-micron fibers are also shown in Fig 6.

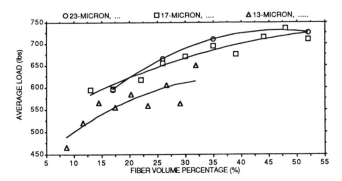

Fig. 6 - Average crush load vs fiber volume percentage in crushing fiber composite rods by a flat plate

It was difficult to pull the fibers through the glass tube at high fiber contents, and there were limitations on the maximum fiber content. For rod specimens with 23-micron and 17-micron fibers, this maximum was 52 vol%. For rod specimens with 13-micron fibers, the maximum fiber volume percentage obtainable was 32%. It is clear from the curves in Fig 6 that the average crush load (or energy absorbed per unit volume) increased with increase in fiber content, but that the trend tended to level off at high fiber contents.

Since all rod specimens had the same cross-sectional area, the interfacial area was inversely proportional to the fiber diameter provided fiber contents were the same. It is seen in Fig 6 that the rod specimens with larger fibers (or smaller interfacial area) had higher average crush load than those with smaller fibers (or larger interfacials area) when the fiber contents were the same. This indicates increasing the interfacial area by using smaller fibers does not give higher energy aborption. It is likely that fracture surface energy from interfacial area is not a major energy sink. Energy absorption fell even with a 77% increment in interfacial area. The interfacial area may not be the only factor which affects the energy absorption of fiber composite rods in this experiment. It is possible that the higher buckling load from larger fibers is a dominant factor which gives higher energy abosrption.

ROD SPECIMENS CRUSHED BY A HEMISPHERICAL TOP - A rod specimen crushed by a concave hemispherical top is shown in Fig 7. It had a mushroom shape with more damage in the peripheral region than that crushed by a flat plate as shown in Fig 3. For the rod specimens with 23-micron fibers, specimens with four different fiber contents were crushed. The average load vs fiber volume percentage plot is shown in Fig 8, compared with those crushed by a flat plate. It is clear that the energy absorption was higher (30-50%) when a concave hemispherical top was used. This may be due to extensive damage in fiber composite materials, caused by the external constraint from a hemispherical top.

Fig. 7 - A rod specimen crushed by a hemispherical top

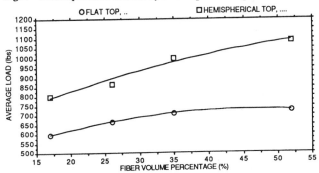

Fig. 8 - Average load vs fiber volume content for rod specimens crushed by a flat plate and a hemispherical top

A more significant result in this experiment is the regular bands found of broken fibers in the crush zone. They are important because the amount of plastic deformation energy in the matrix can be estimated. For a rod specimen crushed by a hemispherical top, four perceptible energy sinks are analyzed as follows.

Plastic deformation in the regular bands - A crushed rod specimen was sectioned and polished and examined under a microscope. Four regular bands with about the same band length (80 microns) appeared next to the uncrushed parts. The inclination angles of bands increased from 45^{0} in the first band to 60^{0} in the fourth band as labeled in Fig 9. Similar kink bands were also found in compression [9-10] and crush [6] of fiber composite materials by other investigators. The rotation of the broken fibers in the first band may have come from release of the elastic strain energy stored in the bent fibers, which may have resulted in plastic work in the matrix. Further compression of the bands resulted in an increase of inclination angle from 45^{0} to 60^{0}. A quasi-steady approach was used to estimate the contribution of plastic deformation energy to the total energy absorption by assuming that the matrix in the regular bands was yielding. Approximately 32% of the total energy expended in crushing is estimated to have been turned into plastic deformation energy in the regular bands. (see Appendix)

Possible plastic deformation in the upper region of the crush zone - No regular bands were found in the region beyond the fourth band in the crush zone. This may have arisen from a well-developed band having lost its regularity as it was squeezed in the crush direction and had more space in the circumferential direction during the process of being squeezed out. The possible plastic deformation in this region is given by the flow of materials in a two-dimention diagram shown in Fig 10.

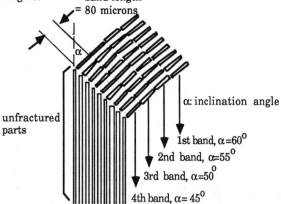

Fig. 9 - Regular bands of broken fibers above the unfractured parts

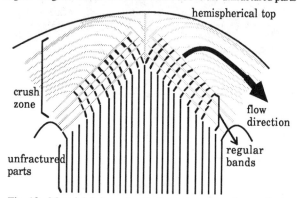

Fig. 10 - Material deformation in the region above the regular band
Friction energy at the interface between the hemispherical top and rod specimens - Friction occurred as fractured materials traveled along the curved surface as shown in Fig 11. An estimation can be made by assuming a uniform load distribution at the contact surface. The traveling distance and the total normal force at certain angle, θ, were r dθ and P cosθ respectively. The friction energy, as fractured materials traveled from point A to point B, was

$$\int_{0^o}^{90^o} P \cos\theta \ f \ r \ d\theta$$

where P : the average load
f : the friction coefficient
r : the radius of the hemispherical top
The total energy absorption was the average load, P, times the traveling distance of the cross-head, rπ/2. Fraction of friction energy to the total energy absorption was

$$\frac{P \ f \ r}{P \ \pi \ r/2} \ 100\% = \frac{2 f}{\pi} \ 100\%$$

The friction coefficient, f, was investigated by different researchers[6,8] and a reported value of 0.41 was used here. This gave a corresponding value of 26% of the total energy.
Friction energy betwen debris in the crush zone - The debris in the crush zone is rubbed together and with partially fractured parts, which has resulted in friction energy dissipation. It is difficult to estimate this friction energy because the local friction force and friction coefficient in a crush zone can not be clearly defined.

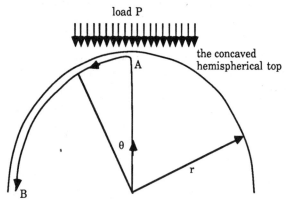

Fig. 11 - Friction between a hemspherical top and the crushed specimen

According to the above analysis, the contribution of various energy sinks to the total energy absorption can be summerized as:
1. plastic deformation energy in the regular bands : 32%
2. possible plastic deformation in the region above regular bands
3. friction between the hemispherical top and rod specimens : 26%
4. friction between debris in the crush zone

The results indicate that plastic deformation energy in the matrix and friction expended during crushing are important energy sinks in crushing fiber composite rods.

PULRUSION TUBES CRUSHED BY A FLAT PLATE - Results from crushing pultrution tubes are presented in this section. Typical load vs deformation length plots for two kinds of pultrusion tubes are shown in Figs 12 and 13. The average loads for these tubes were

	average load
pultrusion tubes with 13-micron fibers	4180 lbs
pultrusion tubes with 23-micron fibers	4650 lbs

When the two kinds of square pultrusion tubes used had the same matrix material, bulk geometry, lay-up, and fiber content (42.6 vol%), the tubes with the larger fibers gave a higher average load. These results were consistent with those from rod specimens.

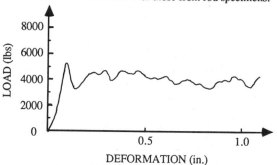

Fig 12 : Load vs deformation plot of a pultrusion tube with 13 μm fibers

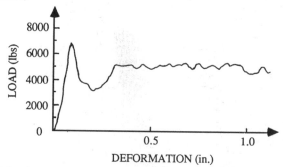

Fig 13 : Load vs deformation plot of a pultrusion tube with 23 μm fibers

Plate specimens were cut from the square pultrusion tube as indicated below.

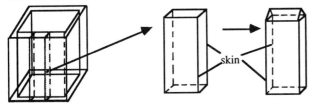

When plate specimens were beveled to 45° and crushed by a flat plate, the load soon fell to zero during the crush. A few cracks parallel with the ridge line penetrated to the bottom, separating the plate specimen into individual thinner plates as shown in Fig 14. When these thinner plates were not lined up with the crush axis, they just fell away and caused the load to drop to zero. In contrast to the stable crush in a pultrusion tube, the crush of this plate specimen cut from a square pultrusion tube was very unstable. The cross-ply layer, as shown in Fig 2, may be important to the crush stability of a pultrusion tube. This cross-ply layer goes around the pultrusion tube acting as hoop fibers. For crush of the plate cut from pultrusion tubes, the instability might have come from the lack of hoop fibers, since the cross-ply layer did not wrap around this plate specimen.

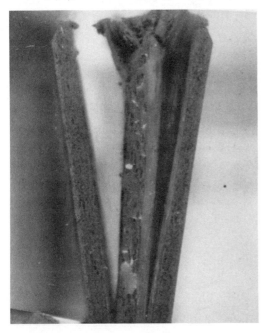

Fig. 14 - A plate specimen crushed by a flat plate

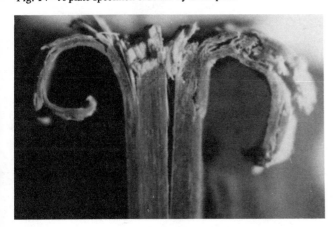

Fig. 15 - A plate specimen crushed by a hemispherical top

The same kind of plate specimen cut from a square pultrusion tube was crushed by a hemispherical top. The crush was stable, and the load did not fall to zero. A plate specimen crushed by a hemispherical top is shown in Fig 15. The hemispherical top therefore seems to function in a similar way to that of the cross-ply layer, offering a clamping force opposing crack opening.

SUMMARY

A crush process can be described as a repeated series of bending and breaking of fiber composite materials accompanied by plastic deformation in the matrix. The stability of crush depends on the adventitious cracks ahead of crush zones. In these experiments, it was found that interfacial bonding, hoop fibers and a hemispherical top were important in stabilizing the propagation of these cracks. Stable crack propagation tended to confine the crush zone in a small region in order to get optimum energy absorption.

The effects of fiber size and content on energy absorption was tested by varying fiber diameters (13 microns, 17 microns and 23 microns) and fiber content. Specimens with the largest fibers (23 microns) gave the highest energy absorption. The average load (or energy absorption per unit volume) increased as fiber content increased, and gradually leveled off at high fiber contents. Test results also showed that increasing the interfacial area by using smaller fibers did not give higher energy absorption. Fracture surface energy from the interface was unlikely to be an important energy sink.

In crushing a rod specimen by a hemispherical top, plastic deformation in the matrix and friction appeared to be major energy sinks. Friction involved both that between debris in the crush zone and between debris and the crushing platen.

REFERENCES

1. J. D. Cronkhite, T. J. Hass, V. L. Berry and R. Winter, Investigation on the crash impact characteristic of advanced air frame structure, USARTL-TR-79-11, Sep. 1979.
2. G. L. Farley, Energy absorption in composite structure, J. Compos. Mater., 17, 267-279 (1983).
3. D. W. Schmuser and L. E. Wickliffe, Impact energy absorption of continuous fiber composite tubes, Trans. ASME, 109, 72-77 (1987).
4. P. H. Thornton, Energy absorption in composite structure, J. Compos. Mater., 17, 247-262 (1979).
5. P. H. Thornton and P. J. Edwards, Energy absorption in composite tubes, J. Compos. Mater., 16, 521-545 (1982).
6. P. H. Thornton, unpublished research, Ford Motor Co., Dearborn, Michigan, 1985.
7. J. N. Price and D. Hull, Axial crushing of glass fiber-polyester composites, Compos. Sci. and Tech., 28, 211-230 (1987).
8. D. Hull, private communication to P. H. Thornton, Univ. of Liverpool, 1984.
9. A. G. Evans and W. F. Adler, Kinking as a mode of structural degradation in carbon fiber composites, Acta Metallurgica., 26, 725-738(1987).
10. C. W. Weaver and J. G. Williams, Deformation of a carbon-epoxy composites under hydrostatic pressure, J. Mater. Sci., 10, 1323-1333(1975).

APPENDIX

PLASTIC DEFORMATION IN THE REGULAR BANDS OF BROKEN FIBERS- The rod specimen gave regular bands was made of 17-micron fibers and 50 vol% vinyl ester. The calculation of plastic deformation energy was based on crushing of one band length. The total energy absorption was the average crush load times the traveling distance of the cross-head (i.e. one band length).

total energy absorption = P l_k = 0.338 J

where P : average crush load = 950 lb_f

l_k: band length = 80 x 10^{-6}m

The plastic deformation energy stored in a band was $\tau \delta V$.

where V is the deformed volume

δ is the shear strain

τ is the shear yield stress

The deformed volume, V, had a conical bottom surface and a height of one band length as shown in Fig 16. V = 1.71 x 10^{-9} m^{-3}

The shear strain, δ, was

$$\delta = \frac{\alpha}{180^o} \pi = 1.05 \text{ (in radian)}$$

where α is the largest visible inclination angle in the bands, i.e. 60°. A shear yield stress of τ = 61.1 MPa was used in this calculation. This gave a value of 0.109 J as the plastic deformation energy stored in the band, which corresponds to 32% of the total energy absorption.

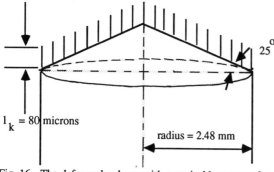

l_k = 80 microns

radius = 2.48 mm

25°

Fig. 16 - The deformed volume with a conical bottom surface and a height of one band length

CRUSH BEHAVIOUR OF SQUARE SECTION
GLASS FIBRE POLYESTER TUBES

J. N. Price
Department of Materials Science and Metallurgy
University of Cambridge
Pembroke St., Cambridge CB2 3QZ   UK

D. Hull
British Petroleum Co. plc,
Research Centre, Sunbury-on-Thames,
Middlesex TW16 7LN, UK

ABSTRACT

Previous studies of the axial crushing of round section composite tubes, to be used as energy absorbing devices, have been extended to cover square cross-section tubes. The results described here relate to compressive testing of various square section tubes made from a random glass fibre/polyester material. The tests were made over a range of speeds from 0.1×10^{-3} m/s to 4 m/s and in simulated crash tests at initial speeds up to 18 m/s (40.3 mph). Failure mode depended on corner radius, wall thickness and sample height. Increasing corner radius and wall thickness, and reducing height, favours stable crushing and results in high energy absorption. Stable crushing is promoted by failure of the corner regions with a crush zone morphology similar to that observed in round section tubes. This morphology extends into the side walls of a square tube if the corner radius and/or wall thickness are relatively large. Specific energy absorption increases with corner radius and wall thickness but values do not equal those for the more efficient round tubes. It is concluded that the stress field which generates a stable crush zone morphology must be understood and related to structural geometry if more complex geometries are to be used as energy absorbing devices.

THE CRUSH PERFORMANCE OF STRUCTURES made from polymer matrix fibre reinforced composite materials has been studied in great detail in recent years [1-8]. This work has been driven by the need for lightweight structures which, when subjected to impact, fail in a progressive manner absorbing significant amounts of energy. The obvious application of such structures is the automobile. Frontal 'chassis' members must support the engine and transmission during normal use, but fail progressively and absorb energy during impact, decelerating the vehicle and protecting the passengers.

It has been shown that cylinders made from a wide range of composite materials fail by progressive crushing if this failure is 'triggered' by a stress raiser, such as a chamfer at one end, provided that the tube geometry lies within certain limits. During failure the crush zone moves down the tube, with the material ahead of the crush zone remaining intact. One important feature is that the energy absorption levels of such tubes expressed as energy absorbed per unit mass of material destroyed, are higher than those of metal structures [1]. This makes composite materials ideal candidates for use in energy absorbing structures for which weight saving is a critical parameter.

Most of the work on crushing of composite structures has concentrated on axisymmetric cylinders, however some work has been done on cones [3] flat plates [4] and square section tubes [4,6-8]. It is obvious that if composites are to replace metals in energy absorbing structures, geometries other than circular cross section tubes will be used. Thornton and Edwards [6] and Thornton [7] have reported the results of crush testing of oriented glass, carbon and Kevlar reinforced square section tubes. They demonstrated that progressive failure will occur provided the relative density of the tube exceeds a limiting value for the materials. Relative density is defined as the ratio of the volume of the tubes to that of a solid of the same external dimensions.

The work described in this paper is part of a study of the crush performance of a range of square section tubes made from random glass reinforced polyester material by resinjection (resin transfer moulding). This material and the fabrication route are representative of the materials and processes which would be used in mass produced components, such as the auto-mobile.

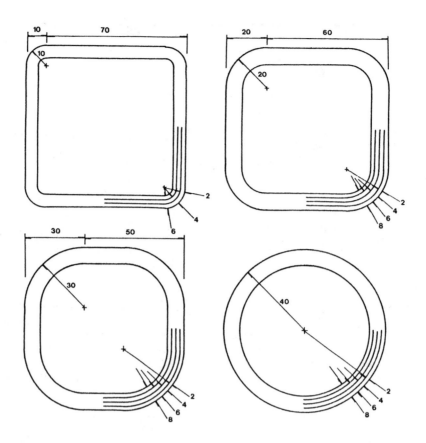

Figure 1. Cross-sectional geometry of tubes (all dimensions in mm)

EXPERIMENTAL METHODS

All specimens were made by resinjection of Crystic 272 polyester resin (Scott Bader Co. Ltd.) into continuous filament random orientation 'spun' E glass mat. The resin was cured by the addition of 0.2% accelerator NL 51P and 3% catalyst Butanox M50. The range of geometries tested is shown in figure 1. Four tube profiles were tested, three square tubes with various corner radii, 10, 20 and 30mm, and round tubes. All square tubes had external widths of 80mm and all round tubes had an outside diameter of 80mm. Tubes with corner radii 10mm were made in three wall thicknesses, 4, 6 and 8mm. Other square tubes and round tubes were made with wall thicknesses of 2, 4, 6 and 8mm. The overall height of all samples was 300mm. The fibre volume fraction was 18%.

The tubes were tested in compression between parallel steel platens using an Instron 8032 100kN servohydraulic machine and a MAND 250kN servohydraulic machine. These constant speed tests were carried out at displacement rates of 0.1×10^{-3} m/s and 50×10^{-3} m/s. High speed constant displacement rate tests were carried out at 4m/s using a Cranfield servohydraulic machine.

In addition to these constant displacement rate tests, tubes were impact tested using a catapult machine which has a fixed available impact energy (12kJ) and an impact velocity upto 18m/s. In this form of testing the sample is fixed to a sled which is driven along horizontal guide rails by elastic cords into an instrumented anvil. Tube failure absorbs the kinetic energy of the sled and sample bringing the impact assembly to rest.

A range of test specimens with various heights were cut from the moulded tubes. All specimens were triggered at one end with a 45° external chamfer.

RESULTS

The failure mode was found to depend on specimen profile, wall thickness and height. Two failure modes were observed, catastrophic shell failure and progressive crushing.

All round tubes failed by progressive crushing regardless of wall thickness or height. Square tubes with wall thickness (t) greater than 2mm failed by progressive crush. The failure mode of square tubes with t = 2mm depended on corner radius (R) and specimen height (h). Increasing R resulted in progressive crush whilst increasing h favoured catastrophic shell failure. These trends are

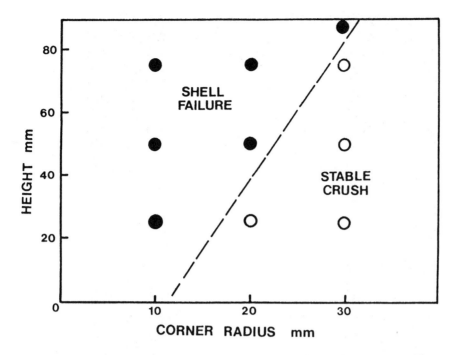

Figure 2. Variation of failure mode with height and corner radius.
Shell failure - full circles; stable crush - open circles.

shown in figure 2. An example of a tube which has failed by progressive crushing is shown in figure 3 and a tube which has undergone catastrophic failure is shown in figure 4.

The morphology of the failure zone produced by progressive crushing has been described in detail elsewhere [1-3]. The main features are internal and external debris fronds separated by an annular wedge of compacted material. In square tubes this morphology does not remain constant around the tube cross-sections, which is in contrast to round tubes. In the corner regions of square tubes the frond/wedge/frond

morphology is well defined for all square tubes which failed by progressive crush. The failure zone morphology of the side regions depended on wall thickness and corner radius. The sides of tubes with R = 10mm, t = 4mm and R = 20 and 30mm, t = 2mm failed by buckling.

An example of this failure mode is shown in figure 5. In tubes with thicker walls (t>6mm for R = 10mm and t>4mm for R = 20 and 30mm) the frond/wedge/frond morphology extended around the complete sample cross section as illustrated in figure 6.

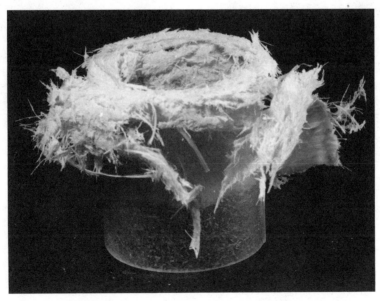

Figure 3. Progressive crush failure. R = 40mm, t = 6mm.

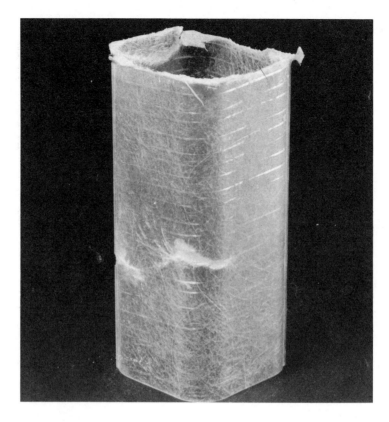

Figure 4. Catastrophic shell failure. R = 20mm, t = 2mm.

Figure 5. Crush zone of square tube showing
frond/wedge/frond failure in corner
regions and buckling failure of side
walls.

Figure 6. Crush zone of square tube showing
'complete' frond/wedge/frond crush
zone morphology.

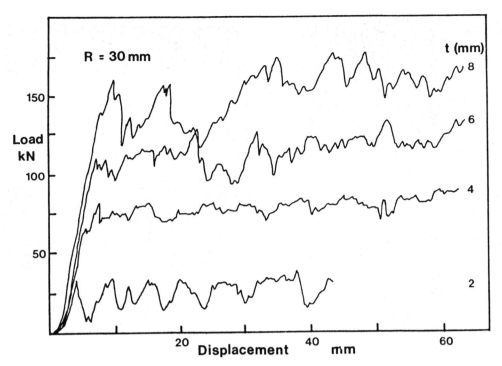

Figure 7. Variation of crush load with displacement, R = 30.

The load-displacement curves for square tubes (R = 30mm) tested at 0.1 x 10⁻³ m/s, which failed by crushing, are shown in figure 7. These curves are typical of progressive crushing; the load rises as the trigger region fails and then fluctuates at approximately constant value as the failure proceeds.

The energy absorbed during crushing is usually quantified in terms of the energy absorbed per unit mass of material that has been crushed. When the crush front moves at the same rate as the testing machine cross-head the size of the crush zone remains constant. If failure takes place at a mean crush load $\bar{P}$, and the density and cross-sectional area of the tubes are ρ and A respectively, the specific energy absorption S_s is given by

$$S_s = \bar{P}/\rho A$$

Values of S_s for a displacement rate of 0.1 x 10⁻³ m/s for the various tube geometries

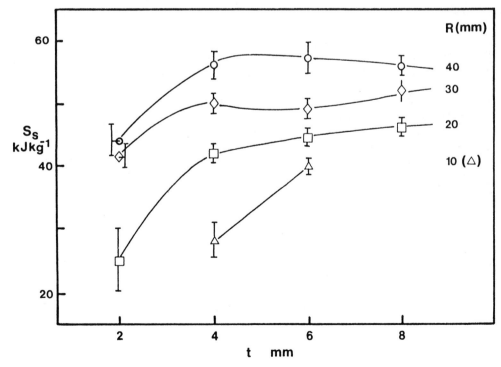

Figure 8. Variation of specific energy absorption, S_s, with wall thickness, t (all corner radii).

Figure 9. Tube with R = 20mm and t = 4mm tested at 0.1×10^{-3} m/s.

Figure 10. Tube with R = 20mm and t = 4mm tested at 4 m/s.

tested are shown in figure 8, plotted against t for each of the corner radii tested. Increasing R resulted in an increase in specific energy absorption for each wall thickness. For tubes with R = 20, 30 and 40mm increasing wall thickness from 2 to 4mm resulted in an increase in S_s. Increasing t further produced only slight increases in S_s. Increasing test speed from 0.1 to 50×10^{-3} m/s produced no obvious changes in mode or crush zone morphology. There were noticeable changes to the crush zone for samples tested at 4m/s, compared with those tested at lower speeds. At high speed (4m/s) the debris fronds were less inverted than at lower speeds. This can be seen by comparing the samples shown in figures 9 and 10. The two tubes shown here have the same geometry (R = 20mm, t = 4mm) but were tested at 0.1×10^{-3} m/s and 4m/s respectively. The reduced degree of frond inversion produced by higher speed crushing is readily apparent in the sample shown in figure 10.

This reduction in frond inversion was also seen in samples which were impact tested using the catapult rig.

The variation in average crush load and specific energy absorption with test speed are listed in the Table. The variation of S_s with speed for tubes with corner radius 20mm is shown graphically in figure 11. Due to the limited number of tests it is difficult to draw

any definitive conclusions from these results, however there is some indication of a peak in S_s at 50×10^{-3} m/s followed by a fall with test speed.

DISCUSSION

For the tubes tested in this work the onset of failure by progressive crush is controlled not only by specimen cross-section (i.e. corner radius and wall thickness), but also by specimen height. These results have been summarised in figure 2.

Thornton and Edwards [6], who carried out similar tests on square tubes made from aligned composites, used the concept of a critical density, Φ_c below which failure is catastrophic. Critical density is defined as the ratio of the volume of the tube to that of a solid of the same external dimensions.

It is readily apparent that this concept of Φ_c is not directly applicable to the tubes tested here, since failure mode can depend on height. In fact the "relative density", Φ, of tubes with R = 20mm and t = 2mm is 0.09, and the maximum height for stable crushing of these tubes is 25mm. This value of Φ (0.09) compares with a Φ_c of 0.2 quoted by Thornton [7] for square glass cloth/epoxy tubes.

The failure of square tubes in a progressive manner does not necessarily involve the formation of a debris frond/debris wedge/debris frond crush zone morphology around the complete tube cross-section. Stable crush

Table: Variation of Mean Crush Load $\bar{P}$ and Specific Energy Absorption S_s with Geometrical Parameters and Test Speed

Corner Radius mm	Wall Thickness mm	Mean crush load $\bar{P}$ kN				Specific energy absorption S_s kJ/kg			
		Test Speed				Test Speed			
		0.1mm/s	50mm/s	4m/s	16m/s	0.1mm/s	50mm/s	4m/s	16m/s
10	4	47.2	56.0	53.5	64.7	28.2	33.5	32.0	38.7
	6	99.0	106.2	95.0	-	40.4	43.4	38.5	-
20	4	66.5	75.0	60.1	64.9	42.3	49.0	38.6	41.3
	6	102.3	107.5	103.7	-	44.5	51.5	45.0	-
	8	138.3	152.5	146.2	-	46.1	53.4	48.8	-
30	4	74.0	75.0	58.0	63.0	50.4	50.7	39.6	44.0
	6	106.5	107.5	106.2	-	49.2	49.7	49.1	-
	8	145.9	145.0	137.5	-	51.9	55.5	48.9	-
40	2	31.1	34.0	26.4	25.8	44.0	48.2	37.4	36.5
	4	76.4	80.0	60.5	66.2	55.8	58.5	44.2	47.0
	6	115.1	108.7	106.3	-	57.4	54.3	52.9	-
		144.8	142.5	140.0	-	55.5	54.6	53.7	-

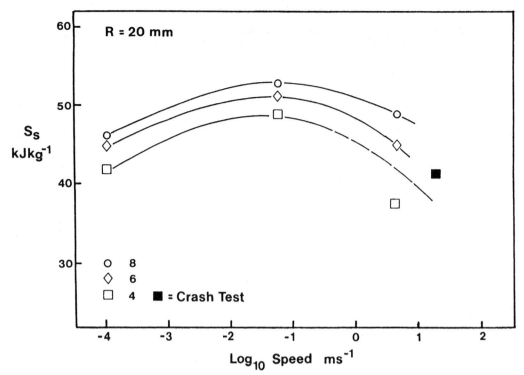

Figure 11. Variation of specific energy absorption, S_s, with testing speed for tubes with corner radius, $R = 20$mm.

first occurs in the corner regions. If corner radius and/or wall thickness is increased this crush zone morphology extends into the flat side walls of the tube. It is apparent that the stress field necessary to generate this type of crush zone is produced by the curved corner regions, but not by the flat side wall of the tubes. Ideally the stress field around the cross-section of a tube during compression should be calculated and compared with that necessary to generate a stable crush zone. This is beyond the scope of the present study and indeed the stress field requirement is not yet fully understood.

It is possible to link the detailed changes in crush zone morphology with cross-section geometry to the recorded variation in energy absorption with geometry (figure 8). As wall thickness, t, is increased the frond/wedge/frond morphology is established around more of the cross-section of the tubes. Correspondingly the energy absorption level increases. Once this morphology is present around the complete cross-section there are only relatively small changes in energy absorption with wall thickness. In figure 8 this can be seen as a steep rise in S_s as t is increased from 2 to 4mm for tubes with $R = 20$mm, followed by only a gradual increase with further increases of t. For tubes with $R = 20$mm and $t = 2$mm only the corner regions failed with a frond/wedge/frond morphology whereas for tubes with $R = 20$ mm and $t = 6$mm the complete cross-section fails in this way.

It can also be seen from figure 8 that increasing R, leads to increases in S_s for tubes with a given wall thickness. This shows that the axisymmetric cylinder (round tube) is the most efficient energy absorbing structure.

This work has shown that there are some changes in failure mode and resultant crush load/energy absorption levels with test speed. Broadly speaking S_s is a maximum at about 50×10^{-3} m/s. Further increases in test speed produce a reduction in the degree of debris frond inversion and corresponding falls in crush load and specific energy absorption levels.

The obvious conclusion is that good frond inversion, which would require a higher level of microdamage, is a more efficient form of energy absorption. The direct relationship between microdamage, crush zone morphology and energy absorption levels have been discussed elsewhere. Whilst this link between damage and energy absorption is readily apparent the reasons for the variation in damage with test speed have yet to be identified.

An impact test during which the crushing sample decellerates itself and the sled from a high initial speed to rest, encompasses all the other test speeds used in constant speed tests. Thus it is difficult to compare the crush load/energy absorption results of these two modes of testing. Ideally instantaneous values of crush load taken from an impact test at points of known sample velocity should be compared with the results of constant rate tests made at the same test speed. This is not feasible because the short term load fluctuations during an impact test tend to dwarf any other variations.

It can be concluded from the present work that the failure mode and presumably failure load of a sample is influenced by its crush history. This is demonstrated by the fact that impacted samples show low frond inversion similar to high constant speed test results. This means that the high speed failure mode (low frond inversion) generated by the initial part of the impact, has been sustained whilst the sample has decellerated to rest. The complete failure zone gives the impression of high speed crush despite the fact that most of it is produced at lower speeds.

With regard to the use of such materials as energy absorbing devices, the significant conclusion is that stable failure still occurs at higher crush speeds and during impact, albeit with some reduction in energy absorption.

The key to the widespread use of composite materials in energy absorbing components must be the ability to use shapes other than axisymmetric cylinders. This work has considered a set of quite simple variations of the axisymmetric cylinder and attempted to determine some of the geometrical factors that lead to progressive crushing. It is apparent that if further advances are to be made in this area then the stress fields that lead to the generation of a stable crush zone morphology must be understood. Furthermore these stress conditions must be linked to the macroscopic structured geometry of component. This type of understanding will reduce the need for widespread testing of many geometries to determine the likelihood of progressive collapse.

CONCLUSIONS

1. Square section glass fibre/polyester cylinders can fail in a progressive manner during compression, absorbing relatively large amounts of energy.

2. The failure mode of square tubes depends on corner radius, R, wall thickness, t, and sample height. Increasing R and t and reducing height favour progressive collapse.

3. For tubes which fail by progressive collapse the crush zone morphology varies around the cross-section and depends on R and t. Curved (corner) sections are first to display the frond/wedge/frond morphology. Side regions fail by buckling unless R and t are large enough to extend the frond/wedge/frond structure into the sides.

4. Specific energy absorption varies with crush zone morphology and hence tube geometry. The most efficient energy absorbing mode is the wedge/frond/wedge morphology. Thus axisymmetric cylinders have the highest energy absorption levels. Reducing R and/or t reduces the energy absorption.

5. Energy absorption and crush zone morphology varies with test speed. Maximum energy absorption occurs at 50×10^{-3} m/s and there is a reduction in debris frond inversion with increasing test speed.

ACKNOWLEDGEMENTS

This work was supported by Ford Motor Company PLC and the Science and Engineering Research Council. The authors are grateful to these organisations and to their colleagues in the Department for their support.

REFERENCES

1. D. Hull, "Axial crushing of fibre reinforced composite tubes", 'Structural Crashworthiness', Eds. N. Jones and T. Wierzbicki, Butterworths, London, (1983) pp.118-135.

2. D. Hull, "Energy absorption of composite materials under crash conditions", Progress in Science and Engineering of Composites, Vol.1, Eds. Hayashi, Kawata and Umekawa, ICCM-IV, Tokyo, (1982), pp.861-870.

3. J. N. Price and D. Hull, "Axial Crushing of glass fibre-polyester composite cones", Comp. Sci. Tech., 28, (1987), 211-230.

4. J. N. Price and D. Hull, "The crush performance of composite structures", Composite Structures, Ed. Marshall, 1987, Elsevier, pp.2.32-2.44.

5. A. H. Fairfull and D. Hull, Effects of specimen dimensions on the specific energy absorption of fibre composite tubes, Sixth Inter. Conf. on Composite Materials ICCM-VI, Ed. F. L. Matthews et al, Elsevier (1987) pp.3.36-3.45.

6. P. H. Thornton and P. J. Edwards, "Energy absorption in composite tubes", J. Comp. Mater., 16, (1982), 521-545.

7. P. H. Thornton, "Energy absorption in composite structures", J. Comp. Mater., 13, (1979), 247-262.

8. G. L. Farley, "Effect of specimen geometry on the energy absorption capability of composite materials", J. Comp. Mater., 20, (1986), 390-400.

IN-PLANE VS. BIAXIAL BENDING
IMPACT OF SRIM COMPOSITES

Dwight A. Rust, Laura K. Gigas
Ashland Chemical Company
Columbus, Ohio   USA

ABSTRACT

Composites are finding increased usage in
energy management and load bearing
applications. These materials are especially
useful where a combination of energy
management, part consolidation, and weight
reduction is required.
Structural reaction injection molding (SRIM)
resins provide unique opportunities for making
composites economically. The SRIM process also
provides a great deal of versatility to optimize the properties of the composite by the
selection of the reinforcement. Reinforcement
materials are generally glass fibers with
woven, chopped, or random continuous strand
construction.
The selection of materials to meet the impact
requirements of the end use products is based
on static and dynamic property tests. Impact
properties are difficult to quantify because
the failure modes are dependent on complex
stress and strain disbributions that develop
within the part.
Impact properties of reinforcements are studied
at various stress states and strain rates to
determine trends that might be useful in selecting materials for specific applications.

TESTS TO MEASURE THE IMPACT PERFORMANCE of
various products are quite involved. Because
of the expense of performing these tests, design engineers have developed techniques such
as finite element analysis to simulate impact
tests and subsequently minimize design
changes(1).

To further help predict part impact performance, material chemists and engineers have
used laboratory impact tests to screen various
materials. One such study of laboratory impact
tests evaluated ten impact test methods on
seven different materials(2). This study
concluded that material screening and selection for impact performance should be made
with a test method having the measured characteristics which match the stress state of the
intended application. In addition to matching the stress state, other studies have
shown the importance of matching the strain
rate(3).

The most common method of measuring in-
plane impact characteristics is notched Izod
(constant strain rate). Another method of
measuring impact is a driven dart test which
measures a biaxial bending mode (varying
strain rate).

The materials chemist and/or engineer
has the problem of defining the proper test
configuration for a particular application by
identifying the stress states and strain
rates involved during impact. Knowing these
parameters, the problem of selecting the
proper reinforcement architecture and matrix
resin can be addressed. It is here where one
is required to make the right choice in
specifying a material for an application.
This choice is more difficult where more than
one mode of impact is operative.

Initial testing of composite materials,
where only the reinforcement was varied, has
shown that increased notched Izod values did
not necessarily predict increased driven dart
impact values(4). Therefore, a study was undertaken to evaluate the broad categories of reinforcements that are currently in use or under
evaluation in composites for SRIM applications. The objective was to compare impact
properties of these reinforcements under
different stress states and strain rates.

MATERIALS

The resin matrix selected for this study
is ARIMAX®* 1100 resin, an acrylamate compos-

---

* Registered trademark of Ashland Oil, Inc.

ition previously described(5).

The type of reinforcement constructions that are available for various applications are almost infinite. In general, glass fiber mats and/or cloths are used rather than carbon, Kevlar, etc. due to the lower costs. The glass type generally used is "E" glass; however, some "A" glass is used in random continuous strand constructions. Properties of these two glass types are shown in Table 1(6). High alkali glass (A) is drawn into fibers for applications where good chemical resistance is advantageous. On the other hand, a high alkali content in the glass is detrimental to electrical properties. A low alkali glass (E) exhibits excellent electrical insulation properties. At present, E glass constitutes the majority of textile fiberglass production.

As parts get more complicated in shape, preshaped fiber construction becomes mandatory. These preshaped fiber constructions can be made by thermoforming continuous random and/or woven constructions or by directing a stream of chopped glass and binder resin against a screen over a large vacuum port. The effects of fiberglass type and binders in preforms used for Structural RIM composites has previously been described(7).

Therefore, a key issue in the Structural RIM process is the choice of fiber reinforcement. This optimization of the type and percent by weight of fibers is important for maximum cost and performance benefits. The reinforcement selection is based on consideration of resin flow, ease of fiber placement, and structural performance requirements of the part. Resin flow is judged by the pressure drop throughout the reinforcement as resin is injected into the part. The introduction of directional material gives the reinforcement package lower resistance to flow than using only continuous strand glass at constant glass levels. Directional reinforcement combined with a random continuous strand mat also yields higher strength properties than a composite made with random continuous strand alone, even though the total glass contents are the same(8).

Four fiberglass reinforcements were chosen to represent the broad categories being used or evaluated. These reinforcements are listed in Table 2.

SAMPLE PREPARATION

All reinforcement materials were molded at random at four different levels of glass content between a range of 20 to 55% glass by weight at a nominal thickness of 3.4mm (0.135 in.). The reinforcements were center injected in a 40 x 51 cm. (16 x 20 in.) plaque mold at an injection rate of 292 gm/sec. (38.5 lb/min.) at a mold temperature of 99°C (210°F). Molding and impact testing were done on the same day to reduce data scatter.

TEST PROCEDURES

Flexural tests were conducted in accordance with ASTM D790, and notched Izod tests were run in accordance with ASTM D256 method D. Figure 1 shows a schematic of the notched Izod and driven dart impact tests.

The equipment used for the driven dart test was a Rheometrics High Rate Impact Tester Model RIT-8000 with a sample dimension of 11 x 11 cm. (4 x 4 in.) to fit a 7.6 cm. (3 in.) ring with a 1.6 cm. (5/8 in.) diameter impacting probe. The impacting probe was driven at a speed of 8 km/hr (5 mph) and 45 km/hr (28 mph).

A typical curve showing the impact force versus impact probe travel at 8 km/hr (5 mph) is shown in Figure 2. The probe contacts the sample at point A. Maintaining a constant speed, the force increases through a relatively linear portion to point B (initial rupture) and then continues upward until it reaches a peak force at point C. Beyond point C, the ram continues to move through the "penetrated" specimen. The amount of energy absorbed by the composite at Point C (peak energy) can be determined by integrating the area under the force-deflection curve.

RESULTS OF IMPACT STUDIES

COMPARISON OF GLASS TYPES - Part of the results of testing the four types of glass reinforcements are summarized in Table 3. One consistent observation has been the better impact properties of the "A" glass versus the "E" glass continuous strand random mat. This is noted for both notched Izod (Figure 3) and driven dart impact at 8 km/hr (Figure 4). These differences could arise from a combination of the softer "A" glass (less brittle) and its highly filamentized nature. However, it is recognized that other factors such as wet out, binder/sizing compatibility, microvoiding, etc. may also contribute to these differences.

The woven biaxial product gives the highest impact overall properties for both notched Izod and driven dart impact. The chopped product has a notched Izod equivalent to the "A" glass; however, it has significantly lower dart impact energy (Table 4).

COMPARISON OF STRESS STATES - While the overall trend shows that higher notched Izods correlate to higher driven dart impact values, the relationship between these two tests and their stress states are different for specific glass types. This is shown by the difference in the rankings in Table 4.

COMPARISON OF STRAIN RATES - At higher strain rates (e.g. 45 km/hr vs. 8 km/hr), the reinforcements do not show large differences in their energy management capability but appear to act similarly (compare Figure 4 vs. Figure 5). This is perhaps due to the catastrophic nature of the failure modes at higher strain rates vs. a more controlled failure at lower strain rates. This can be seen by comparing the impact curves of Figure 2 and Figure 6.

The amount of energy absorbed at the higher strain rates is 5-10% higher than that at the lower strain rates. Increased strain rate of impact speed generally results in a decreased peak force and in increased deflection at the peak force (compare Figure 2 vs. Figure 6). The composites show more ductility at the higher strain rates.

SUMMARY

The composition, construction and level of fiberglass used in SRIM composites does play a role in not only the amount of impact energy managed, but the manner in which it is adsorbed and dissipated.

The type of fiberglass used does contribute to impact resistance. Although not all types have been screened (A, C, E, S, etc.) there is a significant difference between A-glass and E-glass in both Rheometrics and Izod results. This is consistent with previously reported results, wherein continuous strand/filament A-glass was clearly higher than E-glass.

In general, long or continuous fibers have been shown to have better energy managing capability (whether of random or directional orientation) than chopped. However, independent impact testing of experimental chopped glass systems has shown that short fibers (2" length) can be competitive with long fibers(9). Therefore, it is not clear that fiber length, per se, plays a role in the ability to manage impact energy, but is perhaps closely associated to other factors such as wet-out, etc.

Data clearly shows that higher levels of reinforcement yields higher impact strengths regardless of fiber length or construction.

The overall results of these tests are consistent with known composite behavior. However, it is clear that the stress state imposed upon the composite can differentiate reinforcements. It is also clear that in-plane vs. biaxial impacts are not correlatable. In this respect, it is agreed that stress states and strain rates be carefully examined prior to assignment of fiberglass architecture to the composite along with process parameters, costs and other considerations. The best case scenario for maximizing impact resistance must include combinations of A-glass CSM and biaxially woven/stitched cloth.

REFERENCES

1. Glance, P.M., "Computer Aided Analysis of Automotive Bumpers", SAE Technical Paper 840222.
2. Kessler, S.L., G.C. Adams, S.B. Driscoll, and D.R. Ireland, "Instrumented Impact Testing of Plastics and Composite Materials," p. 144, ASTM Special Technical Publication 936, Philadelphia, Pennsylvania (1986).
3. Scammell, K.L., "Comparison of Strain Rates of Dart Impacted Plaques and Pendulum Impacted Bumpers," SAE Technical Paper 870106.
4. Rust, D.A. and P.R. Williams, "SRIM Composites for Energy Management Applications," 43rd Annual Technical Conference Proceedings, Reinforced Plastic/Composite Institute, The Society of Plastics Industry (1988).
5. Wilkinson, T.C. and J.H. Eckler, "Advances in Structural RIM," SPI Polyurethane Proceedings, 29th Annual Technical/ Marketing Conference, October, 1985 (1985).
6. Lubin, G., "Handbook of Composites," pp. 139-140, Van Nostrand Reinhold Company, New York (1982).
7. Williams, P.R., B.D. McGregor, and T.C. Wilkinson, "The Effects of Fiberglass Types and Binders in Preforms Used for Structural RIM Composites," 42nd Annual Technical Conference, Reinforced Plastic/Composite Institute, The Society of the Plastics Industry (1987).
8. Eckler, J.H., and D.A. Rust, "Development of Flow Models for SRIM Process Design," ASM/ESD Advanced Composites Conference Proceeding, pp. 109-111, Detroit, MI (1987).
9. Private Communication, P.R. Williams

Table 1 - Properties of E and A Glass

	"A" Glass	"E" Glass
Chemical Composition		
Silicon oxide	72.0	54.3
Aluminum oxide	0.6	15.2
Calcium oxide	10.0	17.2
Magnesium oxide	2.5	4.7
Sodium oxide	14.2	0.6
Boron oxide	--	8.0
Miscellaneous	0.7	--
Mechanical Properties		
Specific Gravity	2.50	2.54
Tensile Strength at 22°C (MPa)	3033	3448

Table 2 - Types of Fiberglass Reinforcements Evaluated

	A	B	C	D
Form	Mat	Mat	Mat	Fabric
Type	Continuous Strand	Continuous Strand	Chopped 2-inch	Woven Biaxial
Composition	E	A	E	E
Fiber Orientation	Random	Random	Random	0°/90°
Filament Size (Microns)	20	20	13	13
Filaments/Bundle	20-40	1-100	150	2000
Referred to As	"E" Glass	"A" Glass	Chopped	Biaxial

Table 3 - Typical Test Results

Reinforcement	% Glass	SPG	Peak Energy 8 km/hr (J) (in-lb)	Peak Energy 45 km/hr (J) (in-lb)	Notched Izod (J/m) (ft-lb/in)	Flex Strength (MPa) (psi)	Flex Modulus (MPa) (psi)
"E" Glass	40.9	1.547	20.2 / 179	21.2 / 192	603 / 11.3	265 / 38410	9080 / 1316
"A" Glass	41.2	1.523	24.6 / 218	26.8 / 237	817 / 15.3	213 / 30970	8290 / 1201
Biaxial	44.6	1.590	22.6 / 200	22.8 / 202	1340 / 25.1	354 / 51260	12400 / 1798
Chopped	38.8	1.478	11.7 / 104	16.3 / 144	689 / 12.9	200 / 29020	8610 / 1248

Table 4 - Summary of Impact Testing

Notched Izod*	Driven Dart 8 km/hr	Driven Dart 45 km/hr
Biaxial	Biaxial/"A" Glass	"A" Glass
"A" Glass/Chopped	"E" Glass	Biaxial
"E" Glass	Chopped	Chopped/"E" Glass

* Best results are at the top of the column

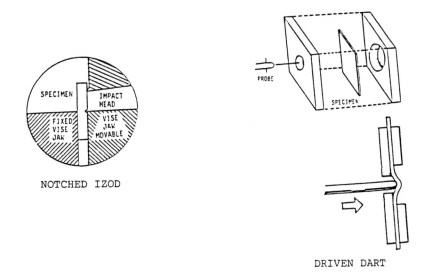

NOTCHED IZOD

DRIVEN DART

FIGURE 1 - SCHEMATIC OF DRIVEN DART AND IZOD IMPACT TESTS

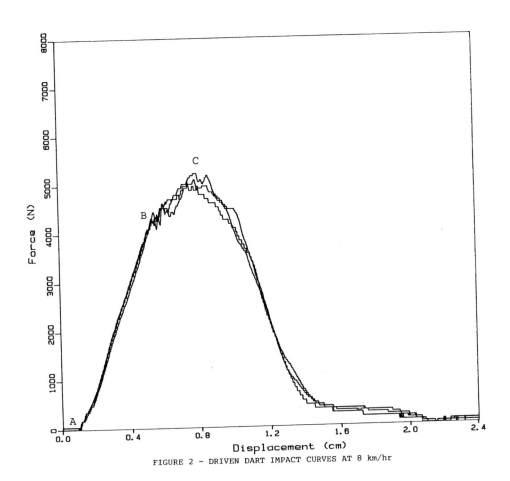

FIGURE 2 - DRIVEN DART IMPACT CURVES AT 8 km/hr

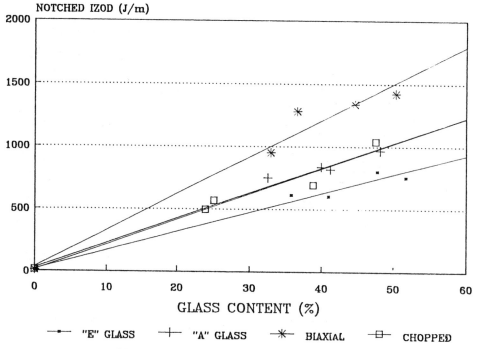

FIGURE 3 - EFFECT OF REINFORCEMENTS ON NOTCHED IZOD

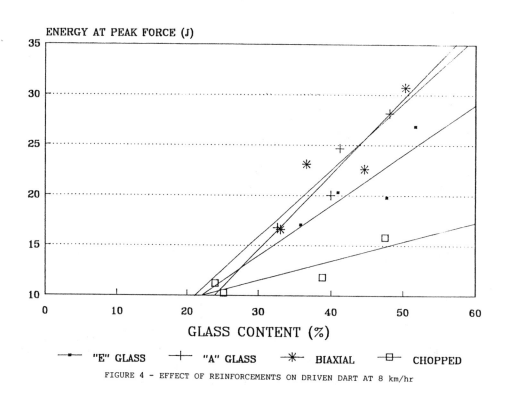

FIGURE 4 - EFFECT OF REINFORCEMENTS ON DRIVEN DART AT 8 km/hr

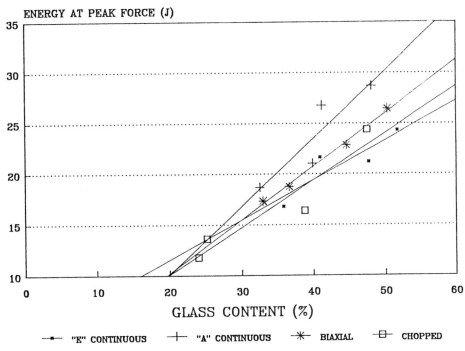

FIGURE 5 - EFFECT OF REINFORCEMENTS ON DRIVEN DART AT 45 km/hr

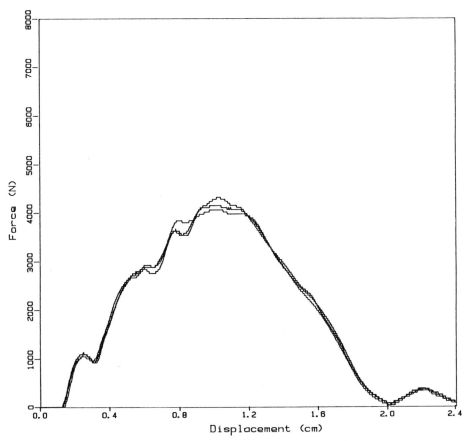

FIGURE 6 - DRIVEN DART IMPACT CURVES AT 45 km/hr

IN-SITU MEASUREMENT OF PROCESSING PROPERTIES DURING FABRICATON IN A PRODUCTION TOOL

D. E. Kranbuehl, P. Haverty, M. Hoff
Department of Chemistry,
College of William and Mary
Williamsburg, Virginia 23185 USA

A. C. Loos
Department of Engineering Science and Mechanics,
Virginia Polytechnic Institute
Blacksburg, Virginia 24061 USA

IN THIS PAPER, we report on our progress in using frequency dependent electromagnetic measurements, FDEMS, as a single, convenient technique for continuous in-situ monitoring of polyester cure during fabrication in a laboratory, and manufacturing environment (1-6). We also report on preliminary FDEMS sensor and modeling work using the Loos-Springer model (7) to develop an intelligent closed loop sensor controlled cure process.

The key is in using the frequency dependence of the complex impedance in the Hz to MHz range to separate and determine parameters which reflect the ionic and the dipolar mobility. These two molecular processes are used to monitor the buildup in viscosity and the extent of cure. The FDEMS measurements have been correlated with frequency dependent mechanical measurements and together used to measure the evolving structure of the polyester network over the entire cure process from a viscous fluid to a crosslinked gel-rubber and finally to a fully cured highly crosslinked glass. The technique is used to measure in-situ the reaction onset, the point of maximum flow, the collapse of the processing window and the post gel buildup in modulus. A correlation is established between Barcol hardness buildup, the mechanical loss peaks and the complex permittivity during the final stages of cure. As an example of one application, the FDEMS technique is used to characterize the processing properties of an unsaturated polyester resin under nonisothermal exothermic cure conditions, the customary manufacturing fabrication environment whose variability often makes controlled laboratory measurements only marginally useful.

EXPERIMENTAL

The frequency dependent complex impedance measurements were made using a Hewlett-Packard 4192A LF Impedance Analyzer controlled by a microcomputer. Measurements at frequencies from 5 to 5×10^6 Hz were taken continuously throughout the entire cure process at regular intervals and converted to the complex permittivity, $\epsilon^* = \epsilon' - i\epsilon''$. Measurements were made with a geometry independent DekDyne microsensor (8). A more detailed description of the equipment and procedures used to calculate the ionic and dipolar mobility parameters has been published (1,2).

Dynamic mechanical measurements were made with a Rheometrics System IV rheometer. G' (dynamic storage modulus) and G'' (dynamic loss modulus) are measured.

Barcol hardness measurements were made with a hardness tester model GYZJ-935 purchased from the Paul N. Gardner Co., Pompano Beach, Florida. This model is one range lower than model GYZJ 934-1 normally used for polyesters and permits significant readings at earlier cure development.

All the polyester resin cure experiments were performed on an orthophthalic general purpose polyester styrene resin provided by the Aristech Chemical Corporation. This resin is typical of the base polyester resin used in marine formulations. Samples were cast in thin sheets to maintain approximately isothermal curing at 24°C and 35°C.

An FDEMS sensor and Loos-Springer model studies were conducted on a Hercules 3501-6 amine cured epoxy resin. Using the results of these studies, a rule based sensor controlled autoclave run was conducted on a one inch 192 ply Hercules AS/3501-6 graphite-epoxy laminate. A 4 foot by 8 foot Tenny production site autoclave was used in this sensor controlled work.

RESULTS AND DISCUSSION

Measurements of the loss component $\epsilon''(\omega)$, of the complex permittivity, $\epsilon^*(\omega) = \epsilon'(\omega) - i\epsilon''(\omega)$, have been made continuously throughout the cure process at 10 frequencies, 50 Hz to 1 MHz, over a range in magnitude of ϵ'' from 10^5 to 10^{-2}. Figures 1 and 2 show the wide variation of $\epsilon''(\omega)$ throughout the cure process as a function of time and frequency under isothermal conditions at 24°C and 35°C. By scaling the values of the loss $\epsilon''(\omega)$ by the angular frequency, $(\omega = 2\pi f)$, all of the experimental values of ϵ'' are conveniently displayed on a single plot. More important, as previously discussed (4,5), plots of $\omega\epsilon''(\omega)$ make it relatively easy to visually determine when the low frequency magnitude of ϵ'' is dominated by the mobility of ions in the resin and when at higher frequencies the rotational mobility of bound charge dominates ϵ''. A detailed description of the frequency dependence of $\epsilon^*(\omega)$ due to ionic, dipolar and charge polarization effects has been previously described (1,2). If we neglect charge polarization effects, which are usually small at frequencies above 10 Hz, the magnitude of the low frequency overlapping values of $\omega\epsilon''(\omega)$ can be used to measure the time dependence of the ionic mobility through the parameter σ where

$$\sigma(\text{ohm}^{-1}\text{cm}^{-1}) = \frac{\omega\epsilon''(\omega)}{\epsilon_o},$$

$$\epsilon_o = 8.854 \times 10^{-14} \, \text{J}^{-1}\text{C}^2\text{cm}^{-1}$$

The dipolar component of the loss, $\epsilon''_{dipolar} = \epsilon'' - \frac{\sigma}{\omega\epsilon_o}$, can then be determined by subtracting the ionic component. The peaks in $\epsilon''_{dipolar}$ (which are usually close to the peaks in ϵ'') can be used to determine the time or point in the cure process when the "mean" dipolar relaxation time has attained a specific value, $\tau = 1/(2\pi f)$, where f is the frequency of measurement. As has been shown previously for tetraglycidyl (TGDDM) epoxy resins, the values of σ, τ and the magnitude of ϵ^* as they vary with time can be used to continuously monitor and measure the viscosity and degree of advancement during the cure process (5,6).

Figures 3 and 4 show the variation of σ and the magnitude of the complex viscosity η measured at 5 rps for the 24°C and 35°C isothermal run. Figures 3 and 4 as well as Figures 1 and 2 clearly show the ability of the onset of change in the magnitude of ϵ'' and σ to detect the reaction onset times as seen both in the onset of the viscosity buildup and the onset of the small exotherm under isothermal conditions.

For unsaturated polyester-styrene resins the gel point, also referred to as the sol-gel transition, occurs very early, approximately 5% conversion, in the reaction. The gel time was measured both using the ASTM D2471 test as well as by using the times at which $\eta = 10^3$ poise and the time at which the real and imaginary components of the complex modulus cross. The observed gel times were 45 minutes at 24°C and 18 minutes at 35°C. The agreement between all three methods was within 10%. The manual ASTM method was inconvenient and less reliable since it required total and constant attention. It is interesting to note that at the gel time the value of σ was 1.5×10^{-9} for 24°C and 3.7×10^{-9} for the 35°C run. It has been shown in

earlier work on TGDDM epoxy resins that the value of σ can be used to quantitatively monitor the viscosity (5,6). Together these observations suggest that the occurrence of a specific value of σ for a particular resin and temperature range can be associated with gel.

Following gel, the collapse of the processing window due to the buildup in the modulus and viscosity can be accurately followed by monitoring the dipolar relaxation times. The dipolar relaxation time monitors the drop in dipolar mobility. It is measured by the time at which peaks occur in $\epsilon''_{dipolar}$ for specific frequencies. The lengthening of the dipolar relaxation times, $\tau = 1/(2\pi f)$ associated with frequencies between 1 MHz and 500 Hz as measured by their corresponding ϵ'' dipolar loss peaks are reported in Table I. The results show that the processing window collapses twice as fast at 35°C relative to 24°C as measured by the elapsed time during which τ drop from $1/(2\pi \times 10^6)$ to $1/(2\pi \times 500)$.

The long time cure properties of the polyester are monitored conventionally using the Barcol hardness number. The measurement, like the ASTM gel procedure requires constant and careful operator attention. Figure 5 reports the buildup in Barcol hardness along with the rate of change in the buildup for the 35°C isothermal cure. Figures 1 and 2 display the sensitivity of the loss to the long time cure process as seen by the corresponding long time decrease in the value of ϵ''. The value of ϵ' displays a similar long time behavior. Figure 6 plots the value of $\frac{d\epsilon'}{dt}$ measured at 5 kHz and 35°C.

The time at which the Barcol hardness is within a few percent of its final value and its rate of change falls to the precision of the measurement can be monitored continuously using the FDEMS technique. For example the times 140 minutes for 35°C and 240 minutes for 24°C at which $\frac{d\epsilon'(5kHz)}{dt}$ drops below .001 are comparable to the times at which the change in the Barcol hardness falls to its measurement precision. These times are also in good agreement with the time to onset of glass formation as determined from mechanical relaxation times. The low frequency 1 Hz measurements of the loss modulus showed the "glass transition" relaxation peaks at 150 minutes for 35°C and 240 minutes for 24°C.

Measurements of the cure properties of the polyesters under the actual nonisothermal fabricating conditions are difficult and the variability in the processing properties is large. A room temperature nonisothermal cure was run as an example of the applicability of the FDEMS techniques to automatically monitor the cure process continuously throughout the cure cycle. Figure 7 displays the variation of $\epsilon''(\omega)$ and the temperature with time. The time to reaction onset, maximum flow, gel, the collapse of the processing window and the onset of glass formation as determined from the FDEMS measurement are reported in Table I.

Finally we turn to work on using the FDEMS sensors and the Loos-Springer model to monitor and model the cure of a one inch thick 192 ply graphite epoxy laminate in a large production size autoclave.

Sensors were placed in the thick laminate at four positions; at the tool surface, at 32 plys, at 64 plys and in the center at the 96th ply (see Figure 8). The bagged 192 ply composite layup was cured in the autoclave using the manufacturer's recommended cure cycle consisting of a 60 minute hold at 116°C and a 120 minute hold at 177°C. The output at each sensor was measured automatically at 2 minute intervals throughout the cure cycle by multiplexing the four sensors through the computer to the impedance bridge. The magnitude of the complex permittivity at each sensor was measured at 10 frequencies: 50 Hz, 125 Hz, 250 Hz, 500 Hz, 5 kHz, 50kHz, 125 kHz, 250 kHz, 500 kHz and 1 MHz. The multiplexed sensor-bridge-computer system made the 40 permittivity measurements in less than the 2 minute interval. Measurements were made continuously throughout the cure cycle without interruption over the 10^5 to 10^{-2} range in magnitude of ϵ' and ϵ'' (see Figure 9). The frequency dependence of ϵ'' was used to determine the conductivity parameter σ. As previously discussed, a plot of $\epsilon''*$ freq, such as Figure 9 shows the frequencies at which ϵ'' is dominated by the ionic mobility.

The magnitude of the ionic mobility as measured by σ is directly dependent on the viscosity of the resin during cure. Calibration curves relating σ to viscosity were generated for representative time-temperature cure cycles by making simultaneous σ and η measurements in the rheometer in which one plate was replaced by the sensor.

Figure 10 is a plot of the viscosity determined by the sensor at each of the four positions in the thick laminate during cure in the autoclave. The sensor data indicates that the middle ply achieves its first viscosity minimum 20 minutes after the surface plys on the tool plate. The middle plys continue to lag the surface ply until the 2nd ramp to 177°C. At this point the exothermic epoxy reaction starts heating the laminate from the inside. As a result, the reaction at the center ply starts to catch up with the extent of reaction of the outer plys. These sensor output measurements of the viscosity at the center and surface suggests that one of the reasons for the success of this recommended cure cycle (undoubtedly determined after many trials), is due to the fact that it causes the center ply to achieve its viscosity minimum at roughly the same time as the surface ply. At all earlier times the center lags the reaction at the surface.

In order to assess both the validity of the Loos-Springer model (7) and the sensor output, the sensor measured values of η were compared with the Loos-Springer model prediction. Figure 11 shows a comparison for the predictions at the 64th ply. The agreement in both the viscosity's time dependence and magnitude is quite remarkable, particularly when one recognizes that the parameters in the Loos-Springer model for 3501-6 were determined on a totally different batch of 3501-6 several years ago.

In the last Figure, 12, we demonstrate the ability of the sensor output to provide useful input for intelligently controlling the cure cycle. We chose to use as a guide the observation that the recommended cure cycle tends to force the 2nd viscosity minima to occur at the same time throughout the laminate. Thus we hypothesize that an efficient and effective cure cycle would endeavor to cause the viscosity minima at the surface and center to occur at the same point, as soon as possible and that the high flow value would continue for as long as possible. Accordingly we proposed raising rapidly the air temperature in the FDEMS sensor controlled run until the exotherm at the inside ply caused the viscosity at the center ply to start to catch up to the surface viscosity. At this point the air temperature would be rapidly lowered in order to hold the surface viscosity in this high flow condition while the center viscosity catches up. At such point as the center viscosity goes through its viscosity minimum and advances beyond the surface plys, the air temperature was set to the final 177°C hold. The 177°C hold would continue until the sensor output indicated through $\frac{d\epsilon''}{dt}$ approaching 0 that the reaction was complete.

The FDEMS sensor measured viscosities from two sensors at the center ply and one sensor at the surface ply are shown in Figure 12. The air autoclave temperature and the temperature at the surface and center ply are also shown. The starting time for the FDEMS sensor controlled autoclave run (Figure 12), and the manufacturer's cure cycle run (Figure 10), was defined as the time at which the tool surface temperature starts to increase.

Figure 12 shows that the FDEMS sensor controlled run significantly reduced the 20 minute lag between the point of the first minimum viscosity for the center ply and the tool surface ply. Similarly the lag between the second viscosity minimum points for the surface and center ply was reduced. The amount of flow as measured by the magnitude of the viscosity minimum was equal or greater in the FDEMS sensor controlled run. Further, the elapsed time to the second viscosity minimum was reduced by 25 minutes. The approach of $\frac{d\epsilon''}{dt}$ and $d\sigma/dt$ to zero indicated cure completion approximately 100 minutes after the 2nd viscosity minimum. Thus the total cure time of

200 minutes was 40 minutes less than the conventional cure cycle.

CONCLUSIONS

Frequency dependent electromagnetic measurements (FDEMS) using impedance bridges in the Hz to MHz region are ideal for automatically monitoring polyester processing properties continuously throughout the cure cycle. Plots of $\epsilon''*$ frequency can be used to determine σ and τ. In turn these molecular parameters can be used to monitor and measure the time-temperature dependence of events such as reaction onset, maximum flow, gel, the collapse of the processing window and completion of cure. The FDEMS measurements can be made in the manufacturing environment in a production tool and used to measure in-situ the processing properties under highly variable fabricating exothermic conditions.

REFERENCES

1. Kranbuehl, D. E., Developments In Reinforced Plastics, Elsevier Science Publishers, LTD, London (1968).

2. Kranbuehl, D. E., S. E. Delos and P. K. Jue, Polymer, 27, 11 (1986).

3. Hoffman, R., J. Godfrey, R. Ehrig, L. Weller, M. Hoff and D. Kranbuehl, J. Polym. Sci: Polym. Symp., 74, 71 (1986).

4. Kranbuehl, D. E., S. E. Delos, E. C. Yi, M. S. Hoff and M. E. Whitham, Proc. of the ACS Div. of Polymeric Materials: Science and Engineering, 53, 191 (1985).

5. Kranbuehl, D., S. Delos, M. Hoff, L. Weller, P. Haverty and J. Seeley, ACS Polym. Mater. Sci. and Eng., 56, 162 (1987).

6. Kranbuehl, D., S. Delos, M. Hoff, P. Haverty, R. Hoffman, J. Godfrey and W. Freeman, Conf. Proc. of the Soc. of Plastics Engineers, 45, 1031 (1987).

7. Loos, A. and G. Springer, J. of Composite Materials, 16, 510 (1982).

8. Questions regarding the sensor and the instrumentation should be directed to D. Kranbuehl.

ACKNOWLEDGMENT

The dielectric work was made possible in part through support from the Center for Innovative Technology of the Commonwealth of Virginia and through support of the National Aeronautics and Space Administration, Langley Research Center NAG 1-237.

PROCESSING PROPERTIES DETERMINED BY FDEMS

Temperature	Reaction Onset $d\varepsilon/dt \neq 0$	Maximum Flow σ max	Gel $\sigma \sim 10^{-9}$	Processing Window τ (1 MHz)	τ (500 Hz)	Glass Formation $d\varepsilon/dt < .001$
35 °C	14 min	13 min	18	31 min	39 min	140 min
24 °C	34 min	26 min	45	53 min	67 min	240 min
exothermic	33 min	25 min	36	42 min	46 min	160 min

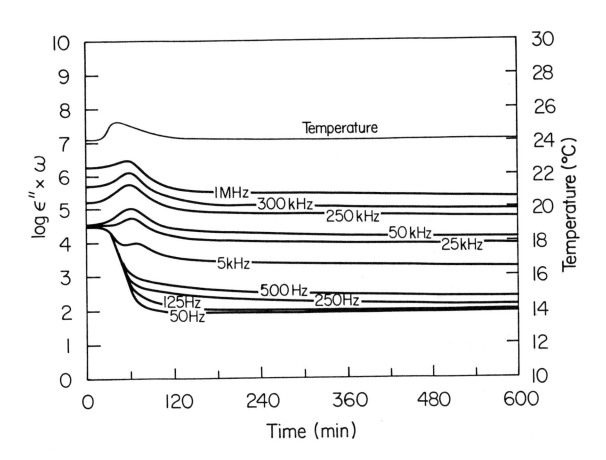

Figure 1. Log($\epsilon''*\omega$) vs time, unsaturated polyester resin, 24°C isothermal cure.

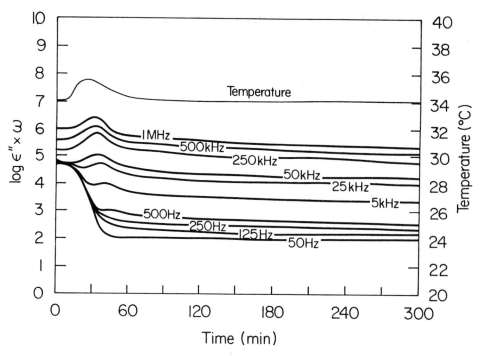

Figure 2. Log($\epsilon"*\omega$) vs time, unsaturated polyester resin, 35°C isothermal cure.

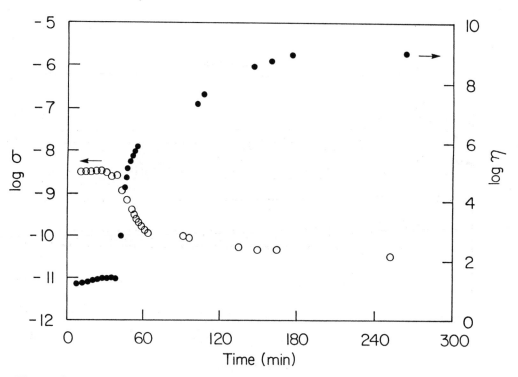

Figure 3. Correlation of ionic mobility, σ, and viscosity, η, 24°C isothermal cure.

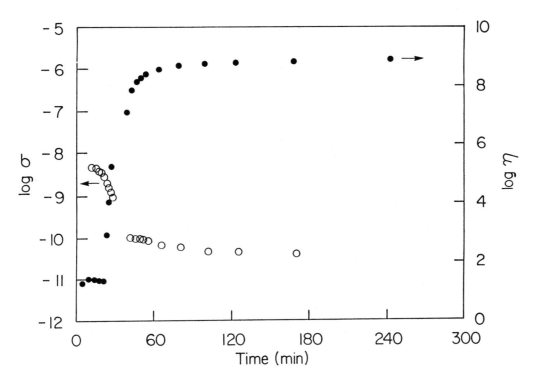

Figure 4. Correlation of ionic mobility, σ, and viscosity, η, 35°C isothermal cure.

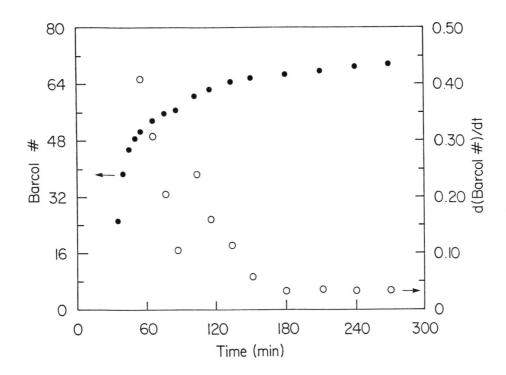

Figure 5. Buildup in Barcol hardness and rate of change in hardness (d Barcol #/dt) vs time, 35°C isothermal cure, unsaturated polyester resin.

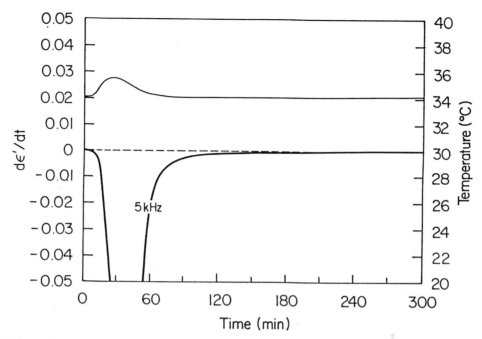

Figure 6. Rate of change in ε' (dε'/dt) at 5 kHz vs time; 35°C isothermal cure, unsaturated polyester resin.

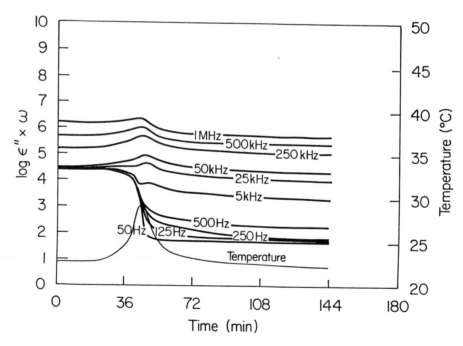

Figure 7. Log(ε"*ω) vs time, unsaturated polyester resin, room temperature non-isothermal cure.

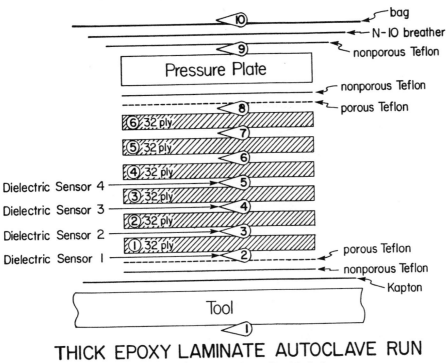

THICK EPOXY LAMINATE AUTOCLAVE RUN
192 PLY 3501-6/AS4

Figure 8. Schematic of the bagged 192 ply graphite epoxy autoclave composite layup.

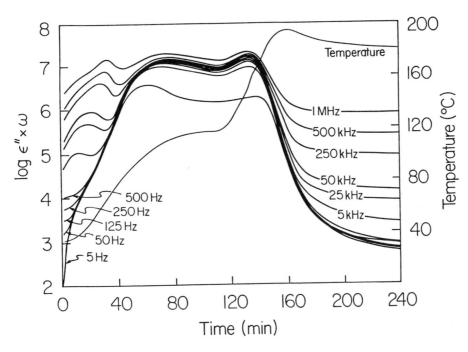

Figure 9. Sensor output of $\log(\epsilon"*\omega)$ at the 64th ply of the composite during autoclave cure.

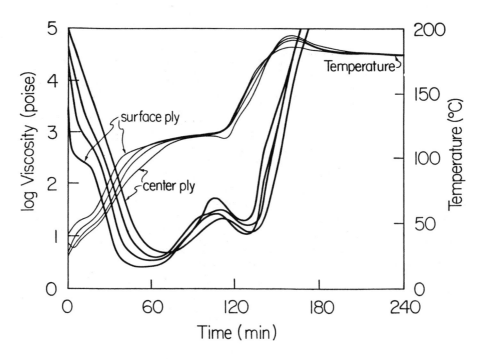

Figure 10. The viscosity at each sensor position as a function of time and temperature as determined by the sensor output.

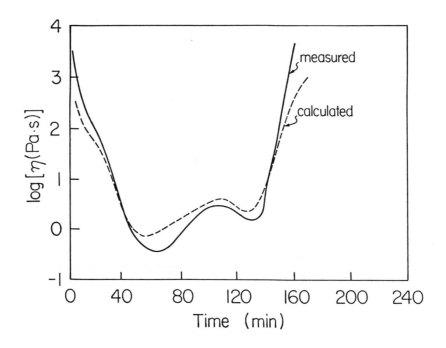

Figure 11. Comparison of the viscosity at the 64th ply of the composite as predicted by the FDEMS sensor and the Loos-Springer model for 3501-6.

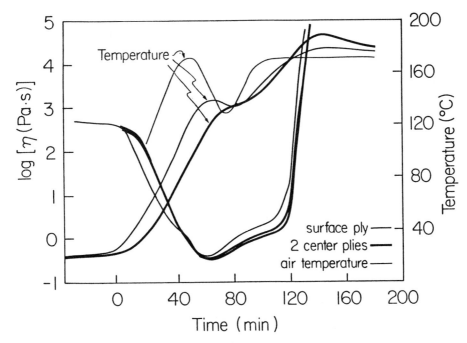

Figure 12.   The viscosity at each sensor position of a 192 ply graphite-
epoxy composite during a FDEMS sensor controlled autoclave cure.

THERMOSET PROCESS CONTROL UTILIZING MICRODIELECTRIC SENSORS

David R. Day
Micromet Instruments, Inc.
Cambridge, Massachusetts   USA

ABSTRACT

Microdielectric sensors are widely used for monitoring reactions in a wide variety of thermosetting and thermoplastic materials. However, software packages for utilizing dielectric response for closed loop control do not yet exist. This paper describes the development of software routines for composite process control under a variety of conditions. The routines measure dielectric loss factor at several frequencies, extract ionic conduction levels, and monitor the first and second derivative of conductivity with respect to time. The software user has the option to specify time and temperature windows in which the process condition under control may be confined. The process under control may be triggered by attaining a user selected first derivative and second derivative. Finally, the ability to average several sets of frequency data has been included. Initial results show that averaging is not necessary for mid-process control points; however, averaging is useful for end of process detection where the first derivative approaches zero. Several examples will be presented ranging from mid-process control (viscosity minimum detection) to endpoint determination.

DIELECTRIC MEASUREMENTS are becoming increasingly important as a means for feedback control in the area of polymer processing. The changes in dielectric response as a function of changing molecular weight or cross link density have been the subject of much research for the last 50 years[1]. Only now is process control through dielectric feedback becoming a reality due to recent advances in dielectric measurement capability[2]. Microdielectric sensors are now available which function down to frequencies characteristic of mechanical measurements (less than 1Hz) and can be inserted directly into curing composite structures. By monitoring the dielectric properties (permittivity and loss factor) at several frequencies, the ionic conductivity can be extracted[3] with the aid of commercial software packages in real time. This paper investigates the use of first and second derivative information (log ionic conductivity with respect to time) for controlling certain key processining steps during composite curing.

EXPERIMENTAL

Dielectric analysis was carried out with a Micromet Instruments Eumetric System II Microdielectrometer. Microdielectric sensors were used to monitor the dielectric response over the frequency range of 0.005 to 10,000 Hz. Materials monitored include a Fiberite 934 graphite epoxy, a USP graphite PMR-15, and a Hexcel 8-297 graphite epoxy. In each case a small piece of glass felt was placed over the sensor to prevent the graphite from coming into contact with the sensor. Temperature was controlled with a Micromet Eumetric Programmable Oven and system software. The commercially available dielectric monitoring software was modified to calculate the derivative of ionic conductivity with respect to time in real time. When the programmed conditions were satisfied, the computer in the first two examples drew shaded areas on the real time plot simulating the initiation of a process control step. This modified software will subsequently be referred to as the Process Control Software (PCS). In the third example, the computer actually controlled the temperature of the oven based on the microdielectric sensor data.

RESULTS AND DISCUSSION

DETECTING VISCOSITY MINIMA—Figure 1 shows a typical industrial cure cycle for a graphite epoxy prepreg, in this case the Hexcel 8-297 material. The left axis of this plot, log conductivity, was extracted from multifrequency loss factor data. The quantity of log conductivity axis may be thought of as a fluidity axis or log (1/viscosity). The peaks in Figure 1 therefore represent fluidity maxima or viscosity minima. It is often the case that pressure is applied near the second viscosity minimum to insure good flow but not too much bleed. In order to trigger pressure application at the second fluidity maximum the software must be able to detect a zero slope of the log conductivity as it passes through a maximum.

In order to set this triggering action the PCS software has the parameter menu shown in Figure 2. The first parameters required are the time and temperature window. The first number represents a value below which the process will not be triggered. For example, if the slope is 0 but the temperature has not yet reached 150C, the process will not be triggered. This prevents the software, for example, from applying pressure at the first fluidity maximum in Figure 1. The second number in the time and temperature window is a value that if attained will automatically trigger the process (if it hasn't been already) no matter what the slope is.

The next parameter is "critical slope." This represents the value of the derivative of log conductivity with respect to time at which the process should be triggered. This is related to the "trend" parameter which defines whether the process trigger should occur if the slope is passing in a positive or negative direction through the critical slope.

The "trend" parameter represents what the sign of the second derivative should be when the critical slope is at-

tained. For a fluidity maxima the "trend" parameter should be set to negative since the slope will be going from positive to negative as it passes through zero.

The final parameter is the number of data sets (n) to average in determining the slope. The PCS takes all the ionic conductivity data from (n+1) frequency cycles and does a least squares fit to determine slope. It has been found that mid-process trigger point detection only requires n to be 1 or 2 for reliable performance. End of cure detection may require more depending on how small a slope is required.

The shaded area in Figure 3 shows where pressure is applied based on the PCS parameters used in Figure 2. Note that pressure is applied almost exactly as the second fluidity maximum is attained.

DETECTING ONSET OF REACTION—Pressure application is again a critical point during the cure of PMR-15, a graphite polyimide material. The ionic conductivity extracted from mulitfrequency loss factor data during a standard PMR-15 cure is shown in Figure 4.

The first large fall off log conductivity is due to imidization along the polymer backbone. During the final ramp up to 600F the chain ends undergo a second crosslinking reaction. This corresponds to the point at about 110 minutes where the ionic conduction bends over with a slightly lower slope. This is precisely where it is desirable to apply pressure (before crosslinking prohibits flow). Although the change in ionic conduction level is minor at this critical point, the change in the slope with respect to time is considerable.

Figure 5 shows the first derivative of ionic conduction with respect to time from Figure 4. The value of the first derivative falls from a value of about 1.1 to lower values at the onset of crosslinking. It is precisely at this point that pressure triggering should occur. In order to set the trigger point appropriately, the PCS parameters are set as follows. The temperature window is set at 450, 600 preventing any pressure application before 450°F is attained. The time window is set to 0, 1000 (not a critical parameter). The critical slope is set to 0.08 so the pressure application will occur just as the slope starts to fall from the 0.13 value. The "trend" parame-

ter is set to negative to insure that triggering occurs as the slope is decreasing (and not during the increase around 95 minutes). Finally the number of point averaged is set to 2.

Figure 6 shows the results utilizing these parameters. The shaded area represents where pressure is applied. Note that as the second temperature ramp was initiated the ionic conductivity level started to increase, due to a steady increase in ionic mobility. However, as the crosslinking starts to occur, the rate of increase if ion mobility, hence ionic conductivity, is impeded and the slope starts to decrease slightly. The PCS detected this decrease and triggered the pressure at exactly this point.

VISCOSITY CONTROL AND END POINT DETECTION—The previous cure control examples were carried out using PCS, a routine intended for generic process application. The next example was carried out by a specialized software routine written specifically for the cure Fiberite 934 graphite epoxy. The program is composed of five main subroutines and was designed to monitor dielectric and temperature information from the microdielectric sensor and automatically control the oven temperature. In Figure 7, the ionic conductivity is converted to resistivity (by inverting) and is labelled as "ion viscosity" (due to the close relation between resistivity and actual viscosity and rigidity). When the condition for each subroutine is met, the program automatically jumps to the next routine. The five subroutines are listed below:

1. heat material to 250°F
2. hold at 250°F until ion viscosity reaches 7.0
3. hold ion viscosity at 7.0 (by controlling temperature) until temperature reaches 350°F
4. hold at 350°F until slope of ion viscosity with respect to time is zero
5. cool down and notify operator

Note that after heat up to 250°F the computer waits for the ion viscosity to increase to 7. Control step 3 is perhaps the most interesting. During this segment, the computer notices

small increases in ion viscosity and automatically increases the temperature to drive in the ion viscosity back to 7. The resulting profile during this period is curved and sucessfully eliminates the second viscosity minimum. In standard cure cycles, a linear temperature ramp is used, thus causing a second viscosity minimum to occur. Note that after a period of time the temperature can not be increased fast enough and the ion viscosity starts to build rapidly. The program then holds the temperature at 350°F until the change in ion viscosity with respect to time reaches zero.

CONCLUSIONS

Microdielectric sensors in combination with the developed software are capable of performing closed loop processing using a variety of control options. The Process Control Software may be instructed to trigger a process step through monitoring of slope of extracted ionic conductivity with respect to time. The additional user selected parameters of time and temperature window, trend, and number of points to average, enable the Process Control Software to very specifically pinpoint the appropriate process trigger point.

More sophisticated control is possible in which the entire cure process may be "intelligently" directed. Hold times may be based on reaching critical values of ionic conduction or slope of conduction with respect to time and viscosity may be controlled in real time by continuously changing temperature.

REFERENCES

1. R. Kienle and H. Race, Trans. Electrochem. Soc., 65, 87 (1934)
2. S.D. Senturia, N.F. Sheppard Jr., H.L. Lee, S.B. Marshall, SAMPE J., 19, 22 (1983)
3. D.R. Day, J. Eng. Sci., 26, 5, 362 (1986)

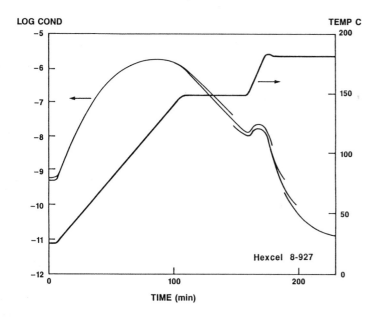

Figure 1. Log conductivity (extracted from 1 to 10,000 Hz loss factor data) during cure of Hexcel 8-927 graphite epoxy.

A. TIME WINDOW= 150 300
B. TEMP WINDOW= 150 200
C. CRIT. SLOPE= 0
D. ON POS OR NEG TREND=NEGATIVE
E. # OF CYCLES TO AVE.= 2

Figure 2. Process Control Software parameter menu.

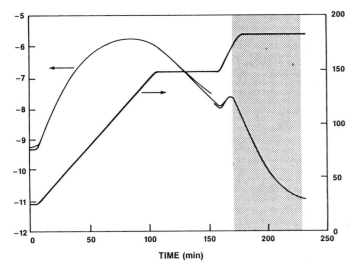

Figure 3. Graphite epoxy cure (from Fig. 1) using control parameters (from Fig. 2). Shaded area is where software applies pressure.

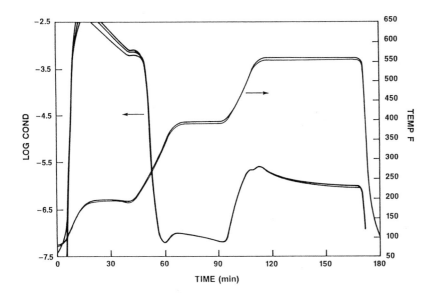

Figure 4. Log conductivity (extracted from 1 to 1000 Hz loss factor data) during cure of USP PMR-15 graphite polyimide.

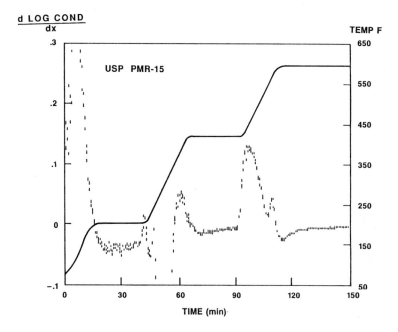

Figure 5. Slope of Log conductivity (from Fig. 4) during cure.

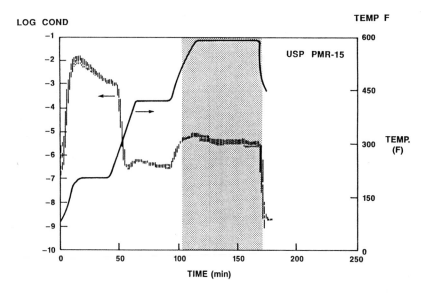

Figure 6. PMR-15 cure using control parameters adjusted to trigger pressure application at onset of crosslinking reaction. Shaded area represents application of pressure.

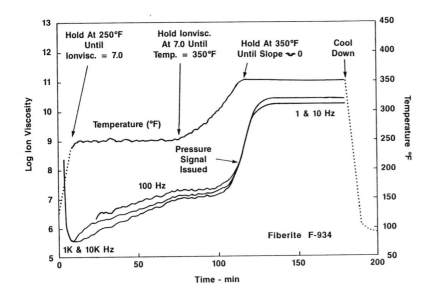

Figure 7. Closed loop cure of Fiberite epoxy graphite
material. Control routines are noted in figure.

CURE MONITORING TECHNIQUES
FOR THERMOSET RESINS

S. S. Chang, D. L. Hunston, F. I. Mopsik
Polymers Divisions,
Institute for Materials Science and Engineering
National Bureau of Standards
Gaithersburg, Maryland 20899   USA

ABSTRACT

The ultimate properties of composite parts made with thermoset resins depend not only on the level of resin/filler consolidation but also on the degree of cure and the cure history. On-line monitoring and control during the cure and post-cure processes can help to ensure uniformity in the final product. To implement this approach, however, requires new monitoring techniques and a better understanding of what the data from these techniques mean. In answer to this challenge, NBS has developed or implemented a large number of cure monitoring techniques that are sensitive to various aspects of the changes that occur during the cure process. As one example of this, the monitoring of cure for a vinyl ester resin by ultrasonic shear wave attenuation and by dielectric time-domain spectroscopy will be shown. These two methods show promise for on-line monitoring and process control. Measurements from these methods are correlated with thermal analysis for degree of cure and curing kinetics.

THE ULTIMATE MECHANICAL PROPERTIES of a cured composite resin depend strongly on the degree of cure and somewhat on the curing path. The cure process may be monitored by measuring any property that change with the degree of cure. NBS has developed or implemented a large number of monitoring techniques, and together they are sensitive to size scales ranging from intramolecular bonds to bulk properties. For example, viscometry, torsional braid analysis, and shear wave propagation techniques are available to monitor changes in mechanical properties from the earliest stages of cure through the post-cure region. On the microscopic level, a number of fluorescence techniques are available to monitor changes in microviscosity in the neighborhood of a probe molecule. Conductance and dielectric measurements can be used to monitor changes in the charge mobility and dipolar relaxations. Neutron scattering techniques can be used to follow structural changes, and size exclusion chromatography can detect increases in molecular weight. At the sub-molecular level, various types of spectroscopy can be used to monitor the formation of specific bonds. Calorimetry and thermal analysis can be used to monitor the energy from bond formation or reaction and to monitor the reaction kinetics and the degree of completion of cure.

All above techniques are available in our laboratories. Since each of the techniques monitors a particular property change and has its own characteristic range of sensitivity, it is the purpose of the NBS program to correlate the results of these techniques to help establish the scientific basis for process monitoring and to explore the on-line monitoring capabilities of those newer techniques where this is appropriate.

The classical methods such as viscosity and calorimetry are not easily implemented for on-line monitoring, but remain indispensible as the basis for comparison and correlation. We report here results for the cure under identical isothermal conditions of a vinyl ester resin measured by ultrasonic, dielectric and calorimetric techniques. Additional studies on curing kinetics and on the cured resin are provided by thermal analysis.

EXPERIMENTAL

MATERIAL - The resin mixture studied is Dow Chemical vinyl ester resin (VER) mixture Derakane 411C50*. The mixture is composed of VER diluted with 40-60% of styrene. The backbone of the VER is the common DGEBA (diglycidyl ether of bisphenol A) type epoxy, which is reacted with propenoic acid to form vinyl ester terminations on the backbone.

The initiator is a difunctional peroxyester,

$$[CH(C_2H_5)(C_4H_9)CO-OO-C(CH_3)_2CH_2-]_2$$

2,5-dimethyl-2,5-bis(2-ethyl hexoyl peroxy) hexane, obtained as Witco, USP-245*. For most of the works studied here, 1 part of initiator per hundred parts of resin (phr) is used.

ULTRASONICS - The techniques used to monitor mechanical properties during cure was ultrasonic shear wave propagation. This technique has the advantage that it can nondestructively monitor the shear mechanical properties of a sample throughout the cure process. A shear technique was chosen for two reasons. First, the shear elastic properties are very sensitive to the liquid to solid transition that characterizes curing; in the limiting case changes in the shear modulus can be 10 orders of magnitude. Second, the shear loss properties are related to the shear viscosity which dictates flow and consolidation during cure. These advantages are most fully realized at low frequencies, but ultrasonics offers the ability to make measurements rapidly, and to characterize thin films or constrained layers. The ability to monitor films makes the technique useful for paints, inks, adhesive, and other coatings as well as composites.

The characteristics of the shear wave technique can be understood by noting that as long as the polymer sample is not completely rigid, the attenuation of high frequency shear waves is very large. This means that wave propagation can not be measured directly since it simply does not travel far enough. They can be characterized in an indirect way, however, by using a low loss substrate,

---

*Certain commercial materials and equipment are identified in this paper in order to specify adequately the experimental procedure. In no case does such identification imply recommendation or endorsement by the National Bureau of Standards, nor does it imply necessarily the best available for the purpose.

such as quartz. A shear wave is generated in the substrate which is pressed against the sample. In the proper geometry, the shear waves in the substrate will generate shear waves in the sample. This alters the waves in the substrate. If these alterations are measured by testing with and without the sample, it is possible to determine what the waves in the sample must be to produce such changes. From this information, the dynamic shear properties of the sample can be calculated. Since the waves in the sample propagate only a very short distance, the thickness of the sample layer does not matter once it is beyond a certain minimum. Film thickness of only a fraction of a mm can be measured.

A number of geometries have been developed [1] and can be used for such experiment [2,3]. In all cases the displacement in the waves must be parallel to the sample-substrate interface so that only shear waves will be present. The differences in the geometries relate to the angle between the direction of wave propagation in the substrate and the sample-substrate interface. Since the sensitivity of the measurement depends on this angle, the proper choice of geometry can facilitate the measurement. If the angle is zero, the wave propagates parallel to and in contact with the interface. This gives the maximum sensitivity. When the angle is not zero, the experiment measures the reflection from the interface with the normal reflection (angle of 90°) giving the lowest sensitivity. The maximum sensitivity is needed when minimum viscosity during cure is 0.1 Poise or less, but the less sensitive geometry provides more information during the later phases of cure. For the samples used here, the minimum viscosities are greater than 1 Poise so the 90° geometry gives the best results.

In these experiments, the wave is reflected from the sample-substrate interface, and the returning signal is measured using a pulse echo technique. The amplitude and phase of the wave are determined relative to those for the reflected wave when no sample is present. The increase in attenuation, Δ, and the phase shift, ϕ, are then used to calculate the shear storage modulus, G', and the shear loss modulus, G'', for the sample using the following equations,

$$G' = C (\Delta^2 - \phi^2)$$
$$G'' = 2 C \Delta \phi$$

where

$$C = Z_Q^2 / 4 \rho_s$$

The density of the sample is ρ_s while Z_Q is the shear mechanical impedance of the substrate.

A complete characterization of the sample requires measurement of both Δ and ϕ. Although this has been done for some samples, the phase shift is quite sensitive to temperature. Since the reactions associated with processing are generally exothermic and the rates of most reactions are rapid, precise control of temperature is difficult. Consequently, a complete characterization of the mechanical properties is usually not possible. On the other hand, the attenuation change, Δ, is not particularly sensitive to temperature, and thus it is a very useful parameters for monitoring the changes that occur during processing even though the exact values of the mechanical properties can not be calculated. Moreover, in most cases the high frequency behavior becomes elastic ($G' \gg G''$) early in the cure and from then on, $\Delta \gg \phi$, so Δ provides a direct measure of G'.

$$G' \approx C \ \Delta^2$$

The experiments performed here utilize a quartz rod with a transducer attached at one end and used to generate and monitor the pulses of shear waves (see Figure 1). The sample is applied to the other end of the rod. A tape is placed around the circumference of the rod and extends past the end to form a cup into which the liquid sample is poured. The sample end of the rod is surrounded by a temperature chamber, and the system is equilibrated at the cure temperature before the sample is introduced. The experiment is started by adding sufficient sample to produce a 2 to 3 mm layer of material on the rod. The layer reaches the cure temperature in about 5 minutes. The attenuation of the shear wave per reflection is measured before the sample is introduced and is monitored continuously throughout the cure.

DIELECTRIC SPECTROMETRY - A time-domain dielectric spectrometer [4] was also used to monitor the cure. The principle of the spectrometer is to apply a step voltage to the electrodes of a capacitor filled with the material under test as the dielectric medium, and to record the change in the charge as a function of time. The charge is directly related to the capacitance and can be Laplace transformed to give data over a range of frequencies. The high frequency limit is determined by the rate of the voltage step, and the low frequency limit by the length of time over which the measurement is made.

In the liquid state, the raw materials are generally quite conductive, due to ionic impurities. Consequently, for the initial phase of the cure a separate instrument was used to monitor the conductance at a fixed frequency, e.g. 50 Hz. This measurement followed the transition of the liquid into a solid or of the low-molecular weight material into a very high molecular weight mass. In the later phases of cure the dielectric spectrometer was used to obtain data over a wide range of frequencies and to extrapolate so the DC conductance could be estimated.

THERMAL ANALYSIS - A Perkin-Elmer DSC-7* power-compensated type of differential scanning calorimeter was used for the thermal measurements. The DSC measurement of the heat of reaction may be obtained in an isothermal mode or in a scanning mode for both curing and post-curing treatments.

The estimation of the heat of reaction for cure at a constant temperature is relatively simple, i.e. the amount of heat released is integrated over the reaction period. Uncertainties in the isothermal measurements may arise from a number of factors: (1) it may take some time for the material to heat up and approximately 1 minute is required to stabilize at the temperature of reaction, (2) a significant reaction rate may already exist by the time the sample is heated to the temperature of reaction, (3) deciding when the reaction "ends" is difficult and (4) there can be drift caused by the inherent instability of the instrument and the instrument's sensitivity toward environmental changes. For each 1 μW of uncertainty in the differential power for a period of 1 hr, there will be an error of about 0.5 J/g for a sample size of 7.2 mg.

The isothermally cured sample is generally post-cured at some higher temperature(s). The sum of all heat released for the cure and post-cure reactions may then be considered as the heat of reaction for a maximum or 100% reacted product. The post-curing may also be obtained by scanning the isothermally cured sample to the highest temperature tolerable by the material and held until the exotherm becomes negligible. A re-scan of the post-cured product is then used to provide a base line as well as to determine the glass transition temperature. The difference between the amounts of heat involved in the two scans is the additional amount of heat released in the post-curing process [5].

The heat of reaction may also be

obtained by scanning the resin mixture first as a reaction run and then as a cured resin. The two-scan computation procedure as described for post-curing may also be used to calculate the heat of reaction at the beginning of the scan temperature [5].

RESULTS AND DISCUSSION

COMPARISON OF ISOTHERMAL CURE AT 90°C - The results of ultrasonic shear wave attenuation are shown in Figure 2. There is a fast increase in the attenuation of the reflected wave at about 7 minutes. The change occurs within about 30 seconds but is followed by a much slower rate of increase in attenuation.

The results of the 50 Hz conductance, Figure 3, also show a rather fast decrease (increase in 1/conductance) at about 9 minutes after an initial increase due in large part to the time required to reach thermal equilibrium. A slower rate of decrease in conductance is observed after about 12 minutes.

In both cases the slower rate of change during the latter part of isothermal curing process indicates that vitrification has caused a slower rate of reaction involving a diffusion controlled process in the glassy state. The changes during this latter stage of cure may also include the consequences of a densification process as the resin is cooling toward the isothermal temperature, as described later.

Calorimetrically, the curing reaction appears to occur rather smoothly over a period of about 8 minutes, Figure 4. Since the material in the DSC reaches thermal equilibrium in about 1 minute, there appears to be an induction period of about 4 minutes.

In looking for an explanation for the different time scales in the various measurements, temperature differences were considered a prime candidate. Because of the small sample size in the DSC (about 10 mg), we believe the temperature gradient across the sample is only in the order of a couple of degrees at an average energy release rate of around 2 W/g. On the other hand, when a thermocouple is placed in the 2-3 mm thick ultrasonic sample (2.5 cm in diameter bounded by air on one side and the quartz rod on the other) or in 2 mm thick dielectric sample (4 cm square bounded on both sides by 2 mm thick fiberglass epoxy circuit boards plus 1.5 mm thick Teflon sheets), the temperature during cure is found to rise 30-40°C above the isothermal control temperature of 90°C, Figure 5.

In both ultrasonic and dielectric measurements, therefore, a runaway reaction is observed because the thermal conduction is not sufficient to remove the heat generated by the self-accelerating autocatalytic reaction. This type of non-isothermal reaction process may have some advantages in reducing fabrication times for sheet molding and other process, as long as the maximum resin temperature is controlled below its decomposition temperature. The temperature differences do, however, complicate the comparisons of interest in this work.

From the temperature profiles in Figure 5, there appears to be time shifts about 6 minutes and 11 minutes, respectively, for the ultrasonic and dielectric experiments. This compares with the faster heating (less than 1 minute) in the DSC. As a reaction may start from any hot spot within the sample, the actual time shift may be shorter than indicated.

ISOTHERMAL CURE BETWEEN 60 AND 100°C - The peak times, $t_{\frac{1}{2}}$, for the exothermic reaction at different temperatures measured by DSC are shown in Figure 6, along with the half-width, Δt, of the reaction. The free radical initiated reaction occurs at a relatively high activation energy of over 120 kJ/mol in comparison to the addition-type of reaction for epoxy systems such as DEGBA and bis(p-amino cyclohexyl) methane, the activation energy is about 55 kJ/mol.

POST CURE - Isothermally cured VER samples are post-cured by scanning the temperature at 10°C/min from 50 to 150°C and then held at 150°C for 2 minutes. No further heat releases are observed by scanning the post-cured samples, Figure 7. There are generally two exothermic peaks observed in the post-cure scanning curves. The lower temperature peak is smaller in magnitude than that of the higher temperature peak. As the isothermal cure temperature, T_c, is increased, the peak temperature of the lower temperature peak is increased from 85 to 105°C but with diminishing magnitudes. At higher T_c, the lower temperature peak may be indistinguishable from the main higher temperature peak. The high temperature (main) peak is less dependent on T_c and occurs between 120 and 130°C, just above the ultimate T_g of vinyl ester resin.

The post-cure energy release, however, diminishes from 30-50 J/g for T_c of 60°C to about 6-7 J/g for T_c of 100°C. The energy releases, however, represent only about 10% to 2% of the total energy of cure, respectively. The sum of the isothermal heat release and the heat released in the post-curing process is about 370 J/g, roughly 10 J/g less than that from scan curing. We believe these smaller values are the consequences of uncertainties associated with long curing time at low temperatures with low signal-to-noise ratio or due to the lost of the beginning of the reaction during fast, high temperature curing.

SCAN CURE - Curing of VER may also be accomplished by scanning up to 150°C at a rate of 10°C/min, Figure 8. If a VER/1 phr initiator mixture is aged at room temperature prior to testing, the cure exotherm starts at lower temperatures, but there is little change in the heat of reaction and the T_g of the cured resin. The exotherm for a VER/2 phr initiator mixture occurs at even lower temperature. T_g of the cured resin with 2 phr initiator is about 5°C lower than that of the 1 phr resin.

DEGREE OF CURE - The heat of reaction at 50°C, $\Delta H_{rxn,50}$ as determined by a two-scan method from 50 to 150°C is 382±8 J/g. The sum, of the heat released at various isothermal cure temperatures T_c from 60 to 150°C plus the post-cure energy by a two-scan method at 50°C, is determined as 366±12 J/g. This sum, although smaller, remains within 5% of the value as determined by the simpler scanning method. The degree of cure is estimated from the ratio of the heat released at T_c from the total energy released, including post-cure energy, for the sample. As shown in Figure 9, the degree of cure for all samples of VER are very high, at above 90%, even for T_c at 60°C.

C_p AND T_g - C_p of the unreacted VER liquid resin mixture and of the cured resin above its T_g (in the "supercooled liquid state") may be connected together and represented approximately by one single linear function for the temperature range 50-150°C:

$$C_{p,l} = 1.95 + 0.001 \ T$$

where C_p is in J/K/g and T in °C. Likewise, C_p of a well-cured VER resin in the glassy state below its T_g may be represented by:

$$C_{p,gl} = 1.43 + 0.004 \ T.$$

$C_{p,gl}$ of less-well cured resin shows a slightly higher value, due to a higher content of liquid-like character.

T_g or fictive temperature is defined as the intercept of enthalpy curves above and below the glass transition region. T_g so defined is independent of the rate of observation while heating, but is dependent only on the thermal history during the glass forming process. However the T_g of fast-cooled or quenched glass which does not exhibit a relaxation peak may easily be estimated as the intercept of the DSC curve with a line representing the average value of $C_{p,gl}$ and $C_{p,l}$ as defined above.

T_g AS A FUNCTION OF T_C AND POST-CURE - T_g of isothermally cured vinyl ester depends strongly on cure temperature, T_c, and degree of cure, Figure 10. At a low T_c, say 60°C, T_g is about 10°C above T_c. At higher T_c, however, T_g is generally near the T_c, Figure 11. After the isothermally cured resin is post-cured by heating to 150°C and held there for 2 minutes, T_g of all post-cured samples fall in the range of 111-116°C, Figure 11. The data are randomly scattered without any discernable trend or inference from prior thermal histories. Although T_g of vinyl ester resin is highly sensitive to the isothermal cure temperature and to the last 10% of the degree of cure the post-cured vinyl ester resin is oblivious to its prior thermal or processing histories.

RELAXATION OF GLASS - Glass transition of cured VER show typical vitreous relaxation phenomena, Figure 12. After slow cooling or soak annealing below T_g, the apparent range of the glassy state is extended above its T_g, due to superheating of the sample with longer relaxation times than the experimental time. Although the actual T_g may be lowered to about 105°C by annealing, a relaxation peak as high as 127°C is observed, as the relaxation time drops to the region of experimental time scale.

CONCLUSION

Peroxyester initiated vinyl ester resin system may be cured in a few minutes at 90°C.

The T_g of post-cured vinyl ester resin appears relatively independent of the curing path.

There is a need to study the different cure monitoring techniques simultaneously on the same sample in order to correlate the results, since the sample would then have a single thermal history.

REFERENCES

1.   G. Harrison, and A. J. Barlow, in "Methods of Experimental Physics," Academic Press, Vol. 19, Chap. 3 (1981).

2.   D. L. Hunston, in "Rev. of Prog.
     Quant. Nondestructive
     Evaluation," D. O. Thompson and
     D. E. Chimenti, Eds., Plenum,
     Vol. 2B, 1711 (1983).

3.   R. K. Elsley, W. J. Pardee, M. J.
     Buckley, F. Cohen-Tenoudji, in
     Rev. of Prog. Quant.
     Nondestructive Evaluation," D. O.

     Thompson and D. E. Chimenti,
     Eds., Plenum, Vol. 5B, 1015
     (1985).

4.   F. I. Mopsik, Rev. Sci. Instrum.
     55, 79 (1984).

5.   S. S. Chang, J. Thermal Anal. in
     press, 1988.

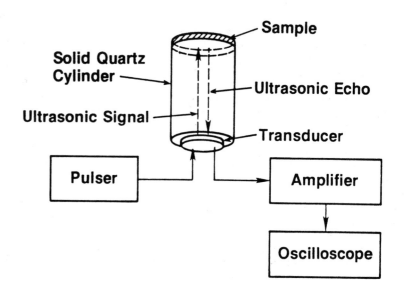

Figure 1.
Schematic of
Ultrasonic
Measurement

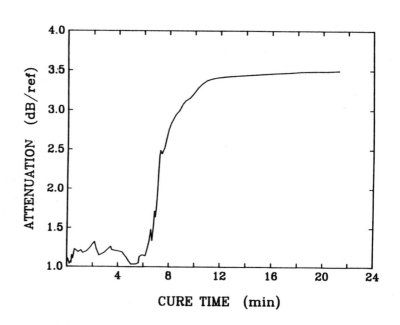

Figure 2.
Cure Monitoring by
Ultrasonic Shear Wave
Attenuation

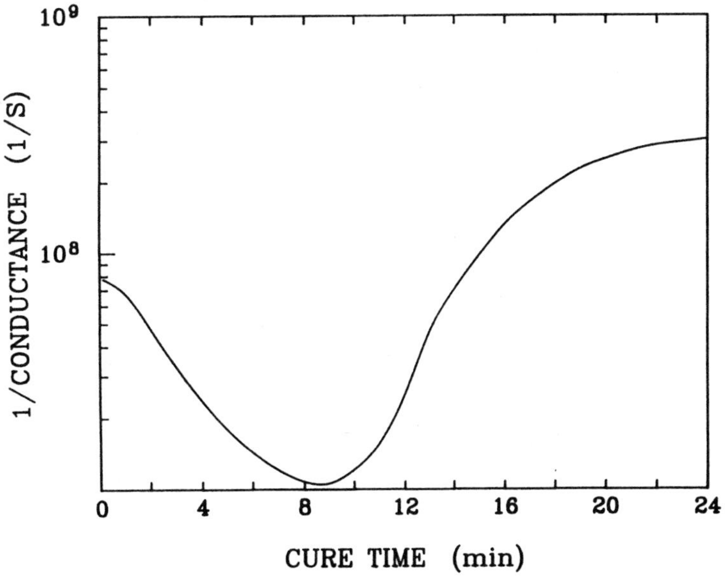

Figure 3.
Cure Monitoring by
Conductance at
50 Hz

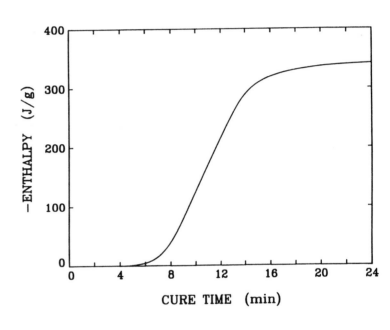

Figure 4.
Cure Monitoring by
Calorimetric
Measurement

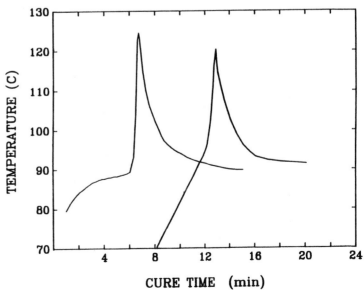

Figure 5.
Temperature Profile
during Cure
Left = Ultrasonic
Right = Dielectric

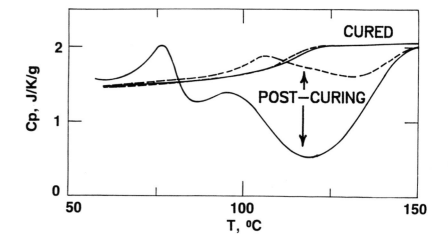

Figure 6.
Peak Time, $t_{\frac{1}{2}}$, and
Half-Width, Δt, as a
Function of
Isothermal Cure

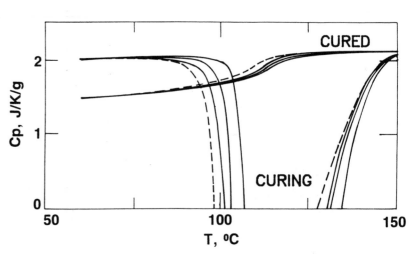

Figure 7.
Post-Curing of
Isothermally Cured
Resins
—————— $T_c = 100°C$
- - - - - - $T_c = 60°C$

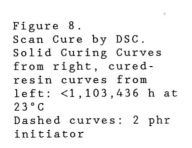

Figure 8.
Scan Cure by DSC.
Solid Curing Curves
from right, cured-
resin curves from
left: <1,103,436 h at
23°C
Dashed curves: 2 phr
initiator

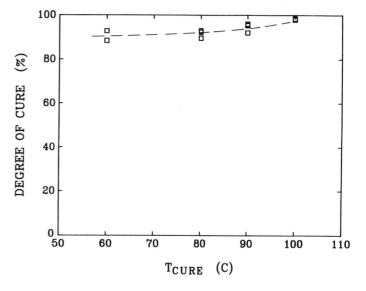

Figure 9.
Degree of Cure as a
Function of
Temperature of cure

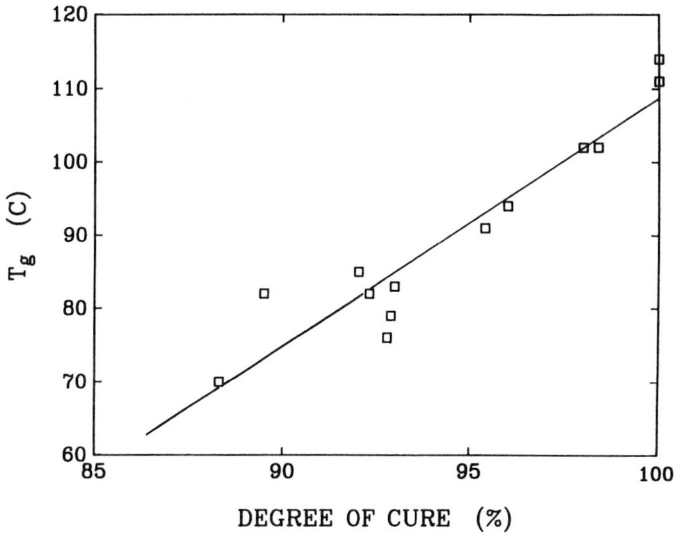

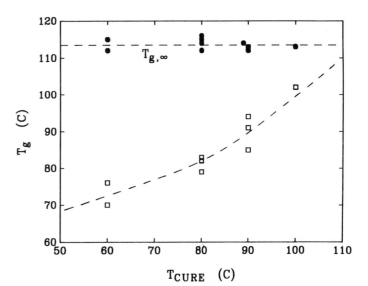

Figure 10.
T_g as a Function of
Degree of Cure

Figure 11.
T_g as a Function of
Temperature of Cure
and after Post-Cure
Open: Isothermal Cure
Closed: Post-Cured

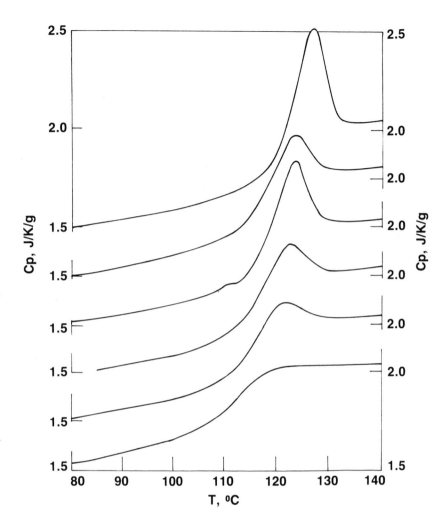

Figure 12.
Relaxation in the
Glass Transition
Region
From bottom: Fast-
cooled,-60°C/h,
-30°C/h and -6°C/h
cooled, cooled with
10°C steps (1 h each
step), cooled with
5°C step (6 h each
step)

MEASURING THE GEL POINT BY DYNAMIC MECHANICAL METHODS

Horst Henning Winter
Department of Chemical Engineering
University of Massachusetts
Amherst, Massachusetts 01003   USA

ABSTRCACT

The universality of the rheological behavior of polymers at the gel point allows a most accurate determination of the instant of gelation. The gel point is reached when the complex rheological behavior reduces to power law relaxation in the terminal frequency range (fractal behavior). The novel technique of Fourier Transform Mechanical Spectroscopy (FTMS) capitalizes on this observation.

INTRODUCTION

The instant of gelation (gel point, GP) of crosslinking polymers is often determined by mechanical test methods.[1] These methods require that the physical properties of the polymer at GP (critical gel) are known in some general way. Mechanical methods are effective as long as the mechanical properties are fully determined by gelation and not by secondary phenomena such as vitrification, crystallisation, etc.. In this presentation, the physical properties will be described briefly and then methods of GP determination will be discussed.

The gel point is defined as the instant at which the weight average molecular weight diverges to infinity. The polymer before GP (sol) is soluble in a good solvent while the polymer beyond GP (gel) is not completely soluble any more. The polymer at the sol-gel transition is called the 'critical gel'. The transition occurs gradually, i.e. the mechanical properties evolve gradually except for the steady state properties as discussed below. Exceptions are polymers in which network junctions form due to a first order transition.

The critical gel is neither a solid, since stress in a deformed critical gel can relax to zero, nor is it a liquid, since stress during flow grows to infinity. At the transition through the gel point, linear viscoelasticity reduces to a simple behavior which is described by the **gel equation** for the stress[2,3]

$$\tau(t) = S \int_{-\infty}^{t} (t-t')^{-n} \; \dot{\gamma}(t') \; dt', \text{ at GP.}$$

Molecular parameters determine the front factor, S, and the relaxation exponent, n. The above power law relaxation behavior seems to be a universal property at GP. It has been found with a large variety of chemically or physically crosslinking polymers, i.e. with all crosslinking polymers which we studied above the glass transition temperature. The relaxation exponent may theoretically adopt values between 0 and 1, while measured values ranged from 0.15 to 0.8.

STEADY STATE MECHANICAL PROPERTIES

At GP, the steady shear viscosity diverges to infinity and the equilibrium modulus is zero. These properties are often used for determining the GP.[4-8] Measurements are fairly easy and of sufficient accuracy for industrial applications. However, it should be kept in mind that steady state measurements are only possible some time before or after GP. The experiment fails in the close vicinity of GP, since the relaxation time diverges (with the molecular weight) to infinity[9] and, in a finite experimental time, no steady state can be reached in flow at constant rate, $\dot{\gamma}$. Attempts of measuring the steady shear viscosity result in breaking of the molecular structure and give an apparent gel point ('gel point' of a broken gel).

DYNAMIC MECHANICAL PROPERTIES

In comparison, dynamic mechanical measurements allow direct determination of GP. In such an experiment, the evolution of $G'(p,\omega)$ and $G''(p,\omega)$ are measured in small amplitude oscillatory shear as a function of extent of crosslinking, p. A consequence of the power law relaxation behavior (gel equation) is that the dynamic moduli follow a power law at GP

$$G' \sim G'' \sim \omega^n.$$

The strength of the critical gel may be determined from dynamic mechanical data as

$$S = G'(\omega) \ \omega^{-n} / [\Gamma(1-n) \ \cos(n\pi/2)],$$

independent of frequency (obviously). The loss tangent $\tan\delta = G''/G'$ depends on the relaxation exponent by

$$\delta = n\pi/2$$

and, most important, $\tan\delta$ is independent of frequency. Therefore, the gel equation suggests several methods to determine the gel point. Two of them will be discussed:

1. The gel point is reached when the loss tangent becomes independent of frequency (in the terminal frequency region). Fourier Transform Mechanical Spectroscopy (FTMS)[10] measures the dynamic modulus simultaneously at several frequencies. It is a most powerful method to find the instant at which the loss tangent, $\tan\delta$, becomes independent of frequency ω. An example is given by Holly et al.[10]. The method applies equally well to chemical[10] and to physical[11] gelation. A narrow critical region is advantageous for GP measurement with FTMS.

2. For some polymers, GP coincides with the G'-G'' crossover,[12,13] referring to the experiment in which G',G'' are measured at constant frequency, ω, (arbitrary value) during the evolution of the crosslinking reaction. There was much dispute whether the above criterion is general and, if not, whether GP occurs exactly at the crossover or just somewhere in its vicinity. Now, the gel equation makes it evident that there exists only one class of network polymers which has the property that GP coincides with the crossover. This class of polymers exhibits, when reaching GP, power law relaxation, i.e. the linear relaxation modulus reduces to

$$G(t) = S \ t^{-1/2},$$

The relaxation exponent has a **specific value n=1/2**. Examples are stoichiometrically balanced network polymers and networks with excess crosslinker, however, only at temperatures much above the glass transition, otherwise the power law behavior would be masked by vitrification. Many polymers have a different exponent value, $n \neq 1/2$, in which case the crossover occurs before GP (for $n<1/2$) or after GP (for $n>1/2$), i.e. the crossover cannot be used for detecting GP.

SUMMARY

Novel methods not only allow the direct determination of the gel point. They also allow, by extrapolation, to predict the gel point when the polymer is close to gelation but has not yet reached it. Mechanical measurements, in general, fail when gelation is interfered by secondary phenomena which alter the molecular mobility (vitrification, crystallization, etc.).

ACKNOWLEDGEMENT - This work is supported by the National Science Foundation grant MSM 8601595 and by the Center of the University of Massachusetts for Industrial Research on Polymers (CUMIRP).

REFERENCES

1. Winter, H.H., Polym. Eng. Sci., **27**, 1698 (1987).

2. Winter, H.H., and F. Chambon, J. Rheol., **30**, 367 (1986).

3. Chambon, F., and H.H. Winter, J. Rheol., **31**, 683-697 (1987).

4. Lipshitz, S., and C.W. Macosko, Polym. Eng. Sci., **16**, 803 (1976).

5. Apicella, A., P. Masi, and L. Nicolais, Rheol. Acta, **23**, 291 (1984).

6. Adam, M., M. Delsanti, and D. Durand, Macromolecules, **18**, 2285 (1985).

7. Bistrup, S. A., PhD Thesis, U. Minnesota (1986)

8. Farris, R.J., and C. Lee, Polym. Eng. Sci., **23**, 586 (1983).

9. Winter, H.H., Progr. Colloid Polym. Sci., **75**, 104 (1987).

10. Holly, E.E., S.K. Venkataraman, F. Chambon, and H.H. Winter, J. Non-Newt. Fluid Mech., **27**, 17 (1988).

11. Te Nijenhuis, K., and H.H. Winter, Macromolecules, 1988, submitted.

12. Tung, C.Y.M., and P.J. Dynes, J. Appl. Polym. Sci., **27**, 569 (1982).

13. ASTM D 4473-85.

CRITICAL ISSUES FOR MRP SYSTEMS IN ADVANCED COMPOSITES FABRICATION

David M. Schneider, William B. Krag
Arthur Young & Company
Detroit, Michigan 48243 USA

ABSTRACT

To maintain both quality and economic manufacturing producing advanced composites many enhancements are required to traditional Manufacturing Resource Planning (MRP) Systems. This paper will elaborate on the modifications required in the software and explore several capabilities required to optimize the often conflicting requirements demanded both by the customer and the physical necessities of the unique and complicated manufacturing process.

INTRODUCTION

THE USE OF COMPOSITE MATERIALS IN AEROSPACE AND AUTOMOTIVE INDUSTRIES has had dramatic growth in the past few years and is expected to increase in the future. These materials have become an increasingly important class of materials because of their unique properties which include a high strength to weight ratio. The manufacture of these components, however, is a labor-intensive, costly process which is difficult to control. This cost, complexity, and need for custom designed equipment has limited the more wide spread use of these materials.

The manufacturing of basic fiber-reinforced polymers -- typically aramid (Kevlar), boron, glass, or carbon (graphite) fibers in an epoxy-resin base or matrix as well as the fabrication of parts for products is very challenging and often results in high variability and subsequent low yields. The high labor content of current methods only adds to the existing high cost of materials. Finally, the disorderliness of the process itself also makes automation difficult. Nonetheless, significant global competition is beginning to emerge in order to take advantage of this new class of materials to produce superior products. Consequently, there exists a critical need both from a cost and strategic viewpoint to improve the fabrication of composite components.

THE MERGING OF TECHNOLOGIES - To economically control the manufacturing of composites an elaborate control system is required. Each of the individual control elements needed is not new, but rather it is the combination of the technologies which is unique. For example, control technologies for producing composites must be adopted from at least the following industries:

technology	industry
lot splitting/merging	computer chip
perishability	food
nesting	garment
kitting	model building
batching	heat treating
tool tracking	NC machining
maintenance	automotive
traceability	pharmaceutical
waste accountancy	nuclear
configuration management	aerospace
computerized process control	chemical, etc.

Thus, one can appreciate the challenge of designing appropriate control systems which contain features and capabilities from all these industries.

COST ISSUES

Because the high cost of raw material, complexity and critical nature of many products, variability of the process, and relatively high content of touch labor, the cost issues span procurement as well as the utilization, control and maintenance of production resources.

PROCUREMENT - In the procurement area the current supplier practice is to manufacture basic composite materials in large, bulk batches. Thus, a purchaser may have no choice but to acquire significantly more composite material than may be needed in the short run. The "Just-In-Time" capability between supplier and consumer is rarely feasible at this point in the development of the composite industry. Because of the perishability of room temperature composites, this leaves the users with little choice but to store the large batch in an expensive, refrigerated environment and exercise careful lot control procedures to assure accountability of the entire batch.

UTILIZATION - Utilization of material, equipment, tooling, and labor are, on the other hand, key cost issues within a composite fabrication center. Shop floor control systems must be able to conveniently portray real time indications of the utilization of raw materials including off-fall/scrap accountability. Likewise, utilization of key equipment such as cutting tables, autoclaves, utilization of tooling including "ready for use" status, as well as labor availability are all critically important contributors to overall costs. Manufacturing control systems for composite operations must maintain visibility of the utilization/ availability of all these factors.

CONTROL - Utilizing enhanced MRP and Shop Floor Control (SFC) systems to maintain visibility over composite operations is essential because of the complexity of the manufacturing process. Most cured piece parts are made up of dozens of pre-preg plys or layers of wound (wet or dry) composite filament/tape. Because each ply has individual part numbers/revision levels assigned, the cured piece part becomes an irreversible (can't uncure) "assembly" part to which positive identification is typically added after curing. Thus thousands of plys in hundreds of kits are flowing through layup and autoclave queue while relying for their identity on shop documentation which may travel with the lot. When plys are damaged, replacements must be recut and quickly brought to the kit to minimize material "out-time". Just a little of this remanufacturing and expediting can set off scheduling impact ripples which can have serious overall impact on the ability of an operation to maintain schedules and complete curing before material shelf life expiration. Thus the SFC systems have to be full featured, real-time, and convenient to use.

QUALITY - The above example of a torn ply requiring replacement is but one example of several potential quality problem areas. Unless rigorous inventory procedures are maintained through out the process, the total

process rapidly becomes chaotic. Each shop work order must be analyzed and nested to determine the smallest practical rectangle from which a group of plies can be cut from the same pre-preg (partially cured, semi-tacky condition) sheet. If pre-preg whisker orientation is important, nesting arrangements may be significantly limited. This has the effect of generating even more off-fall from the cutting operation. Plys with narrow, long finger shapes are very fragile and careless handling can easily tear them. Layup on autoclave or press tooling is another area where handling is required with the same exposure to damage. Tape laying operations are another potential problem area. Although much device programming has been done to determine "natural paths", still the potential exists for wrapping gaps or overlaps. Variations in pre-preg material thickness can even influence curing cycles. The cured part, as hard and sometimes brittle as it is, presents an entire new list of potential quality problems characteristic of difficult to machine materials. Ceramic or diamond coated cutting tools may be required to minimize tool wear and spindle deflection. Ultrasonic inspection is typically used to check for internal part continuity and suspect areas can be verified using radiographic inspection. The results of all quality inspections, process cycles (ovens, autoclaves, etc.) must be archivally captured and related to the appropriate work order/part number/lot number. Thus, the role of quality systems in composite manufacturing can not be separated from the main line production control systems.

MAINTENANCE - Traditional maintenance systems have been installed to minimize down time. A more progressive philosophy is to install a maintenance system to be able to actually improve process capability. This can be accomplished by reducing variation -- variation in machine operating characteristics, in tooling, in gaging, and in energy used for the operation. Because composite operations are typically accomplished in clean environments (sometimes class 100,000 clean rooms), maintenance of building related equipment is as important as the production equipment and tooling. Thus, composite manufacturing requires the installation and utilization of a full featured maintenance system.

SPECIFIC CONSIDERATIONS

RECEIVING INSPECTION - The control of refrigerated rolls of pre-preg requires unusually disciplined receiving inspection procedures. Each lot of received material must be "quarantined" until sample pieces (coupons) have been fully inspected and tested. As composites are evolving in material composition, likewise the inspection/testing procedures must change. Thus, the receiving environment must not only contend with "hold for testing" situation, but the test procedures themselves must be carefully related to revision levels of ordered material and part configuration. As distributed network systems merge text and graphics, on-line systems are being implemented which contain the latest documentation for all such testing as well as archivally retaining the results of each test.

PERISHABILITY - Some composite materials have limited useful life even if stored within a refrigerated environment. Most, however, have very limited lives at room temperature until they are cured in hot presses or autoclaves to "stop the clock". Consequently, an detailed shelf life system is mandatory which can track shelf life while cold (in storage) at one rate and then shelf life while hot (room temperature) at a much accelerated rate. In the event that a ply is damaged and a kit of plies has to be held up awaiting a replacement ply, the system must also be able to maintain shelf life for individual parts (plies) not just kits. After cutting unused material is often returned to refrigerated storage. In this case, the shelf life accumulator must record shelf lives under more than one condition to be able to accurately indicate remaining useful life.

NESTING AND WASTE ACCOUNTABILITY - Nesting, obviously, may only occur for plies made from the same pre-preg material. Thus, several work orders may have to be combined to accumulate a sufficient number of plies for economic cutting. The other scenario encourages product design to utilize as many plies as possible within any given design from the same type of material to make nesting by vehicle set more effective. Nesting by vehicle set can be done as a static exercise and the NC tape saved for reutilization. In the cases, however, where combinations of work orders are assembled, more dynamic nesting is required. More full featured nesting programs consider mirror images as candidate possibilities in the nesting solution. To maintain fiber orientation, plies made from woven material can be rotated only in increments of 90 degrees; those from unidirectionally reinforced material, only in 180 degree increments. The off-fall is valuable, approaching $100/#. It must be weighed, packaged, labeled, deleted from inventory according to lot identifications, and shipped much like product itself.

KIT ORGANIZATION - The plies lifted from the cutting table (urethane-surfaced table with chisel-point cutters, bristle top tables with textile saws, water jet, etc.) must be ordered in the reverse sequence from the part

itself. For example, the first ply the layup technician places is the bottom ply in the part. Conversely, the last ply which is applied is the top ply on the part. The engineering drawings should indicate ply order and the kits arranged accordingly. As plies are removed from the cutting table, they could each be placed on a "bakers' tray" and slid by level into a rolling cart which, in turn, can easily be transferred to the layup area. Manufacturing control can now begin to be maintained by kit number, providing that individual plies are still identifiable in case replacement plies are required.

TOOL STATUS - Each time a tool endures a curing cycle it may encounter buildups of parting materials and/or dimensional distortion. Thus, it becomes necessary to control tools by individual serial number and superimpose a repeating tool "clean and check" operation after a predefined number of curing cycles. For this reason, the shop floor control system must be able to recognize not only the quantity and location of tools, but their condition (availability for use). Because tool size can vary greatly, for large parts the aircraft manufacturers have had special "stretch" Automated Guided Vehicles (AGVs) created to automatically move large, long tools from layup to autoclave to trim & drill to tool queue, and tool refurbishing. In short, an entire system should be created just to control form tools. The system could include automatic storage and retrieval systems integrated with AGVs for automated handling combined with random access to needed tooling of practically any size.

LAYUP - The layup operation combines the OK tool with the kit of plies to form the layered composite part. For accurately locating plies on the tool several tricks have been used including making special tabs or locators in the plies and photographically projecting the ply image on the tool. Completed build ups which are going to be cured in an autoclave are then compacted by vacuum bagging. Hot press formed parts may not require bagging. Once assembled, the laid up tool can be controlled by part number and lot number.

CURE QUEUE - Bagged parts are next transferred to the curing areas which typically include autoclaves. Hot press cycles are relatively short. Autoclave cycles extend to several hours. Queuing for autoclaves must be accomplished on the basis of overall part size and cure time. Thus, the order of production is rescrambled by cure time to more effectively utilize the autoclaves. Once a part is committed to a curing operation, the shelf life accumulator can be stopped and the total cold storage and hot (room temperature) time profile recorded archivally against each ply in the assembly. Manufacturing control systems thus need access to basic part descriptive information such as cure time and composite thickness and overall size to be able to effectively queue jobs for curing.

IDENTIFICATION - Prior to curing the flimsy, material like composites typically rely upon external documentation for their identification. This can take the form of Kit Bills of Material and shop Traveler packets and/or CRT resident likenesses associated with a tool serial number. Although some experimentation has occurred imbedding part number identification in the layup of plies, more commonly positive identification is added to the part after curing. Traceability systems must recognize this problem with positively marking uncured composites and provide high confidence systems to control the kit from the cutting table through to curing.

TOUCH LABOR - Accurate cost for a labor intensive operation obviously depend upon accurate measurements of this component. Recently, the industrial engineering community has combined computer aided time standards systems with computer graphic systems to provide a method by which the method definition can include a graphic representation of the work situation. Using this approach, any changes to a work place description automatically generates a new time standard. This level of measurement provides the highest level of confidence in work measurements and thereby exceeds the requirements of MIL-STD-1567A. Manufacturing routings which include this level of definition for both operations and setups will provide an excellent basis for accurate costs. This level of documentation also defines motions and tasks so easily that non-value adding work can easily be identified and eliminated.

SUMMARY

The superior performance of composite materials is significantly enhancing the capabilities of products in many industries -- including defense, marine, automotive and recreation. Composites offer an opportunity to design products with far fewer numbers of parts which are strategically lighter and stronger. Thus, they represent a very important emerging industry. For the many reasons reviewed in this paper, the manufacturing process is, however, complex and to some degree unpredictable. Thus, the overall success of the manufacturing process is to a large degree dependent upon the effectiveness of the scheduling, quality, and shop floor control systems to positively control all inventory and do so maximizing utilization of critical resources to minimize cost.

AUTHORS

This paper has been prepared by Messrs. David M. Schneider and William B. Krag of Arthur Young and Company.

Mr. Schneider has had an extensive career with Grumman and General Electric managing leading edge composite manufacturing operations. Mr. Krag participated on a design team to design a major computer automated composite aircraft parts manufacturing center.

EQUIPMENT FOR PROCESSING STRUCTURAL RIM

Bruce C. Mazzoni
Cannon, U.S.A., Inc.
Mars, Pennsylvania   USA

NEW DEVELOPMENTS FOR STRUCTURAL RIM PROCESSING are occurring daily. Current activities include prototyping and identification of hundreds of parts for both automotive and non-automotive applications. Parts as large as a truck pick-up bed have been made completely out of Structural RIM.

Its light weight properties, high flex modulus, unique design flexibility, and adaptation to high volume production make SRIM ideal as a replacement for structural steel and other composites.

The goal of this paper is to update the reader with the current equipment being used for SRIM development and to detail the next generation of SRIM production equipment.

STRUCTURAL RIM

Initial SRIM developments were conducted on standard RIM and RRIM metering machines and clamps. Soon after these initial steps, Cannon and manufacturers who were doing these developments determined that special designs for both metering units and clamps were necessary for SRIM processing.

Major chemical, manufacturing, and development companies then decided on metering units with variable output capabilities and closed loop control, including controlling pressure. The clamps were unique to the RRIM market because of the higher tonnage requirement (as high as 600 tons) and large platen size (8 feet x 12 feet). Since the molds were so large, traditional 45 degree booking of the platen would not give full access to the part. It was necessary to have the bottom platen shuttle to the side for up to four-side access to the bottom mold. The upper platen was designed with the unique feature of tilting a full 90 degrees so that the operator could have full access to the upper platen.

These new metering units and clamps are currently used in development and have now led to a new generation of SRIM equipment necessary for providing parts in 60 seconds or less. The new turnkey system called COMPOTEC (for Composite Technology) is an integrated plant for the production of Structural RIM parts, which includes all aspects of processing including automatic mat preforming.

The new COMPOTEC system is the newest member of the integrated turnkey plants that Cannon manufactures for the automotive markets.

Like our other turnkey plants, the Cannon COMPOTEC system is the latest technology to reduce manpower requirements and gives faster cycle times, which improves profitability and productivity for the manufacturer.

STRUCTURAL RIM CLAMP

The SRIM clamp has been proven ideal for SRIM processing because of the ability to book the upper platen 90 degrees for full operator access to the mold (Figure 1). Platen size for a SRIM clamp is designed for up to 12 feet by 8 feet with tonnages as high as 600 tons. The clamp can handle total mold weight of 40 tons.

The clamp is designed using four cylinder strokes for high pressure closing via the lower platen. The booking of the upper platen is obtained by two hydraulic cylinders. THe bottom shuttle platen(s) is operated by two electrically driven pinions engaging two gear racks fitted on the lower platen's bottom side that drives the platen along the rails to the service station (Figure 2).

The complete automatic cycle sequence is as follows:
1. The mold is closed.
2. Bottom platen lowers and engages to shuttle.
3. Bottom platen shuttles to service station
4. Upper platen books 90 degrees to vertical position.
5. Bottom platen shuttles back to clamp station.
6. The bottom platen rises; high tonnage occurs.

Also designed is a SRIM clamp with two bottom platen shuttle beds. While one bottom platen is in the clamp station for injection and curing, the other bottom platen is in one of the two service stations for part removal and to prepare the next preform. The goal is to accomplish the capability to rapidly produce parts by automating a task or doing several tasks at once (i.e., part curing-preparing the next bottom platen).

The clamp hydraulic system will be supported on top of the clamp structure. Typical of all of our clamps, ventilation systems are compatible with the design of the clamp.

The SRIM clamp includes closed loop position of control and variable tonnage to help with the distribution of chemicals through the mat.

STRUCTURAL RIM METERING UNIT

The SRIM metering unit has been designed for the high pressure generation caused by the resistance of the glass mat and large liquid distribution. Typical of most SRIM machines, a Cannon Model HE closed loop cylinder machine is utilized. The machine incorporates variable output capability with closed loop control on injection pressure during the length of the shot.

The SRIM Model HE machine with closed-loop control is capable of controlling and correcting all parameters including injection pressures with a response time of the whole system of 28 milliseconds and correction adjustment of less than 0.2 seconds (Figure 3).

The component metering is operated by hydraulically driven metering pistons. The cylinder's speeds are controlled via the proportional valves which dispense the amount of oil necessary according to the output requirements. An encoder on the cylinder measures the speed and position, and with the high resolution of the digital system gives the output with extreme precision.

The position control card (CCP) compares this value with the value entered into the machine by the operator. Should a difference be noted between the theoretical and actual data, the servo drive corrects the proportional valve controlling the output.

The unit eliminates the need for calibration and can calculate the specific gravity of the chemical at any process condition. The metering machine is supplied with a diagnostic system which controls all operational functions and will alert the operator to a possible malfuction.

The only interface between the operator and machine consists of a hexadecimal keyboard and monitor. A printer will record all vital information including date, time the part was made, number of parts made that shift, set and actual values of all parameters. Also, other intelligent peripheral units can be connected for data exchange and Computer Integrated Manufacturing (CIM).

SRIM MIXHEAD

The processing of SRIM depends on a proper flow distribution of chemicals throughout the part. The use of a self-cleaning mixhead with the ability to control the injection pressure of the chemicals proves to assist for good flow distribution. With the pressure control nozzles it is possible to adjust or vary injection pressure as a function of time to assist with chemical distribution.

The pressure control nozzles are accomplished via pneumatic/hydraulic actuator. The pressure via a transducer is communicated to the Metering Unit where it compares the set value and corrects it accordingly. It permits automatic adjustment to allow low pressure recirculation through the mixhead.

SRIM - CANNON COMPOTEC

The goal of Cannon was to combine all aspects of processing SRIM and to combine each of them to provide a modular turnkey system. This goal is now realistic with the introduction of the Cannon COMPOTEC system.

The COMPOTEC process integrates the clamping system, metering unit, computer closed-loop control system, and the automatic mat preform system.

The different techniques of preforming can incorporate IR heating of the mat with pressure forming plug (Figure 4) or a vacuum drawn membrane to form deeper profile parts. For complex parts a spray-up of the preform may be necessary.

The operating sequence of a turnkey Cannon COMPOTEC system is shown on Figure 5, utilizing a double shuttle system.

CONCLUSION

The advancements in Structural RIM development have now made it a viable production composite for the replacement of other structural supports. With the introduction of the Cannon Structural RIM COMPOTEC turnkey system, the process steps can now be obtained automatically with only an operator(s) to remove parts.

FIGURE 1. Upper platen book 90 degrees for easy access.

FIGURE 2. Example of a bottom shutting platen.

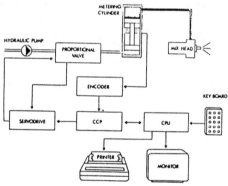

FIGURE 3. Cannon HE closed loop control.

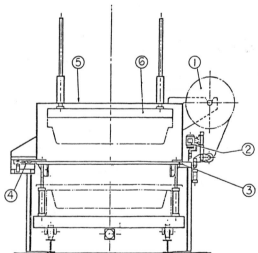

FIGURE 4. Automatic mat preform (front view). (1) mat/fiberglass, (2) mat cutter (rotary knife), (3) mat support frame holder, (4) mat unwinding grippers, (5) enclosure, (6) forming plug (heated), (7) antifolding device, (8) bottom shuttle platen.

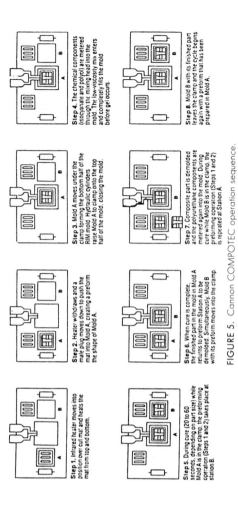

Step 1. Infrared heater moves into position over cut mat and heats the mat from top and bottom.

Step 2. Heater withdraws and a male plug moves down to push the mat into Mold A, creating a preform the shape of Mold A.

Step 3. Mold A moves under the clamp forming the bottom half of the RIM mold. Hydraulic cylinders raise Mold A to clamp onto the top half of the mold, closing the mold.

Step 4. The chemical components (isocyanate and polyol) are metered through the mixing head into the mold. The low-viscosity mix enters and completely fills the mold before gel occurs.

Step 5. During cure (20 to 60 seconds, depending on part size) while Mold A is in the clamp, the preforming operation (Steps 1 and 2) takes place at station B.

Step 6. When cure is complete, the finished part in the mold in Mold A returns to preform Station A to be demolded. Simultaneously, Mold B with its preform moves into the clamp.

Step 7. Composite parts demolded and the polyurethane components are metered again into the mold. During cure while Mold B is in the clamp, the preforming operation (Steps 1 and 2) is repeated at Station A.

Step 8. Mold B with its finished part leaves the clamp and the cycle begins again with a preform that has been prepared in Mold A.

FIGURE 5. Cannon COMPOTEC operation sequence.

RAPID COMPOSITE ARMOR FIELD REPAIR PATCH/KIT CURED BY THE SUN OR UV

Novis Smith, Mark Livesay, Emiro Castaneda
Sunrez Corporation
El Cajon, California 92020   USA

ABSTRACT

Sunrez Corporation has developed a rapid field repair patch/kit for composite armor, composites, and related materials. This system is based on an unique photoinitiator which is activated by UVa (350-400 nm) present in sunlight and in commercially available UV lights. These patch systems incorporate unsaturated polyester and vinyl ester resins containing either 50% E or S-2 glass fiber. The cure time is less than 10 minutes at 9ºF. The properties of the resins, the forms of the repair patch, and the variation in the types of resins used will be discussed. Due to its ease of use, excellant results can be obtained by untrained individuals. This system has wide application for composite fabrication and repair.

THE USE OF FIBER REINFORCED PLASTIC composite armor structures is now and important thrust by the U.S. Army for the enhancement of the performance and survivablility of American combat troops. Parallel developments of composite armor are also being carried out by other branches of the U.S. Military. In order to maximize the effectiveness of composite armor or any composite vehicle structure subject to battlefield damage, it must be repaired as soon as possible after receiving damage to at least restore environmental integrity(sealing of all holes). Those personnel carrying vehicles which maintain a protective environment must be resealed in case of chemical, biological or radioactive weapons being used.

Additionally, in many cases, the interior of the vehicle and the crew must be protected from weather extremes such as rain, wind, cold heat dust and humidity. The restoration of full structual strength is also desirable in the field, but this is a function of what tools are available to remove the interferring damaged resin/fiber.

In order that this necessary rapid repair be performed on the vehicle or structure as soon as possible after being damaged, the repair procedure must be: 1. Simple; 2. Quick; and 3.Reliable as possible, in order that the reinforced armor structure be operationally serviceable again including providing protection for its crew and passengers. Because the repair must be done under battlefield conditions, the repair procedure must perform under almost all weather conditions that may be encountered from desert to arctic to tropical climates. Another consideration in the development of a repair system is that the damage to be repaired can consist of irregular holes and cracks and jagged, torn edges. It is also necessary to be able to repair the ehicle quickly from either the outside and/or the inside.

A summary of the requirements for this new field repair system for reinforced plastic composite armor (and possibly any other armor or military vehicle structure) should be the following:
1) Simple - A crew member can apply it with minimum training and equipment;
2) Rapid - Total time for preparation, application, and cure must be a matter of minutes - under ten if possible;

3) Reliable - The system should have a high tolerance to moisture, temperature variation, preparation and application time, require no mixing, and yet produce good properties when cured;
4) Conformable - Easily applied to any shape of crack or hole;
5) Storable - Shelf life should be a minimum of six months under normal temperature conditions;
6) Usable over wide temperature range, -40°F to 120°F;
7) Compact and lightweight - It will be stored aboard the vehicle;
8) Low cost - This permits greater latitude in deployment and spares for a more effective repair system;
9) Minimum power - Requires no external heat or can operate off the vehicle battery;
10) Minimum of noxious fumes - Requires that the fumes will not cause a problem when being used in a confined space;
11) Satisfactory physical properties - The patch should be structurally sound and perform until the vehicle is back in a rear area for more extensive repair.

All of these requirements have not been met by any single existing repair system up to now. Current RTV silicone patching systems do not develop enough strength, are slow to cure and can have adhesive bonding problems. Sunrez is developing a field repair system to meet all of these requirements under contract to U.S. Army Materials Technology Laboratories (MTL).

TECHNICAL APPROACH

Currently, all practical reinforced resin or plastic armor systems are based on thermoset resins, i.e., epoxy, unsaturated polyester. phenolic, vinyl esters, and certain cross-linked polyurethane systems. The vehicle armor or structure which is being repaired or patched is a large heat sink. A thermally dependent cure of a resin is difficult to attain under all conditions. If a very exothermic self-curing resin system is used, difficulties in storage and subsequent use by the crew will result. Thermosets are usually two part systems or are premixed and require cold storage prior to use, and usually require external heating to initiate and cure. On the other hand, the thermoplastics which have the necessary properties require high heat and are difficult to conform to the damage holes and cracks because their high melt viscosity.

Another alternative is the use of a UV cured resin system. The usual disadvantages of an UV curable resin system are that only thin sections can be done, usually expensive UV lights are required, and reinforcement fiber glass interferes with or prevents curing.

The required new field repair patch system will also require a thermoset to produce the necessary high structural properties after curing in place and to form a patch which conforms to the irregularities of the hole or crack with good adhesion.

An ideal rapid structural repair system for composite armor and other structures would consist of an easily applied, moderate viscosity, glass fiber reinforced, putty-like material which requires no mixing(one part) and cures in place without the addition of heat. Usually exothermic thermoset resins are two part systems requiring careful measuring and mixing, and rapid application. This mixing is difficult to do under battlefield conditions, much less having to wet out out reinforcement. Additionally, systems which require heat for curing whether generated internally or externally added, often do not cure satisfactorily at the edges, especially in cold weather.

Sunrez has chosen to work with an ultraviolet thermoset resin system which meets the requirements listed previouly for a field repair patch system for armor and structures including high strength properties, rapid cure, nonthermal cure, one part, good storability, easily reinforced as a prepreg, and conformable.

The UV resin systems which we have chosen to develop for the field repair patch are based on both vinyl ester and unsaturated polyester resins utilizing a new and novel photoinitiator from BASF. (Sunrez has an agreement with BASF and the appropriate PMN to import this initiator.) This system cures with long wavelength UVa(360-400nm) light. This corresponds to standard tanning light, but sunlight is equally effective. The system can be handled under standard indoor lighting or in the shade for several hours with no problem. Although there has been considerable experience in Europe with this photoinitiator, it was necessary to reformulate and screen this photoinitiator with American resins and determine the limits on the system for application to a field repair patch for armor and structures. Additionally, it was also necessary to determine in what reinforcement configurations could this resin system be used.

RESULTS

EVALUATION OF AMERICAN RESINS-Samples of most readily available American unsaturated isophthalate esters and vinyl ester resins which were listed as lighter in color and having good tensile strength were obtained from commercial producers. Upon receipt, the resin was examined to make sure it was relatively light in color. Samples were then taken and blended with less than 1% photoinitiator by weight. A 1/4-inch layer of the resin was poured into a cup and irradiated with a 400-watt UVa lamp at a fixed distance of 12 inches. Those samples which had a satisfactory cure rate under the UVa light were noted along with the time to achieve curing all the way through the formed resin disk in the cup. Many of the resins exhibited a surface tackiness when cured in air, but this tackiness did not appear when cured with a protective clear PVA film or under nitrogen. This surface tackiness disappeared in a few hours when it did form. The test ended if the curing time exceeded 20 minutes.

We found a total of 14 American resins could be used for the patch system. There are probably a number of other resins that might also work, but we already have found a significant number which offer a great latitude in resin selection. OCF E-780 was among the resins which were satisfactory and is related to the initial resin selected for development work with composite armor by FMC and OCF.

The tensile strength(without reinforcing glass fiber) was measured on three of the more interesting resins which were photocured and are reported in Table 1. We selected Sunrez 9000(an unsaturated isophthalate ester) as the resin of choice for the final patch samples of our Phase I SBIR program to be sent to U.S.Army Materials Technology Laboratory, Watertown, MA (MTL).

TABLE 1

Resin	Min.to Cure	Tensile,N/mm² (psi)
A	5	52.6 (7626)
B	5	39.8 (5773)
Sunrez 9000	5	61.7 (8946)

The most important property is rapid cure rate because this relates to how thick a practical patch can be with a given resin. Resin B was the fastest evaluated although the spread in cure time for most of the resins was from 4 to 9 minutes with all other conditions being equivalent. We found that the physical properties of MEK-peroxide cured resins are equivalent to this UV cured system.

The listed properties (MEK-peroxide cure) of the resins selected for evaluation fell within a range of 25% for tensile and modulus.

PREPARATION OF PATCH –

Determination of Maximum Practical Patch Thickness - A number of layers of E glass woven fabric were layered up in 1/4- and 1/2-inch thicknesses and wetted out thoroughly with resin. These were cured under a 400-watt UVa lamp at 12 inches in the same manner as the initial resin evaluation procedure. The 1/4-inch sample cured in 8 minutes. The 1/2-inch patch cured in 25 minutes. It was faster to cure two 1/4-inch patches on top of each other to obtain a 1/2-inch patch than to start with a 1/2-inch patch. We have made one sample up to two inches thick for ballistic testing.

Effect of S-2 Glass - The initial attempt to cure a resin impregnated 24-oz. S-2 glass fabric layer resulted in a slow and incomplete cure due to the opacity of the 463 sizing. Leaching the fabric with solvent(methyl ethyl ketone) improved the rate of cure significantly, but it still was not satisfactory. A developmental sample of 24-oz. S-2 fabric with 363 sizing was found to be not only clearer in appearance but actually cured satisfactorily in a 1/4-inch thickness.

Weight % Glass Content - Resin impregnated E-glass woven fabric samples were made up with a squeegee used to pull away the excess resin. After curing the glass content was 50%. From experience this would be the most convenient and practical level to formulate a patch system. The problem is that on storage, the resin poor system starts to show dry spots or less tacky surfaces and this will decrease bonding. In a field situation, the patch must provide its own tack, and there is little time to work the resin back into place if it has shifted or wicked around in the patch.

In other words the glass fabric is "full" of resin at 50%. Vacuum compression or applied pressure in a stack of fabric layers will compress the fiber glass fabrics permitting a higher glass loading while maintaining a good wet out of the fabric. On the other hand when we fabricated a thick armor section (2.1 inches thick) using standard wet layup techniques, the glass content rose to 68%. In this case, the layers are rolled and com-pressed by hand and the excess resin is removed with a squeegee. The use of a vacuum bag would increase the glass content more.

Surface Adhesion - A ballistically damaged thin plate of DAP/fiber glass armor and a thick section of standard polyester/S-2 fiber glass armor was received from MTL for developing good adhesion of the patch system to these surfaces.

Initially we studied the peel strength/adhesion of the patch system cured to itself. We found that the isophthalate resins gave better adhesion than the vinyl ester systems. When laminating one patch on top of another, the first one was cured in air until it was hard but with a slight tackiness. Then the next one was placed on top of the previous one. This one was subsequently cured. The adhesion was excellent and both layers had bonded and showing preferred failure though the whole sample and not at the interface when broken.

The DAP/S-2 thin armor section with two bullet holes which had been sent to us by MTL was wiped with MEK and lightly sanded with #80 sandpaper, and finally wiped off. A repair was made using two patches consisting of one patch applied on top of the other. Each patch consisted of one ply of mat followed by 4 plies of 24-oz. woven S-2 glass fabric all preimpregnated with resin and cured with a hand held UVa lamp. The other side was done in the same manner with E-glass fabric in place of the S-2 fabric. The adhesion of both of the 1/2-inch thick patches was excellent. This plate was returned to MTL with the patches in place. Similar patches were also cured in the sun with approximately the same cure rate. A patch was also placed on the flat section of the large 2-inch section of armor which we had received from MTL. In this case, the surface was not prepared and the solvent action of the styrene and the tackiness of the wet patch were all 'that contributed to the wet out of the surface. Satisfactory adhesion was also obtained on curing.

We conclude that the Sunrez patch system will be applicable to most dry structural and armor surfaces within reasonable limitations.

Spectra and Kevlar - A sample of Spectra fabric was impregnated with resin and was cured with the UVa lamp. The temperature was controlled by switching off the light when the exotherm began approaching 200°F, and switching the light back on when the sample started to cool. The curing cycle was about twice as long as that of the E-glass even with the cooling pauses. The resulting totally cured composite was satisfactory in appearance and toughness. In a patch system we could adjust conditions to maintain the temperature below 200°F without having to interrupt the cycle.

We attempted to cure resin B with 1% by weight Kevlar or carbon fiber added. In both cases, the curing was significantly inhibited.

A group of potential promoters for the photoinitiated reaction were evaluated with the hope of enhancing the reaction in the presence of the opaque fibers such as Kevlar. These chemicals included: anthracene, fluoranthene, dimethyl benzylamine, several fluorescent dyes, laser type dyes, and vinyl imidazole, vinyl pyridine. None seem to speed the cure of the pure resin and several decreased the rate of cure.

Thickening the Patch - If needed we have utilized 1% magnesium oxide (as an oil paste) to thicken the resin by heating at 250°F for 15 minutes. A five ply E-glass patch was made, thickened, and successfully cured.

TESTING OF PATCHES - A patch kit was made of 2.0-oz. nonwoven mat, and four plies of woven E-glass roving and impregnated with resin to 50% by weight and then encased with PVA film on both sides and placed into a opaque plastic bag and sealed to prevent styrene vapor from escaping. This patch was approximately one foot square and 1/4-inch thick and was sent to W. Haskell(MTL). It was placed in the direct sunlight in Watertown, MA sunlight (1315 hours) at 9°F. The patch hardened in 9 minutes and demonstrated that it could be used at even lower temperatures.

The patched thin armor plate which was sent to MTL also was satisfactory. We have made up patches with long continuous roving, chopped fiber, and nonwoven mat with Sunrez 9000 which were sent to MTL as the final patches of Phase I. All of these appeared to behave satisfactory with respect curing and handling. As long as the patch system is transparent and not too thick, there appear to be few limitations on the form the patch takes.

CONSTRAINTS OF THE PHOTOINITIATOR - The reaction is strictly dependent upon light and not temperature. Evidently, the reaction is the absorption of UVa light (360-400 nm) followed by a mild exotherm of the actual polymerization.

OTHER CONSIDERATIONS - The stability of the resin system appears to be at least six months at 72°F. At 140°F, the resin was fluid for nine days before gelling. However, the photoinitiator still reacted as normal causing curing in 4-5 minutes in sunlight. The problem in obtaining long term high temperature stability will focus strictly on the stabilization of the polyester resin to prevent high temperature gelling. A problem will be the difficulty of the photoinitiator in overcoming the large amount of inhibitors normally used to effect this stability.

Phase I of this program demonstrated feasibility of this field repair patch system based on a photocure using sun or UVa light. This patch system is versatile enough that it should be adaptable for the repair of all armor systems to achieve environmental integrity of any damaged vehicle. Not only does this patch system appear to work well for field repair, but it may perform very well for fabricating the composite armor itself.

DISCUSSION

Over twenty American resins were evaluated. Significant variation was found between these resins with respect to rate of cure and extent of hardness. Many of these resins were found to be satisfactory for making a patch kit. Single patch thicknesses were made up to 1/2 inch thick. A multilayer laminate was made that contained 68% by weight glass and was readily cured.

Patches or prepregs have been made in a number of forms including woven and knit fabric, nonwoven mat, and chopped fiber. Techniques have been developed to control the viscosity of the resin in the prepreg. Both a DAP and a FMC S-2/OCF resin panel with ballistic damage were satisfactorily repaired. One of these panels was repaired using a sunlight cure on one side and a handheld UVa lamp on the other with equal results. The patches were 1/2-inch thick and were done in two successive layers.

With the currently available U.S. resins, the maximum thickness of one layer with woven roving reinforcement was about 1/2 inch. However, a thicker layer might be achievable with even clearer resins. The best way to thicker patches was found to be the use of 1/4-inch patches and just cure one layer on to another. Thicknesses of at least 2 inches are readily built up.

As presently configured, the Sunrez field patch system is simple and can be readily used by the combat vehicle crewman. It is also practical for ready field deployment in vehicles. This has been a very successful program in demonstrating the feasibility and practicality of the Sunrez light-cured reinforced-plastic field repair system for armor and other structures.

Acknowledgement: We wish to thank the U.S. Army for their support of this program though a Phase I SBIR award, and William Haskell (MTL) for his direction and guidance in the performance of the program.

FRP MILITARY COMBAT VEHICLE HULL AND TURRET DESIGN

P. R. Para, D. E. Weerth
FMC Defense Systems Group
San Jose, California   USA

ABSTRACT

Research of thick laminated Fiber Reinforced Plastic (FRP) composite materials for use on military tracked vehicles is increasing. The increased emphasis is in response to a major Army thrust directed toward increasing mobility by weight reduction. Performance improvements such as equivalent ballistic protection with elimination of spall, increased operational life by reduced corrosion and stress cracking, and potential sound and vibration reduction may be realized with this technology.

FRP materials require the combat vehicle designer/analyst to focus on the inherent anisotropic characteristics of the material when considering its use. FRP combat vehicle structures are being designed with fibers of lower matrix bond strength resulting in a combination of properties not obtainable in many other material systems. Design by compromise is necessary to exploit maximum material benefits. Studies related to design, analysis, and testing have demonstrated, that for armored military vehicles, ballistic performance will remain the limiting design consideration. This paper will examine other design considerations and performance advantages achieved through the use of these new materials.

SIGNIFICANT RESEARCH IS CURRENTLY underway in the application of FRP materials in military tracked vehicles as a substitute for aluminum and steel. As mission requirements become more difficult to satisfy, the following performance improvements will draw increased attention towards the use of FRP composites.

- Weight Reduction with equivalent ballistic performance
- Reduced Life Cycle Cost
- Increased Corrosion Resistance
- Survivability Enhancements
- Reduced Noise and Vibration

The utilization of FRP materials for structural applications has been perfected by numerous aerospace companies. The designs are driven by the need to develop high quality load bearing structural laminates which have high fiber/resin bond strengths. FRP laminates for military vehicle applications are required to be structural and capable of resisting impact due to ballistic projectiles. Previous research and development programs have demonstrated that FRP laminates, with intentionally degraded interfacial bond strengths, ballistically out-perform laminates with high fiber/resin bond strengths.

FMC Corporation has successfully designed, analyzed, fabricated, and tested hull and turret structures using FRP materials. One example of this technology is the composite portion of a hybrid Bradley Fighting Vehicle (BFV) turret (Figure 1) developed for the U.S. Army Materials Technology Laboratory (AMTL). Conclusions revealed that hybrid turret fabrication is straight forward and economically feasible. A 15-20% weight savings was realized in areas where the aluminum was replaced with FRP [1]. Another example is the composite hull of an M113 Armored Personnel Carrier (Figure 2) developed for the David Taylor Naval Ship Research Development Center. Three years of field testing have thus far revealed that an E-glass fabric with vinyl-ester and epoxy

Fig. 1 - Composite Turret for Bradley Fighting Vehicle

Fig. 2 - Composite M113 Armored Personnel Carrier

thermosetting resins may be used effectively as a highly loaded structure with equal or improved performance characteristics.

The success of the above mentioned programs have led to the award of the Composite Infantry Fighting Vehicle (CIFV) Program by AMTL to FMC Corporation. This multi-year program will design, analyze, fabricate, and field test two (2) demonstrator vehicles with FRP hulls. Structural and ballistic performance requirements for the CIFV demonstrators conform to existing aluminum BFV M2/M3 A1 vehicle requirements. The conceptual phase has been completed and the detailed design and fabrication of the 1st hull demonstrator is in progress. An artist's rendition of the CIFV demonstrator is illustrated in Figure 3.

MATERIAL CONSIDERATIONS

The use of FRP materials provides the designer flexibility in maximizing strength and stiffness by the preferential placement of fibers. The vehicle designer is faced with identifying appropriate material characteristics which simultaneously meet both the structural and ballistic performance needs.

The selection of an S-2 Glass™/semi-compatible polyester resin system offers this needed compromise between mechanical properties and ballistic performance. The material systems for combat vehicle structures require characterization throughout an environmental operating range of -70°F to 180°F with moisture contents

ranging from 0% to 100% Relative Humidity (R.H.). The minimum acceptable room temperature mechanical properties for the selected S-2 Glass™ (24oz-yd^2 woven roving) polyester resin were determined from a 4-ply [0]$_T$ laminate and are shown in Table 1.

The importance of fatigue behavior in FRP structures has been thoroughly researched in the aerospace industries. The application of FRP ballistic laminates requires additional characterization to assess the materials fatigue resistance due to reduced interfacial bond strengths. Displacement controlled flexural fatigue performance is illustrated in Figure 4. Fatigue testing was conducted under elevated environmental conditions, 43°C (110°F) and 100% relative humidity. The Cycles to Failure (N) reported in Figure 4 represent the point at which the stiffness of the specimen is reduced to 70% of the initial level.

BALLISTIC CONSIDERATIONS

The most important design consideration for the hull structure is the level of ballistic protection needed to meet the mission requirements. Ballistic testing of various laminates comparing ballistic efficiencies revealed that S-2 Glass™ is superior to Kevlar™ 29, E-glass, and 5083 aluminum [2,3]. This is illustrated in Figure 5.

Parametric studies performed by Owens-Corning Fiberglas (OCF) and AMTL investigated the influence of fiber type, fiber finish, and resin on ballistic performance [2,3]. Findings from this study revealed that a fabric treated with a non-compatible sizing exhibits a higher ballistic efficiency than fabric with compatible sizing. This however, has an inverse effect on the mechanical properties (Table 2). The ballistic performance of FRP laminates is also strongly influenced by the resin system

Fig. 3 - Composite Infantry Fighting Vehicle Concept

Table 1 - Mechanical and Physical Properties

Property	Requirement	ASTM Test Method
Resin Content, % by wt	32 ± 3	D2584
Density, g/cc	1.84 - 1.98	D792
Tensile Modulus, (min avg)	18.6 GPa (2.7 msi)	D638
Tensile Strength, (min avg)	496 MPa (72 ksi)	D638
Flexural Modulus, (min avg)	20.7 GPa (3.0 msi)	D790
Flexural Strength, (min avg)	276 MPa (40 ksi)	D790
Compressive Strength, (min avg)	124 MPa (18 ksi)	D695
Short Beam Shear Strength, (min avg)	17.2 MPa (2.5 ksi)	D2344
Glass Transition Temp, (min)	160°C (320°F)	D4065

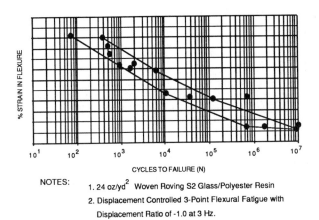

NOTES:
1. 24 oz/yd^2 Woven Roving S2 Glass/Polyester Resin
2. Displacement Controlled 3-Point Flexural Fatigue with Displacement Ratio of -1.0 at 3 Hz.
3. 8-Ply Laminate [0°,90°,± 45]s

Fig. 4 - Strain (ε) Versus Cycles to Failure (N)

[3,4]. Figure 6 illustrates the ballistic efficiency (i.e., weight merit rating, E_m, relative to MIL-12560 steel) of S-2 Glass™ laminates using various resin systems.

Protection for the hull and crew for light-medium combat vehicles has been defined for threats from artillery, small arms, kinetic energy (KE) projectiles, and mine blast. Fragments from artillery detonation impact the hull surface at various velocities and obliquities; therefore, this ballistic threat effects the largest surface area of the vehicle. Ballistic performance assessment with respect to artillery bursts is accomplished using Fragment Simulating Projectiles (FSPs). FSPs are hardened steel projectiles with performance characteristics similar to shell fragments of the same weight and hardness. The fragment simulating projectile test is the simplest method to screen various armor systems and rank relative ballistic performance.

Structure surfaces may utilize a monocoque FRP armor system and/or a system consisting of FRP as the backup material with applique/stand-off armor. Ballistic laminates employ very high fiber contents, (typically 70% by weight) and a semi-compatible fiber/interface. Tuning of the material system promotes a penetration mechanics defeat mechanism for the monocoque system which combines transverse fiber shear, fiber tension, and delamination to absorb the projectile kinetic energy. Fragments of higher mass and/or velocity require additional energy absorbing material (hull thickness) or increased obliquity to defeat the projectile. The most effective system against an FSP threat is the monocoque S-2 Glass™/polyester system when ballistic performance, fabrication complexity, and cost are considered.

In addition to protection against artillery fragments, structure surfaces are also exposed to KE projectiles. Ballistic protection from KE threats (at shallow obliquities) requires higher areal densities than required for FSPs and a different approach in defeating the projectile. A monocoque FRP system is not a weight efficient system for defeating KE projectiles (at shallow obliquities). The ballistic efficiency is greatly increased by utilizing the laminate as a backup material with a suitable armor capable of fracturing the projectile. High-hardness steel (either spaced or applique) and ceramic applique have proven to be weight efficient alternatives to the monocoque system. The front surface of the steel/ceramic armor fractures the armor piercing projectile, and the FRP catches the resulting steel and ceramic fragments.

At higher obliquities, the monocoque FRP system becomes more efficient at defeating armor piercing projectiles. Rather than fracturing the projectile, as does the steel/ceramic FRP system, the projectile is deflected and turned within the laminate. A monocoque FRP system at obliquity exhibits a similar defeat mechanism as that seen in aluminum by deflecting and turning an armor piercing projectile. The critical angle defines the obliquity at which a projectile will be defeated, while impact velocity and laminate thickness remains constant.

Knowing the defeat mechanisms and ballistic performance levels for various FRP armor systems, the general configuration of the combat vehicle may be defined. Structure shapes may be developed which consider surface obliquity, surface area, and internal volume in such a way as to reduce weight. Figure 7 illustrates the CIFV demonstrator which optimized the armor system and

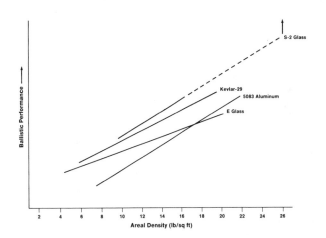

Fig. 5 - Fiber Effect on Ballistic Performance

Table 2 - Ballistic Performance and Flexural Strength as a Function of Fiber/Resin Compatibility

Material	Relative Ballistic Performance	Flexural Strength MPa (ksi)
S-2 Glass™/Polyester Non-Compatible (Starch Oil)	100 %	113.0 (16.4)
S-2 Glass™/Polyester Semi-Compatible (Epoxy Compatible)	93 %	171.5 (24.9)
S-2 Glass™/Polyester Compatible	85 %	333.4 (48.4)

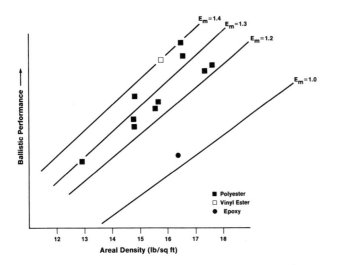

Fig. 6 - Resin Effect on Ballistic Performance

geometry to meet the ballistic requirements and program objectives (i.e., reduced structure weight at equivalent or improved ballistic performance).

STRUCTURAL CONSIDERATIONS

Design practice has relied extensively on field testing to identify and correct structural inadequacies to meet mission requirements. Field failures identify items requiring redesign but do not identify items with excess safety margins (i.e., excess weight). The traditional approach in evaluating structural integrity of hulls has been to impose conservative loading conditions which may be encountered during the life of the vehicle. The following categories are generally the most severe:

- Ground Transport Loads
- Rough Terrain Impact Loads
- Airlift Transport Loads
- Impulse Loads (Mine blast, ballistic impact, and reactive armor detonation)

The above loads have been established empirically and are conservative estimates of the maximum expected loads. Static acceleration (g) loading conditions have historically been used as a method to establish structural design loads. The increased utilization of FRP for combat hull structures warrants the development of realistic load spectra based on the performance characteristics of the vehicle. Dynamic load spectra may be determined from simulated terrains using mobility models to establish force time histories for the structure. The resulting loads should be utilized to establish a realistic design criteria for the structure. Variables such as system performance (i.e., propulsion system, suspension system, etc.) as well as crew and driver discomfort are limiting criteria in defining maximum loads. Such revised loading criteria replaces conservative static 'g' loads used in the past.

The increased emphasis of FRP hull structures will have a significant impact on meeting the military objectives of mobility and rapid deployment. Lighter weight vehicles will respond to external loading with increased dynamic flexibility.[5] This increased flexibility is not only

attributable to lighter weight, but also reduced elastic properties (i.e., tensile and in-plane shear modulus). The 24 oz/yd^2 woven roving S-2 Glass™/polyester resin laminate has a tensile modulus and an in-plane shear modulus which are 33% and 18% of aluminum respectively. The in-plane shear modulus has the most significant impact on reducing the 1st torsional frequency of the hull structure. The mode shape of the 1st torsional frequency is twisting around the longitudinal axis of the structure which occurs when the vehicle impacts an obstacle on one side and introduces an asymmetric load on the structure. The use of FRP materials allows the selective placement of discrete fibers to effectively satisfy the strength and stiffness requirement. This advantage of using FRP materials is a cornerstone in developing weight efficient designs.

The built up layers of high strength fibers increase the flaw tolerance of the structure when compared to metals. The flaw which is developed during service may be mitigated by the woven roving fabric. Preliminary damage tolerance models have indicated that the critical flaw size is large and therefore elaborate Non-Destructive Testing (NDT) hardware and methods are not required. Currently, work is in progress to evaluate other fiber resin interface failures to further characterize strain energy release mechanisms.

Understanding the limitations of the FRP ballistic laminate will serve to minimize the structural weight while maintaining adequate structural integrity and durability in considering the trade off between weight reduction and increased hull flexibility.

JOINING CONSIDERATIONS

The use of FRP materials for combat hull structures will increase parts consolidation by replacing many structural elements with one piece moldings. In hybrid metal/FRP combat vehicle structures, adequate attention to the proper transfer of mechanical loads across joints is essential in optimizing performance.

The BFV composite turret design retains aluminum in regions that are stiffness critical and incorporates a one piece molding to minimize the joints as shown in Figure 8. This method of assembly offers the maximum advantage in utilizing the FRP material. Recent investigations have

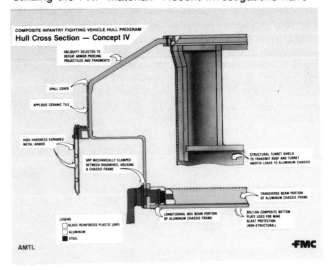

Fig. 7 - CIFV Concept Hull Cross Section

shown that through the thickness constraint, as provided by a bolted connection, improves the static strength of an FRP laminate by 100-180% and the fatigue limit by as much as 100% over a pin bearing connection [6,7]. Bolted connections are subjected to triaxial stress with the bolt preload acting as a lateral constraint. The additional constraint increases the fiber buckling stress which results in higher laminate strength.

Typical bolted joints sandwich the thick laminate between two metal surfaces, namely the aluminum substructure and the metal clamping plate. In a simple joining arrangement, the mechanical fastener is perpendicular to the surface of the laminate as shown in Sections A-A through C-C of Figure 9. These types of joints are susceptible to bolt preload loss due to the visco-elastic behavior of the resin. The behavior of the material is further aggravated by the extreme environmental conditions (temperature and moisture content) which may be encountered during the life of the vehicle.

In order to characterize the visco-elastic behavior of FRP materials, stress relaxation tests were conducted on E-glass/vinyl ester resin specimens. The specimens were subjected to varying bolt torques (i.e. transverse preload) at 49°C (120°F) over a period of 1000 hours. Preload data versus time was collected for each of the six (6) initial preload levels. Test results [8] showing the extent of preload relaxation are illustrated in Figure 10. As expected, the relaxation rate is dependent on the initial preload magnitude. Once the visco-elastic behavior of an FRP material has been determined, designs which consider this behavior may be developed. Preload in a connection serves to retain proper clamping between mating parts and to reduce the effects of fatigue during the application of a cyclic working load.

A bolted connection is subjected to fatigue by the application of a working load which alternates between a maximum and minimum level. In connections where flange

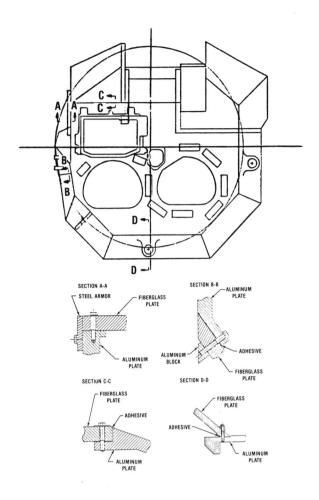

Fig. 9 - Hybrid Turret - Joining Configurations

stiffness exceeds bolt stiffness, the cyclic portion of the load is small relative to the total working load. This is typical in metal connections where a larger portion of the load is carried by the flange. In composite joints, the bolt stiffness exceeds the flange stiffness, therefore the majority of the cyclic load is carried by the bolt. The importance of preload and flange stiffness on joint fatigue behavior is illustrated in Figure 11.

An example of a critical connection subjected to an initial preload and a working load is the roadwheel support housing/FRP hull connection of the CIFV hull demonstrator. The cross-section of the connection is illustrated in Figure 7. This connection clamps the thick laminate between the steel roadwheel housing and the aluminum chassis frame and is subjected to cyclic loading. The terrain induced loads transmitted through the connection are a function of the vehicle weight, vehicle speed, and terrain conditions. A typical duty cycle for this connection which considers cross country and paved surfaces is shown in Table 3.

Testing of this connection was conducted under the CIFV demonstrator program to verify the connection design and its ability to withstand the equivalent of 6,000 miles of combat vehicle operation. A 36" specimen was fabricated and assembled to simulate a segment of a combat vehicle hull. Cyclic loads were applied to the connection using the loading from the Duty Cycle (Table 3). The setup for the roadwheel support housing fatigue test is illustrated in Figure 12. The fasteners which join the

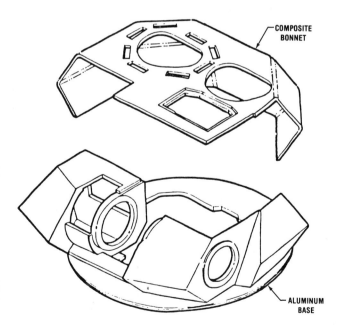

Fig. 8 - Hybrid Turret - Exploded View

roadwheel housing to the hull were instrumented with strain gages to measure bolt load fluctuations. Bolt strain data was continuously monitored throughout testing and the subsequent evaluation revealed that no failure or degradation occurred. The success of this test indicates that efficient joining configurations may be developed using FRP material which are capable of withstanding significant cyclic structural loads.

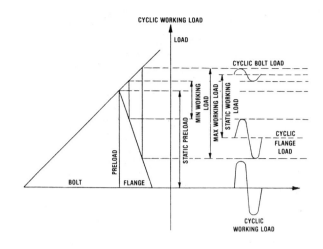

Fig. 11 - Bolted Joint Subjected to a Cyclic Working Load

Fiberglass tooling was selected for the CIFV hull demonstrator program. Similar coefficients of thermal expansion for the tool and FRP structure will reduce cure cycle stresses. Female tooling was selected to produce a high quality surface finish on the structure exterior and requires that careful attention be exercised in consolidating the FRP plies in female radii during the lay-up process.

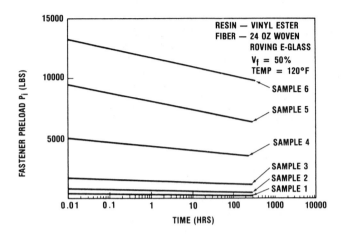

Fig. 10 - Experimental Bolt Load Relaxation for Vinyl Ester/E-glass Composites

FABRICATION/ASSEMBLY CONSIDERATIONS

Metallic structures require that surfaces be defined using a series of flat plates assembled to the desired shape. Reducing the number of piece parts will reduce fabrication costs [9] and may be easily accomplished by consolidating the items into larger but more manageable FRP elements. The selected assembly method should minimize the extent of joining in the structure. The approach used for the CIFV demonstrator consists of left and right-hand side molded structures which are joined along the vehicle centerline. This method takes advantage of a parting line containing many cut outs (cut outs for the turret, cargo hatch, exhaust grille, etc.) and reduces the extent of joining.

Table 3 - Roadwheel Duty Cycle

ROADWHEEL TRAVEL	NO. CYCLES @ TERRAIN CONDITION			
	" IMPROVED ROADS " 23.3 % OF TOTAL USAGE 35 mph AVERAGE SPEED	" SECONDARY ROADS " 23.3 % OF TOTAL USAGE 20 mph AVERAGE SPEED	" CROSS COUNTRY " 53.4 % OF TOTAL USAGE 12 mph AVERAGE SPEED	TOTAL
2 In.	50,500	81,500	228,000	360,000
4 In.	10,800	29,300	139,900	180,000
6 In.	3,500	7,000	61,500	72,000
8 In.	2,100	2,100	31,800	36,000
10 In.	-	-	24,000	24,000
12 In.	-	-	12,000	12,000
14 In.	-	-	6,000	6,000
TOTAL NO. CYCLES	66,900	119,900	503,200	690,000

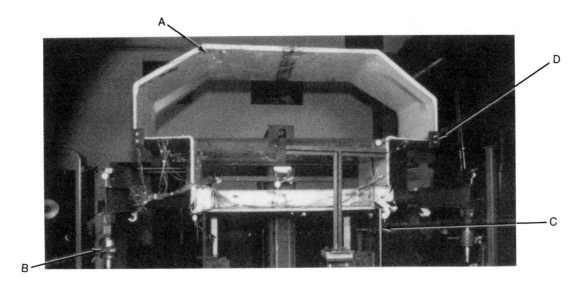

NOTE: Letters indicate the following:

A-Test Speciment I   C-Vertical Restraint Fixture
B-Hydraulic Actuator   D-Horizontal Sponson Restraint

Fig. 12 - Roadwheel Support Housing Fatigue Test Setup

CONCLUSIONS

Two fundamental barriers explained the slow acceptance of polymer composites in the military market place; namely, a lack of material data/design experience and the need for new manufacturing technology and facilities. The benefits of using these more expensive materials, in overall cost and performance, are becoming more pronounced as mission requirements become more difficult to satisfy. Drivers generating increased attention toward the military application of FRP composites are:

- Weight reduction
- Reduced noise and vibration
- Improved ballistic performance
- Enhanced survivability
- Corrosion resistance
- Reduced life cycle costs
- Reduced production costs
- Reduced signature
- Improved blast load performance
  (i.e., mines and reactive armor)

Research in the last five years has resulted in the development of ballistic thick laminate FRP composites that will compete structurally and ballistically with steel and aluminum, in addition to providing the added benefits listed above.

The major programs funded to date have demonstrated various aspects of using FRP material for combat vehicle structures. Phase I of the CIFV program has established proof of material principle in the laboratory. Phases II and III of the program will establish proof of principle under realistic operating conditions.

REFERENCES

1. FMC Corporation, "Reinforced Plastic Turret for M2/M3, "Final Report AMTL TR 87-39, August, 1987.

2. Discover S-2 Glass™ Fiber, Pub. No. 5-ASP- 13101, Owens-Corning Fiberglas Corporation, Toledo, Ohio (1985).

3. W.E. Haskell, III and L.J. Dickson, "Reinforced Plastics for Military Ground Vehicle systems," 40th Annual Conference, Reinforced Plastics/Composites Institute, The Society of the Plastics Industry, Inc., Jan 28-Feb 1, 1985.

4. A. Vasudev and M.J. Mehlman, "A Comparative Study of the Ballistic Performance of Glass Reinforced Plastic Materials," Sampe Quarterly, Volume 18, No. 4, July 1987.

5. D.E. Weerth and C.R. Ortloff, "Computer Analysis Methods for the Composite Bradley Fighting Vehicle," Proceedings of the Army Symposium on Solid Mechanics, MTL MS 86-2, U.S. Military Academy, West Point, NY, October, 1986.

6. Mattews, F.L., and Kalkanis, P., "The Strength of Mechanically Fastened Joints in Kevlar Fiber Reinforced Epoxy Resin" Proceedings of the Fifth International Conference on Composite Materials ICCM-V, San Diego, California, July 29, 30, August 1, 1985.

7. Crews, J.H. Jr., "Bolt-Bearing Fatigue of a Graphite/Epoxy Laminate" <u>Joining of Composite Materials ASTM STP 749,</u> American Society for Testing and Materials, copyright 1981.

8. D.E. Weerth, "Creep Considerations in Reinforced Plastic Laminate Bolted Connections," Proceedings of the Army Symposium on Solid Mechanics, MTL MS 86-2, U.S. Military Academy, West Point, NY, October, 1986.

9. D.E. Weerth, C. Huffman, and S.A. Ellery, "The Engineering Role of Composite Materials in Reducing the Cost of Ordnance Products," 41st Annual Conference, Reinforced Plastics/Composites Institute, The Society of the Plastics Industry, January, 1986.

FRP MATERIAL CONSIDERATIONS FOR MILITARY COMBAT VEHICLE HULL AND TURRET APPLICATIONS

G. E. Thomas

FMC Corporation
Central Engineering Laboratory
Santa Clara, California 95052   USA

D. E. Weerth

FMC Corporation
Defense Systems Group
San Jose, California   USA

Abstract

Many composite material applications employ high strength fibers which are well bonded to the polymer or metal matrix surrounding the fibers. Fiber Reinforced Plastic (FRP) composite military combat vehicle structures are being designed employing fibers with lower matrix bond strengths requiring a combination of properties not obtainable in many other composite material systems. These include performance as armor and structure, very low flammability and toxicity, long term environmental resistance, and the ability for thick composite sections to be joined to other materials for dynamic structural load transfer.

This paper reviews the development of material systems employing high strength S-2 Glass® in a polyester thermoset matrix which meet the property requirements for this specific application. The selection criteria utilized to determine the best material system will be discussed, along with the material properties achieved to date.

THE DEVELOPMENT OF FRP MATERIALS for armored combat vehicle structures (1) has addressed two military performance needs that are difficult to achieve simultaneously, namely, vehicle weight reduction and improved ballistic performance. Early parametric investigations were conducted by Owens-Corning Fiberglas (OCF) and the U.S. Army Materials Technology Laboratory (AMTL, formerly AMMRC) to study the effects of resin cure, fiber weave, fiber type, and fiber finish (sizing), on the ballistic performance of polyester resin matrix flat panel laminates (2)(3). The results of that study showed that S-2 Glass® treated with a non-compatible sizing exhibited superior ballistic performance to material treated with moderately compatible or compatible sizings. However, the inverse relationship was found for the mechanical

properties of the three types. S-2 Glass® woven roving (24-oz/sq yd) treated with a moderately compatible sizing (OCF 463) was determined to yield the best compromise of ballistic and mechanical properties for combat vehicle structures.

To demonstrate the feasibility of using FRP materials for a combat vehicle structure, AMTL contracted with FMC Corporation to design, analyze, fabricate, and test hybrid Bradley Fighting Vehicle (BFV) turrets (4). The roof, right side, and rear sections of the welded aluminum armor turret were replaced with a single FRP laminate structure (see Figure 1).

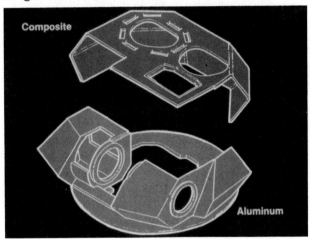

Fig. 1 - FRP/aluminum turret concept

Fabrication of the FRP turret section was accomplished by a wet lay-up process. The woven roving was wet-out on a mechanized impregnator, placed on the mold, consolidated, vacuum bagged, and cured in a hot air recirculating type oven. After removal from the mold, the FRP section was machined and attached to the aluminum base plate to complete the assembly (see Figure 2). Ballistic, structural, and field tests were successfully completed on the hybrid turret. A 16.5% weight saving was realized in the sections replaced by the FRP laminate.

Fig. 2 - Assembled hybrid turret

The success of the FRP turret program resulted in the award of the Composite Infantry Fighting Vehicle (CIFV) hull contract by AMTL to FMC Corporation. This is a 4-year program to design, analyze, fabricate, and test two demonstrator vehicles having FRP hulls. The performance requirements for the demonstrator vehicles are equivalent to the current aluminum hull BFV-A1.

Phase I, which included material selection, conceptual design, ballistic qualification, structural analysis (5), test panel/section fabrication, and process development, was completed in December 1987. Final detailed design and fabrication of the first demonstrator vehicle commenced this year in Phase II.

It is important to note that ballistic laminate technology should not be confused with thin aircraft laminates which require high consolidation and react differently during curing than thick ballistic laminates. This technology is also significantly different than ship building techniques which are restricted to low discontinuous fiber volume fractions unsuited to armored vehicle applications.

BALLISTICS

Ballistic laminates differ uniquely from structural laminates, which require high fiber-to-resin bond strengths normally achieved through autoclaving. High quality ballistic laminates employ very high fiber contents, typically 70% by weight, wherein the fiber-to-resin interfacial bond strength is intentionally degraded. The proper tuning of the interfacial bond strength promotes a penetration mechanics ballistic defeat mechanism illustrated in Figure 3. This defeat mechanism results in fiber shearing in the first third of the laminate, fiber tension in the middle third of the panel, and ply delamination in the last third of the plate thickness. This unique defeat mechanism explains how tuned composite laminates can be ballistically more weight efficient than their metal counterparts. By altering the defeat mechanism, the notion that more mass improves ballistic performance is dispelled. Both fiber and resin play an integral role in ballistic performance, as does the fiber finish (sizing).

Figure 4 illustrates the superior ballistic performance of an S-2 Glass® laminate compared to 5083 aluminum and Kevlar® 29 (2). Only at lower areal densities (approximately 5-7 lbs/sq ft) does Kevlar® outperform S-2 Glass®. Figure 5 illustrates the effect that resin has on ballistic mass efficiency, E_m (6). Variations in resin performance are based on the degree of interfacial bond strength developed between fiber and resin.

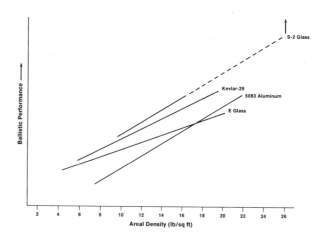

Fig. 4 - Fiber effects on ballistics

Fig. 3 - Typical example of a ballistic impact on an FRP thick laminate

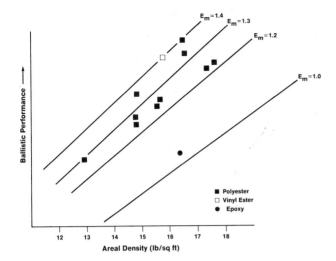

Fig. 5 - Resin effects on ballistics

- AMTL propane burner flammability
  - Time for smoke generation to cease
  - Weight loss
- In-coming material condition
  - Weave distortion
  - Resin wet-out uniformity
- Processability
  - Tack/drape
  - Cure
  - Compaction
- Safety
  - OSHA requirements
- Physical properties
  - Density
  - Glass transition temperature (T_g)

CIFV HULL MATERIAL SELECTION

Certain processing and performance improvement needs were identified with the resin system used during the FRP turret program. Specifically, these were cure reproducibility and environmental/safety issues involved with wet lay-up of a styrene type polyester resin. To avoid these problems in the CIFV Hull Program, several other polyester resin systems were investigated in prepreg form. From an original list of fourteen promising systems (6), four material candidates were selected for further evaluation. The four candidate material suppliers were requested to prepreg dry 24-oz/sq yd S-2 Glass® woven roving supplied by FMC.

The selection criteria established for the material evaluation were the following, in order of importance:

- Ballistic performance
  - Fragment simulating projectile V$_{50}$
- Ballistic (incendiary round) flammability
  - Time to extinguish

Mechanical tensile properties were not evaluated since there is a strong direct correlation with ballistic performance.

Prepreg material returned from the four selected material suppliers was used to fabricate 300-lb test sections, which had a shape simulating the complex geometry of the proposed hull structure. Processing differences between each of the materials were monitored, and after final testing, two prepregs were qualified.

Upon completion of the material selection, a specification was established for material procurement. This material specification was adopted by AMTL and will be issued as MIL-L-46197(MR), 23 DECEMBER 1987.

MATERIAL MECHANICAL AND PHYSICAL PROPERTIES

The mechanical and physical properties for in-coming material are listed in Table 1. These properties are determined from a 4-ply $[0_4]_T$ laminate cured at 250°F for 2-hours in a vacuum bag. The mechanical values represent minimum acceptance standards. Consistently, the strength of the material has exceeded these requirements by 10% or more.

Table 1 - Mechanical and Physical Properties

Property	Requirement	ASTM Test Method
Resin Content, % by wt	32 ±3	D2584
Density, g/cc	1.84-1.98	D792
Tensile Modulus, (min avg)	18.6 GPa (2.7 msi)	D638
Tensile Strength, (min avg)	496 MPa (72 ksi)	D638
Flexural Modulus, (min avg)	20.7 GPa (3.0 msi)	D790
Flexural Strength, (min avg)	276 MPa (40 ksi)	D790
Compressive Strength, (min avg)	124 MPa (18 ksi)	D695
Short Beam Shear Strength, (min avg)	17.2 MPa (2.5 ksi)	D2344
Glass Transition Temp, (min)	160°C (320°F)	D4065

FLAMMABILITY AND TOXICITY

The acceptance of FRP materials for combat vehicle applications requires a thorough understanding of its performance when subjected to burning conditions, such as an incendiary round, burning fuel, electrical fires, or other battlefield hazards. Ballistic flammability and the AMTL Propane Burner Flammability Test were used for the initial material screening. Currently, work is in progress to comprehensively measure the flammability and toxicity characteristics of the qualified FRP materials. The appropriate tests are listed below:

- Ballistic (incendiary round) flammability
- AMTL Propane Burner Flammability Test
- Oxygen Index (ASTM D2863)
- Smoke Density (ASTM E662)
- Toxic gas analysis
- Flame spread
- Halon 1301 extinguishment

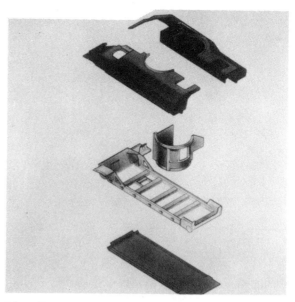

Fig. 7 - CIFV concept exploded view

VEHICLE DESIGN

Five different conceptual designs were developed for government review, each with varying ballistic, structural, and fabrication trade-off alternatives. Figure 6 illustrates the selected concept for the first demonstrator vehicle, which offered the maximum weight savings (27%) of the five candidates when compared to the aluminum BFV-A1 hull.

Fig. 6 - CIFV selected concept

The hull structure will be composed of an FRP left and right hand side joined (7) to an aluminum box beam chassis (see Figure 7). The turret shield is a structural member which torsionally stiffens the hull and transmits turret inertial loads to the aluminum chassis. A non-structural FRP bottom plate is bolted to the bottom of the chassis for mine blast protection.

The thickness of each region of the vehicle has been ballistically designed to afford maximum weight savings, but at the same time provide the proper level of structural integrity. The roof and glacis region of the FRP vehicle are designed for overhead artillery bursts, whereas the bare glacis of the roof defeats armor piercing (AP) rounds at obliquity. Ceramic tile is appliqued over the FRP vertical sidewalls to defeat zero degree obliquity AP threats. The horizontal sponson has no ballistic requirements, consequently a thinner section is sufficient to carry the structural loads. The vertical side wall behind the roadwheels employs a stand-off steel perforated armor to defeat the AP threat, rather than applique tile.

PROCESSING

Process development for Phase I was done on an 11-foot long female tool of fiberglass construction. It represented the left front quadrant of the proposed vehicle concept. The surface area of the tool was approximately 100-sq ft.

A three-man team was used for the lay-up. Four foot wide material was cut to length, oriented on the tool, and pressed into place by working it from one end to the other (see Figure 8). In order to avoid distortion of the woven roving, the separator film was removed as the material was applied. Particular attention was given to consolidation in all the female radiuses to avoid bridging. In this fashion a complete ply coverage was applied to the tool by butt splicing the various pieces of material. Debulks were performed every five plies to consolidate the material.

Fig. 8 - Lay-up procedure

Due to the complexity of the tool geometry, the tight radiuses, and the loft in the material, "high elongation" bagging materials were used in the final bagging stage to avoid any possibility of bridging. The bleeder was engineered to remove approximately 2% of the resin from the lay-up, to yield a final resin content in the laminate of 32 ±3% resin by weight. This weight fraction of resin provides maximum ballistic performance, without compromising material processing. A vacuum of approximately 28-in Hg was maintained on the lay-up during the entire cure cycle (see Figure 9). Curing was accomplished in a hot air recirculation type oven. Thermocouples were placed in various positions throughout the lay-up to monitor the cure.

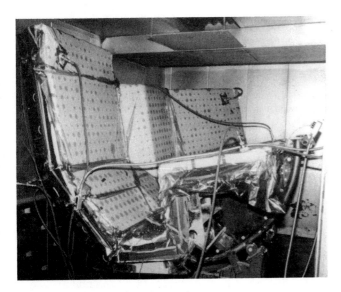

Fig. 9 - Vacuum bag cure in oven

The cure cycle was designed to avoid an excessive rise in temperature or exotherm within the laminate in order to avoid part damage. Careful control of the cure cycle is an important aspect in processing thick laminate composites made with thermosetting resins. The cycle shown in Figure 10 is a three step cure, with the leading thermocouple indicating the temperature of the lay-up surface and the lagging thermocouple representing the temperature at the center of the lay-up. The first step heats the resin to reduce viscosity without initiating a reaction. This allows consolidation of the material and removal of volatiles or trapped air. The second step increases the temperature of the material sufficiently to cause the desired chemical reaction to occur, without causing an excessive exotherm. Once the exotherm has subsided, the temperature is raised to complete the final cure of the resin in the third step. Further work is being done to optimize the cure cycle to maximize ballistic performance.

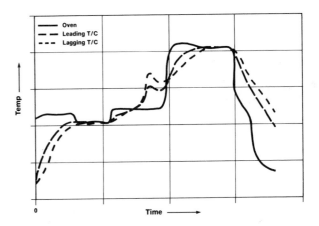

Fig. 10 - Cure cycle for thick laminates

After curing, the 1400-lb laminate was removed from the tool with an overhead hoist (see Figure 11). Figure 12 shows the completed hull section. For structural testing purposes, a three foot cross-section of the proposed vehicle concept was machined and assembled (see Figure 13). This test section was subjected to a 10.5-g vertical static load test and dynamic fatigue loading induced by roadwheel motions that simulate a 6000-mile field durability test. Test results were used to validate the structural finite element analysis. The design has passed all test requirements, and work is proceeding rapidly to complete the final design, fabrication, and field testing of the first FRP hull demonstrator vehicle by early 1989.

Fig. 11 - FRP hull laminate being extracted from tool

Fig. 12 - Completed FRP hull laminate

Fig. 13 - CIFV concept cross-section

CONCLUSIONS

Two fundamental barriers have explained the slow acceptance of polymer composites in the military market place, namely, a lack of material data/design experience and the need for new manufacturing technology and facilities. The benefits of using these more expensive materials, in overall cost and performance, are becoming more pronounced as mission requirements become more difficult to satisfy. Drivers generating increased attention toward the military application of FRP composites are:

- Weight reduction
- Reduced noise and vibration
- Improved ballistic performance
- Enhanced survivability
- Corrosion resistance
- Reduced life cycle costs
- Reduced production costs
- Reduced signature
- Improved blast load performance (i.e., mines and reactive armor)

Research in the last five years has resulted in the development of ballistic thick laminate FRP composites that will effectively compete structurally and ballistically with steel and aluminum, in addition to providing the added benefits enumerated above. Phase I of the CIFV program has established proof of material principle in the laboratory. Phases II and III of the program will establish proof of principle in the field.

REFERENCES

1. D.E. Weerth, C. Huffman, and S.A. Ellery, "The Engineering Role of Composite Materials in Reducing the Cost of Ordnance Products," 41st Annual Conference, Reinforced Plastics/Composites Institute, The Society of the Plastics Industry, January, 1986.

2. Discover S-2 Glass® Fiber, Pub. No. 5-ASP-13101, Owens-Corning Fiberglas Corporation, Toledo, Ohio (1985).

3. W.E. Haskell, III and L.J. Dickson, "Reinforced Plastics for Military Ground Vehicle Systems," 40th Annual Conference, Reinforced Plastics/Composites Institute, The Society of the Plastics Industry, Inc., Jan 28-Feb 1, 1985.

4. FMC Corporation, "Reinforced Plastic Turret for M2/M3," Final Report MTL TR 87-39, August, 1987.

5. D.E. Weerth and C.R. Ortloff, "Computer Analysis Methods for the Composite Bradley Fighting Vehicle," Proceedings of the Army Symposium on Solid Mechanics, MTL MS 86-2, U.S. Military Academy, West Point, NY, October, 1986.

6. A. Vasudev and M.J. Mehlman, "A Comparative Study of the Ballistic Performance of Glass Reinforced Plastic Materials," Sampe Quarterly, Volume 18, No. 4, July 1987.

7. D.E. Weerth, "Creep Considerations in Reinforced Plastic Laminate Bolted Connections," Proceedings of the Army Symposium on Solid Mechanics, MTL MS 86-2, U.S. Military Academy, West Point, NY, October, 1986.

S-2 Glass® is a registered trademark of the Owens-Corning Fiberglas Corporation.

Kevlar® is a registered trademark of the DuPont Company.

FIRE HAS TO BE CONFINED: COMPARTMENTALIZATION IS A KEYWORD

Kevin H. Kramer, Jan M. G. Peper
Mobay Corporation
Pittsburgh, Pennsylvania USA

ABSTRACT

Fire-resistant intumescent materials and intermediates have been developed which enable the design and manufacture of effective barriers to help prevent the spreading of fire, smoke and gas in various types of structures. These products are available in different forms such as: a putty-like material, a one-component coating, and several two-component systems. The two-component systems allow for the production of parts and profiles to meet specific requirements.

These fire-resistant, intumescent materials:
-- are activated in the case of fire or extreme temperatures;
-- build a foam which fills and closes joints, cracks, cavities;
-- build an insulating carbonized foam to help prevent penetration of the flames and the spread of smoke and gases;
-- achieve a cooling effect; have heat-consuming properties to help prevent the transfer of heat towards the backside of the barrier;
-- do not contain halogens, asbestos, phenol or formaldehyde.

Fire barriers and fire stops of various configurations are applied in power plants, commercial buildings, hospitals, and industrial plants. Depending upon the type of intumescent material used and the design of the system, fire rates of up to four hours can be achieved. Also, most of these systems can be easily handled, installed and maintained.

FIRE PROTECTION is a very important aspect of our lives today. One of the best forms of fire protection is prevention. This, however, is not always possible. Although we do not expect it, fires do occur and we must be ready.

In case of fire, steps must be taken to confine the fire; compartmentalization becomes the keyword. One way to compartmentalize a fire is to make each area an independent fireproof structure. This, however, is not practical as ventilation ducts and cable passes are present in most structures. Therefore, we must find a way to seal or close these openings or penetrations in the event of fire. In this way, the fire, smoke or gases generated do not spread through these openings.

A few problems become evident, however, when you try to accomplish this. They include:

1) A ventilation duct must be open for the movement of air under non-fire conditions. To use a damper in the event of fire could require some mechanical means for closing.
2) In the cable passes or penetrations, any system design or product would have to take into account that eventually the cables could burn, leaving voids in the fire protection system where fire, smoke or gases could penetrate.
3) The inclusion of additional cables after the protection system is in place could also leave voids in the system's protection, and
4) Around any opening, no matter how small, cracks could occur which could lead to the spread of fire, smoke or gases.

With this in mind, we need to find a system which can be used and addresses these needs. One such product classification that can achieve this is intumescent products.

In developing these products, one approach has been more the functionality of the resulting materials rather than the "pure chemistry."

Parallel to the growing experience with

new fire protection materials, certain aspects have been established to govern consideration of future fire protective materials:
-- The chemistry of fire protection materials
-- The physics of the fire protection
-- The handling and processing of the materials

By chemistry, we do not mean the actual composition of the fire protection system because the finished part or configuration can contain metal, wood, concrete or other materials.

Of more importance is that a fire protection material does not contain halogens. In the case of fire, halogens develop halogen/hydrogen products, and these components very often cause considerable secondary damages.

Another important consideration of fire protection material composition is the presence of a plasticizer. These could attack cable sheathing or any hygroscopic component which could cause draining off or blooming and should be avoided.

Furthermore, in developing fire protection materials, hazardous chemicals like asbestos, phenol, formaldehyde and halogens must be avoided.

The possible release of toxic gases, especially from incomplete burning, is also growing in importance. When such a release cannot be avoided completely, the amount of gases should be as minimal as possible.

By physics, we could mean how the fire protection is established or created. This can be simply achieved by using fire clay. However, the formation of cracks enables the spread of fire gases and the heat transport through a layer of fire clay happens relatively quickly. Additionally, all inorganic materials like cement or concrete, gypsum, asbestos, etc., show similar disadvantages. Better results are already being obtained with more elastic, porous and endothermic materials for they seal off and have some insulating properties.

Also of importance is the pressure of the intumescent foam. This pressure should not be so high as to destroy the construction. However, enough foam should be formed to fill the cracks and cavities sufficiently.

The temperature at which the intumescent material starts its activity is also important. If the activating temperature is too low, the intumescent behavior starts too early and can be exhausted after a period of time. In practice, it has been shown that at temperatures between 300 and 605°F, various organic materials found in buildings release combustible gases thus requiring protection.

As for handling, it is necessary that any materials should avoid complex construction and handling of the end products. Appropriate application methods are: spraying, casting, using cartridges, plastering, etc. Moreover, these systems should be adjusted easily, without destroying the fire stop or needing a labor intensive procedure.

Knowing these items, the key developmental objectives were:
1) The material must be activated in the case of fire. In this way, the system being protected would not be hindered until the protection is needed.
2) The material must build an insulating and fire rejecting foam. This foam will fill and close existing and future penetrations and openings, thus preventing the spread of the fire, smoke and gases.
3) The material should consume heat during the building of the foam layer. This should have a cooling effect which prevents the transfer of heat toward the backside of the fire barrier.

Once the development stages were completed, the mechanism of the intumescent products could be evaluated. Unlike non-foaming inorganic materials, intumescent products undergo a number of reactions when exposed to a flame or high temperature. These reactions, such as polymerization, decomposition and expansion function, all contribute to the formation of the carbonized foam structure. Simply put, we have a displacement of fire, the dilution of gases, a breakdown of the radicals, decomposition, and the formation of an insulating layer (Figure 1).

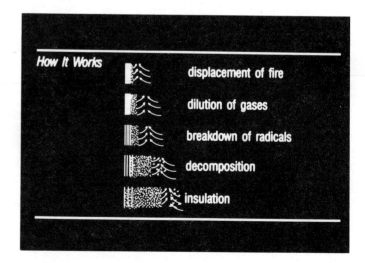

Figure 1: How it Works

In general, the reactions take place at or just beyond the surface of the intumescent material. Although described as single and separate, the reactions overlap. They ultimately result from the several processes which, when synchronized, occur in the intumescent mass.

In the process, we have a number of reactions and processes which occur. Initially, the process begins by a heating-up phase which cases dehydration to begin. This results in the displacement of the hot fire gases from the surface by the release of decomposition products. At this stage, melting begins to occur which leads to further decomposition of the surface. Also, this begins the formation of the expansion promoters and the polymerization reaction. These result in the fixation of potential flammable organic components by carbonization at the surface, resulting in a considerable reduction of flammability. During this process, the reaction of pyrolysis, release of pyrolysis gases, and the crosslinking process takes place. This results in carbonization or the build-up of a rigid carbonized foam skeleton. This foam insulates the materials from continued attack by flames.

Figure 2 shows this process in a slightly different manner. In the first stage, the material is rigid (or in its original form); it then softens or begins to melt. The expansion promoter then begins to react with further decomposition and crosslinking. And finally, you get the blown up, rigid, cross-linked carbonized foam.

Figure 3 (Photo: CSD International)

an intumescent material being subjected to a heat source. As you can see at the point where the heat source comes in contact with the coating, an insulating foam is formed.

The proper running of these reactions and processes heavily depends upon the formulation of the intumescent material. For example, the formation of the expansion promoter should not take place before the mass is melting and should not be released too quickly or in excessively large quantities.

The formulation of an intumescent material can contain four functional ingredients where each ingredient can consist of several components:
1) The carbonifier or skeleton builder
2) The catalyst
3) The expansion promoter
4) Additives or fillers

The skeleton builder is the ultimate building source of the carbonized foam. When exposed to fire, it should not become volatile due to decomposition. On the contrary, flames and heat should cause a further polymerization to obtain the macromolecule for the fire-resistant foam skeleton. Polyhydrocarbons could be used.

The catalyst has three functions:
1) Prevent decomposition of the skeleton builder into smaller volatile molecules.
2) Create unsaturated bondings by splitting off side chains in the molecules, enabling crosslinking to occur.
3) Aid in the generation of the char through the crosslinking and polymerization.

"FORMATION OF CARBON SKELETON"

| T;t | | cold |
| melt |
| expansion |
| cross-linking |
| rigid cross-linking |
| carbonized |
| foam |

Figure 2: Formation of Carbon Skeleton

Figure 3 shows a substrate coated with

Ammonium phosphate, polyphosphate and/or borates can be used as a catalyst.

The expansion promoter should exhibit several properties; not only should it adequately blow-up the skeleton builder, but, if possible, it should release gases like ammonia to bind halogen-hydrogen compounds and break down radicals. The timing of this expansion promoter release is important. After blowing up the macromolecule skeleton, it should end so as not to cause cracking off of the insulating foam. Nitrogen-containing compounds can be used for this purpose.

The additives can include pigments for color, fillers or even resins to change the physics. It should be noted, however, that fillers or resins can influence the intumescent behavior of the formulated product.

Having understood the need for an intumescent product and then developing this product, one must now determine how it performs. Table 1 shows the results of fire tests performed on one such range of intumescent products.

Figures 4 and 5 show a mock-up of a cable penetration, which used an intumescent product, after exposure to fire. The fire side of this penetration has extensive damage, yet the intumescent product has prevented the damaging effects from affecting the other side of the penetration.

Figure 5

Additional systems where intumescent products are used are shown in Figures 6-8. These systems can be designed to achieve specific fire rates, depending upon the materials used. In fact, fire rates of up to four hours are possible in certain circumstances with some of the intumescent products available today.

Figure 4

Figure 6 (Photo: CSD International)

138

Figure 7 (Photo: CSD International)

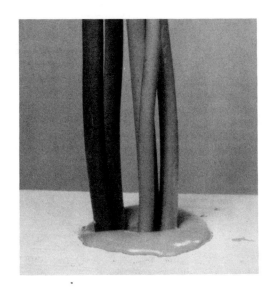

Figure 8 (Photo: CSD International)

The successful development of a complex
technology makes products available with the
intumescent properties or behavior. These
products can be used for the design and
manufacture of effective fire barriers and
fire stops.

ACKNOWLEDGEMENTS

We wish to thank Dr. Wolf von Bonin (Bayer
A.G., West Germany) for his assistance in the
realization of this presentation. Also, we
thank Mr. William J. Eicher (Mobay Corporation)
for the fire test results.

TABLE 1

SUMMARY OF SOME TEST RESULTS CONDUCTED BY
MOBAY CORPORATION FOR VARIOUS FOMOX® PRODUCTS

	UL-94 Rating*	Oxygen Index	Radiant Panel Flame Spread Index	Mobay Tunnel Flame Spread Class	Smoke Density
Putty (Retains Elasticity)	5V	67.6	1	16	176
Putty (Dries Rigid)	5V	>68	2	13	450
Two-Component Rigid	5V	40.7	9	ND**	382
Two-Component Medium Soft	5V	46.2	7	ND**	319
Two-Component Heat Absorbing	5V	>68	2	14	196
One-Component Coating	5V	>68	3	19	284

* The flammability results are based on small-scale laboratory tests and do not reflect the hazard presented by this or other materials under actual fire conditions.

** ND - Not Determinable

The conditions of your use and application of our products and information (whether verbal, written or by way of production evaluations), including any suggested formulations and recommendations, are beyond our control. Therefore, it is imperative that you test our products and information to determine to your satisfaction whether they are suitable for your intended uses and applications. This application-specific analysis at least must include testing to determine suitability from a technical as well as health, safety and environmental standpoint. Such testing has not necessarily been done by Mobay. All information is given without warranty or guarantee. Mobay Corporation disclaims any liability, in negligence or otherwise, incurred in connection with the use of our products and information. Any statement or recommendation not contained herein is unauthorized and shall not bind Mobay Corporation. Nothing herein shall be construed as a recommendation to use any product in conflict with patents covering any material or its use. No license is implied or in fact granted under the claims of any patent.

CONE CALORIMETER METHOD
FOR DETERMINING THE FLAMMABILITY
OF COMPOSITE MATERIALS

James E. Brown
Center for Fire Research
National Bureau of Standards
Gaithersburg, Maryland   USA

Cone Calorimeter Method for Determining
the Flammability of Composite Materials

By

James E. Brown
Center for Fire Research
National Bureau of Standards
Gaithersburg, MD 20899

ABSTRACT

A study was undertaken to evaluate the fire
performance of composite materials using the
cone calorimeter as the bench-scale method of
test simulating the thermal irradiance from
fires of various magnitudes. Parameters were
derived from the calorimetry measurements to
characterize the ignitability and flammability
of the composite materials.

The parameters are, to some extent,
empirical since radiative heat losses from the
samples were unknown. These parameters are: 1)
minimum external radiant flux (MERF) required to
produce pilot ignition in a predetermined
exposure time; 2) thermal sensitivity index
(TSI) which indicates the burning intensity
dependence on external heat flux; and 3)
extinction sensitivity index (ESI) which
indicates the propensity for continued flaming
combustion without an external heat flux. MERF
values at 300 s for 3 mm composite panels of a
FR epoxy resin and of a poly(phenylene sulfide)
(PPS) resin were about 18 and 28 kW/m^2,
respectively. The TSI of the PPS resin
composite revealed that it had the greatest
dependency on external flux. Additionally, the
ESI of the PPS composites was the only one to
indicate an external flux requirement to sustain
combustion during the first 60 s after ignition.

1.0   INTRODUCTION

The purpose of this project is to assist the
Navy with selection criteria for ship components
made of synthetic resins or composites of fiber-
reinforced resins by systematically assembling
data on the flammability properties of these
materials. The ultimate use of such a data base
assembly is input for a method for accurately
predicting the real fire performance and
flammability characteristics of composites and
resins from bench-scale tests. This is an
interim report.

The primary objective of this phase of the
study is to determine the rate of heat release
(RHR) of selected composites and resins burned
under various levels of radiative heating. The
rate of heat release, $\dot{q}$, of fires, especially
the peak $\dot{q}$, is a primary characteristic
determining the size, growth, and suppression
requirement of a fire environment [1]. Combined
with other measurements, such as flame spread
rate and ease of ignition, the fire hazard
contribution of the object can then in principle
be calculated. Some such computational methods
are already under development for other
applications [2].

The calculations of Huggett [1]
demonstrated that the average heat of combustion
for a broad range of materials is a constant
13.1 kJ/g oxygen consumed, with an accuracy of $\pm$
5% or better. Thus the RHR can be determined
with good accuracy from two simple measurements,
the flow rate of air through the combustion
system and the oxygen depletion in this flow
[3]. A bench-scale apparatus, the Cone
Calorimeter, developed and described by
Babrauskas [4], utilizes this oxygen consumption
principle to determine the RHR per unit area $\dot{q}''$.
Data from this instrument can be used in
research to predict the full-scale fire behavior
of certain furnishings and wall lining materials
[5].

The Cone Calorimeter has a well characterized and controlled irradiance source. Babrauskas and Parker [6] deduced that the spectral distribution of this source approximates the irradiance in compartment fires, where radiation is the primary process for energy transfer. Thus it was concluded that the instrument suitably determines the ignitability of upholstery and wood in fire-like environments.

We report the results of Cone Calorimetry measurements on a variety of composites. The primary measurements are RHR as a function of time and ignitability under pre-selected, constant levels of irradiance simulating radiant fluxes from real fires of various magnitudes.

2.0 MATERIALS

The materials used in this study are listed in Table 1. For the most part, only the generic classification of the resin and a general classification of the fiber reinforcement are known. Where greater detail of the materials is available, it will be indicated when the test results are presented. The generic classification of the resin and fiber identification were provided by the indicated sources. The resin classifications are epoxy, polyester, bismaleimide (BMI), and poly(phenyl sulfide) (PPS). In general, the resin reinforcement was a glass fiber fabric except for a sample set of Ryton** PPS panels and laboratory prepared panels in which carbon fibers were used.

Test specimens were cut into 10 cm by 10 cm squares at the full thickness of the supplied product. The specimens were maintained at laboratory conditions (approximately 40-45% RH and 25°C) prior to testing.

3.0 MEASUREMENT TECHNIQUE

The data reported here were obtained using the NBS Cone Calorimeter shown schematically in figure 1. The Cone Calorimeter and its function have been previously described [5,7]. An external radiant energy flux of up to 110 kW/m^2 may be applied to the surface of a sample by a temperature-controlled radiant cone heater. An electric spark igniter mounted above the specimen was used to ignite the pyrolysis products generated by a specimen exposed to a preset irradiance. The ignition delay time for each composite material usually was measured at three different radiant flux levels. In general, specimens are exposed to flux levels of

---

**Certain commercial materials are identified in this report in order to adequately specify the experimental procedure. In no case does such identification imply recommendation by the National Bureau of Standards nor does it imply that the material identified is necessarily the best available for the purpose.

35, 50, and 75 kW/m^2. Ignition delay time measurements were made by an operator with a stop watch as part of the standard test procedure used to determine the rate of heat release. Samples were placed under the cone heater and the amount of time necessary to ignite the evolving decomposition products by the spark of an electric arc was recorded.

A load cell was used to continuously measure changes in sample mass. Data from all the measurement devices were collected every five seconds by a digital data acquisition system and stored for later data reduction.

4.0 RESULTS AND DISCUSSION

4.1 IGNITION - The objective here is to obtain a measure of the resistance of materials to pilot ignition under radiative heating.

The times to pilot ignition are reported in Table 2. Where duplicate measurements were made, the average values are reported (one half of the difference between the two measurements is shown in parentheses). As expected, the data show that as the incident flux increases the ignition delay time decreases. It should also be noted that the range of ignition delay times for all materials tested at a given radiant flux increased with decreasing radiant flux, with ranges of 21 s to 54 s at 75 kW/m^2 and 92 s to infinity (no ignition) at 35 kW/m^2. Therefore, in very large and incipient fires, one might expect similar ignition-delay times for all composites, that is, a factor of two or more in the range of ignition-delay times depending on the chemical composition of the resin and configuration of the material.

There are two possible modes of radiative ignition: one is auto-ignition and the other is pilot ignition. For auto-ignition to occur, the fuel/air ratio above the material substrate must be within the flammability limits and the gas phase temperature must be above its ignition temperature. Because of the existence of an intentionally supplied hot spot, pilot ignition only requires that the fuel/air ratio above the material substrate in the vicinity of the pilot source be within the flammability limit.

Kashiwagi [7,8] investigated the effect of external radiant flux on the surface temperature at ignition and the ignition delay time of red oak and poly(methyl methacrylate) (PMMA) in auto- and pilot-ignition modes. These tests showed that for PMMA the surface temperature of the substrate at ignition was relatively constant, approximately 400°C, for both auto- and pilot-ignition modes. The surface temperature of red oak, a char former, at ignition increased as the external flux decreased.

Atreya [9] reported results of tests on different types of wood and found that, for the pilot-ignition mode below an external flux of 25 kW/m^2, the wood surface temperature at ignition increased with decreasing external radiant flux.

Above 25 kW/m^2 the surface temperature at ignition was constant at approximately 350°C.

If it is then assumed that this trend is applicable in general to all composites (i.e. that the pilot ignition occurs at a fixed surface temperature, dependent only on the thermal gasification characteristics of the substrate), one can use the simple thermal heat-up model derived by Carslaw and Jaeger [10] to suggest a form or trial function that ignition-delay time might have with respect to the heat absorbed from the constant external radiant flux. Heat losses are ignored in this simple model. Furthermore, this model assumes that the substrate can be treated as an inert, thermally thick, and opaque solid. Then the delay time to reach the ignition temperature is given by:

$$ t_{ig} = \frac{\pi k \rho c}{4} \left(\frac{T_{ig} - T_o}{\dot{q}} \right)^2 \tag{1} $$

where

t_{ig} = ignition delay time (s)
k = thermal conductivity (kW/m/K)
ρ = density (kg/m^3)
c = heat capacity (kJ/kg/K)
T_{ig} = ignition temperature (K)
T_o = initial temperature (K)
$\dot{q}$ = external radiant flux absorbed at the surface of the solid (kW/m^2)

Equation (1) shows that the ignition delay time, t_{ig}, is proportional to $1/\dot{q}^2$ provided, however, that the temperature wave does not reach the back surface of the sample before ignition. If, however, the material is thermally thin, that is, if the thermal wave hits the back surface before ignition occurs, the solid material will have an ignition delay proportional to about $1/\dot{q}$. Additionally, equation 1 provides an approximate basis for determining the influence of k, ρ, and c on ignition behavior. Therefore, a log-log plot of ignition time versus flux should give a straight line with a slope of minus 2 for an inert, thermally thick, and opaque material. Figure 2 shows the results for the Koppers Dion 6692T panel 25 mm thick and a Corflex panel, 3 mm thick. A linear regression line drawn through these data results in negative slopes for the two lines of 2.3 and 1.7, respectively, for the Koppers and Corflex panels. Table 3 is a listing of the slope of the regression line for each composite. It is recognized that derived values are essentially empirically based since heat losses are not known. However, plots of the data appear to be linear as was shown by typical examples in figure 2. Moreover, ignition apparently occurs before the thermal wave penetrates to the back surface.

Plotting ignition-delay data for all Cone Calorimeter experiments as illustrated in figure 2 allows one to extrapolate the regression line to some chosen location. Extrapolation to 600 s represents the minimum external flux (MERF) necessary to produce pilot

ignition after a protracted exposure to radiative heating. The MERF values may not be completely realistic because of the extensive extrapolation (for some materials) from a limited data range and the neglect of heat losses. A more practical application of the ignition times would involve determining the required irradiance to cause ignition at some other time, such as 300 s, for example. The values obtained at 300 s are also listed in Table 3 and are called MERF$_{300}$. Although the data for ignition are very limited, the maximum deviation of the regression slope from the theoretical value of minus 2 towards a value of minus 1, was 0.3 (15%) for the 3 mm Corflex epoxy panel. This difference, in part, may be due to a rapid heat up of the entire thin sample. This material indicates the lowest MERF values for ignition at 12 and 18 kW/m^2, respectively, for MERF$_{600}$ and MERF$_{300}$.

4.2 RATE OF HEAT RELEASE - We present in this section the Cone Calorimeter results on the heat released as measured by the oxygen consumption method during the combustion of several materials under radiative heating in a fully ventilated atmosphere (21% oxygen).

Profiles of the rates of heat release with respect to time, under constant external irradiance, were found to have features which are unique to the composite studied. The profiles appear to depend primarily on the chemical composition of the resin and the thickness of the composites. Figure 3 shows the RHR with respect to time of 3 mm thick PPS/glass fiber (Ryton) panels which were radiatively heated at 35, 50, and 75 kW/m^2. These curves demonstrate typical variations observed in the RHR-time profiles of composites panels.

In general, all of the curves exhibit at least two maxima for RHR. The initial peak is due to surface volatilization slowed by char formation. The second peak is a result of an increase in the gasification rate of the unburned substrate caused by an increase in the bulk temperature of the substrate. The bulk temperature increases because the unburned substrate is no longer thermally thick. Back surface temperatures should increase as the second peak of RHR is approached. While these measurements were not made in this investigation, Parker [11] has shown the same phenomenon with wood, a char forming material. The second peak increased as the back surface temperature increased.

Since the RHR, in most cases, changes quite significantly with time, it appears that more meaningful information may be gained about the fire behavior of the composites under radiative heating if the rates of heat release are averaged over periods of time during the burning process. Not only are the advantages of curve smoothing brought forward to clarify trends in the heat release data, but Babrauskas and Krasny [12] demonstrated that the rate of heat release averaged over 180 s could best be used to predict the fire performance of upholstered furnishings in large-scale experiments. Kanury and Martin [13] also have used average values for deducing physicochemical properties of essentially homogeneous materials in fire environments. The proposed ASTM Method P 190 [14] specifies that average $\dot{q}"$ values for the first 60, 180, and 300 s after ignition, or for other appropriate periods, be included in the report of the cone calorimeter results. In addition to the recommended practice in ASTM P 190, the average RHR over 60 s intervals from ignition are reported here.

Figure 4 illustrates the behavior of the RHR of Ryton composite averaged over the post-ignition periods as recommended by the ASTM method. This composite shows the greatest sensitivity to irradiance level. The effect of irregular volatilization of fuel from the surface is reduced. The lowest irradiance level, 35 kW/m^2, as was seen in the ignition data, provides barely enough energy to promote combustion. On the other hand, the average RHR at irradiances of 50 and 75 kW/m^2 increases until 300 and 240 s, respectively, when the panels are burned out. Table 4 summarizes the average RHR of the Ryton (PPS) panels. We note that ignition did not occur in one specimen reinforced by a chopped mat of glass fibers. Overall, the RHR at the 35 kW/m^2 flux level is low, always less than the irradiance.

Figure 5 shows a plot of the RHR also averaged as recommended from ignition up to 360 seconds for 25 mm thick polyester (Koppers) panels which were irradiated at four flux levels. A reduced effect of flux level on the average RHR values, caused at least in part by greater material thickness, may be seen by comparing figure 5 with figure 4. In every case this includes the initial peak RHR's but may not include the second maximum. Table 5 summarizes the results obtained for the 25 mm thick Koppers polyester composite. The $\dot{q}"$ value averaged at 60 s for one of the specimens tested at 35 kW/m^2 appears to be larger than expected. An explanation for this behavior is not known.

Next, the average $\dot{q}"$ values are summarized for two epoxy resin composites. The average $\dot{q}"$ values of a fire retardant epoxy are listed in Table 6. These results show that there are only small differences between the average RHR values of the thermally thin panels (3 mm) and thick assemblies (about 37 mm) and suggest that the heat loss from the back surfaces of the thin samples is relatively small as a result of effective insulation in the sample holder. The average $\dot{q}"$ results for the second resin type, a high performance epoxy resin composite prepared at the Navy's David Taylor Research Center (DTRC), are shown in Table 7. Although the composition of the resin is not known beyond it being an amine-cured epoxy resin, it appears that the fire performance of this latter resin closely resembles that of the FR epoxy composite shown in Table 6. The average $\dot{q}"$ results of another 3 mm panel composite prepared at DTRC

from a bismaleimide (BMI) and graphite fibers are listed in Table 8.

It may be seen in reviewing the average $\dot{q}"$ data in Tables 4 through 8 that the composites with polyester and epoxy resins generally show maximum $\dot{q}"(t)$ values in the first 60 s post ignition. The $\dot{q}"(t)$ values generally decrease with time after the first 60 s which suggest that the peak RHR is associated with initial surface burning of the composite rather than subsequent combustion of the pyrolyzate from the interior of the composite. For irradiances of 50 kW/m^2 or more, the composites with PPS and BMI resins show maxima at times greater than 60 s. For these samples, the maximum $\dot{q}"(t)$ is not the initial peak.

4.3 FIRE SENSITIVITY INDICES - Kanury and Martin [13] and Kanury [15] reported simplified models by which heat release rates (peak, instantaneous, and average values) may be related to basic properties of materials in fire environments using the Spalding B number concept. These authors and Tewarson and Pion [16] deduced from energy conservation at the sample surface that the heat release rate $\dot{q}"$ may be expressed by the following equation

$$\dot{q}" = (\Delta H_{c,eff}/L)[\dot{q}_T" + \dot{q}_e" - \dot{q}_\ell"] \qquad (3)$$

where

$\Delta H_{c,eff}$	=	effective heat of combustion
L	=	heat of gasification (pyrolysis)
$\dot{q}_T"$	=	heat transferred from flame to material surface
$\dot{q}_e"$	=	imposed external flux
$\dot{q}_\ell"$	=	heat flux loss by the surface to ambient

The slope $(\Delta H_{c,eff}/L)$ of a plot of the measured RHR against the external radiant flux provides one measure of the flammability of materials; it is a key determinant of the B number value. This parameter, termed the thermal sensitivity index (TSI) [13], provides a basis by which the fire performance of the materials may be indexed and compared over a broad range of external irradiances, simulating different fire environments. The intercept of such a plot in principle indicates whether the flame is self-sustaining in the absence of an external radiant flux for the time period under consideration. We will call this parameter the extinction sensitivity index (ESI); Kanury and Martin [13] called this parameter the limiting thermal index. Equation 3 can be expressed as

$$\dot{q}" = (TSI) \cdot \dot{q}_e" + (ESI). \qquad (4)$$

We illustrate the dependence that the average RHR has with respect to imposed heat flux levels by plotting the average $\dot{q}"$ at 60 s versus external flux, $\dot{q}_e"$. Using 60 s average $\dot{q}"$ minimizes the effect of sample thickness and conductive heat losses. Figure 6 shows the

144

results for composites whose resins are polyester, FR epoxy, PPS and BMI. This plot illustrates the TSI and ESI at 60 s interval on resin composition.

Table 9 summarizes the slopes, intercepts, average effective heat of combustion. The ESI values (slopes) are estimates of the sensitivity of the combustion intensity to variations in external irradiance and show that the Koppers composite, Corflex Panel Assembly, and BMI Panel had about the same sensitivity to variations in $\dot{q}_e"$. Because of differences in sample thickness these samples should not be compared to each other without caution. However, the TSI values indicate that the rate of heat release of these samples, although not the same in magnitude, would be fairly insensitive to small changes in external irradiance. This suggests that in a real fire the decay in an external fire imposing energy on a target material made from one of these composites would not be reflected as rapidly in a reduced heat release rate of the target material as compared to the materials with higher TSI values. For example, the Ryton Panels, which ranged in value from 1.3 to 1.8, would be expected to respond most strongly to variations in source irradiance.

The Ryton Panels also exhibited a negative intercept, ESI. This suggests that the heat loss from the flame is greater than its flux to the surface. With the removal of an external heat source these materials can be expected to self-extinguish, while the other materials with a positive ESI would be expected to continue burning at least for the first 60 s. The intercepts indicate that the epoxy matrix composite exhibits the most potential for sustained combustion with an external radiant flux following ignition.

In Table 9, the effective heat of combustion values are averages taken from each exposure over the entire measurement; they are computed from the ratio of $\dot{q}"$ to mass loss rate, $\dot{m}"$. These values fall into two groups, the lower one (about 12 kJ/g) where the resin is flame retarded and the upper values (20-25 kJ/g) where it is unretarded.

6.0 CONCLUSIONS AND RECOMMENDATIONS

The Cone Calorimeter can measure an array of flammability parameters for composite materials. These include the external radiant flux requirements for pilot ignition and sustained flaming combustion and the flux dependency of the rate of heat release. Since these measurements are based in the fundamentals of fire science, we expect they will correlate well with the larger-scale compartment fire behavior.

Thus, this development represents the first step towards predicting the performance of composite materials in various fire environments. Therefore, we recommend investigating the use of these parameters from the Cone Calorimeter and, additionally, a radiant-flux-based flame spread index to provide the basic data from bench-scale measurements for correlation with larger-scale (quarter scale and full scale, for example) compartment fire measurements. The bench-scale parameters may then serve as input for hazard prediction models of the type already under development at the NBS Center for Fire Research.

ACKNOWLEDGMENTS

This work was supported in part by the U.S. Department of the Navy, Naval Sea Systems Command, (SEA 05R25), Washington, D.C. The authors are grateful for the cooperation of Mr. George Wilhelmi, Mr. Usman Sorathia, and Mr. James Morris of DTRC, Annapolis, MD in obtaining and providing test materials.

REFERENCES

[1] Huggett, C., "Estimation of Rate of Heat Release by Means of Oxygen Consumption Measurements", Fire and Materials, 4, 61 (1980).

[2] Bukowski, R.W., Jones, W.W., Levin, B.M. Forney, C.L., Stiefel, S.W., Babrauskas, V., Braun, E., and Fowell, A.J., "Hazard I. Volume 1: Fire Hazard Assessment Method", NBSIR 87-3602, U.S. Dept. of Commerce, National Bureau of Standards (1987).

[3] Parker, W.J., "Calculation of Heat Release by Oxygen Consumption for Various Applications", J. Fire Sci. 2, 380 (1984).

[4] Babrauskas, V., "Development of the Cone Calorimeter - A Bench-Scale Heat Release Apparatus Based on Oxygen Consumption", Ibid 8, 81 (1984).

[5] Babrauskas, V., "Bench-Scale Methods for Prediction of Full-Scale Fire Behavior of Furnishings and Wall Linings", Technology Report 84-10, Society of Fire Protection Engineers, Boston, MA.

[6] Babrauskas, V. and Parker, W.J., "Ignitability Measurements with the Cone Calorimeter", NBSIR 86-3445, U.S. Dept. of Commerce, National Bureau of Standards (1986).

[7] Kashiwagi, T., "Experimental Observation of Radiative Ignition Mechanisms", Combustion and Flame, 34, 231-244 (1979).

[8] Kashiwagi, T., "Effects of Sample Orientation on Radiative Ignition", Combustion and Flame, 44, 223-245 (1982).

[9] Atreya, A., "Pyrolysis, Ignition and Fire Spread on Horizontal Surfaces of Wood", NBS-GCR-83-449, U.S. Dept of Commerce, National Bureau of Standards (1984).

[10] Carslaw, H.S. and Jaeger, J.C., Conduction of Heat in Solids, 2nd Ed., Oxford University Press (1959).

[11] Parker, W.J., "Development of a Model for the Heat Release Rate of Wood - A Status

Report", NBSIR 85-3163, U.S. Dept. of Commerce, National Bureau of Standards (1985).

[12] Babrauskas, V. and Krasny, J., "Predictions of Upholstered Chair Heat Release Rates from Bench-Scale Measurements", Fire Safety Sci. Eng. ASTM SP-882, Harmathy, T.Z., Ed., American Society for Testing and Materials, Philadelphia, 1985, pp.268-284.

[13] Kanury, A.M. and Martin, B.M., "A Profile of the Heat-Release Rate Calorimeter", International Symposium on Fire Safety of Combustible Materials, University of Edinburgh, UK (1975).

[14] Proposed Test Method for "Heat and Visible Smoke Release Rates for Materials and Products Using an Oxygen Consumption Calorimeter", P 190, American Society for Testing and Materials, Philadelphia, Pa.

[15] Kanury, A.M., "Modeling of Pool Fires with a Variety of Polymers", Fifteenth Symposium (International) on Combustion, The Combustion Institute, Pittsburgh, PA (1975).

[16] Tewarson, A. and Pion, R.F., "Flammability of Plastics - I. Burning Intensity", Combustion and Flame, 26, 85-103 (1976).

TABLE 1
Composite Materials

Material	Resin Classification	Fiber Reinforcement	Source
Koppers Dion Panels	Polyester, Brominated	Glass Woven Roving	Koppers Co., Inc.
Corflex Panel Assembly	Epoxy filled with aluminum silicate	Glass	Corflex Corp. per DTRC
Ryton Panels	Poly(phenylene sulfide)	Glass/Graphite	Phillips Petroleum Co.
Lab. Epoxy Panels	Epoxy	Graphite	DTRC
Lab. BMI Panels	BMI	Graphite	DTRC

TABLE 2
Ignition Delay Times for Composite Materials
Exposed to Various External Flux Levels

Ignition Delay Time (s)

Material (thickness)	Incident Flux (kW/m^2)			
	25	35	50	75
Koppers 6692T (25 mm)	263	120(10)*	60(2)	21
Corflex Panel (3 mm)	---	92(5)	54(2)	25
Corflex Assembly (37 mm)	---	122	70(0)	30(2)
Ryton Panels (3 mm):				
Glass Mat (Chopped)	---	183[+]	66	27(2)
Glass Woven Mat Prepreg	---	154(7)	75	29(1)
Lab. Epoxy Panel (3 mm)	---	116	76	40
Lab. BMI Panels (3 mm)	---	211	126	54

* Numbers in parentheses indicate range about a mean of duplicate measurements were made.

[+] Duplicate tests performed; only one specimen ignited.

TABLE 3

Minimum External Flux for Long Exposure Time and for 300 s
Exposure To Cause Ignition Computed from the Regression of Ignition
Delay Time and External Flux

Material	Regression Slope	MERF$_{600}$[a] (kW/m^2)	MERF$_{300}$[b] (kW/m^2)
Koppers 6692T (25 mm)	-2.3	18	24
Corflex Panel (3 mm)	-1.7	12	18
Corflex Assembly (37 mm)	-1.9	15	22
Ryton Panels (3.2 mm):			
Glass Mat (Chopped)	-2.5	21	28
Glass Mat (Swirl)	-2.6	23	31
Glass Woven Mat Prepreg	-2.1	18	25

a: Minimum External Radiant Flux necessary to cause ignition after 600 s exposure.
b: Minimum External Radiant Flux necessary to cause ignition after 300 s exposure.

TABLE 4

Results of Post-Ignition Averaging of the Rate of Heat
Release of 3 mm Thick Ryton Panels (Reinforced Poly(phenylene sulfide))

Fiber Reinforcement	Flux (kW/m^2)	Average Rate of Heat Release (kW/m^2)					
		60 s	120 s	180 s	240 s	300 s	360 s
Chopped Mat	35	50	30 (10)*	20 (5)	20 (10)	20 (20)	20 (25)
		N.I.	--	--	--	--	--
	50	75	65 (60)	70 (75)	85 (135)	90 (105)	80 (30)
	75	110	115	130	130	115	100
	75	100	95 (95)	105 (125)	120 (155)	110 (75)	100 (45)
Woven Mat	35	10	5	<5	<5	<5	<5
	35	5	5	<5	<5	<5	<5
	50	35	45 (60)	50 (64)	55 (95)	55 (110)	55 (45)
	75	85	90	95	100	95	80
	75	80	85 (85)	90 (110)	95 (110)	90 (60)	80 (25)

*Values in parentheses are averaged over the previous 60 s interval
N.I. = No ignition during a 600 s exposure.

TABLE 5
Results of Averaging the Rate of Heat Release of 25 mm (1 in)
Koppers Dion 6692T Panels (FR Polyester/Glass Fiber
Composite)

Flux (kW/m²	Average Rate of Heat Release (kW/m²)					
	60 s	120 s	180 s	240 s	300 s	360 s
25	50	40(35)*	35(25)	30(5)	25(5)	20(<5)
35	55	65(60)	55(40)	70(35)	65(25)	60(20)
35	70	55	45	40	40	35(25)
50	60	50(40)	45(35)	45(35)	40(35)	40(35)
50	60	45(35)	40(25)	35(25)	35(25)	30(25)
75	80	80(75)	70(55)	65(40)	60(40)	55(40)

*Values in parentheses are averaged over the previous 60 sec interval.

Table 6
Average Rate of Heat Release of 3 mm Corflex Panels
(FR Epoxy-Fiberglass Composites)

Sample	Flux, (kW/m²)	60 s	120 s	180 s	240 s	300 s	360
Corflex Panel	35	170	155 (140)*	160 (175)	140 (7)	--	--
Corflex Panel	35	170	170 (170)	160 (145)	130 (30)	105 (15)	90 (10
Corflex Panel	50	175	190 (205)	155 (90)	120 (20)	100 (10)	--
Corflex Panel	50	175	180 (185)	180 (180)	145 (45)	120 (20)	105
Corflex Panel	75	215	215 (215)	165 (75)	130 (25)	--	--

*Values in parentheses are averaged over the previous 60 s interval.

TABLE 7
Averaged Rates of Heat Release of 3 mm
Thick Laboratory Samples of Epoxy-Graphite Fiber
Composite Panels

Flux (kW/m²)	Average Rate of Heat Release (kW/m²)					
	(60 s)	(120 s)	(180 s)	(240 s)	(300 s)	(360 s)
35	150 (160)*	155 (50)	120 (20)	95 (≈0)	75	
50	185 (155)	170 (60)	135 (15)	105 (10)	85 (10)	75
75	210 (165)	190 (60)	145 (25)	150 (≈0)	100	

* Values in parentheses are averaged over the previous 60 s interval.

TABLE 8

Results of Post-Ignition Averaging of the Rate of Heat Release of a 3 mm Thick Laboratory Sample of BMI-Graphite Fibers Composite[a]

Experiment Number	Flux (kW/m^2)	Average $\dot{q}''$ (kW/m^2)					
		(60 s)	(120 s)	(180 s)	(240 s)	(300 s)	(360 s)
2296	35	105	130 (130)*	135 (140)	120 (90)	105 (40)	90 (10)
2308	50	120	145 (170)	145 (150)	130 (90)	110 (35)	96 (15)
2313	75	140	170 (200)	165 (155)	145 (75)	125 (30)	105 (25)

(a) Taken from Reference [12]
* Values in parentheses are averaged from the previous 60 s interval.

TABLE 9

Comparison of Inferred Flammability Indices of Composite Materials

	$\Delta H_{c,eff}$ (kW/m^2)	TSI*	ESI[+] (kW/m^2)
Koppers Dion 6692T (25 mm)	12 ± 2	0.6	30
Corflex Panel (3 mm)	12 ± 0.9	1.1	125
Corflex Panel Assembly (37 mm)	12 ± 0.4	0.6	100
Lab. Epoxy Panel (3 mm)	20	1.4	100
Lab. BMI Panel (3 mm)	20	0.9	75
Ryton Panels (3 mm)			
Chopped Mat	25 ± 1.6	1.3	5
Swirl Mat	22 ± 2.0	1.6	-55
Woven Mat	23 ± 2.2	1.8	-40
Graphite Woven Mat	23 ± 0.03	1.6	--
Average	23 ± 1.3	1.6 ± .20	

* TSI = Thermal sensitivity index.
[+] ESI = Extinction sensitivity index.

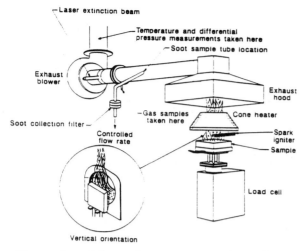

Figure 1. Schematic representation of the cone Calorimeter.

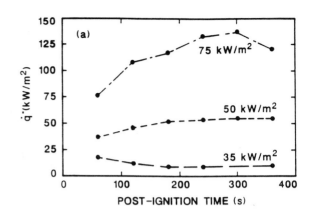

Figure 4. Average $\dot{q}''$ of a PPS composite at various times after ignition.

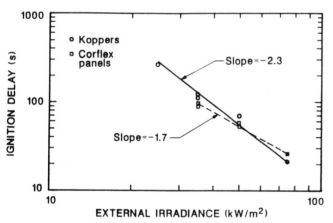

Figure 2. Effect of irradiance on ignition delay of two composite materials.

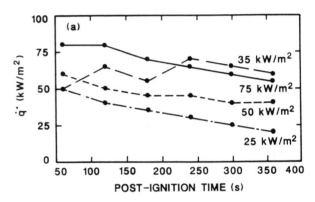

Figure 5. Average $\dot{q}''$ of a polyester panel at various times after ignition.

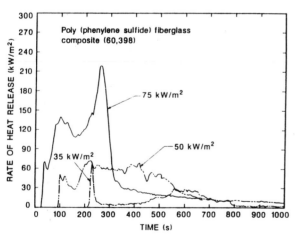

Figure 3. Effect of irradiance on the rate of heat release with respect to time.

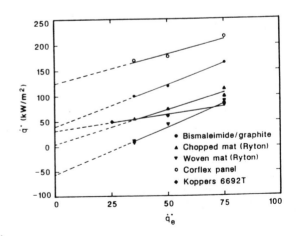

Figure 6. Average $\dot{q}''$ at 60 s post ignition versus external flux.

FIRE SAFETY CONSIDERATIONS IN THE SELECTION OF COMPOSITES AND PLASTICS FOR USE IN GROUND COMBAT VEHICLES

Rex B. Gordon
Ground Systems Division
FMC Corporation
San Jose, California   USA

Abstract

Land combat vehicle designers find many potential advantages in applying composites and other polymeric materials in place of traditional metals. These new materials have the capability of providing improved performance and increased crew survivability at reduced manufacturing costs. However, as vehicles become increasingly loaded with non-metallics, concerns arise in the area of fire safety which require attention. These concerns include ignition, flame spread, smoke and toxic products of combustion; and cover both the combat and non-combat use phases.

The establishment of practical and realistic fire safety criteria is needed in order to realize the full potential benefits from the application of polymeric materials in land combat vehicles. Failure to initiate a timely fire safety criteria increases the possibility of catastrophic fire related accidents, and subsequent imposition of overly restrictive safety regulations and attitudes which could deter subsequent development in this area.

This paper provides a review of the current status of fire safety issues in combat vehicles, and suggests some approaches for consideration by those involved in the selection of non-metallic materials for application in ground combat vehicles, relative to fire safety considerations.

GROUND COMBAT VEHICLE FIRE SAFETY is a significant concern to those involved with both the use and development of these increasingly complex systems. In peace time, vehicle fires cause extensive loss of material resources as well as severe injury and health hazard potential. During combat, the ability to successfully extinguish a fire without loss of fighting capability can be the key to survival for the crew involved, and for success of the mission.

It has been reported that the U.S. Army experienced some 213 non-combat tracked vehicle fires during the period 1974-1984. These resulted in 2 fatal and 30 serious injuries. The material losses were estimated at $12.7 million.[1] Further investigations have indicated that these reported accidents represented perhaps only a fourth of the actual fire incidents occurring in the field. Subsequent sampling of the accident data base indicates that the frequency and cost impact of these non-combat fires are increasing as more sophisticated and expensive vehicles are entering the inventory. [2]

The impact that the increasing application of composites and other polymers may have on vehicle fires' safety, and considerations for limiting any adverse consequences through effective testing and selections of materials during development are addressed in this paper.

COMPOSITES AND VEHICLE FIRE SAFETY

The traditional fuel sources for vehicle fires include heated fuel and hydraulic fluids, on-board ammunition, personal gear and other stored combustibles. As their application increases, polymers will become a significant addition to this listing.

To meed the increasing demands for battlefield performance, technology development efforts are focusing on new materials that increase ballistic protection, reduce radar signature, lower weight, enhance producibility and crew safety. Composite material systems, comprised of high strength fibers in organic resin matrices are finding increasing acceptance as viable solutions to these requirements.

Composites are being evaluated for various applications. These include alternate designs for vehicle components such as battery trays, ammunition containers, turret baskets, doors and ramps; as well as major structures such as all composite hulls and turrets. In addition, numerous parts, such as seats, containers, controls, etc. are made of plastics, fabrics and foams. Additional polymer containing parts are under development for cost reduction replacements for existing difficult to manufacture metal components.

To help frame a basis for developing fire safety criteria guidelines for non-metallic material selection, a brief overview of the various fire related hazard considerations is needed.

FIRE RISK ASSESSMENT - An evaluation of the potential impact these non-metallic materials present to ground combat vehicle safety is a complex issue. A number of factors require consideration:

- (a) Ignition sources
- (b) Flame spread dynamics
- (c) Fire detection and crew response
- (d) Fire suppression provisions
- (e) Effects of smoke and toxins
- (f) Hazardous chemicals used for field repair.

Some Basics - Given sufficient oxygen and heat input, all organic polymers will burn. Therefore, absolute fire safety of polymeric materials does not exist. There are always trade-offs in safety, utility, and costs. In general, these materials do not present an unmanageable fire safety problem. It is primarily when applied in a new environment, where increased fuel loading occurs in limited, inhabited spaces, does the fire safety issue become acute. Some examples of this are high rise buildings, manned space craft, ships, rapid transit vehicles, and aircraft. As will be discussed later, various fire safety codes and standards have been developed for these applications. A study of these are useful in establishing a framework for practical fire safety guidelines for ground combat vehicles.

Many synthetic organic polymers burn in a manner differing from that of the more familiar natural polymers such as wood, paper, cotton, or wool. Some synthetics burn much faster, some give off much more smoke, some evolve potentially noxious and toxic gases, and some melt and drip. Others burn less readily. Fire retarding additives are often applied to polymers. In general, these are effective in resisting small ignition sources or low thermal fluxes in large scale-scale fires. However, in high thermal fluxes, these systems may burn readily, often with increased smoke and toxic gas in the combustion products.

Fire Safety Aspects - When considered from a "systems" viewpoint, the fire safety evaluation of ground combat vehicles is seen to have two primary aspects: (a) combat survivability, and (b) non-combat safety. Both the nature of potential ignition sources, and potential crew response actions are markedly different. For the past 10 years, most U.S. Army experience and exposure has been in the non-combat safety aspect. However, since combat vehicles are designed primarily for battle, both aspects must be considered in the formulation of an effective fire safety criteria.

Combat Survivability Aspects - In combat, ground vehicles are vulnerable in varying degrees to fire causing threats. These include incendiary, shaped charges, fragmentation, blast kinetic energy and nuclear type weapons. Increasingly effective advances in armor, firepower, and mobility are minimizing the vulnerability to these threats. In addition, fixed fire extinguishing systems (FFES) are installed on most current ground combat vehicles. These systems incorporate very fast reacting optical sensors, which upon detecting a hydrocarbon fuel fire, will send a signal that results in the release of Halon gas into the crew compartment, all within 250 milliseconds. These automatic fire suppression systems are designed to control an explosive fire ball within the crew area resulting from a projectile or hot gas jet penetration of the external armor and on-board diesel fuel tank. If the resulting fireball can be controlled at its initial formation, it is likely that the crew will not receive incapacitating burns. This system can also be manually activated by the crew for the control of slow growth fires. The limitations, if any, regarding the ability of the automatic optical sensor systems to either detect slow growth or smoldering type fires (such as those experienced in aircraft and building fires involving composites), have yet to be fully defined.

The combat related fire safety mission of composites in this environment is to: (1) minimize fire ignition potential of external threats, through absorption of hot metallic spall, etc., (2) minimize spread of secondary and slow growth fires, and (3) minimize generation of hazardous combustion products (smoke and fumes). In a combat engagement situation, the ability to remain under armor within a vehicle, rather than having to evacuate due to fire and smoke conditions is a desirable option for a crew to have.

Non-Combat Aspects - Although not as severe from a fire threat standpoint, the non-combat environment presents significant challenges also. Typically, vehicles will spend thousands of operating hours in training exercises, and undergo extensive maintenance and modifications. Army accident statistics indicate that faulty repairs and modifications on vehicles constitute the primary underlying cause of vehicle fires .(1) The most common non-combat ignition sources are leaking fuel lines near hot surfaces, electrical short circuits due to improper repairs, and improper repair or operation of personnel heaters.

Since most fire ignition sources are not a direct function of a design defect, attention to fire spread control will be a continuing requirement in combat vehicle design, independent of combat survivability considerations. This requires a basic understanding of how polymers burn, and existing criteria for fire hazard control.

OVERVIEW OF POLYMER COMBUSTION

Combustion of polymers typically occurs in four stages - heating, decomposition, ignition and combustion. Initial application of heat raises the temperature of the polymer at a rate dependent on the size of the item being heated, and the heat flux present, normally from a flame, spark, hot object or some combination. As the polymer heats, thermoplastic materials soften and melt, while thermoset plastics retain their shape. When sufficient heat is induced, vaporization occurs.

During the decomposition stage an endothermic process occurs when sufficient energy is produced to overcome the binding energy of the individual atoms. Decomposition is the result of free radical chain reactions initiated by oxygen or oxygen laden contaminants embedded in the polymer during manufacture. Gases such as monomers and saturated and unsaturated hydrocarbons are produced during decomposition. Materials not turned to a gaseous state remain as carbonaceous residues.

Ignition occurs when the gases produced during decomposition mix with atmospheric oxygen or embedded oxidants. When temperatures are high enough the polymer ignites, either from flame or self ignition. When sufficient heat and oxygen are available an exothermic reaction occurs, which if sufficient energy is available, overrides the endothermic reaction of decomposition and initiates flame spread. Once the gases are burning, decomposition increases through thermal feedback, which increases the speed and energy of gas production and flame spread. The burning process produces toxic gases, smoke, heat, and residue.

FIRE RELATED HEALTH HAZARDS

Fire related fatalities and injuries occur from oxygen deficiency, toxic poisoning, skin burns, heat, damage to the respiratory tract, and entrapment caused by lack of visibility and fear. While carbon monoxide poisoning is considered the major cause of fatalities and injuries, all of the above noted hazards have contributed to fire related casualties. Toxic gases, including HCl, HF, and HCN are often produced during the combustion of polymers. There is some consideration that combustion products, and reduced available oxygen for breathing (under 17%) may cause disorientation and debilitating effects, so that normally expected fire fighting and escape actions are not taken by those involved, thus increasing injury and death potential. Pulmonary irritants, such as oxides of nitrogen and ozone, may cause cardiac edema and increased respiration rates. Smoke particles can restrict respiratory passages, reduce vision, and contribute to the onset of panic. (3)

Carbon dioxide is usually the most prevalent of the various products of combustion. Although not toxic, except in very high concentrations, as little as a 3% concentration can cause the respiration rate to double. The result of increased respiration is inhalation of greater quantities of other, more toxic gases. (4)

Reducing Smoke and Flammability Characteristics - As a means to mitigate the above noted fire related hazards in polymers, various additives are often used. This is a growing technology which has resulted in ever increasing beneficial results. Experience has demonstrated that increasing demands contained in performance criteria in these areas has spurred the polymer industry to find cost-effective solutions to retardation challenges; which a few years previous were felt impossible.

Fire retardants, primarily halogens, phosphorus and antimony, inhibit or suppress the combustion process by altering the gases produced during thermal decomposition. Methods of altering the flammable gas production include production of non-combustible gases, production of radicals of molecules which interfere with the combustion process, formation of non-combustible char or fluid barriers, and generation of fine particles which reduce the flame by altering the course of gas phase reactions. There is a potential down side to this approach which requires careful evaluation; fire retardants can increase the level of smoke and toxic gas release.

Smoke suppressants are of special interest in applications where there is a need to have time and spatial orientation to escape from fire. Polymers have a very high smoke potential in comparison to other materials. While present in all fires to some extent, the quantity and composition of the smoke produced is a function of the combustion temperature and introduction of contaminates such as fire retardants. Smoke includes all airborne products of combustion, including water, soot particles, and toxic gases. Inhalation may result in immediate, and possible permanent pulmonary impairment.

Smoke suppressants can act either physically or chemically and in a variety of ways. In some cases by forming glassy coatings or intumescent foams; or by dilution. In thermosets, soot-forming cross linkers such as styrene, are replaced with non-aromatic compounds. Additives which form char layers are also used.

RELATED FIRE SAFETY STANDARDS

Due to the limited use of composites and other polymers in combat vehicles to date, little exists at present in the way of consensus or formal fire safety criteria for the selection of non-metallic materials for vehicle components. Although considerable research and criteria exists for fire safety properties of fuels, hydraulic fluids, and explosives used in combat vehicles, an overall, systems oriented, fire safety criteria guideline has yet to be developed. A U.S. Army contract for the preparation of a Handbook on Design of Combat Vehicles for Fire Survivability may address some aspects of this issue when published, currently scheduled for late 1989.

A survey of the literature will uncover an extensive volume of documentation on a wide variety of subjects associated with the fire safety aspects of composites and other polymer materials. The coverage of combustion product toxicity is particulary expansive, but difficult to correlate to applications not specifically addressed. In addition to numerous laboratory environment studies, various specific applications problems have been studied which are of related interest to our subject. In the United States the role of materials in fire safety is governed by a complex hierarchy of test methods, standards, specifications, codes and related regulations. These tend to fall into general or specific application categories. An overview of the most pertinent of these existing testing criteria and selection standards follows:

TEST METHODS - Numerous test methods have been established for measuring a property or behavioral characteristic of a material, product or assembly as an aid to predicting its performance in application. In certain cases, Government agencies such as the DOD, NASA, FAA and NRC have established product specific flammability testing requirement for material selection. Most commonly used methods, however, are of the general type which have been documented by the American Society for Testing and Materials (ASTM), the National Fire Protection Association (NFPA), Underwriters Laboratories (UL), or by a technical society or trade association. Unless referenced in an applicable code, contract, specification or similar document, these carry no legal obligation to a designer or builder. However, since they may often be used as a material suitability evaluation tool during development of a product, it is important to recognize the types and limitations of such test methods.

Test methods differ widely in purpose, scale, and degree of sophistication, making systematic classification difficult. Some are primarily for research and others for acceptance screening. For combat vehicle applications the following fire hazard characteristics and standard test methods appear vivable candidates for testing:

- Flammability: How a sample burns in contact with flame, (ASTM D 635 & 668, UL-94, MVSS302).
- Ignitability: Ability to burn from exposure to an electric spark or radiate heat, (ASTM D 3874 & 3638, UL 746A para 24, 25, 42 & 43).
- Oxygen/Temperature Index: A determination of the minimum oxygen content needed to sustain burning, and the minimum temperature at which a material will burn at in normal air, (ASTM D-2863).
- Flame Spread Index: Rate of surface flame spread in radiant heat environment, (ASTM E-162).
- Smoke and Toxic Gas Generation: Measurement of light transmission reduction and concentration of selected gases generated from a burning sample, (ASTM E 662).
- Rate of Heat Release: Becoming recognized as the single most important measure of fire safety of a material. It provides a means for predicting the rate of fire growth and its effect within a given environment. The average heat release rate is used to estimate the heat contribution of the composite or plastic material to an enclosed space fire. Smoke and offgas evaluations can be conducted in the same test. Most referenced procedures are the OSU (ASTM E 906) and the NBS Cone Calorimeter (ASTM P-190).

Most standard flammability test methods involve repeatable, small scale material sample testing in a laboratory apparatus designed to reduce potential environmental variables. A typical flammability test involves preparing a strip of material sample in a prescribed orientation (horizontal, vertical or at an angle), placing a controlled heat source at one end for a specified time and noting the burn length, duration, and melting characteristics of the sample. In some test methods, (i.e. the UL 94 series) the accept-reject criteria is included, but in most ASTM test methods this is an open issue.

Although repeatable and fairly inexpensive, there is a serious shortcoming of these types of test methods. They are not able to predict or describe the burning characteristics of plastics products under actual fire conditions, with any reliability. In addition to the important considerations of actual product size, orientation, ventilation, etc., within an actual fire environment, the energy feedback issue is most critical. In the combustion of a solid fuel, energy feedback from the high temperature gaseous combustion zone pyrolyzes the fuel surface to provide a continuing supply of gaseous fuel to the flame. The rate of burning is directly related to the magnitude of this energy feedback and the intensity of combustion. In the typical small scale test method as described above, most of the energy of combustion is dissipated in the rising convective plume and through radiation to the cool surroundings. In a real fire, on the other hand, energy exchange between adjacent fuel surfaces and radiation from the heated surroundings greatly increases the energy feedback and the intensity of combustion. (5)

To overcome this inherent shortcoming, activities concerned with establishing specific fire safety material criteria for a given application are tending toward testing methods that provide a radiant flux input, combined with full scale testing to better evaluate the effectiveness of the acceptance criteria selected. A recently implemented revision to the Federal Aviation Administration (FAA) airworthiness standards for materials used in aircraft interiors (14 CFR 25.853) provides a good illustration in an application somewhat related to fire safety issues associated with ground combat vehicles.

The selection of an improved flammability test method was made from correlation studies of data from candidate material testing and full scale fire testing. Studies of actual aircraft fire incidents indicated that a post crash landing fuel fire located external to an opening in the aircraft passenger cabin provided the most likely severe fire accident scenario. Full scale testing of alternate composite interior surface materials (partitions, sidewalls, stowage bins) was conducted in a C-133 wide body crew compartment coverted for fire testing use. A large fuel fire was initiated external to the cabin and cameras monitored that reaction of the composite test panels. It was found that the different composite materials presented significant differences in both delay times to flashover and toxic gas (HF) levels. This was not as evident from the normal laboratory burn rate testing results. Experimentation showed that the best correlation between actual full scale fire testing findings and laboratory testing methods was through use of a modified version of the Ohio State Univ. (OSU) rate of heat release apparatus used in ASTM E 906, Test Method for Heat and Visible Smoke Release Rates for Materials and Products. This is basically a flow through device that measures the heat release rate produced as a function of time by a material subjected to a preset level of radiant heat flux.

Acceptance criteria adopted by the FAA for the OSU testing results were set, at least in part, on findings of the full scale testing. These tests found that panels made of phenolic glass had about 3 times the delay to flashover characteristics than those made of epoxy glass or phenolic kevlar, which are excluded for use by the acceptance level

criteria. Panels made of non-commercially available resins developed by NASA did not ever result in a flashover condition. This gives promise of even further advances in fire safety once a sufficient demand is present for commercial marketing of such fire safe materials. (6)

Other vehicle related specific testing criteria for material selection can be found in NASA publication NHB 8060.1B for use in space craft applications, MIL-STD-1623D for naval shipboard applications and a draft standard for submarine applications. These are comprehensive fire safety criteria documents which typically establish a variety of testing methods and acceptance criteria based on the fire risk category assigned to a given polymeric item. The more severe the risk to safety of the crew and passengers, the more severe the flammability and off gasing criteria.

To date, the only material selection criteria directly applicable to ground combat vehicles is found in MIL-STD-1180(AT) - "Safety Standards for Military Ground Vehicles" dated July 1976. This references Federal Motor Vehicle Safety Standard (FMVSS) No. 302. "Flammability of Interior Materials". This was issued by the DOT in 1975 to provide a minimum standard for interior materials used in passenger vehicles. It was based on a scenario in which a seat cover fire was initiated by a dropped cigarette, and all passengers exit the vehicle within a minute or less. The test will pass a material sample strip, which when held horizontal in a holder, and contacted with a flame at one end, burns no more than 4 inches a minute.

SUMMARY AND CONCLUSIONS

Providing complete fire safety in modern combat vehicles is an impossibility. This does not preclude the desirability of reducing the fire risk to crew safety and vehicle damage wherever feasible. In addition to the active fire suppression systems previously discussed, various passive fire safety provisions are utilized in vehicle design. These include optimum armor protection, compartmentalization and minimizing such potential ignition sources as high temperature surfaces, fuel leakage sources, and electrical short circuits. Rapid crew emergency exiting is also a fire safety feature of these vehicles.

The prevention or reduction of flame spread within a vehicle is also a basic concern, since accident history demonstrates that fires will still occur despite utmost care in design. The selection of material which have suitable fire safety provisions is a key part of this goal. In developing a selection criteria guideline, the following considerations appear appropriate for ground combat vehicle application:

1. Because of the minimal fire hazard risk presented by the low quantities of polymer materials currently present in production combat vehicle, the primary attention should be oriented toward the selection evaluation and control procedures for engineering changes on current vehicles, and new vehicle development programs.

2. The fire safety acceptance criteria for a candidate material should be based on the extent of hazard risk involved when the item is part of an operating vehicle. This risk determination to be based on both combat and non-combat worst case forseeable fire scenarios. As an initial approach four primary fire safe categories of materials could be defined as:

   Type A. Material which have suffcent fire safety characteristics that their use in a vehicle is unrestricted.

   Type B. Materials which can be used within a crew compartment only in limited quantities as specifically authorized.

   Type C. Materials which are restricted from use within a crew compartment, but can be used in isolated unmanned compartments or externally, when specifically authorized.

   Type D. Materials authorized for use as electrical cable insulation.

3. In the absence of specific ground vehicle full scale testing data, an initial criteria for being a Type A material could be meeting the standards established by the FFA for aircraft interiors as defined in 14 CFR 25.853, or equivalent. Type B materials should pass the 94V-1 and FMVSS 302 flammability tests, and not exhibit excessive smoke or toxic gas generation. Type C materials should pass the FMVSS 302 test, and Type D the applicable UL746A tests.

4. Material selection must include a balance between performance, producibility and safety. Where a trade off between alternate material candidates is involved, weighting should be given to resistance to ignition and low flame spread, over smoke and toxic gas characteristics.

5. Due to the specialized nature of the test method selection and evaluation process, and efficiencies in avoiding duplication of effort, a centralized activity within the organization should be provided the charter and resources to provide technical fire risk evaluation, test method and acceptance criteria. This activity should also maintain data base management of test results and a material usage log on all products. This fire safety activity would provide support to the various responsible project engineering groups within the organization.

6. As the application of composites and plastics grows, initial test methods and acceptance criteria values should be reevaluated periodically by conducting full scale fire testing based on realistic worst case ground vehicle fire scenarios.

Acknowledgement - The author wishes to acknowledge the technical help received from the following fire safety and polymer material specialists who provide valuable insights and comments on an earlier draft of the material used for errors or omissions herein: J. Hill, A. Vasudev, G. Thomas, and C. Crebs of FMC; D. Macaione, U.S. Army Materials Technology Laboratory; P. Zabel, of SwRI, C. Sarkos of the FAA Research Center, and LTC F. Sisk of the USASC. The Committee on Fire Safety Aspects of Polymeric Materials of the National Research Council (Ref 5) was extensively used for the technical overview on combustion dynamics, retardants and test methods.

References:
(1) U.S. Army Safety Center, "Tracked Vehicle Fire Analysis: 1984 Update. USASC TR 84-2, June 1984
(2) Sisk, LTC F. USASC, Private Communications
(3) G. Hartzell, "Combustion Products and Their Effects on Life Safety", Fire Protection Handbook, NFPA 1976
(4) Anon. "Fire and Smoke: Understanding the Hazards", Committee on Fire Toxicology, National Research Council, National Academy Press, Washington D.C., 1986
(5) Anon. "Elements of Polymer Fire Safety and Guide to the Designer", National Research Council Publication NMAB 318-5, National Academy of Sciences, Washington, D.C., 1979
(6) Sarkos, C., ASTM Standardization News, December 1988

APPLICATION OF RESPONSE SURFACE METHODS TO STRUCTURAL SMC COMPOSITE CHARACTERIZATION

Allan B. Isham
Owens-Corning Fiberglas Corp.
Granville, Ohio   USA

Douglas L. Denton
Battelle Memorial Institute*
Columbus, Ohio   USA

Edwin S.-M. Chim
Shell Development Company*
Houston, Texas   USA

ABSTRACT

The application of response surface methods to material characterization experiments is an efficient approach to produce and present reliable mechanical property data. This methodology is used to study a system of structural SMC composites containing continuous unidirectional (C) fibers and/or random chopped (R) fibers. Modulus and strength in tension, compression and shear, as well as Poisson's ratio are reported for five SMC-C/R composites. Response surface models are presented that relate material composition and fiber configuration to composite properties for the entire family of composite materials.

RELIABLE MATERIAL PROPERTY DATA are required for the efficient design of parts used in structurally demanding applications. Fulfilling the need for data on fiber reinforced composite materials presents a particular challenge. The properties of composite materials can be tailored to specific part geometry and loading conditions through controlled modification of material composition and fiber configuration. This allows a designer to select composites with the desired properties from a myriad of candidate material systems, but significantly increases the amount of property data that is needed. The generation of reliable design data for composites has not kept pace with the rate of new applications of these materials primarily due to the high cost of testing programs required to fully characterize composite materials.

Typically, the approach to providing reliable data is detailed characterization studies of specific materials. Several such studies have been performed on sheet molding compound (SMC) composites [1-8]. However, the expense and time required to study these individual programs restricts the number of materials that can be fully characterized. An alternative approach is to use designed experimental techniques to systematically investigate related materials. Data on these materials can then be used to develop empirical, response surface models to describe the behavior of an entire family of composite materials.

In the present study, SMC materials and processing were used to examine the effects of glass fiber content and configuration on molded composite properties. Compression molding of SMC can produce composites with a wide range of glass fiber contents and fiber configurations without significant changes in constituent materials or process conditions. Thus, unconfounded effects of the variables under study can be identified. While SMC materials and processing are used in this particular investigation, the experimental methods illustrated, and to some extent, the mechanical property models produced can be generalized to other material/processing systems such as resin transfer molding (RTM) and structural reaction injection molding (SRIM).

EXPERIMENTAL DESIGN

The specific design selected for this experiment includes five SMC composite materials that contain different amounts of continuous unidirectional (C) and random chopped (R) fibers. This basic design, which includes two variables at two levels each and a "center point," is one of many that can be applied to response surface experiments [9]. Economic constraints prevented additional formulations or replicated materials from being included in this test program. However, the design is adequate to provide an estimate of the effect of the variables on the measured properties over the experimental region.

* Current Address

A graphic representation of the experimental design is shown in Figure 1. The material variables are expressed as the total weight percent of glass fibers in the composite (horizontal axis) and the percent of the total glass fiber content that is continuous (vertical axis). A secondary vertical axis shows the percent of the reinforcement that is randomly oriented chopped fiber. By definition the percent continuous and percent random fiber must sum to 100 percent.

The levels of variables are chosen to bound the range of SMC compositions generally considered to be useful for structural applications. The total glass fiber content is varied among three different levels, 30, 50, and 70 percent by weight. At the same time, the portion of the unidirectional continuous glass fibers are either 100, 50 or 0 percent of the total glass fiber weight. The remainder of the reinforcement in each composite is randomly oriented chopped fibers (0, 50, or 100 percent, respectively). Thus, the five nominal compositions in the design are SMC-C70, SMC-C30, SMC-R70, SMC-R30 and SMC-C25/R25.

SMC generally contains a particulate filler such as calcium carbonate. As the glass fiber content is increased in SMC formulations, the filler content must be reduced. However, since the resin demand of the filler and the glass fibers are not the same, the ratio of the filler to the resin must also be changed along with the glass fiber content to maintain adequate fiber wetting in the sheet molding compound. This effect is illustrated in the SMC composition diagram which shows the three nominal glass contents used for the design of this study (Figure 2).

MATERIAL PREPARATION

The SMC formulations are presented in Table 1. The SMC paste composition varied depending on the nominal glass fiber content of the SMC. The same glass fiber reinforcement was used for both the unidirectional continuous and the random chopped glass fibers. The SMC areal density was approximately five ounces per square foot.

Each SMC mold charge was four plies thick and was dimensioned to cover 93 percent of the mold surface area. All of the continuous fibers were oriented in same direction in each panel. SMC containing both continuous and random fibers had a stacking sequence as follows: (R/C)(C/R)(R/C)(C/R). Flat sheets measuring 21 in. X 24 in. X 1/8 in. were molded at 1,000 psi and 295°F for three minutes. Panels were postcured at 320°F for ten minutes to achieve full cure as indicated by dynamic mechanical analysis.

Quality assurance testing was performed to determine the actual composite compositions. Results are presented in Table 2. The SMC containing both continuous and random fiber was found to approximate SMC-C25/R20 more closely than SMC-C25/R25. Therefore, this material is designated as SMC-C25/R20 throughout the remainder of this paper. The measured glass contents for the other composites were fairly close to the desired compositions so the original nominal designations are retained.

MECHANICAL TESTING

Six molded panel of each material were sampled for mechanical property measurements. Specimens were cut in both the longitudinal (L) and transverse (T) directions. The longitudinal direction is parallel to the continuous fiber and the SMC machine direction. A description of the procedures for each type of test is presented in the following paragraphs.

Straight edge tensile test specimens measuring 10 in. X 1 in. were tabbed and tested according to the specifications of ASTM D3039. Tensile tests were performed at a crosshead speed of 0.2 in./min. An LVDT extensometer provided strain measurements for the modulus calculations.

Compression tests were performed on 6.5 in. X 0.5 in. X 0.125 in. coupons in an ITTRI fixture. End tabs were bonded to the specimens which gave a gage section measuring 0.5 in. X 0.5 in. at the center of the rectangular strip. A rosette with 1/16 in. strain gages oriented parallel and perpendicular to the loading direction was mounted on one face of each specimen. Stress-strain data were collected with a microcomputer interfaced to a data acquisition/control unit.

Poisson's ratio measurements were made in both tension and compression. A biaxial strain gage extensometer was attached to the tensile specimen and a small load was applied. Stress-strain data were acquired with a digital processing oscilloscope, which calculated the Poisson's ratio. Each specimen was loaded three times to measure Poisson's ratio before the specimen was tested to failure in the tensile test. In the compression tests, strain data from the axial and lateral gages were collected simultaneously. Poisson's ratio was determined from regression lines fitted to the initial portion of the lateral strain-axial strain curves.

The inplane shear modulus and shear strength were determined using the University of Wyoming version of the Iosipescu shear test [10]. The test specimens were 2 in. long and 0.5 in. wide. A 90° notch was cut 0.1 in. deep on each edge of the specimen at the midlength.

A pair of 1/16 in. strain gages were oriented at +/-45° relative to a line connecting the notch tips.

The shear modulus was calculated by regression of the initial "linear" portion of the stress-strain curve for each strain gage. The averaged value of the modulus calculated from each pair of strain gages is reported. For SMC-R70 and SMC-R30 the average secant moduli at 0.5 percent strain are reported because the stress-stain relationships of these materials show significant non-linearity.

The shear strength of each material was calculated by dividing the failure load by the cross-sectional area of the specimen between the notches. Failure was defined as the point where the first major loss of load occurred. Shear strengths of the SMC-R70 and SMC-R30 were determined from a second set of specimens that had notch depths of 0.125 in. and were tabbed to prevent failure outside of the gage area.

EXPERIMENTAL RESULTS

The tensile, compressive and shear moduli, and Poisson's ratios measured for the five SMC-C/R composites are summarized in Table 3. Tensile, compressive and shear strengths of these materials are similarly presented in Table 4. The mean, standard deviation and number of specimens tested are listed in these tables. The coefficient of variation (100 x mean/standard deviation) generally ranges from five to 15 percent, which is reasonable for these materials. Data for the longitudinal (L) and transverse (T) directions are given for the composites containing continuous unidirectional fibers. Averaged values for the two directions are reported for SMC-R30 and SMC-R70, but it should be noted that process induced fiber orientation can result in significant anisotropy of SMC-R properties [11].

The data presented on individual composites in Tables 3 and 4 are intrinsically valuable. However, their use as input data to create response surface models provides a wealth of additional information in a very accessible format.

RESPONSE SURFACE MODELS

An empirical model was developed for each of the mechanical properties measured. The mean value observed for each composite material was fit to a regression model of the form

$$z = ax + by + cxy + d \qquad (1)$$

where z is the predicted property in units of 10^3 psi or 10^6 psi, x is the total glass fiber content in weight percent, and y is the continuous unidirectional fiber content expressed as a percentage of the total amount of glass fiber in the composite. The cross product term (xy) represents "curvature" in the model that arises from the interaction of the independent variables. Regression coefficients are represented as a, b, and c, while d is the "intercept" of the model.

The experimental design provides one degree of freedom for the estimation of error. Ideally, a better estimate of error is needed to judge the statistical significance of the regression coefficients. Since the error is not well-known, all four terms are included in the models used to generate the response surface contours. In general, the models fit the experimental data very well (+/-10%) and appear to be physically meaningful.

The response surface contours provide an excellent means of visually presenting the relationships between the material variables and the material properties. Interactions among variables are more easily understood in such plots. In light of the statistical variation of the composite mechanical properties, values read from the contour plots are sufficiently accurate for most engineering purposes. Some of the important features of the response surface models for the SMC-C/R properties are discussed in the following sections.

TENSILE PROPERTIES - The model for longitudinal tensile strength shows an increase as the composition moves from SMC-R30 to SMC-C70 (Figure 3). Longitudinal tensile strength increases faster with glass content when all the fibers are continuous. Likewise, replacement of random fiber with continuous fiber produces a greater increase in longitudinal tensile strength at higher total fiber contents.

In contrast, transverse tensile strength is the lowest and the least sensitive to total fiber content when most of the fibers are continuous (Figure 4). This is expected since the glass fibers provide little reinforcement in the transverse direction. Within the range of compositions examined, SMC-R70 shows the highest transverse tensile strength.

SMC-C70 provides the maximum longitudinal tensile modulus (Figure 5). The longitudinal modulus of composites with a 70 percent glass content is predicted to be very dependent on the proportion of continuous fibers, whereas at 30 percent total fiber content, substitution of random fibers with continuous fibers only provides modest increase in the longitudinal modulus. The model reflects the experimental data (Table 3) in estimating the modulus of SMC-R30 to be higher than SMC-R70. This can be explained, at least in part, by the higher resin content of the SMC-R70 composite compared to SMC-R30 (Table 2, Figure 2).

Transverse tensile modulus is predicted to decrease about 40 percent over the composition range of SMC-R30 to SMC-C70 (Figure 6).

COMPRESSIVE PROPERTIES - The model clearly shows the longitudinal compressive strength for this system of composites is independent of the total glass fiber content (Figure 7). Only by increasing the proportion of continuous fiber can the strength be increased. Apparently the highly filled resin matrix in composites with low fiber content provides as much support to the fibers to resist microbuckling as the more highly reinforced composites, which contain unfilled resin.

The compressive strength in the transverse direction also exhibits little dependence on total glass fiber content (Figure 8). In this case, however, increasing strength is predicted with increasing random fiber content.

The general effects of composition and fiber configuration on the longitudinal and transverse compressive moduli (Figures 9 and 10) are similar to those of the tensile moduli (Figures 5 and 6). The magnitude of change in the longitudinal compressive modulus over the composition range studied is much greater than for the transverse compressive modulus.

INPLANE SHEAR PROPERTIES - SMC-R70 composite provides the greatest inplane shear strength among the range of materials considered (Figure 11). Decreasing the total glass content or the proportion of random, chopped fiber results in a loss of shear strength. As the blend of continuous and random fibers approaches 100 percent continuous, the shear strength becomes less dependent on total the fiber content. These relationships are essentially the same as for the transverse tensile strength. This is not unexpected since the underlying mechanisms of failure are dependent on the same factors.

There is relatively little change in the shear modulus of the composites over the range of this study (Figure 12). The effect of random fiber content is more pronounced at 70 percent fiber than at 30 percent fiber.

POISSON'S RATIO - Good agreement is observed between Poisson's ratio measured in tension and compression (Table 3), so averaged values were used to generate regression models. For the range of SMC-C/R composites examined, the longitudinal Poisson's ratio is independent of the factors studied and can be approximated as 0.31. Thus, no contour plot is presented for this property.

In contrast, the transverse Poisson's ratio changes significantly over the range of materials studied and exhibits a response surface much like those of the transverse compressive and tensile moduli (Figure 13). This is not unexpected as the transverse Poisson's ratio (ν_T) is not an independent property, but is directly related to the transverse Young's modulus (E_T) by the expression

$$\nu_T = \nu_L \cdot (E_T / E_L) \qquad (2)$$

where E_L is the longitudinal Young's modulus and ν_L is the longitudinal Poissons's ratio [12]. In this case ν_L is considered to be a constant (0.31). Young's modulus from the models (Figures 5, 6, 9 and 10) can also be substituted into Equation (2) to estimate the transverse Poisson's ratio. Agreement between the model predictions and values calculated for the transverse Poisson's ratio from Equation (2) is very good.

SUMMARY AND CONCLUSIONS

Response surface methods are a cost-effective approach to the characterization of related composite materials. Carefully designed experiments provide reliable property data on specific materials and can be used to develop empirical models that demonstrate the effect of compositional changes on composite properties. Graphic representation of these models makes the information readily available and clearly illustrates important relationships and interactions.

While demonstrating the advantages of response surface methods, this study also provides mechanical property data on five structural SMC-C/R composites with different compositions and fiber configurations. Empirical models based on these data provide the following general conclusions:

- Longitudinal tensile strength, tensile modulus and compressive modulus increase with total glass fiber content and with the portion of continuous fibers.

- Longitudinal and transverse compressive strengths are independent of the total fiber content of the composite.

- Transverse tensile modulus, transverse compressive modulus and transverse Poisson's ratio decrease with increasing total fiber content, but increase with increasing random fiber content.

- Transverse tensile strength and inplane shear strength increase with the amount of random fiber in the composite.

- The longitudinal Poisson's ratio is essentially the same for all of the SMC-C/R composites modeled.

These results demonstrate that continuous, unidirectional fibers provide high properties in the fiber direction, but also show the beneficial effect of random chopped fibers in improving transverse and shear properties.

ACKNOWLEDGEMENT

This study was conducted entirely at the Owens-Corning Fiberglas Technical Center in Granville, Ohio. Appreciation is expressed to the Owens-Corning Fiberglas Corporation for supporting this work and for permitting the publication of this paper.

Other Owens-Corning Fiberglas employees who provided valuable contributions to this work are S. H. Munson-McGee, R. J. Cannizzaro, D. E. Musick, H. C. Gill, D. E. Wise, L. A. Bowman, B. J. Hankins, and A. Majoy.

REFERENCES

1. R. A. Heimbuch and B. A. Sanders, Composite Materials in the Automobile Industry, ASME, 110 (1978).

2. R. B. Jutte, SAE Technical Paper Number 780355 (1978).

3. D. L. Denton, The Mechanical Properties of an SMC-R50 Composite, Owens-Corning Fiberglas, Toledo, Ohio (1979).

4. J. H. Enos, R. L. Erratt, E. Francis and R. E. Thomas, Polymer Composites, 2(2), 53 (1981).

5. D. A. Riegner and B. A. Sanders, Proc. SPE 37th ANTEC, (1979).

6. J. F. Kay, Composites Technology Review, 4(4), 110 (1982).

7. N. S. Sridharan, Short Fiber Reinforced Composite Materials, ASTM STP 772, 167 (1982).

8. C. T. Sun, R. L. Sierakowski and S. K. Chaturvedi, Polymer Composites, 4(3), 167 (1983).

9. G. E. P. Box, W. G. Hunter and J. S. Hunter, Statistics for Experimenters, John Wiley & Sons, Inc., New York (1978).

10. D. F. Adams and D. E. Walrath, Experimental Mechanics, 23(1), 105 (1983).

11. D. L. Denton, Proc. 36th Ann. Conf. RP/C Inst., SPI, 16A (1981).

12. R. M. Jones, Mechanics of Composite Materials, McGraw-Hill, New York (1975).

Table 1

SMC Formulations

Paste Constituents	Parts by Weight		
	30% Fibers[a]	50% Fibers[a]	70% Fibers[a]
Polyester Resin[b]	100	100	100
Calcium Carbonate Filler[c]	200	90	0
Magnesium Oxide Thickener[d]	2.5	3.0	3.5
Zinc Stearate Mold Release	2.5	2.5	2.5
TBPB Initiator	1.25	1.25	1.25

(a) Continuous and/or one inch chopped OCF 433-113 roving.
(b) Low viscosity version of OCF E-987 resin.
(c) Georgia Marble Calwhite II.
(d) Plasticolors PG9033 (38 weight percent active MgO).

Table 2

Measured Composition of SMC-C/R Panels

Nominal Designation	Percent Resin[a] (Std. Dev.)[d]	Percent Fiber[b] (Std. Dev.)[d]	Percent Filler[c] (Std. Dev.)[d]
SMC-R30	22.6 (0.20)	33.1 (0.42)	44.3 (0.26)
SMC-C30	23.1 (0.20)	32.1 (0.52)	44.8 (0.35)
SMC-C25/R20	28.8 (0.47)	45.9[e] (0.68)	25.3 (0.31)
SMC-R70	27.4 (0.95)	72.6 (0.95)	No Filler
SMC-C70	29.1 (0.42)	70.9 (0.42)	No Filler

(a) Measured as loss-on-ignition (LOI).
(b) Measured as LOI residue or as acid wash residue.
(c) Measured as loss from acid wash of LOI residue.
(d) Mean and standard deviation based on six observations.
(e) 24.9% continuous and 21.0% random fiber based on three measurements.

Table 3

Moduli and Poisson's Ratios of SMC-C/R Composites

Material/ Orientation[a]	Statistic[b]	Modulus (10^6 psi)			Poisson's Ratio	
		Tensile	Compression	Shear	Tensile	Compression
SMC-R30	$\bar{x}$ s n	2.87 0.216 12	2.34 0.203 15	1.18[c] 0.184 13	0.31 0.026 12	0.301 0.0238 15
SMC-R70	$\bar{x}$ s n	2.54 0.253 12	2.33 0.326 18	1.33[c] 0.210 14	0.33 0.032 12	0.301 0.0354 15
SMC-C25/R20 (L)	$\bar{x}$ s n	3.54 0.225 12	2.95 0.304 16	0.789[d] 0.0899 13	0.32 0.029 12	0.307 0.0293 15
SMC-C25/R20 (T)	$\bar{x}$ s n	2.01 0.185 12	1.56 0.177 16	0.789[d] 0.0899 13	0.19 0.029 12	0.158 0.0241 16
SMC-C30 (L)	$\bar{x}$ s n	3.78 0.449 10	3.76 0.437 17	0.901[d] 0.118 13	0.30 0.022 12	0.302 0.0305 17
SMC-C30 (T)	$\bar{x}$ s n	2.17 0.219 8	1.98 0.363 10	0.901[d] 0.118 13	0.16 0.027 12	0.136 0.0269 10
SMC-C70 (L)	$\bar{x}$ s n	5.80 0.171 10	5.19 0.237 17	0.589[d] 0.0916 13	0.31 0.026 12	0.312 0.0323 17
SMC-C70 (T)	$\bar{x}$ s n	1.61 0.193 11	1.25 0.327 10	0.589[d] 0.0916 13	0.10 0.021 12	0.056 0.0103 10

(a) L = longitudinal, T = transverse.
(b) $\bar{x}$ = mean, s = standard deviation, n = number of specimens
(c) Secant modulus at 0.5% strain.
(d) Shear modulus is the same in longitudinal and transverse directions.

Table 4

Strengths of SMC-C/R Composites

Material/ Orientation[a]	Statistic[b]	Strength (10^3 psi)		
		Tensile	Compression	Shear
SMC-R30	$\bar{x}$ s n	16.6 1.09 12	31.4 1.91 18	18.4 1.49 20
SMC-R70	$\bar{x}$ s n	31.5 4.71 12	34.2 3.73 18	23.8 2.00 20
SMC-C25/R20 (L)	$\bar{x}$ s n	46.6 2.33 12	62.6 5.14 18	14.6[c] 1.07 17
SMC-C25/R20 (T)	$\bar{x}$ s n	10.3 0.518 12	25.9 1.73 18	14.6[c] 1.07 17
SMC-C30 (L)	$\bar{x}$ s n	49.8 4.60 12	97.5 6.71 18	9.02[c] 0.422 18
SMC-C30 (T)	$\bar{x}$ s n	2.48 0.552 8	18.8 1.37 15	9.02[c] 0.422 18
SMC-C70 (L)	$\bar{x}$ s n	113.5 8.77 12	96.4 14.4 11	7.94[c] 0.239 17
SMC-C70 (T)	$\bar{x}$ s n	1.85 0.446 11	15.2 0.906 11	7.94[c] 0.239 17

(a) L = longitudinal, T = transverse.
(b) $\bar{x}$ = mean, s = standard deviation, n = number of specimens
(c) Shear modulus is the same in longitudinal and transverse directions.

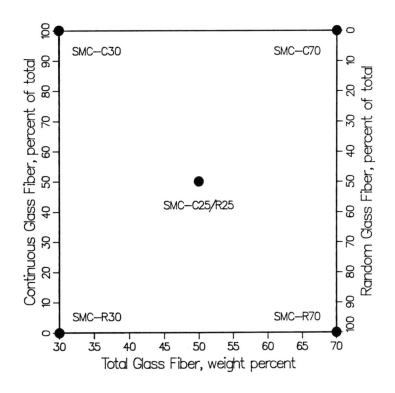

Figure 1. Experimental Design for the SMC-C/R
Characterization.

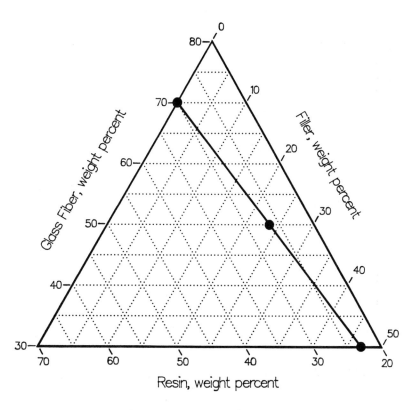

Figure 2. Composition Diagram for the SMC-C/R Composites.
Formulations containing 30, 50 and 70 percent
glass fibers are noted.

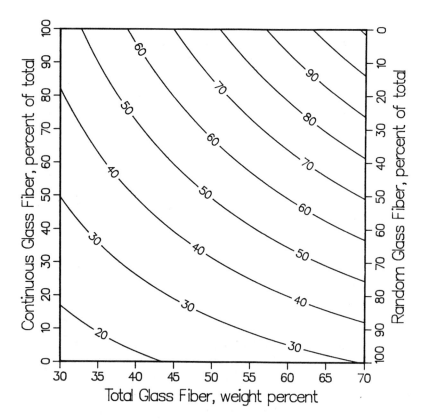

Figure 3. Longitudinal Tensile Strength (10³ psi).

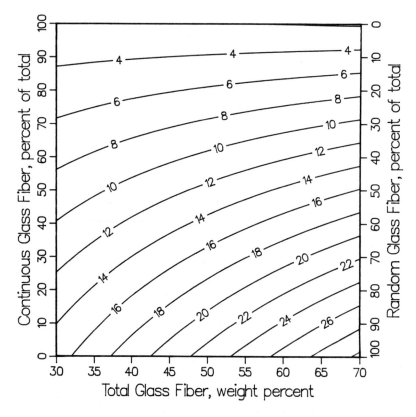

Figure 4. Transverse Tensile Strength (10³ psi).

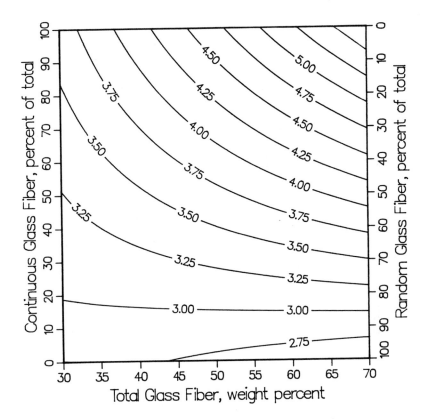

Figure 5. Longitudinal Tensile Modulus (10⁶ psi).

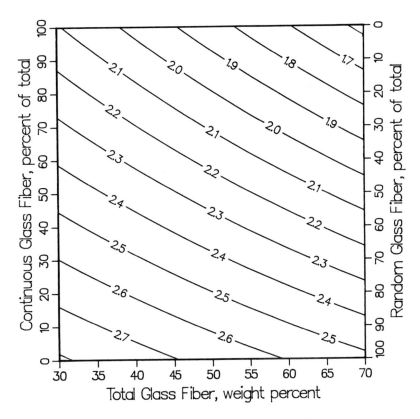

Figure 6. Transverse Tensile Modulus (10⁶ psi).

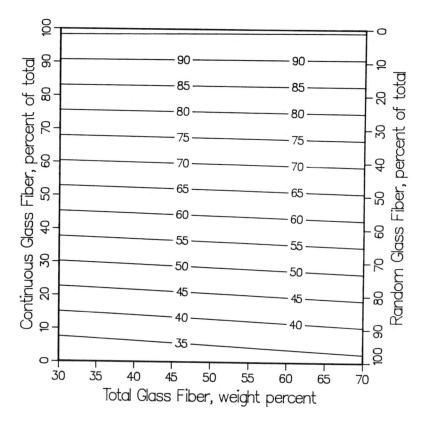

Figure 7. Longitudinal Compressive Strength (10^3 psi).

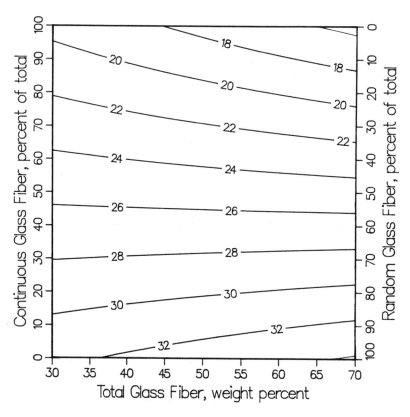

Figure 8. Transverse Compressive Strength (10^3 psi).

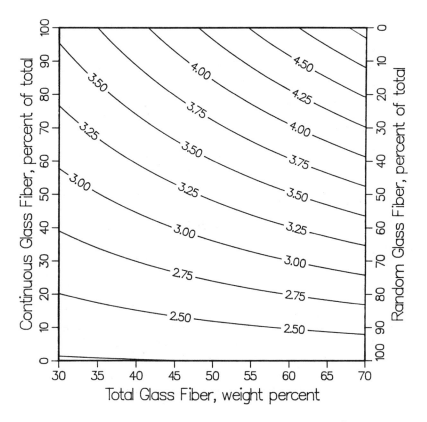

Figure 9. Longitudinal Compressive Modulus (10⁶ psi).

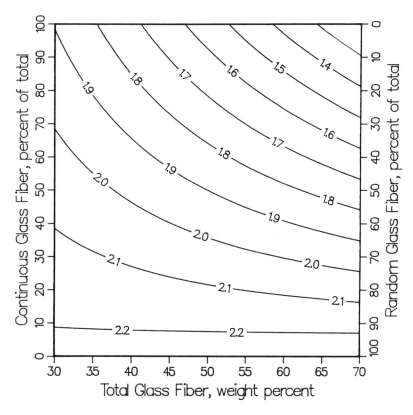

Figure 10. Transverse Compressive Modulus (10⁶ psi).

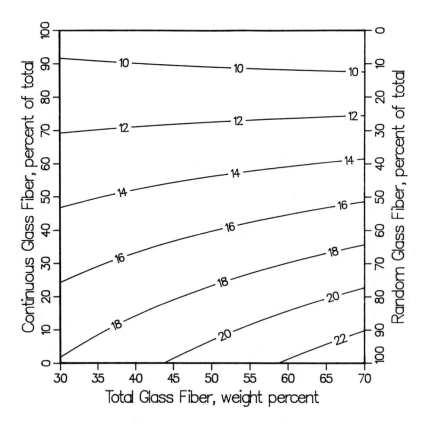

Figure 11. Inplane Shear Strength (10³ psi).

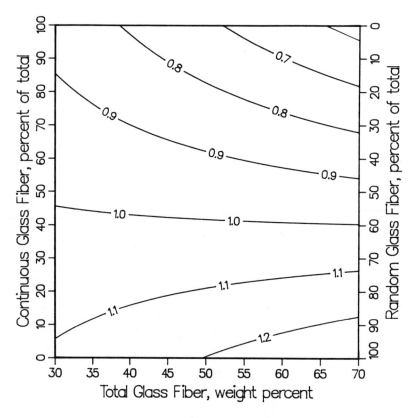

Figure 12. Inplane Shear Modulus (10⁶ psi).

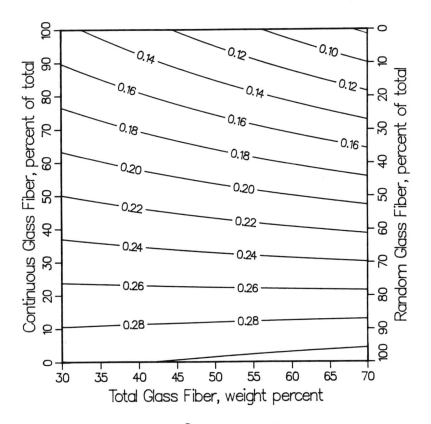

Figure 13. Transverse Poisson's Ratio.

VAMAS—A FORUM FOR INTERNATIONAL COOPERATION ON THE DEVELOPMENT OF COMPOSITE STANDARDS

Donald Hunston, Herzl Chai
National Bureau of Standards
Gaithersburg, Maryland 20899   USA

ABSTRACT

In 1982 representatives and Heads of State from Europe, Japan, Canada, and the United States met in Versailles for the first Joint Economic Summit. One item addressed at that Meeting was the need to encourage international trade through technical collaboration. This led to the establishment of VAMAS, the Versailles Project on Advanced Materials and Standards. The VAMAS Steering Committee then established Technical Working Groups in specific areas such as composites. The goal of these Working Groups is to facilitate the development of the science and technology base to accelerate the generation of needed standards. This is certainly critical to the composites field where the lack of standardized tests for many of the important properties is a major handicap. The VAMAS Working Group on Composites is providing a forum for information exchange and international cooperation so the efforts of scientists addressing questions related to test methods can be encouraged. The focus in the VAMAS program is on the basic research that must precede the establishment of standards. By working closely with organizations such as ASTM, the time frame for establishing standards can therefore be shortened.

INTRODUCTION

The standardization of definitions, terms, and test methods plays a vital role in the economic utilization of any technology. This is increasingly becoming important in the field of polymer composites where deficiencies in standardization represents a major area of concern. In response to this, activities have been initiated by a number of groups in addition to the ongoing programs of the American Society for Testing and Materials (ASTM). The Department of Defense with MIL Handbook 17, the Suppliers of Advanced Composite Materials Association (SACMA), and the Aerospace Industries Association of America are just three of many organizations with current activities. In these programs the focus is on specifying standards and obtaining data in areas where adequate knowledge already exists.

In addition to these efforts, however, there is a need to conduct basic research which helps resolve areas of controversy and lays the groundwork for new or improved standards in areas where the current knowledge is not adequate. These issues are also being addressed with programs at government laboratories, such as the National Bureau of Standards (NBS), Universities, and some industrial laboratories. To achieve success in these efforts, however, there is a need to increase intercommunication among the groups involved. This is one of the goals in a recently initiated international program which provides a form for information exchange and cooperative research. This effort is part of the Versailles Project on Advanced Materials and Standards (VAMAS). A complete description of VAMAS is given in Reference 1, but a brief summary from that paper will be presented here to provide the appropriate historical perspective. The paper then goes on to discuss the initiation and status of the polymer composites work in VAMAS and finishes with an example of these studies.

BACKGROUND

In the early 80's the importance of novel materials was becoming increasingly apparent to scientists, manufacturers, and governments. Many of the advances in technology depended on novel materials. Moreover, the rapid development and incorporation of advanced materials in these technologies was clearly an international phenomenon. Materials developed in one location, such as the United States, would be produced in a second, such as Japan, incorporated in products in a third, such as Europe, and used throughout the world.

Rapid growth in the manufacture and use of highly engineered materials, together with the promise of even greater advantages in the future, was placing severe strains on existing materials performance standards, both national and international. On the other hand, the new technologies were presenting entirely new demands on materials characterization and performance measurements, requiring new testing concepts and procedures. At the same time, sophisticated materials responded to existing tests in ways that differed from the response of traditional materials. New materials therefore mandated novel approaches to measurement and performance evaluation. The rapid development of new technology made it impossible to wait for the long basis of experience that has traditionally preceded the development of standards. Furthermore, since these problems were international in nature, a new, broadly based, international effort was essential to ensure that the prerequisite measurement basis would be in place as the development of internationally accepted standards became necessary.

INITIATION OF VAMAS

At the same time these needs were becoming clear, the Heads of States from the 7 countries listed in Table I and a representative from the Commission of the European Communities (Common Market) met at the 1982 Economic Summit in Versailles to

consider the various technical sources of expansion in the world economy. The leaders identified a number of specific

Table I: Participants in the Economic Summits
Canada
France (1982)*
Germany (Federal Republic) (1985)
Italy (1987)
Japan (1986)
United Kingdom (1984)
United States (1983)
Commission of the European Communities (Common Market)

*Year indicates summit meeting in host country since founding of VAMAS

technologies and a group of critical generic areas on which the establishment of these specific technologies is based. The Heads of State selected 18 of these technologies (Table II) for the inauguration of international collaboration within the framework of economic summits, under the general heading of Technology, Growth, and Employment. One of the

Table II: Versailles Projects in Technology, Growth, and Employment
Advanced Robotics
Aquaculture
Biotechnology
Controlled Thermonuclear Fusion
Fast Breeder Reactors
Food Technology
High Speed Trains
Photosynthesis and Photochemical Conversion of Solar Energy
Photovoltaic Solar Energy
Remote Sensing from Space
Advanced Materials and Standards
Basic Biology
High Energy Physics
Solar Systems Exploration
Housing and Urban Planning in Developing Countries
Impact of New Technology on Mature Industries
New Technologies Applied to Culture, Education, and Vocational Training
Public Acceptance of New Technologies

project areas identified was Advanced Materials and Standards.

After selection of the project areas, scientific representatives of the Heads of State were identified for each of the projects in which there was a national interest. One or two of the participating countries were identified as project leaders or coleaders in each of the 18 areas. For Advanced Materials and Standards, initial leadership by the United Kingdom and the United States was designated, to take advantage of strong materials research interest in these two countries. This effort was designated as VAMAS (Versailles Project on Advanced Materials and Standards), and now operates independently under a Memorandum of Understanding signed by Cabinet-level representatives of the Heads of State.

The VAMAS structure has two tiers, a steering committee and a set of technical working parties. The Steering Committee is made up of representatives for each of the Economic Summit participants and has the responsibility to initiate, review, and terminate the individual VAMAS projects. Dr. Lyle Schwartz, the Director of the Institute for Material Science and Engineering at NBS, is currently the Chairman of the VAMAS Steering Committee. The United States participants on the Steering Committee are appointed by the Office of Science and Technology Policy in the White House.

The Technical Working Parties carry out specific VAMAS programs and are established at the discretion of the steering committee. Technical Working Party members are appointed by the steering committee with the concurrence of the chairman of the respective working parties. Both in its program development and in the resulting research and intercomparison, the Technical Working Parties are actively encouraged to coordinate their activities with existing standards organizations.

VAMAS has now launched 13 Technical Working Parties (see Table III). Activities in these areas include establishment of priorities for prestandards activity, consultation and collaboration in prestandards research, and intercomparison of specific

---

Table III: VAMAS Technical Working Areas

Winter 1983-1984*

Wear Testing Methods (Germany+)
Surface Chemical Analysis (United Kingdom)

Summer 1984

Ceramics (France)
Polymer Blends (Canada)

Winter 1984-1985

Bioengineering Materials (United Kingdom)
Hot Salt Corrosion Resistance (United Kingdom)
Polymer Composites (France)
Superconducting Cryogenic Structural Materials (Japan)
Weld Characteristics (United Kingdom)

Fall 1985

Materials Databanks (United States & CEC‡)
Creep Crack Growth (United Kingdom)

Spring 1986

Efficient Test Procedures for Polymers (United Kingdom)

Fall 1986

Low Cycle Fatigue (CEC)

*Time of activity initiation
+Lead country
‡Commission of the European Communities

---

measurements in a number of governmental and commercial laboratories.

VAMAS WORKING GROUP ON COMPOSITES

The area of polymer composites was added as a Working Party in the winter of 1985. The lead country was France with the United States and the United Kingdom also playing major roles. The effort in the U.S. is lead by the National Bureau of Standards. Although there was early agreement concerning the importance of standards work in the composites field, considerable discussion was devoted to selecting the particular areas where effort could be focused productively. It was important to select topics where the work could be coordinated with ongoing research since special funding was often unavailable. Delamination testing was selected as the first area for study with fatigue testing added more recently and creep testing now being considered.

Delamination Testing: The area of delamination testing was chosen first because there were ongoing programs in this area, and the work could be used to augment the ASTM round robin that was underway. The ASTM program involved static testing of samples made from unidirectional, graphite-fiber, reinforced samples so the VAMAS project selected glass-reinforced samples made with both unidirectional tape and woven cloth. The studies will begin with static tests but will examine fatigue tests as well. Potential areas for study include:

The effects of specimen size and initial crack size.
Initiation of crack growth and methods of measurement.
Strain rate.
Means of measuring displacement and load.
Methods of analyzing results.
Presentation of results.

Laboratories in Canada, France, Japan, the United Kingdom, and the United States are currently involved in this project with Sweden expected to join soon. In the U.S. the initial studies were conducted at NBS with cooperative efforts donated by three other laboratories (NASA Langley Research Center, Texas A & M University, and BASF Structural Materials). The program has now reached the point where opportunities are available for additional participants who might wish to donate time for cooperation. Inquiries from interested parties should be directed to the authors at NBS.

Fatigue Testing: The second area selected for investigation was fatigue testing. The objective was to establish reliable specimen specifications and testing methods for use in predicting fatigue limits of glass and carbon fiber composites. Several different comparisons among shear, tension, and flexure are being made. Fatigue test parameters are:

Loading conditions: frequency, wave type, R ratio.
Specimen design.
Mode of Loading: Tension and bending.
Load control or stroke control experiments.
Environment effects.

Participants in this program include Laboratories in Canada, France, Japan, and the United Kingdom. Although the U.S. is not currently involved in these tests, the program is seeking groups in the U.S. who would be interested in cooperation with this effort. Any such organization should contact the authors at NBS to pursue this topic further.

Creep Testing: A third area of study, creep testing, has been identified and is currently under discussion. The final program has not yet been developed, but as indicated above, parties who have any interest in this area should contact the authors.

EXAMPLE OF RESULTS

The VAMAS composites program currently has ongoing activities in the areas of delamination and fatigue. Since most of these studies have just begun, however, very few results are currently available. Consequently, to illustrate the type of work being conducted, some data will be presented here from one aspect of the delamination research at NBS. A brief summary of the overall VAMAS delamination effort will be given first to provide a general perspective.

The VAMAS program is focusing initially on two types of tests: mode I loading where a tensile force is applied perpendicular to the crack plane, and mode II loading where a shear force is applied parallel to the crack plane (see Figure 1). The specimens involve layers of continuous fibers oriented parallel to the crack plane, and the crack propagates between the layers, i.e. interlaminar crack growth. Fracture energies are then calculated from the loads required to propagate the crack, and the results represent one measure of the resistance to delamination.

Within the VAMAS program, participating laboratories are investigating many of the parameters mentioned above, i.e. specimen design, crack initiation methods, etc. In addition, a round robin has been organized using samples donated by the French and Japanese. The round robin is examining 4 specimen designs, 2 for mode I and 2 for mode II (see Figure 2). The data that follows are from the NBS effort on mode II testing.

Most research to date on interlaminar fracture has focused on mode I loading, which proved to be the critical component in first-generation, brittle-matrix composites. Use of tough matrix systems was shown to greatly improve the composites damage tolerance, but unfortunately the accompanying reduction in G_{IIC}/G_{IC} ratio for such materials necessitates the additional consideration of shearing modes in the fracture analysis.

Although research on mode II interlaminar fracture is extensive, there are a number of controversial issues, the most pressing being perhaps a meaningful interpretation for the observed nonlinearity in the load-deflection curve. The majority of work to date has largely neglected this effect, treating the fracture problem in accordance with linear elastic fracture mechanics. Recent mode II fracture tests on adhesively-bonded joints [2] have shown, however, that this nonlinearity can be due to the development of a plastic deformation zone in the resin region ahead of the crack tip. It is only after the damage zone becomes fully-developed that crack growth in the usual sense can take place. Consequently, it is desirable to develop a test specimen that provides stable crack growth over as large a range of crack lengths as possible so that steady state behavior can be studied.

The most widely used mode II interlaminar fracture geometries is the End-Flexure Notch (ENF) specimen shown in Figure 3a. One of its main drawbacks in the above context, however, is the limited range of crack lengths over which the crack growth is stable, i.e. a/L > 0.68 [3], where a and L are the crack length and the half-span length, respectively (see Fig. 3a). A greater range of stable growth, i.e. a/L > 0.55 (Eq. 5 of Ref. 3, with c/b → 0), is provided by the split-beam specimen shown in Fig. 3b. This test geometry (see Ref. 4) was therefore examined in the VAMAS program, and the initial results are

reported here.

Both glass-fiber and graphite-fiber composites were examined in this work. The glass reinforced specimens that have been available to date were not stiff enough to meet the requirements for the tests so all of the results reported here are for graphite-reinforced composites. The mode II fracture specimens were supplied by Toray Industries of Japan. The material* was a 24 ply, unidirectional T300/Epoxy #3601. The resin was a 177°C cure epoxy based on MY720-DDS formulation. The fiber content was 60% by weight. The specimens contained a 37 μm thick, 35 mm long midplane insert (polytetrafluoroethylene film). Prior to testing, the crack was advanced several mm beyond the insert border by wedging open the end of the specimen slightly, thus generating a natural crack.

The tests were run on an Instron* testing machine at a loading rate of 2 mm/min (see Figure 3b for loading geometry). The displacement, δ, under load, P, was obtained using an LVDT mounted directly under the loading pin. The crack tip position was recorded during the test using a video recording system. Fig. 4 exemplifies the loading history. The tic marks on the chart correspond to instantaneous crack length measurement, obtained from the video records.

The fracture energy was calculated as a function of crack length from

$$G_{IIC} = \frac{9(aP)^2 C}{2b(L^3 + 3a^3)}$$

where b is the beam width and C ($=\delta/P$) is the compliance. The calculations start from the peak load and continue through the region of stable crack growth. The compliance C was evaluated at each tic mark by connecting a straight line from the origin to marked point on the load deflection curve. Table IV summarizes the results. As shown (specimen #118-2-14), the

_____

*Certain commercial materials and equipment are identified in this paper in order to specify adequately the experimental procedure. In no case does such identification imply recommendation or endorsement by the National Bureau of Standards, nor does it imply necessarily the best available for the purpose.

fracture work was quite independent of crack extension, and is consistent among the test specimens.

CONCLUSIONS

The preliminary work described above suggests that the split beam specimen has an advantage over the ENF specimen since the former has a larger range of crack lengths over which stable crack growth can be obtained. This allows more data to be taken so variations within the sample have less influence. Moreover, the data can be obtained in the region where steady state growth has been achieved. This allows the true fracture work, corresponding to a self-similar crack growth, to be determined more easily.

This is only one example of the many studies currently underway as part of the VAMAS activity on polymer composites. Many more results will become available in the near future. This will provide valuable information for efforts to establish new or improved test methods and standards. The participation of additional laboratories in this effort to advance the field of composite standards is most welcome, and interested parties are invited to contact the authors.

REFERENCES

1. L. Schwartz and B. W. Steiner, ASTM Standardization News, (October, 1986), 40-44.

2. H. Chai, "Shear Fracture", in press, International J. of Fracture.

3. H. Chai and S. Mall, International J. of Fracture, R3-R8 (1988).

4. K. Kendall, J. of Materials Science (1976), 1263-1266.

Table IV: Mode II Fracture Data

Specimen	Critical Load p(N)	Crack Length a(mm)	Specimen Length L(mm)	b(mm)	2h(mm)	C(mm/N)	G_{IIC} (N/m)
118-2-14	256.7	43	78	25.03	3.42	60.0	1838
	174.1	57				92.5	1593
	141.5	66				122.8	1453
	136.1	67				137.0	1488
118-2-9	234.9	43	61.5	24.95	3.39	46.8	1838
118-2-10	276.7	38	61.5	25.00	3.40	37.7	1890
118-2-11	283.8	37	61.5	25.03	3.38	36.5	1890
118-1-2	234.2	38	81.5	25.00	3.20	69.1	1400
117-1-4	158.0	63	106	24.94	3.28	163.6	1505
						Avg. (SD)	1686(214)

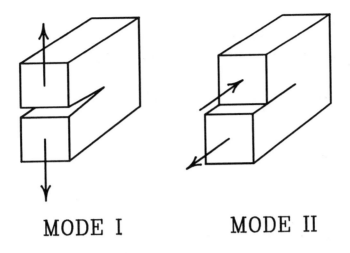

MODE I   MODE II

Figure 1: Schematic illustration for mode I and Mode II loading.

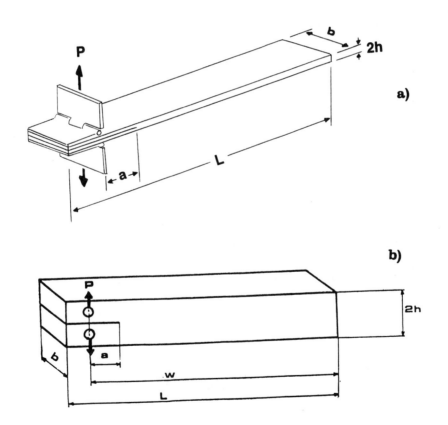

Figure 2: Two geometries for mode I interlaminar fracture testing.

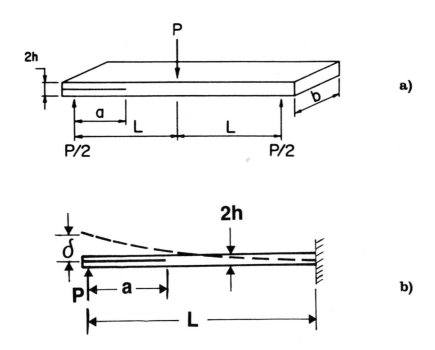

Figure 3: Two geometries for mode II interlaminar fracture testing, a) edge notched flexure (ENF), and b) split beam specimens.

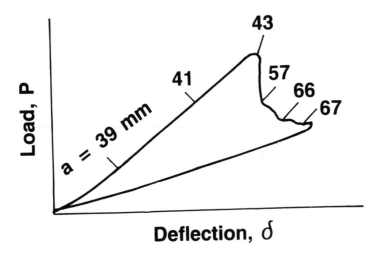

Figure 4: Loading history for split beam fracture test. Tic marks indicate instantaneous crack length measurements.

THE EFFECT OF HEATING RATE ON THE SHEAR STRENGTH OF A ROOM TEMPERATURE CURED EPOXY RESIN

M. R. Mathews, L. J. Doctor
Raytheon Company
West Andover, Massachusetts 01810   USA

S. P. Petrie
Plastics Engineering Department, University of Lowell
Lowell, Massachusetts 01854   USA

Abstract

During the determination of engineering design data for the shear strength of a room temperature cured epoxy resin at elevated temperatures, some interesting effects were observed. When the rate of heating was varied, the shear strength, at a given temperature, was found to vary greatly. Since this effect had not previously been reported in the literature, an experimental investigation was made to determine the cause and magnitude of the effect.

The decrease in the shear strength of the epoxy resin was determined for equilibrium and dynamic heating conditions using thick lap shear specimens. Differential scanning calorimetry (DSC) experiments were used to show, in addition to adhesive tests, that the observed differences were due to differences in the extent of cure of the resin.

The effect of increasing the heating rate was found to accentuate the difference between the dynamic and equilibrium shear strength values.

THE RADOME FOR THE PATRIOT MISSILE is a slip cast fused silica ceramic and it is bonded to an epoxy–Kevlar[R] composite thread ring using a room temperature cured epoxy resin. During flight, the nose of the missile sees elevated temperatures due to frictional heating with air and high mechanical stresses produced in maneuevers. For this reason, it is desirable to know the shear strength of the adhesive under dynamic heating conditions.

The chemistry of epoxy resins has been widely investigated. It has been reported in the literature (1) that three reactions can occur during the cure of an epoxy resin with an aliphatic amine curing agent. It was proposed (1) that the reaction of the primary amine with an epoxy group occurs first yielding a secondary amine. The secondary amine can then react with another epoxy group which will then yield a tertiary amine, which is not capable of any further reaction. The hydroxyl group, which is produced by opening the epoxide ring, can also react with an epoxy group to produce an ether. The authors proposed that the amine reactions predominate at room temperatue while the etherification reaction does not occur to any significant extent unless a tertiary amine catalyst is used. The reaction of the secondary amine is responsible for the branching and crosslinking of the epoxy resin system.

Numerous articles have appeared in the literature (2, 3, 4) detailing various test specimens available for evaluating adhesive properties. The most commonly used test is ASTM D1002 (the thin lap shear specimen). It is frequently reported that the thin panels tend to deform under high loads. The problems encountered with the thin lap shear specimen have led to the development of new test specimens which reduce the peel stress on the adhesive joint. Namely, the double lap shear specimen and the thick lap shear specimen. It was reported by Aker (2) that for the thick lap shear specimen a thicker panel reduces the normal stress and shear stress concentration factors due to greater rigidity and less deformation of the test specimen under load.

Articles have appeared in the literature concerning the curing kinetics and extent of cure of epoxy systems (5, 6). The use of a differential scanning calorimeter (DSC) has been shown to offer a practical method for the determination of residual cure in an epoxy system as shown by Fava (7).

It was hypothesized that as the extent of cure of a room temperature cure epoxy system increases the shear strength of the epoxy will also increase. It has been observed that the majority of the data in the literature which deals with the strength of epoxy with increasing temperature involves the use of equilibrium heating conditions. Thus, when the test specimen is brought to the desired temperature, further reaction of the epoxide groups with the curing agent can take place. This phenomenon of strengthening the epoxy by increasing the cure or extent of reaction, will be opposed by the normal degradation of strength associated with an increase in test temperature. The theory that two opposing effects are occuring simultaneously introduces the time element into the determination of epoxy strength.

This paper examines the effect of relative heating rates on the shear strength of an epoxy resin using an improved test specimen. The DSC was used to correlate the effect of the extent of reaction on the strength of the epoxy. In order to reduce problems associated with improper surface preparation, commercially prepared test panels were purchased and used in this investigation.

EXPERIMENTAL

The adhesive used in this study is manufactured by the Hysol Company; a division of Dexter-Midland Corporation and is sold under the trade name of "EA-934". Table I lists the reported tensile shear strength of this adhesive when bonded to 2024-T3 Alclad 1/16 inch thick aluminum and then tested per ASTM D1002 (7).

Table 1 - Vendor Reported Test Data - EA934

Temp (deg. F)	Stress (psi) **	ln Stress
77	3100	8.04
100	2200	7.70
200	1800	7.50
250	1500	7.31
300	1000	6.91
400	760	6.63
500	400	6.00

** std. dev. was not listed in vendor data

Before using accelerated heating rates to evaluate this adhesive under dynamic conditions, it was desired to reproduce the equilibrium test data reported in the vendor literature. The adhesive was mixed according to the manufacturer's specifications and all assemblies were made within the 40 minute pot life. During the application of the adhesive to the panels, care was taken so that excessive amounts of adhesive did not leave a large fillet along the overlap of the panels. The panels were placed into a bonding fixture and clamped for cure using spring loaded clamps. The panels were left undisturbed for 16 hours minimum to cure at room temperature.

Upon completion of the cure cycle the panels were removed from the bonding fixture and allowed to cure for an additional two days prior to machining. The finished size of the test specimens was obtained by machining the strips, which were cut from the plate, to a one inch width.

The purpose of using the thick lap shear test was to reduce the difficulties associated with the thin lap shear specimens. Plates of aluminum, treated with the Boeing process, were obtained from the Limco Manufacturing Company of Glen Cove, New York. This company is a Boeing licensed facility for surface pretreatment of aluminum and they were chosen as the source of supply for the prepared thick lap shear panels. The purchased panels were 1/4 inch thick unclad 2024-T351 aluminum 7 3/4 inches by 8 3/4 inches. After a cleaning cycle they were phosphoric acid anodized and primed with BR-127 primer which is manufactured by the American Cyanamid Company of Wayne, New Jersey.

Again the adhesive was mixed according to the manufacturer's specifications and all assemblies were made within the pot life. Both mating surfaces were covered with adhesive using a wooden applicator stick. Prior to assembling the panels together, .009 inch shim stock was placed on one panel in four places to allow for a uniform bond line thickness. The location of the shim stock was such that it did not come near the test area of the specimens when they were cut to size.

The panels were bonded together in a hydraulic press at twenty pounds per square inch of pressure. Figure 1 is a detailed diagram of a finished thick lap shear specimen.

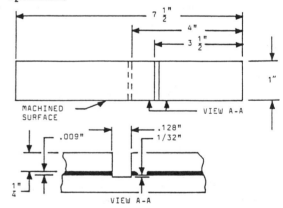

Fig. 1 - Thick Lap Shear Specimen

The ASTM D1002 standard specifies that for testing at elevated temperatures, an oven or furnace should be used capable of maintaining the specimens at the desired test temperature. To accomplish this for the equilibrium temperature tests, an Instron[R] portable oven and controller were used. The specimen was kept in the oven for thirty minutes (minimum) while the oven controller kept the temperature in the oven at the set point temperature.

For the fast heating technique of this study, radiant quartz heaters were used because they were capable of supplying immediate intense heat to the specimen. The quartz heaters selected were manufactured by Research Incorporated of Minneapolis Minnesota. The model 4083 "Pyropanel" offers the flexibility to operate with between one and six lamps per panel. It has ceramic reflectors to help direct the radiant heat to the specimen. Two pyropanel heaters were used for the testing so that the heat would be applied evenly from both sides of the specimen. The lamps that were used in the panels, were model 500T3/CL which are capable of supplying 500 watts per lamp at 120 volts A.C. The power controller used was model 663F-12-331, it was rated for 120 volts A.C. and 45 amps. It has a ten (10) turn potentiometer dial to control the power output to the lamps. Both the lamps and the power controller were also supplied by Research Incorporated.

In order to monitor the heating characteristics of this system, dummy thermocoupled specimens were prepared, using type J thermocouple wire. Figure 2 is a diagram of a finished dummy thermocouple specimen.

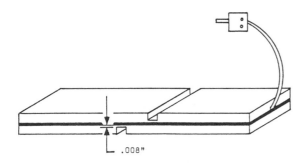

Fig. 2 - Dummy Thermocouple Specimen

The quartz heaters were mounted to an angle iron support frame so the centerline of the test specimen when loaded into the jaws of the Instron[R] was in line with the centerline of the heaters. The quartz heaters were attached to the frame with a threaded rod so that the distance the panels were held from the specimen could be adjusted. Figure 3 is a photograph of the heating apparatus attached to the support frame.

All specimens were tested with an Instron[R] Tensile Test Machine using the 10,000 lb and 50,000 lb load cells. The specimens were loaded into the jaws approximately 7 inches apart and pulled to failure while the force required was recorded on chart paper.

The sequence of events for testing the thick lap shear specimens at the various

heating rates was as follows: First, the equipment was allowed to heat up for thirty minutes minimum with a slave specimen. Then the power to the pyropanel quartz heaters was shut off and the slave specimen was removed from the grips. A new specimen was quickly inserted and the power to the heaters was turned on. Once the power was turned on a timer was started to track the test time.

Fig. 3 - Quartz Heating Apparatus

For the thin lap shear tests the oven temperature was recorded as the equilibrium test temperature. For the dynamic thick lap shear tests the test temperature was determined using the dummy thermocouple specimen.

A Perkin-Elmer Model DSC-IV differential scanning calorimeter (DSC) was used to determine the extent of cure for this epoxy system. A starting temperature of 25 deg. C was used and the scan rate was 20 deg. C per minute. The final temperature for all the DSC runs was chosen as 275 deg. C. The DSC samples were prepared by filling preweighed DSC pans approximately 1/4 full, assuring that the epoxy did not extend over the top, and then covering the pan with a platinum cover.

RESULTS AND DISCUSSION

Figure 4 is a plot of the data reported by the vendor as listed in Table 1. As can be seen the relationship follows a smooth curve.

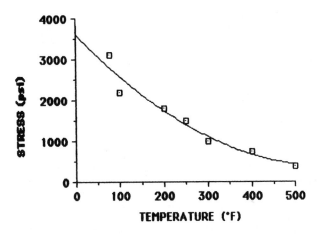

Fig. 4 - Plot of Vendor Data

When the data are plotted on a semi-log graph, the relationship is linear with a negative temperature dependence as shown in Figure 5.

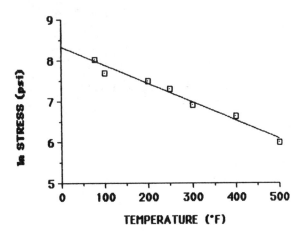

Fig. 5 - Semi-log Plot of Vendor Data

Table 2 lists the initial set of data obtained in this study from the thin lap shear tests along with the mean, standard deviation, and coefficient of variation.

Table 2 - Initial Room Temperature Test Data Thin Lap Shear Specimens

Type	Load (lb)	Stress (psi)
thin	1350	2700
thin	1470	2940
thin	1630	3260
thin	1350	2700
thin	1470	2940

AVE STRESS = 2910 psi, s = 230 psi, v = 8%

Similar data obtained for the thick lap shear specimens are given in Table 3.

The first portion of this paper was to duplicate the published values of the lap shear data. Figure 6 is a semi-log plot of strength vs. temperature for all of the previous conditions.

Table 3 - Initial Room Temperature Test Data Thick Lap Shear Specimens

Type	Load (lb)	Stress (psi)
thick	2500	5000
thick	2340	4680
thick	2450	4900
thick	2450	4900
thick	2200	4400
thick	2510	5020
thick	2570	5140
thick	2480	4960
thick	2475	4950
thick	2630	5260

AVE STRESS = 4925 psi, s = 240 psi, v = 5%

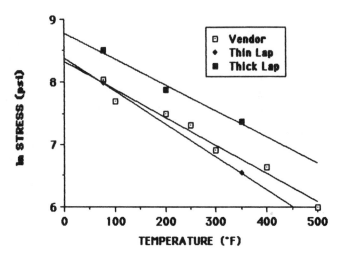

Fig. 6 - Semi-log Plot of Strength Versus Temperature for All Conditions

As can be seen in Figure 6, all of the curves appear to have the same shape - they are straight lines. The slopes and intercepts for the different sets of results were obtained using a programmable calculator and linear regression analysis to determine the "best fit". The slope, obtained from the analysis, indicates the temperature dependence of the shear stress. An increase in the value of the slope would result in a greater loss of strength with increasing temperature. The y-intercept, obtained from the analysis, is the shear at zero degrees Fahrenheit. This value can be used to compare the magnitude of the strength which would be produced by using different test geometries. For all practical purposes the vendor data have been reproduced.

An examination of Figure 6 shows that an upward shift is observed for the thick lap shear specimens. This is believed to be due to the reduction in peel stress located at the ends of the lap joint when the thick specimen is used (2). The slopes of the results are similar which shows that for equilibrium conditions, over the temperature range investigated, that the relationship between the stress and temperature is well defined. It is apparent however, that the test method affects the magnitude of that relationship.

A series of photomicrographs were taken on a scanning electron microscope (SEM) to more closely look at the failure mode of the epoxy joint in order to see if the reduced shear stress values associated with the thin lap shear specimens could be explained.

Figure 7 is a photomicrograph at 500X of a typical thin lap shear specimen at a point where the failure mode is primarily cohesive. However, as can be seen in the upper portion of the photomicrograph, there is evidence of slick or adhesive failure.

An examination at higher magnification (1000X) in Figure 8 shows that the surface appearance of the adhesive failure portion is very smooth.

A magnified view (1000X) of the cohesive failure portion is shown in Figure 9 and already the increased surface texture is plainly visible. It is interesting to note the asbestos fiber running along the right hand side of the photomicrograph.

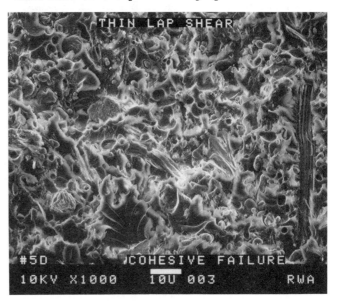

Fig. 9 – SEM Photomicrograph at 1000X

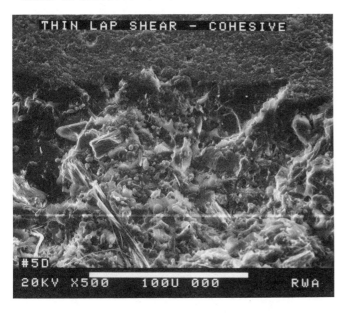

Fig. 7 – SEM Photomicrograph at 500X

For the thick lap shear specimens there was never any evidence of adhesive failure to the substrate. Figure 10 is a 1000X magnifaction of the fracture surface of a typical thick lap shear specimen pulled at room temperature.

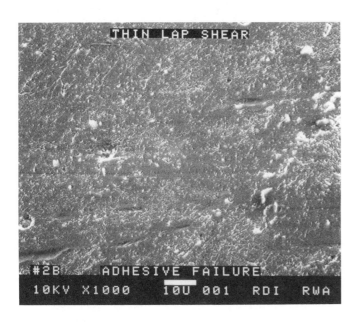

Fig. 8 – SEM Photomicrograph at 1000X

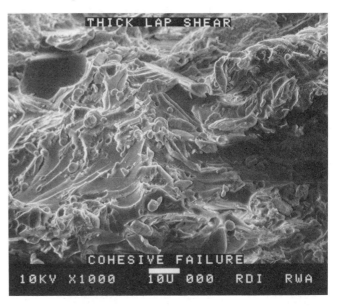

Fig. 10 – SEM Photomicrograph at 1000X

As can be seen, the surface appearence in Figure 10 is very rough when compared with Figure 8. It can be seen that the cohesive surfaces of the thick and thin lap joints are very similar. It is also interesting to note that in all the photomicrographs it can be seen that the orientation of the fiber in the matrix is totally random.

Based on the SEM photomicrographs, it appears that the reduction in strength of the thin lap shear specimens is due to the mixed mode of failure. The thick specimens offer a more defined state of shear with less peel and cleavage resulting in higher shear stress values.

The second topic to be investigated was to examine the effect of different heating rates on the shear strength of the epoxy at elevated temperatures. Figure 11 is a plot of log shear stress versus temperature at various heating rates.

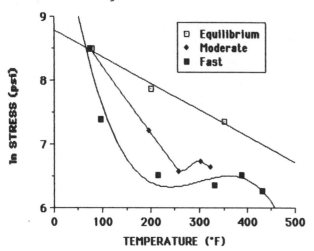

Fig. 11 – Semi-log Plot for
Various Heating Rates

As can be seen the temperature dependence of the log shear strength is no longer linear. From the results, it appears that the mechanism of the dynamic heating failure process over the range of temperatures investigated in this study has changed from the equilibrium heating situations. The data were subsequently plotted as shear stress vs. temperature in Figure 12 along with the thick lap shear equilibrium data.

It can be seen in Figure 12 that as the heating rate is increased the curves shift away from the equilibrium curve.

It is interesting to look at the pattern of the curves for the various heating rates. Initially the shear strength falls off sharply and then begins to level off, however as the time to reach temperature increases to five or six minutes the curves show a slight increase. It is suspected that this is due to the increase in strength associated with an increase in the cure of the epoxy resin. Then at a time of ten minutes the shear stength falls off again apparently due to the lessening of strength associated with elevated temperatures.

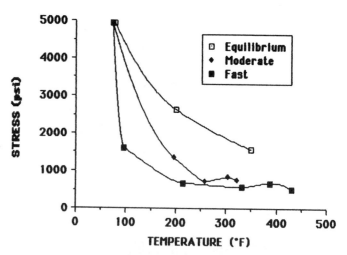

Fig. 12 – Plot of Various Heating Rates

In order to further look at the effect of the heat curing, or increased degrees of crosslinking and to test the hypothesis, it was decided to post cure two sets of thick lap shear specimens. The two post cure conditions used were 200 deg. F for four hours and 350 deg. F for two hours. The average values of the data for the room temperature tests of the post cured samples are given in a bar graph in Figure 13, where A is 7 days at room-temperature, B is 5 days at room-temperature plus 4 hours at 200°F, and C is 5 days at room-temperature plus 2 hours at 350°F.

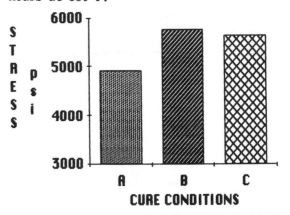

Fig. 13 – Bar Graph of Post Cure Results

It can be seen that the post cured specimens show an increase in shear stress over the room temperature cured specimens.

The previous discussion and results have focused on the macroscopic level. In an attempt to understand what was happening on the molecular level the extent of cure of the epoxy was examined by DSC. Figure 14 shows a typical dynamic DSC scan for the epoxy resin. Using the DSC integrator, the area under the curve was determined by calibration. From the mass of the epoxy resin, the exotherm of the EA-934 was determined as 85.31 cal/gram.

Samples from the same batch of adhesive were allowed to cure for five days and the scan was run again as shown in Figure 15.

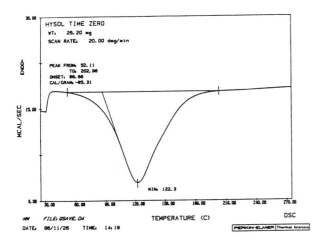

Fig. 14 - DSC Scan for the Epoxy Resin

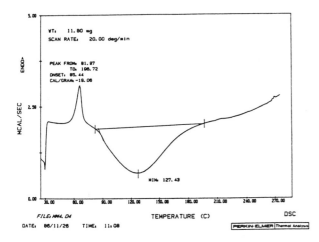

Fig. 15 - DSC Scan After Five Day Cure

The residual cure (exotherm) was determined to be 19.06 cal/gram.

For the sample, post cured at 200 deg. F, the residual cure was determined to be .42 cal/gram. The residual cure for the sample post cured at 350 deg. F was negligible which indicates that the reaction had gone to completion.

The extent of cure can be calculated from the DSC results. After five days at room temperature, the EA-934 was found to be about 75% reacted. Following the 200 deg. F post cure, the extent of reaction was almost complete. The evaluation of extent of cure by the DSC method reinforces chemically what has been observed on the macroscopic level.

CONCLUSIONS

Based on the previous results and discussion presented in this work, the following conclusions can be made:

1. The vendor reported shear stress data at elevated equilibrium temperatures are reliable for the range of temperatures investigated.

2. The shear stress vs. temperature profile of this adhesive obeys a logarithmic relationship for equilibrium heating conditions.

3. The test specimen configuration affects the magnitude of the shear stress vs. temperature relationship as evidenced by a comparison of the thin and thick lap shear data.

4. The reduction in the thin lap shear stress values compared to the thick lap shear values can be explained by the mixed mode of failure common to the thin lap specimens. Evidence of this was shown in the SEM photomicrographs.

5. Increasing the heating rates of the test specimens above the equilibrium rate causes a drastic reduction in the shear stress values at the same test temperatures. Additionally, under dynamic heating conditions, the shear stress vs. temperature profile no longer obeys a logarithmic relationship which suggests that the mechanism of the failure process has changed.

6. The increased shear stress values of the equilibrium conditioned specimens over the dynamic heating specimens is due to the increase in the extent of chemical reaction or degree of cure of the epoxy resin. This was confirmed by the differential scanning calorimetry (DSC) analysis of post cured samples.

7. When selecting an adhesive for a particular application involving elevated temperatures, it is important for design engineers to use test data which most closely simulate the end use heating conditions.

ACKNOWLEDGMENT

The authors would like to acknowledge the support of the Raytheon Company in this work.

REFERENCES

1. Shechter, L., Wynstra, J., Kurkjy, R., Industrial and Engineering Chemistry, Vol. 48, No. 1, 94-97, (1956).
2. Aker, S.C., Journal of Applied Polymer Science, Symp. 32, 313-314, (1977).
3. Marceau, J.A., McMillan, J.C., Boeing Commercial Airplane Co., Seattle, WA, Tech. Report D6-41317-1.
4. Lin, C.J., Bell, J.P., Journal of Applied Polymer Science, Vol. 16, 1721-1733, (1972).
5. Fava, R.A., Polymer, Vol. 9, 137-151, (1968).
6. Crane, L.W., Dynes, P.J., Kaelble, D.H., Journal of Applied Polymer Science, Polymer Letters Ed., Vol. 11, 533-540, (1973).
7. Technical Data Sheet, EA-934, Epoxy Adhesive, Hysol Company, Pittsburg, California.

STANDARD DATA BASES REQUIRE STANDARD TEST PROCEDURES

Gary E. Hansen
Hercules Aerospace
Magna, Utah   USA

IT IS IN EVERYONE'S BEST INTERESTS to define and use a standard data base. The advanced composites industry, in conjunction with various government agencies, are participating in the writing of Mil-HDBK-17, Polymer Matrix Composites. Part of this effort is to define what properties of composites should be measured, and how this data should be presented. This endeavor will result in considerable cost savings if it is adopted throughout industry. The "proof of the pudding" will be measured only by whether the major airframe suppliers use and require composite test data which has been obtained and reported in accordance with Mil-HDBK-17 guidelines.

The subject discussed herein is complimentary to this effort. What I would like to discuss is just how important - and just how complex - mechanical testing of advanced composite materials is. I will cite several examples of observations I have made over years of progressing down the learning curve.

FINITE DEFINITION FOR MODULUS CALCULATION

A typical stress-strain curve on unidirectional carbon fiber laminates is non-linear (Figure 1).

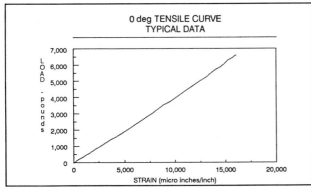

FIGURE #1

Interestingly, the modulus gets higher with increasing stress. This creates a problem when trying to report the modulus. The calculated modulus is highly dependant upon the method used to obtain the slope of the curve (Figure 2). As can be seen, a difference of 8 Msi (or 20%) can be measured on the same tensile curve. Therefore, it is imperative that the method and strain level for obtaining modulus should be standardized.

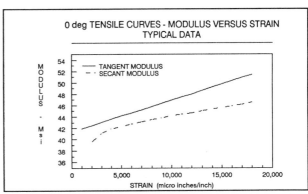

FIGURE #2

A typical ±45° stress-strain curve for carbon fiber epoxy laminates is shown in Figure 3. As can be seen, no initial straight line portion of the curve exists. The shear modulus is defined as:

$$G_{12} = \frac{M}{2 \cdot b \cdot d \, (\epsilon x - \epsilon y)}$$

where   M is Load at strain
          b is Width
          d is Thickness
          ϵx is Strain in x direction
          ϵy is Strain in y direction

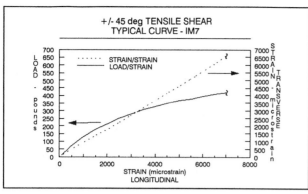

FIGURE #3

Figure 4 shows the myriad range of shear modulus values available from a single curve. The designer must understand what is needed and define what is required when asking for this property. Again, this is a strong argument for a well defined, secant modulus to reduce subjectivity and its inherent error.

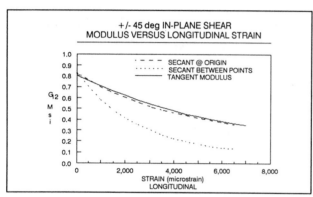

FIGURE #4

Another area of concern is the precision with which modulus is measured. A recent study presented to ASTM showed the hazards of defining modulus as "the initial straight line portion of the stress-strain curve". Sample traces were sent to 8 different laboratories for measurement of the slope of the lines. There were 3 separate curves: Linear, a ±20° Layup; Stiffening, a 0° Layup; Softening, a ±45° Layup (FIGURE 5). The error shown in this type of subjective analysis is not acceptable. We must rely on more precise means for measuring modulus. I would suggest using secant modulii based on two finite points (i.e., 6000 microstrain minus 1000 microstrain) for the following reasons:

1) The "noise" in a stress-strain trace takes place in the first 500 microstrain and is primarily caused by grip seating and alignment.
2) Measuring loads at specified standard strains is much more precise than attempting to draw a tangent line to a non-linear curve.

VARIABILITY OF MEASURING SLOPES OF TENSILE CURVES		
LINEAR (+ -20 deg)	STIFFENING (0 deg)	SOFTENING (+/-45 deg)
+/- 1.4%	+/- 3.4%	+/- 20%

FIGURE #5

EFFECT OF RESIN IN COMPOSITE TENSILE STRENGTH

While tensile strength is a fiber dominated property, it is in fact dictated to some degree by the resin. For example, the tougher (softer) the resin, the higher the strength (Figure 6).

EFFECT OF RESIN ON TENSILE PROPERTIES		
RESIN	TENSILE STRENGTH (ksi)	TENSILE MODULUS (Msi)
TOUGH	527	32.1
NORMAL	500	32.7
BRITTLE	435	32.4

FIGURE #6

EFFECT OF LAMINATE SURFACE CHARACTERISTICS ON TENSILE STRENGTH

A recent study of 0° tensile testing on IM6/3501-6 laminates showed a strength difference of greater than 10% when comparing peel plies used in the fabrication of the laminates (Figure 7). This difference was noted using nominal panel thickness rather than actual thickness to avoid any data shifts due to differential bleeding. An examination of the specimens showed the surfaces to be obviously different (Figure 8). The panels having the higher tensile strength had a consistent resin rich surface while the sub-standard laminates had areas of surface fibers. It is significant to note that the panels having the highest resin bleed had the resin rich surface, which was counter to what would have been anticipated.

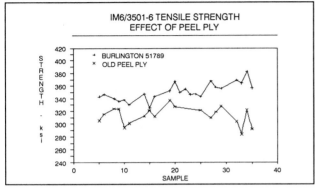

FIGURE #7

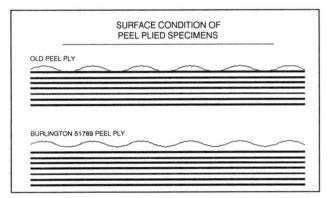

FIGURE #8

A similar observation relating to surface condition was made by another test laboratory. Samples of laminates were grit blasted prior to tab bonding and compared to lightly hand sanded specimens. A reduction in tensile strength of 15% was observed (FIGURE 9). We have seen reductions as high as 30% due to grit blasting test laminates versus peel ply only.

EFFECT OF SPECIMEN MACHINING

A major aerospace contractor used to cut test specimens using a diamond embedded band saw to rough cut and achieving a final finish with a diamond embedded surface grinder. When they changed to a 200 grit diamond embedded wafering saw they measured a 10% increase in tensile strength and 50% decrease in variability (Figure 10).

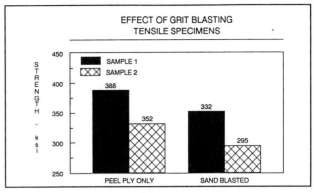

FIGURE #9

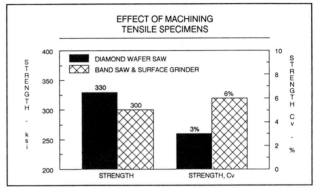

FIGURE #10

COMPRESSION TESTING

There is much we do not yet understand about compression testing - which is why there are dozens of "standard compression tests". In ASTM D3410 alone there are 3 different methods called out, none of which can be relied upon to give comparable test data. Furthermore, most composites companies rely on a different method, which does not even have ASTM approval. We as an industry need to agree on one standard method and to "fine tune" that procedure to allow consistent measurements.

The examples cited have, I hope, identified the need for better defined and controlled test methods. I would like to introduce to you the efforts that SACMA (Suppliers of Advanced Composite Materials Association) is taking to reduce the confusion. SACMA is presently writing recommended test methods for the following tests.

SACMA RECOMMENDED METHODS

SRS 1-88   SACMA Recommended Test Method for COMPRESSIVE PROPERTIES OF ORIENTED FIBER-RESIN COMPOSITES.

SRS 2-88   SACMA Recommended Test Method for COMPRESSIVE AFTER IMPACT PROPERTIES OF ORIENTED FIBER-RESIN COMPOSITES.

SRS 3-88   SACMA Recommended Test Method for OPEN-HOLE COMPRESSION PROPERTIES OF ORIENTED FIBER-RESIN COMPOSITES.

SRS 4-88   SACMA Recommended Test Method for TENSILE PROPERTIES OF ORIENTED FIBER-RESIN COMPOSITES.

SRS 5-88   SACMA Recommended Test Method for OPEN-HOLE TENSILE PROPERTIES OF ORIENTED FIBER-RESIN COMPOSITES.

SRS 6-88   SACMA Recommended Test Method for FLEXURAL PROPERTIES OF ORIENTED FIBER-RESIN COMPOSITES.

SRS 7-88   SACMA Recommended Test Method for INPLANE SHEAR STRESS-STRAIN PROPERTIES OF ORIENTED FIBER-RESIN COMPOSITES.

SRS 8-88   SACMA Recommended Test Method for APPARENT INTERLAMINAR SHEAR STRENGTH OF ORIENTED FIBER-RESIN COMPOSITES BY THE SHORT-BEAM METHOD.

SRS 9-88   SACMA Recommended Test Method for BEARING STRENGTH PROPERTIES OF ORIENTED FIBER-RESIN COMPOSITES.

SRS 10-88   SACMA Recommended Method for CALCULATION OF FIBER VOLUME CONTENT OF COMPOSITE TEST LAMINATES.

SRS 11-88   SACMA Recommended Procedure for CONDITIONING OF COMPOSITE TEST LAMINATES.

These are the tests required in Mil-HDBK-17, standard testing matrix.

These methods were compiled with the following philosophy in mind.

- They are not standards.
  - They will be submitted to ASTM for standardization.
- They are written based on methods "most commonly used in industry".
- All modulii are based on secants taken from two distinct points, avoiding initial "noise".
- They include guidelines for better specimens, (i.e., surface finish criteria, tab compliance).
- They have been agreed upon by 15 + task force member companies.

We are convinced that in the advanced composites industry it is hopeless to pool data until more exact test procedures are followed. This is happening. The payoffs from this endeavor are obvious - the data will be poolable; hence, larger data bases will be available which will result in lower costs of qualification. In addition, reduced costs from test fixtures, paperwork, coordination of variances, and of testing will result from test standardization.

These methods are presently in the final approval loop. After this, round robins are to be conducted to verify repeatability. Methods and results will be available to the public and will be submitted to ASTM in November of 1988.

We feel this effort is a giant step in the establishment of reliable data bases for composite materials. As is evidenced by the examples given, it is impossible to get there without this type of effort.

It is SACMA's intent to reduce costs of raw material. Reduced testing and testing problems will be a major contributor to this effort.

GAS PLASMA AND THE TREATMENT OF PLASTICS

S. L. Kaplan, P. W. Rose, D. A. Frazier
Plasma Science, Inc.
Belmont, California   USA

ABSTRACT

Cold gas plasma, though not a new technology, has only recently begun to be accepted as an important surface preparation method to enhance adhesion and decoration permanency of plastics. During the past several years, acceptance of this technology has been growing rapidly because of its inherent virtues of operator and environment safety. There are no hazardous chemicals to be used or to be disposed of. In addition, the process is very forgiving— it is very rare that materials are damaged or over-treated.

Plasma surface treatment provides surprising efficiency. After plasma treatment, bond strengths using conventional epoxy, urethane or acrylic adhesives usually are limited by the cohesive strength of either the substrate or adhesive. Plasma surface treatment appears to be a "universal" treatment— efficacious with all plastics, polymers and elastomers.

Results on a variety of engineering materials, including polyacetals, liquid crystal polymers, and other engineering thermoplastics are discussed. Typically the treatment results in a several fold increase in bond strength with the locus of failure moved from the surface into the bulk of the engineering material.

BACKGROUND

PLASMA TREATMENT IMPROVES THE ADHESION OF COATINGS AND STRUCTURAL ADHESIVES in two ways: cleaning and activation. Even within the confines of clean rooms in the semiconductor and medical industries, materials to be bonded or decorated are covered with organic contamination. The contamination may be from oil mists in the environment, common in manufacturing areas where there is pneumatic equipment, or transferred by contact with lubricants and hydraulic oils from mechanical equipment. With plastics the contaminant may be from processing additives such as internal lubricants and mold release. These contaminants become weak boundary layers on the surface of the parts. Achieving good bond strengths and permanent coatings is greatly enhanced by removing these contaminants.

Simply put, a plasma is an electrically excited gas consisting of ions, electrons, free radicals, metastables, atoms and with a significant amount of high energy ultra-violet. The mechanics of a plasma and the chemical changes that occur to materials placed within a plasma are complex. It is suggested that those interested in a more through explanation explore the recommended readings appearing at the paper 's end.

PLASMA SURFACE TREATMENT- AN EXAMPLE

In a cold gas plasma four competing processes occur: ablation, chain scission, crosslinking, and activation. Process parameters can favor one or more processes over another. Since plasma treatment is conducted in a vacuum chamber the process conditions are controllable, and thus producing highly reproducible results.

What does occur in a cold gas plasma? Let 's examine the effect of an oxygen plasma on polyethylene. Low molecular weight hydrocarbon contaminants, such as air borne oil mists or wax lubricants, are readily oxidized and removed in the oxygen plasma. The by-products are water, carbon dioxide and low molecular weight fragments. These contaminants are thus quickly removed via the vacuum system and are safely exhausted out onto the roof. The bulk material, long chain polyethylene, is simultaneously attacked by the oxidizing plasma. Its longer chain length, however, shifts the predominant effect on the surface from one of removal to one of substitution. The plasma is efficient at abstracting hydrogen and cleaving chains to generate free radicals where the subsequent substitution of various oxygen functional groups occur. These oxygen moieties increase the surface energy and provide reactive functional groups for interaction with adhesives and decorating media.

In an earlier study on Spectra™ composites[1], the authors showed a very significant improvement in interlaminar shear and flexural strengths of Spectra/ epoxy composites when the fiber or fabric was treated in an oxygen plasma prior to resin impregnation. ESCA (Electron Spectroscopy for Chemical

Analysis) results demonstrated that the top 50 Å of the plasma treated fiber contains an excess of 18% bound atomic oxygen, a significant portion of which is present as hydroxyl and carboxyl functional groups. The carbonyl and hydroxyl groups are reactive with epoxy resins forming covalent bonds between the resin and the fiber. This covalent bonding provides greater adhesion of the fiber resin interface than can be accomplished by simple wetting. Molded or extruded polyethylene can be expected to respond similarly to plasma.

Interestingly, the ESCA measurements of the total oxygen content of the untreated Spectra indicated an equivalent amount of oxidation as the plasma treated fiber showed. Since virgin polyethylene should have no oxygen, the results were initially perplexing. The answer came when the samples were subsequently examined by employing derivatization techniques.

Derivatization is done prior to ESCA analysis by exposing materials to be analyzed to functional group specific reagents. One set of samples was immersed in a pentafluoro-hydrazine solution, and a second set immersed in a solution of a fluorinated chloro- silane. The hydrazine reacts with carbonyl but not hydroxyl groups, whereas the reverse is true for the silane reagent. Derivatization of the untreated fiber resulted in the loss of most of its oxygen groups. It quickly became obvious that the oxygen was present as a contaminant— in this case a fatty acid used as part of the fiber sizing composition. Thus, the plasma treatment resulted in not only the oxidation of the surface but also in the removal of nearly all of the weakly bound contaminates. Evidence has also been reported that the treatment of Spectra fiber can increase the surface area by a micro-etching of the amorphous areas of the plasma[2].

[1] Kaplan, Rose, Nyugen, and Chang, SAMPE Proceedings, 1988.

[2] Nguyen, et al, SAMPE Proceedings, 1988.

THE MECHANICS OF TREATMENT

In a cold gas plasma the plastic to be treated is bathed in an gas activated by RF energy. This activated gas (plasma) contains free electrons, ions and other activated species which bombard the plastic causing various covalent bonds to rupture. As discussed earlier, hydrogen abstraction and chain scission are the predominant reactions resulting in free radicals at the ruptured bond sites. These free radicals are thermodynamically unstable and and quickly react with other species in the plasma adding new or additional chemical groups to the polymer.

The nature of the chemical groups can be varied by the gas selected for the plasma. For example, if carboxylic acid and carbonyl functional groups are desired, as in the example above, an oxygen plasma would by employed. However, if amino functional moieties are desired, an ammonia plasma may be preferred.

Cold gas plasma treatment is conducted at reduced pressure in a vacuum chamber; thus all process conditions are controlled, providing excellent reproducibility. Plasma equipment is available for automated, continuous plasma treatment of fiber, film and for the batch treatment of discrete parts.

EXPERIMENTAL

The effect of the plasma treatment on adhesion strength was evaluated by testing the lap shear strength of two different commercial adhesives: 3M Scotch-Weld #3549 urethane and Scotch-Weld #2216 epoxy adhesives. Both systems are two component room temperature curing adhesives. The engineering plastics were supplied by their manufacturers as injection molded dog bone tensile specimens. Each specimen was cut mid-way through the gauge length, plasma treated (except for the control) in a Plasma Science PS0500 treatment system, and bonded with the referenced adhesives to provide a 1.27 x 1.27cm (1/2 x 1/2 in) overlap. Lap shear tests were conducted on a universal testing machine in a tensile mode, using a cross-head speed of 1.27 cm/minute (0.5 inch/minute). Values are the average of at least 3 specimens at each condition. Control specimens, i.e. not plasma treated, were solvent wiped with ethanol to remove contaminants before bonding.

Since new materials or process chemistries may produce a different chemistry in the plasma, it is often difficult to predict what will be the exact effect of a new process on the surface. Thus it is natural that an empirical approach to investigating new processes is used. For this set of trials two plasma processes (gases) are usually chosen as starting points: Oxygen, because it has proven to be of excellent value in prior work; and ammonia, because it is reported to aminate[3] the surface of polymers exposed to its plasma. Since the epoxy and urethane adhesive have functional activity toward both amino and carboxylic groups, these gases were of great interest.

3 Allred, Merrill, and Roylance, Proceedings of the ACS Symposium on Composite Interfaces, 1983.

DISCUSSION OF RESULTS

Without exception, at least one of the treatments provides significant improvement of adhesive lap shear strength over the control (Table I).

Upon close examination of the data, two types of cohesive failure was observed. The desired failure mode was tensile fracture of the adherents (plastic dog bones). Such a failure was obtained with each engineering material after plasma treatment, however, the corresponding stress on the plastic adherent at the time of cohesive failure was calculated to be 8,000 to 10,000 psi, some what lower than published ultimate tensile strength values.

Unfortunately, uncut samples were not retained for tensile testing and so independently values were not measured in this study. Since we do not expect that the tensile strength will differ much from the published values we attribute the differences to stress concentrations due to single lap shear geometries.

Plasma cleaned aluminum to aluminum lap shear joints were made as a control for each adhesive. Lap shear strengths of 2398 psi and 1700 psi were obtained for the epoxy adhesive (3M#2256) and for the urethane adhesive (3M#3549), respectively. These values are in excellent agreement with those reported for epoxy in the literature, 2350 psi, but somewhat lower than the 2000 psi reported for the urethane. Since no material failures were observed from within the adhesive, this shortfall had little effect on the results.

Comparison of plasma treatment gases within one adhesive type results, in general, in great improvements in adhesion. Lexan™ is an exception. A significant decrease in the adhesive strength with both adhesives occurs when ammonia plasma was employed. Perhaps the ammonia plasma generates a weak boundary on the surface of the Lexan, limiting adhesion.

Delrin™ gave uncharacteristically low adhesion results. At the adhesive failure load of 2.79 Mpa (724 psi) the stress in the adherent was only 11.16 Mpa (1622 psi), far below its reported tensile strength and the bond failed adhesively. Delrin™ is known to have at least one degradation mechanism which involves depolymerization evolving formaldehyde. If this occurs the depoymerization would result in removal of modified polymer thereby negating the effects of treatment. Considerable room exists for improvement and work on this system continues.

SUMMARY

The data presented clearly shows the significant improvements in adhesive bond strengths for engineering thermoplastic materials which have been plasma surface treated. The treatment is equally effective at imparting better wetting properties, increasing coating quality, providing dramatic interfacial fiber adhesion, and generally improving the composite performance of fibers which have been treated prior to impregnation.

The importance of the technology lies in its ability to economically treat a broad range of materials effectively, cleanly, and with low operator sensitivity. Additionally, the option of tailoring surface chemistry for application specific systems allows a means of optimization which usually results in material failure within the adherand.

TABLE I

Lap Shear Strength Vs. Adherent Pre-Treatment

MATERIAL	TREATMENT	LAP-SHEAR STRENGTH [MPa (psi)]			
		EPOXY	failure mode	URETHANE	failure mode
Valox™ 310	none	3.59 (522)	adh.	1.32 (192)	adh.
polyester resin	oxygen	**11.33 (1644)**	**adher.**	1.48 (215)	adh.
	ammonia	9.78 (1420)	adher.	**6.58 (955)**	**adh.**
Noryl™ 731	none	4.25 (617)	adh.	1.54 (223)	adh.
polyphenylene	oxygen	10.23 (1485)	adh.	**13.02 (1890)**	**adher.**
ether	ammonia	**12.40 (1799)**	**adher.**	6.63 (962)	adh.
Durel™	none	1.72 (250)	adh.	0.83 (125)	adh.
polyaryl ester	oxygen	**14.89 (2161)**	**adher.**	**6.13 (890)**	**adh.**
	ammonia	13.65 (1981)	adher.	2.34 (340)	coh.
Vectra™ A625	none	7.17 (939)	adh.	0.87 (191)	adh.
liquid crystal	oxygen	**11.01 (1598)**	**adher.**	6.72 (976)	adher.
polymer	ammonia	8.54 (1240)	adher.	**7.22 (1048)**	**adher.**
Ultem™ 1000	control	1.27 (185)	adh.	n/a	
polyetherimide	oxygen	13.36 (1939)	adher.	n/a	
	ammonia	**14.17 (2056)**	**adher.**	**13.32 (1933)**	**adher.**
Lexan™ 121	control	11.74 (1705)	adh.	3.74 (537)	adh.
polycarbonate	oxygen	**15.45 (2242)**	**adher.**	**7.83 (1137)**	**coh.**
	ammonia	4.73 (686)	adh.	3.16 (458)	coh.
Delrin™ 503	control	1.14 (141)	adh.	0.22 (32)	adh.
polyacetal	oxygen	**4.46 (743)**	**adh.**	**9.30 (1350)**	**adh.**
	ammonia	3.97 (576)	adh.	2.70 (393)	adh.

adher. = cohesive failure of adherent
coh. = cohesive failure within the adhesive
adh. = adhesive failure

ACKNOWLEDGMENTS

The authors would like to acknowledge Chuck Orlando of Plasma Science for his assistance in sample preparation, and Robert Cormia of Surface Science Labs, Mountain View, California for his expert aid with reference to ESCA characterization.

REFERENCES

Kaplan, S. L., Rose, P. W., Nguyen, H. X., Chang, H. W., "Gas Plasma Treatment of Spectra Fiber", SAMPE 33rd International Conference, Anaheim, CA, March 1988.

Nguyen, H. X., Riahi, G., Wood, G., Poursartip, A., "Optimization of Polyethylene Fiber Reinforced Composites using a Plasma Surface Treatment" SAMPE 33rd International Conference, Anaheim, CA, March 1988

Allred, R. E., Merrill, E. W., and Roylance, D. K., "Surface Aminated Polyaramid Filaments", Proceedings of the ACS Symposium on Composite Interfaces, Seattle, WA, 1983.

Chapman, B., Glow Discharge Processes, John Wiley & Sons, New York, 1980.

Hollahan, J. R. and Bell, A. T., Techniques and Applications of Plasma Chemistry, John Wiley & Sons, Inc., New York, 1974.

PLASMA TREATMENT OF COMPOSITES FOR ADHESIVE BONDING

John G. Dillard, Ionel Spinu
Chemistry Department
Virginia Tech
Blacksburg, Virginia 24061   USA

Abstract

Sheet molded composites based on different polymers (polyester, polycarbonate, polyurea) were treated in an oxygen (O_2) and an argon (Ar) radio frequency plasma to alter the surface chemistry and topography. The plasma treatment increased the oxygen-containing functional groups on the SMC surface when each gas was used. The SMC surfaces became rougher as a result of chemical reactions and sputtering processes. Adhesive bonding and testing of lap shear specimens were carried out for polyester based SMC. Adhesive bonding was enhanced for the plasma treated samples. The lap shear mode of failure is dependent on the nature and the time of pretreatment. Characterization of the treated and failed surfaces was accomplished using X-ray photoelectron spectroscopy, scanning electron microscopy, and photoacoustic infrared spectroscopy.

Introduction

Surface treatment of polymers in a gaseous plasma enhances adhesion (1-4). The principal changes produced by exposing a polymer to a plasma occur in surface wettability, molecular weight of the surface layer, and the chemical composition of the surface (3). It was found that the plasma treatment of polyethylene improved its adhesive properties (1,2,5). Similar results were found for other polymers such as PTFE (4).

Argon plasma treatment generally introduces oxygen functionalities into the polymer structure and nitrogen plasma treatment incorporates nitrogen and oxygen functionalities into the surface (6). Generally the bulk properties of the treated polymers are unaltered by plasma treatment,

the effect of plasma treatment being confined to a surface layer 1-10μm in depth (3,4).

The objective of this research was to obtain fundamental surface chemical and physical information regarding the alterations that accompany the plasma surface pretreatment and the formation of adhesive substrate bonds in composites. Different types of SMC (polyester, polycarbonate, polyurea) were plasma treated under various conditions. Treated polyester-based SMC samples were bonded with a urethane adhesive and tested using lap shear specimens. Improved adhesion was found for plasma treated SMC samples. The failed sample surfaces were studied in order to explore the failure mechanism. Surface chemistry was evaluated using X-ray photoelectron spectroscopy (XPS) and photoacoustic infrared spectroscopy, and alterations in surface topography were monitored by scanning electron microscopy (SEM).

Experimental

Oxygen and argon plasma treated SMC

Coupons for XPS studies were cut to approximately 15 x 17 mm. The samples were handled in such a manner that contamination of SMC surfaces was avoided. A Plasmod (Tegal Corporation) was used to treat the samples. All plasma treatments were accomplished at 50W and 13.56 MHz. The treatment times were 1, 5, 10 and 30 min. The SMC treated samples were characterized by XPS, SEM, and FTIR.

XPS surface analysis was accomplished using a Perkin-Elmer PHI Model 5300 X-ray photoelectron spectrometer. Photoelectrons were generated using Mg K_α radiation ($h\nu$ = 1253.6 eV). Ejected photoelectrons were analyzed in the hemispherical analyzer and detected using a position sensitive detector. The area of the specimen sampled by the analyzer electron optics was approximately 2.0 x 10.0 mm. The spectra were accumulated under control of a Perkin-Elmer Model 7500 computer. In the presentation of the elemental analysis results photoelectron spectral areas were measured and subsequently scaled to account for ionization probability and an instrumental sensitivity factor to yield results which are indicative of surface concentration in atomic percent. The precision and accuracy for the concentration evaluation are about 10% and 15%, respectively. The XPS results represent an average of two separate determinations on two samples from different plasma treatments, except for those marked with an asterisk (*) in the table which represent only one measurement. The energy scale was calibrated by setting the CH_n carbon 1s binding energy at 285 e.v. (7)

Scanning electron micrographs were obtained using an ISI SX-40 scanning electron microscope. The SMC samples were coated with a film of sputtered gold. The infrared spectra were obtained using a Nicolet 5DBX FTIR spectrometer and a MTEC Model 100 photoacoustic cell.

Oxygen and plasma treated/bonded SMC

Polyester based SMC samples were bonded using a two component urethane adhesive obtained from Ashland Chemical Co., Ashland, OH. SMC coupons 2.5 x 10 cm (1" x 4") were cut and treated in an oxygen or argon plasma for 5, 10, and 20 min. The SMC coupons were treated in a variety of ways prior to bonding as noted a) non-treated SMC sample bonded in "as received" condition, b) oxygen or argon plasma treated and, c) primer treated SMC sample wiped three times with a Kim-wipe tissue saturated with an isocyanate primer solution (Ashland Chemical Co.)

Lap shear specimens were prepared following the treatment procedures of a, b, and c noted above. The bonded area was 1 in^2 (1" x 1"). Bond thickness was controlled by addition of 0.030 in. glass beads to the bonded area. The bonding of SMC specimens was done immediately after pretreatment. After bonding the samples were cured one hour at room temperature and then for 30 min at 150°C (302°F) in an oven and excess adhesive was removed.

Lap shear measurements were obtained using an Instron apparatus (Model 1123) operating at a cross head speed of 1.27 cm/min. with a thermally controlled testing chamber (Model 3116). The samples were conditioned at 82°C (180°F) for 30 min before testing and were tested at 82°C (180°F). At least five measurements were made for each kind of sample.

In the failure experiments, two specimens were obtained one specimen surface was principally adhesive and the other SMC. From the failure specimens, samples were cut for XPS and SEM studies. For these measurements the side showing bulk adhesive is designed the adhesive side (a) and the corresponding SMC side, (b).

Results and Discussion

Oxygen and Argon Plasma Treated SMC

XPS Results

X-ray photoelectron spectroscopic analysis results for argon and oxygen plasma treated samples are summarized in Tables 1a, 1b, and 1c. The most important results are that following the plasma treatment the percent oxygen increases dramatically and the carbon percent decreases for all three types of composite. The concentration of oxidized carbon constituents on the surface (-COR, >C=O, -CO_2R) increases and the CH_n content decreases for all treated samples. In the case of polyester based SMC an increase in the percent of Zn, Ca, Mg, and Al is found after treatment, and an increase in nitrogen content is found for oxygen plasma treated samples. The increase for metallic elements is higher for samples treated in the oxygen plasma than for samples treated in the argon plasma. For polyurea based composite the plasma treatment produced an increase in Na, Si, and N content. An increase in Na content was observed after plasma treatment of polycarbonate based materials.

The treatment of SMC in the plasma alters the chemical nature of surface carbon. When the C 1s spectra of the plasma treated samples are compared with those for the untreated samples it is noted that the intensity for carbon photopeaks with binding energies at 286.5, 288.0, and 289.1 eV increases. Carbon at these bonding energies corresponds to -COR, >C=O and -CO_2R type species. The increase in these carbon species is accompanied by a relative decrease in CH_n carbon content (BE = 285.0 eV). Representative C 1s spectra for plasma treated samples and untreated samples are shown in Fig. 1a, 1b, and 1c.

To investigate the changes in the surface chemistry with aging the plasma treated samples were stored under vacuum (10^{-6} torr) and were measured again using XPS after different time intervals, the results being compared with those obtained by measuring the samples immediately after the plasma treatment. Generally an increase in carbon concentration and a decrease in oxygen concentration was found. For example, a 24 hrs aging produced a decrease in oxygen content in the case of polyester, polyurea, and polycarbonate samples treated 10 min.

with oxygen plasma from 35.2%, 36.8%, and 28.2% to 31.1%, 32.9, and 23.9%, and an increase in carbon content from 53.5%, 48.5%, and 67.0% to 57%, 54.2%, and 73.4%, respectivelty. Generally the aging of the plasma treated samples produces a decrease in intensity of the peaks associated with -COR, >C=O, and -CO$_2$R species. The concentrations of the other elements remained essentially unchanged.

SEM Results

SEM photomicrographs of the untreated samples show relatively uniform and smooth surfaces. SEM micrographs of the treated samples show roughened surfaces (Fig. 2a, 2b, and 2c).

FTIR Results

The infrared spectra of the plasma treated samples are equivalent to those for the untreated samples except for the polyester based SMC samples treated in an oxygen plasma for 20 and 30 min. This result suggests that only in the case of the polyester based SMC are changes in the surface chemistry indicated by XPS profound enough to be detected by photoacoustic infrared spectroscopy. Comparing the spectra of the polyester samples treated the oxygen plasma for 20 min and 30 min with those of the untreated samples, an increase in the following peaks is observed 1441 cm^{-1}, (CO$_3^{2-}$, carboxylic acids, aldehyde); 1795 cm^{-1} (ester, aldehyde, ketones, carboxylic acids, ethers); 2516 cm^{-1} (ketones). These findings can be seen in Fig. 3 for polyester based SMC where spectra for as received SMC and SMC treated for 5 and 30 min in the oxygen plasma are compared. These findings correlate with XPS data which showed an increase in Ca concentration and in the -COR, >C=O, -CO$_2$R species for oxygen plasma treated samples.

Oxygen and Argon Plasma Treated/Bonded Specimens

The lap shear test results are summarized in Table 2. The average failure force for the untreated samples is low (~0.4 MPa) and the failure mode is 100% adhesive. The samples treated with primer gave good values for the failure force (~2.4 MPa). The failure mode was mixed. The oxygen plasma treatment for 1 min and 5 min produce only moderate increases in failure force, giving 1.1 and 0.8 MPa, respectively, but the failure mode remained 100% adhesive. The 10 min oxygen plasma pretreated samples showed a much larger average failure force, 1.7 MPa and showed a mixed mode failure. In the case of the samples treated with oxygen plasma for

20 min and 30 min the average failure force was 2.6 MPa and 2.5 MPa, respectively and mixed mode failure was found.

The treatment of SMC samples with argon plasma also produced an appreciable increase in the failure force with values at 1.2, 1.2, and 1.7 MPa for samples treated in the plasma for 5, 10, and 20 min, respectively. The failure mode was 100% adhesive for all samples.

The failed sample surfaces were examined by XPS and SEM confirmed the failure modes of the samples. In the case of samples for 10, 20, 30 min. in the oxygen plasma the shape of C 1s spectra for the SMC side is different from that of the treated unbonded SMC samples. The results indicate the presence of adhesive on the SMC side. The SEM micrographs of the same samples show particles of adhesive on the SMC failed side.

Summary and Conclusions

The plasma treatment of SMC produces important changes in physical and chemical properties of the SMC surface. The plasma treatment of SMC transforms the relatively smooth surface of SMC into a rough surface, a fact revealed by SEM.

The most important change in chemical properties of the SMC surface is an increase in oxygen percent after plasma treatment and a decrease in carbon concentration. Accompanying the change in oxygen content is an increase in the concentration of oxidized carbon constituents on the surface -COR, >C=O, -CO$_2$R, and a decrease in CH$_n$ content.

Lap shear test results show that the plasma treatment of SMC surfaces promotes adhesive bonding using a urethane adhesive.

The important increase in the adhesive properties of SMC after plasma treatment may be explained by the changes in the physical and chemical properties of SMC surfaces and reflected in a roughening of the surface, and an increase in the polarity of functional groups.

Acknowledgments

The authors gratefully acknowledge the Ashland Chemical Co. and General Motors for financial support. Instrumentation used in this study was purchased with funds provided by the National Science Foundation, the Virginia Center for Innovative Technology, and the Commonwealth of Virginia.

References

1. D. M. Brewis, D. Briggs, Polymer, 22, 7 (1981).

2. B. Westerlind, A. Larsson, M. Rigdahl, Int. J. Adhesion and Adhesives, 7, 141 (1987).

3. J. R. Hollahan, A. T. Bell, **Techniques and Applications of Plasma Chemistry,** John Wiley, New York, (1974) pp. 116-140.

4. L. Mascia, G. E. Carr, P. Kember, 3rd International Conference, Adhesion 1987, York University, p. 22/1.

5. D. Briggs, C. R. Kendall, Int. J. Adhesion and Adhesives, 2, 13 (1982).

6. H. Yasuda, H. C. Marsh, S. Brandt, C. N. Reilley, J. Polym. Sci. Polym. Chem. Ed. 15, 991 (1977).

7. D. Briggs, "Application of XPS in Polymer Technology", in **Practical Surface Analysis by Auger and X-Ray Photoelectron Spectroscopy,** D. Briggs and M. P. Seah, Editors, Chapt. 9, John Wiley, New York (1983).

8. J. G. Dillard, C. E. Burtoff, Surface Characterization of Sheet Molded Composite (SMC) as Related to Adhesion. **Proceedings: 2nd Conference on Advanced Composites,** Dearborn (1986).

9. J. G. Dillard, C. Burtoff, F. Cromer, A. Cosentino, T., Rabito, G. MacIver, J. Saracsan, D. Cline, "Surface Chemistry and Adhesion in Sheet Molded Composite (SMC)". **Proceedings: 3rd Conference on Advanced Composites,** Detroit (1987).

10. C. Burtoff, **Surface Analysis of Sheet Molded Composite (SMC) Material as Related to Adhesion.** MS Thesis, VPI & SU, 1985.

Table 1a Plasma Treated Polyester Based SMC

(atomic percent)

Element	Untreated sample	Argon plasma				Oxygen plasma			
		1 min*	5 min	10 min	30 min*	1 min*	5 min	10 min	30 min*
C	82.75	66.10	63.60	63.02	61.37	59.58	58.19	53.43	27.93
O	14.20	30.50	32.00	31.96	31.06	35.34	35.77	35.20	44.21
N	<0.1	0.55	0.38	0.35	0.55	0.32	1.15	1.55	2.26
Si	0.40	0.58	0.48	0.60	0.44	0.29	0.18	0.34	1.20
Zn	0.38	0.44	0.67	0.81	1.22	0.84	0.67	1.28	3.86
Ca	0.30	0.45	0.60	0.53	0.90	0.69	0.40	1.16	4.55
Mg	1.09	1.00	1.82	2.11	3.16	2.07	2.70	5.34	10.42
Al	0.76	0.38	0.43	0.61	1.12	0.86	0.91	1.26	4.08
Na	<0.1	<0.1	<0.1	<0.1	<0.1	<0.1	<0.1	<0.1	1.47
Cl	<0.1	<0.1	<0.1	<0.1	0.17	<0.1	<0.1	<0.1	<0.1

* Measurements on one sample only.

Table 1b Plasma Treated Polycarbonate Based SMC

(atomic percent)

Element	Untreated sample	Argon plasma				Oxygen plasma			
		1 min	5 min	10 min	30 min	1 min	5 min	10 min	30 min
C	80.52	70.96	67.60	67.18	67.67	72.98	70.17	69.97	67.90
O	19.05	27.88	30.72	30.61	30.35	26.00	28.43	28.17	29.07
N	<0.1	0.34	0.53	0.13	0.30	0.60	0.53	0.34	0.31
Si	0.30	0.39	0.51	0.60	0.19	0.20	0.18	0.25	0.36
Zn	<0.1	<0.1	<0.1	0.26	<0.1	<0.1	<0.1	<0.1	<0.1
Ca	<0.1	<0.1	<0.1	<0.1	<0.1	<0.1	<0.1	<0.1	<0.1
Mg	<0.1	0.15	<0.1	<0.1	<0.1	<0.1	<0.1	<0.1	<0.1
Al	<0.1	<0.1	0.31	<0.1	<0.1	<0.1	<0.1	<0.1	<0.1
Na	<0.1	0.44	0.72	1.00	1.47	0.20	0.67	1.27	2.30

Table 1c Plasma Treated Polyurea Based SMC

(atomic percent)

Element	Untreated sample	Argon plasma				Oxygen plasma			
		1 min	5 min	10 min	30 min	1 min	5 min	10 min	30 min
C	67.41	55.02	55.33	54.60	54.40	58.02	56.18	48.48	42.42
O	22.70	34.01	32.72	33.64	32.94	31.45	32.14	36.77	40.0
N	5.15	5.81	6.57	6.50	6.51	6.80	6.99	7.88	7.46
Si	3.75	3.66	2.92	3.64	3.55	2.42	6.24	5.37	7.31
Zn	0.13	<0.1	0.20	0.11	<0.1	0.13	<0.1	<0.1	<0.1
Ca	0.16	<0.1	0.12	0.22	<0.1	<0.1	<0.1	<0.1	0.12
Mg	<0.1	<0.1	<0.1	<0.1	<0.1	0.20	<0.1	<0.1	<0.1
Al	<0.1	<0.1	<0.1	0.10	<0.1	<0.1	<0.1	<0.1	0.17
Na	0.53	1.30	2.20	1.99	2.56	0.90	1.41	1.95	2.47

Table 2 Lap Shear Test Results

	Untreated sample	Argon plasma			Oxygen plasma					Primer Treated Sample
Time (min)		5	10	20	1	5	10	20	30	
Failure Force (MPa)	0.4	1.2	1.2	1.7	1.1	0.8	1.7	2.6	2.5	2.4
Failure Mode	100% Ad	100% Ad	100% Ad	100% Ad	100% Ad	100% Ad	Mixed	Mixed	Mixed	Mixed

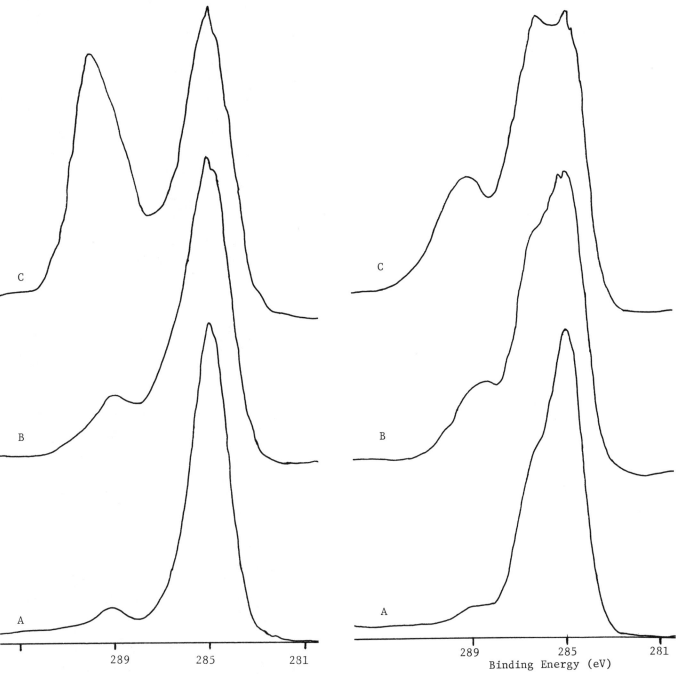

C

B

A

289   285   281

Binding Energy (eV)

Fig. 1a C 1s Spectra for Polyester
   A: As Received SMC
   B: Argon Plasma Treated SMC 30 min
   C: Oxygen Plasma Treated SMC 30 min

C

B

A

289   285   281

Binding Energy (eV)

Fig. 1b C 1s Spectra for Polyurea
   A: As Received SMC
   B: Argon Plasma Treated SMC 30 min
   C: Oxygen Plasma Treated SMC 30 min

As Received SMC

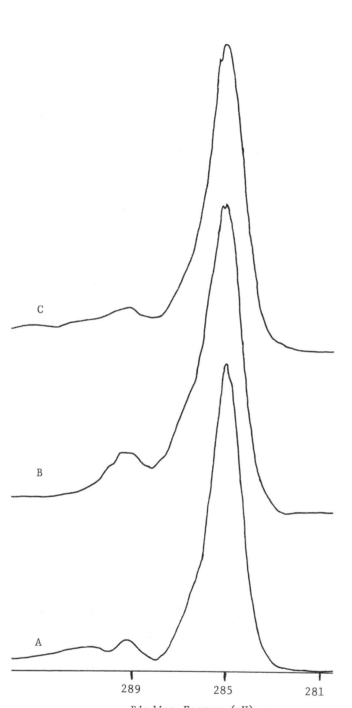

Binding Energy (eV)

Fig. 1c C 1s Spectra for Polycarbonate
A: As Received SMC
B: Argon Plasma Treated SMC 30 min
C: Oxygen Plasma Treated SMC 30 min

Oxygen Plasma Treated SMC 30 min

Fig. 2a SEM Photos for Polyester

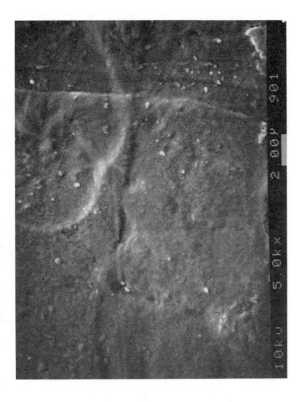

As Received SMC

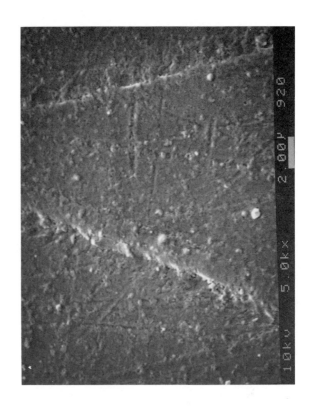

As Received SMC

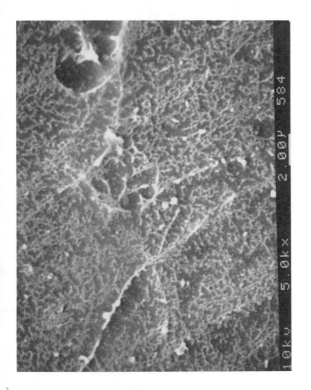

Oxygen Plasma Treated SMC 30 min

Fig. 2b SEM Photos for Polyurea

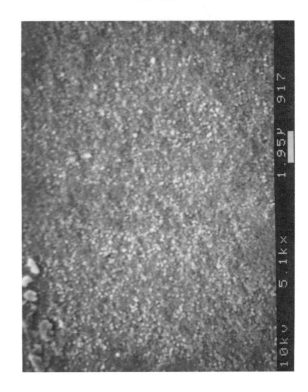

Oxygen Plasma Treated SMC 30 min

Fig. 2c SEM Photos for Polycarbonate

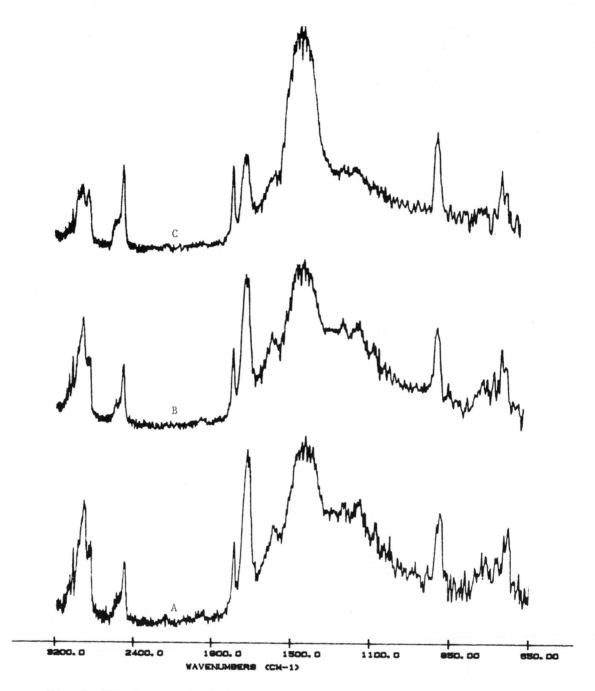

Fig. 3 FTIR Results for Polyester

     A: As Received SMC
     B: Oxygen plasma treated SMC 5 min
     C: Oxygen plasma treated SMC 30 min

JOINING SECTIONS OF THICK FIBERGLASS LAMINATE

N. Keith Young
FMC Corporation
Ground Systems Division
Santa Clara, California USA

C. C. Chen
FMC Coporation
Central Engineering Laboratories
Santa Clara, California USA

W. L. Cheng
FMC Corporation
Central Engineering Laboratories
San Jose, California USA

Joining Sections of Thick Fiberglass Laminate

ABSTRACT

Thick laminated composite hulls for armored vehicles are being developed to provide lighter, quieter vehicles that will meet all service requirements. Making large vehicles in one piece places severe constraints on manufacturing. Qualified joints make it possible to fabricate the vehicle in partial sections.

In this study, several designs of thick laminated joints are described. Two joints were studied, analyzed by the finite element method, and tested to validate their performance. Further investigation is being conducted to verify they would carry the required loads under all possible conditions.

Test programs were used to determine data for use in the finite element models for both the adhesives and composites. Standard failure tests, tests after exposure to long periods of water immersion, to high and low temperatures, and to other conditions are described in this paper.

A high-performance adhesive was selected to withstand loads that caused failure in the composites before the adhesives failed in nearly all tests. The joint design is expected to meet all of our requirements.

COMPOSITE STRUCTURES CAN BE very large. This makes it necessary to divide the assembly into sections with connecting joints. Composites for ballistic protection must be thick. Joining such thick composite sections to each other requires a joint that when placed in tension causes each part to align as the joint deforms so that bending will be minimized. Any bending results in peel stresses for bonded parts. Adhesives are excellent in shear but have a low tolerance for peel stresses. Yet, bonded structures have many advantages if a successful design can be found.

Study objectives were to establish designs:
- Capable of high loads.
- Resistant to chemical degradation.
- Resistant to ballistic attack.
- Resistant to environmental heat, cold, fungus, and etc.

Two joints were designed, analyzed, and tested in tension and bending. Joints have so far been tested for resistance to high and low temperature extremes and to water immersion for over six days.

DISCUSSION

Advantages and disadvantages are found when thick laminate joints formed by adhesives are compared to those made by bolts. The advantages of structural adhesive bonding over mechanically joined structures are:
- Improved fatigue resistance.
- Resistance to shock.
- Resistance to vibration.
- Resistance to thermal cycling.
- Increased or inhibited heat transfer.

Any part of the outer shell on an armored personnel carrier must protect the personnel from an expected ballistic threat. This is also true of any structural joint that is a part of this outer shell. Joints without bolts have an advantage in ballistic attack. Any projectile striking a bolt can shear off the head and the shank can continue as a secondary projectile to injure personnel or damage any objects it strikes. Bonded joints furnish no internal projectiles but must stop the projectile.

The disadvantages of structural adhesive joints are possible bond degradation by heat, moisture, poor joint preparation, and chemical attack.

Once bonded, joint integrity cannot visually be determined nor can the joint be disassembled for inspection. However, non-destructive testing can determine if there are voids in the bond.

Stress analysis is more difficult to perform for bonded joints than for mechanical joints. Critical areas require generation of more finite elements. Different laminate mechanical properties in all three directions must be used. Adhesive mechanical properties, though the same in all directions, must also be determined and used.

Future tests are planned to evaluate ballistic performance, fatigue, chemical degradation, tolerance for adhesive voids, creep, and allowable variations in bonding preparation.

JOINT DESIGN

Several common joints for composites are shown in Figure 1[1]. Both the simple lap-joint or single-strap joint with one cover plate as shown in Figure 1, beveled or unbeveled, introduce bending when tensile loads are applied. The double-lap joints do not evidence the same bending as simple-lap joints but are not acceptable for exterior appearance.

Ballistic tests of a recessed-metal mechanical joint were not acceptable, thus the bonded double-strap with beveled-cover plates became the preferred joint. A variation of this joint is to replace the butt joint with a scarfed lap. The scarf and butt joints of Figure 2 were selected for testing and analysis. Total joint strength can be varied by changing dimensions of the two cover plates.

Delaware University experiments "...have shown that significant improvements can be made in the efficiency of single lap joints through several modifications. Tests have been conducted in which the adherends were linearly tapered from full panel thickness to practically zero thickness over the length of the bondline in the load direction..."[2] This gave a significant improvement in joint efficiency under either static or fatigue loads.

Beveling the ends of the joint cover plates concentrates the load into fibers near the bond where the best load exchange can take place. The U. S. Army Handbook[1] page 4-34 (see Figure 1) selected a radius at the bondline edge as a better configuration to decrease local stress but such a radius is difficult to machine. Angles less than 45 degrees also are difficult to cut. Radiuses and shallow angle cuts both adversely affect producibility.

Increasing the thickness of the adhesive or adding rubber to the adhesive to decrease stiffness in the end of the lap area are additional approaches to reduce stress in critical areas.

Both butt and scarf joints were found to have good structural characteristics; they are simple to make, and easy to inspect before bonding to assure a good fit. The latter feature helps to lower cost.

MATERIAL SELECTION AND TESTING

The composites were fabricated from S-2 polyester preimpregnated cloth (prepregs). Glass content of the laminate was maintained at around 70 percent by weight. The composite joint was made with 64- and 32-ply laminates for the main and cover plates, respectively.

American Cyanamid FM 73 film adhesive with some rubber content was used to make the joints. Properties of this film adhesive were determined by standard laboratory testing and are listed in Table 1. The adhesive shear properties as a function of the shear strain and temperature were compiled from test data provided by the adhesive supplier.

Joints were tested to the following specifications:

Tensile Properties: ASTM D638.
Adhesive strength: ASTM D2095,
ASTM D2094.

Fracture toughness: ASTM D3433,
 contoured double-cantilever beam
 (CDCB).

Table 1 - Adhesive Property
 Test Results (Average)

Tensile Properties (ASTM D638)

E	350,000 psi	(Deviation	18,000)
T	6,800 psi	(Deviation	210)
e	3.3 %	(Deviation	0.4)

Strength (ASTM D2094)

T	5,200 psi	(Deviation	600)

Fracture Toughness (ASTM D3433)*

G_{1c} 660 lbf/in. (Deviation 42)
G_{1a} 560 lbf/in. (Deviation 28)

Shear Strength 5,900 psi
Shear Strain 0.87
Shear Modulus 116,000 psi

* Five data points on one sample
 were taken. Fifteen total data
 points from samples tested.

COMPOSITE JOINT FABRICATION AND TESTING

The ply orientation for the 64 ply
laminate was [(-45/45/0/90)$_8$]s and was
[(-45/45/0/90)$_4$]s for the 32 ply
laminate. The laminates were hand
laid-up, vacuum bagged, and oven
cured. Bonding surfaces were sanded
lightly with fine grit sand paper
followed by solvent cleaning (ASTM
D2093).

The FM 73 film adhesive was
applied to joint surfaces after
cleaning. Bondline thickness was
controlled to 0.02 inches thick. When
the joint was assembled, a vacuum bag
was made to cover all of the surfaces
and to apply pressure while the
laminate joint was oven cured.

Tensile and flexural specimens
were cut from the large laminated
joint panels for testing. Specimens
were 1.5-inch wide by 1.65-inch thick
plus the cover plates each half the
1.65 thickness.

Testing was performed at
different temperatures and the mode of
failure of each specimen was
identified. For the testing at high or
low temperatures, the joint samples
were kept in the environmental chamber
until the temperature was stabilized.
Table 2 shows that all of sample
failures involved the substrate to
some significant degree indicating
that the bonding was successful.

Table 2 - Tested Properties of
 Scarf and Butt Joints.
 (Average results)

Butt Joint	Test Temp.	Stress (psi)	Failure Mode
Tensile,	80°F	15,400	1
Tensile,	-65°F	15,200	2
Tensile,	180°F	14,600	3
Flexural	80°F	3,650	2
After 160 hours at 100° in water			
Tension		14,730	2

Scarf Joint

Tensile,	80°F	21,760	4

Notes:
Three samples used each test.

Failure Modes:
1. 80% Substrate, 20% Adhesive.
2. 100% substrate.
3. 97% substrate, 3% adhesive.
4. 90% substrate, 10% adhesive.

JOINT ANALYSIS

FINITE ELEMENT MODELING - Different
approaches are available for
determining the adequacy of a joint to
sustain service loads. MARC, a general
purpose finite element analysis
program (FEA) was used in this study
to evaluate the strengths of the butt
and scarf joints and as an appropriate
means of analyzing failure modes.
First, the FEA analysis models were
made to simulate the structure and
probable sites of joint failures.

Plane strain models were used to
determine the stress distributions of
joints subjected to the test loads.
Symmetry of the butt joint enabled
analysis with only one quarter of the
joint modeled as shown in Figure 3A. A
whole model was required for the scarf
joint shown in Figure 3B.

Fine meshes were placed to capture the high stress gradients of material and geometrical discontinuities such as bonded regions and re-entry corners. The composites were modeled using transversely-isotropic material properties found by testing. The adhesive was considered isotropic having material properties given in Table 3. Linear behavior was assumed for both materials.

Table 3 - Transversely-isotropic
Material Properties

Composites
_____

E_x = 2.315 x 10^6 psi
E_y = 0.470 x 10^6 psi
E_z = 2.315 x 10^6 psi
μ_{xy} = 0.300
μ_{yz} = 0.061
μ_{zx} = 0.250
G_{xy} = 0.1810 x 10^6 psi
G_{yz} = 0.1810 x 10^6 psi
G_{zx} = 0.2634 x 10^6 psi

Adhesives
_____

E = 0.350 x 10^6 psi
μ = 0.400

FAILURE MODES - Adhesive, cohesive, and sub-laminate failures were considered as possible failure modes. Failure is assumed to be controlled by the internal load distribution resulting from the applied loads. It is assumed that edge effects do not essentially contribute to failure. Therefore, for comparison purposes, edge effects can be ignored.

Potential failure sites in both designs are the bond edges that form re-entry corners between the main and cover plates. The 45 degree bevel at the edges of the cover plate reduces the peel stress at this re-entry corner for both designs. For the scarf joint, the peel stress is not as dominant at this failure initiation site because of the bending moment introduced as it is for the butt joint. Sublaminate failure, initiated as a delamination mode, can occur if the applied load exceeds a critical value yielding high transverse-normal and shear stresses.

Stresses in the composites along the fiber direction for both designs shown in Figures 4A and 4B are well below the strength at the applied load. The transverse normal and shear stresses for both designs are high in comparison with their strengths. A simplified criterion is proposed to determine when delamination will begin:

$$\frac{\sigma_y}{S_y} + \left|\frac{\sigma_{xy}}{S_{xy}}\right|^2 = 1$$

Figures 5 and 6 show high stress gradients at the re-entry corners. As in a notch problem, the stresses used to determine the failure were calculated based on an average scheme up to a distance of a ply thickness. The average stresses normalized by the strengths are tabulated in Table 4 for the composites.

Using stress ratios as a performance index, it was found the scarf joint design is 1.45 times better than the butt joint at initiation of failure based on the proposed strength criteria. The well known Tsai-Wu failure criteria[3] is less conservative. For the adhesives, equivalent stress/allowable equivalent stress ratio equals 0.8604 for the butt joint and 0.7162 for the scarf based on an applied load of 15,400 lb.

Stresses along the material principal directions were used in the strength calculations for the composites but Von Mises equivalent stresses were used for the isotropic adhesives. In both designs the equivalent stresses in the adhesives are below the allowable strength; therefore failure will not occur in the adhesives. This is an adequate approximation as long as there is no flaw (void) in the adhesive and the adhesive/composite interfaces. Thus for this configuration, the joint strength is dependent on the stresses and strengths of the composite laminates.

Table 4 - Stresses near Re-entry
Corners in Laminates*

	σ_x/S_x	σ_y/S_y	$[\sigma_{xy}/S_{xy}]^2$
Butt	0.454	1.178	0.609
Scarf	0.371	0.691	0.425

SUMMARY AND CONCLUSIONS

Testing has shown that two joint designs for thick composite laminates can meet design goals. The scarf joint is the better of the two designs based on the analytical results using the proposed criterion. Ultimate failure occurred at a load about 1.5 times higher than the butt joint. The analysis method predicted failure initiation at 1.45 times the butt joint load capability. Predicted laminate failure was the actual failure mode in these joints. The adhesive scarf joint is an effective alternative to mechanically fastened joints in thick laminate vehicle structures. More testing to further qualify the joints for other conditions can proceed with confidence.

References

1. "Joining of Advanced Composites", Engineering Design Handbook, p. 4-43, DARCOM Phamplet No. 706-316, 5001 Eisenhower Av., Alexandria, Va (March 1979)

2. Renton, W.J., Flaggs, D. L., Vinson, J. R., "The Analysis and Design of Composite Material Bonded Joints.", Delaware Univ., Newark, N.J., Report No. AD-A060624 (July 1978)

3. Tsai, Stephen W., "Composite Design" pg 11-6, 3rd Ed., Think Composites (Co.),Dayton, Ohio (1987)

Nomenclature

e	Elongation
E	Modulus of elasticity
G	Modulus of rigidity
G_{1c}	Strain energy release rate, mode I fracture mechanics
G_{1a}	Strain energy release rate, mode II fracture mechanics
S	Allowable strength*
T	Tension*
σ	Stress*
μ	Poisson's ratio

* Subscripts denote direction. See figures for stress directions.

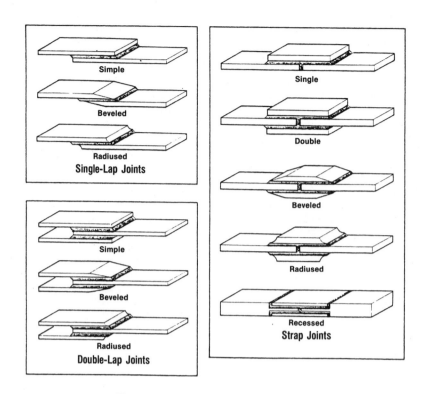

Figure 1. Common Composite Joints

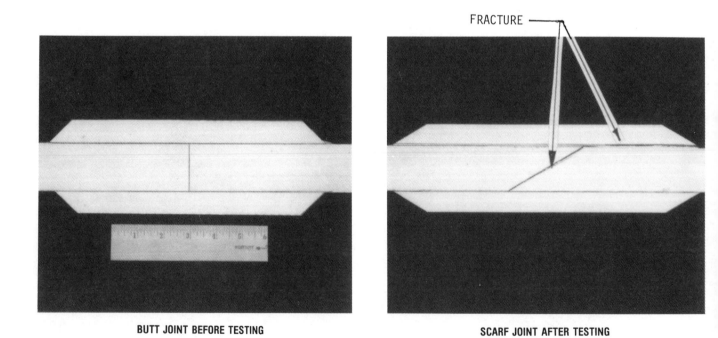

BUTT JOINT BEFORE TESTING

SCARF JOINT AFTER TESTING

Figure 2. Joints Tested

Figure 3A.   Finite Element Model of the Butt Joint

Figure 3B.   Finite Element Model of the Scarf Joint

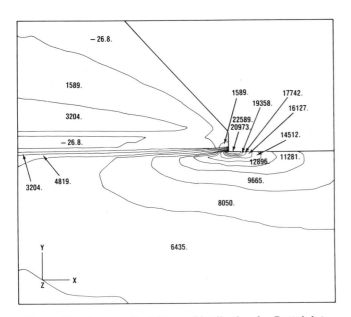

Figure 4A.   Longitudinal Stress Distribution for Butt Joint

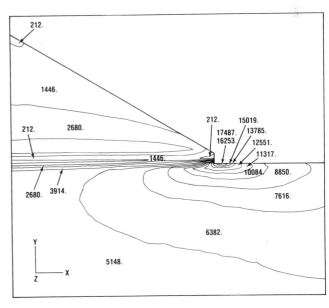

Figure 4B.   Longitudinal Stress Distribution for Scarf Joint

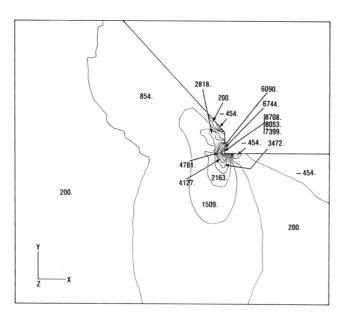

Figure 5A. Transverse Normal Stress Distribution for Butt Joint

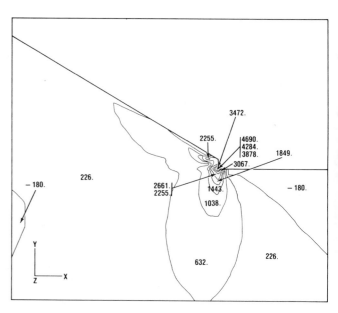

Figure 5B. Transverse Normal Stress Distribution for Scarf Joint

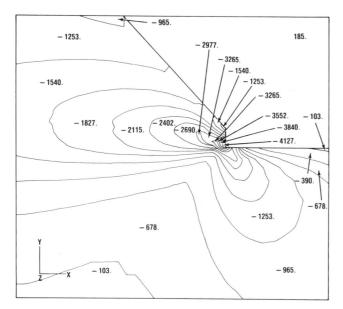

Figure 6A. Transverse Shear Stress Distribution for Butt Joint

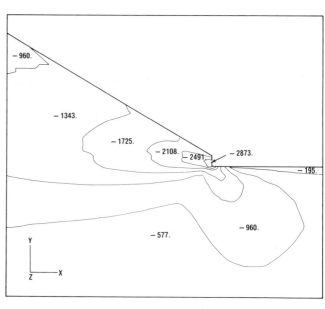

Figure 6B. Transverse Shear Stress Distribution for Scarf Joint

USING STRUCTURAL ADHESIVES TO BOND COMPOSITES MADE BY STRUCTURAL REACTION INJECTION MOLDING (SRIM)

Tai Ming Liang, Timothy A. Tufts, Gary E. Kimes
Ashland Chemical Company
Columbus, Ohio USA

ABSTRACT

Composites made by structural reaction inject-ion molding (SRIM) can be bonded satisfactorily with existing commercial structural adhesives. However, proper selection of the mold release agent, surface treatment, and adhesive material is critical. Bonding of SRIM materials was found to require different surface preparation and a stronger adhesive when compared to SMC.

FIBER-REINFORCED COMPOSITES have gained broad and growing use in automotive, truck, marine, and aerospace industries. The primary driving force to use composites include parts consolidation, lighter weight, and corrosion resistance.

The bonding of composites to themselves or other substrates has been widely practiced for more than 15 years. Corvette exterior body panels and Mack Truck Ultraliner cabs are two well known examples of bonding SMC composites with structural adhesives. The bonds are so strong that the SMC composites fail, rather than the adhesives, when they are subjected to shear and wedge tests. Structural adhesives also have good fatigue resistance and good resistance to some detrimental environments. In addition, they are adaptable to robotic operations for fast production speeds.

The SRIM process is a cost effective method to make structural composites for both automotive and non-automotive applications. Its advantages include fast speed, and low tool and equipment costs. It produces very strong composites with a high loading of strong fibers. Generally, long glass fibers in the form of continuous strand mats, uni- or bi-directional mats, or any combination of these are used in SRIM. However, other fibers, like carbon or Kevlar®*, can be used also. Typical physical properties of SRIM composites made from ARIMAX®* 1150 and ARIMAX 2100 resins are shown in Table 1.

Two major driving forces to study the bonding of SRIM parts by adhesives are: many applications require adhesive bonding rather than mechanical fastening; and it is important to know the impact of the bonding step on the total cost.

The objective of this paper is to present some results we have generated in our laboratory using commercial structural adhesives to bond ARIMAX composites. The effects of mold release agents, matrix polymers, structural adhesive materials, and substrates other than those made using ARIMAX resins are discussed. Finally, a current application is presented in which ARIMAX parts are assembled using structural adhesives.

TEST PROCEDURES

ARIMAX composites used for bonding tests contained 30 ± 2% by weight of OCF M-8610 continuous strand random glass mat. Standard lapshear test samples, consisting of two 1" x 3" x 0.125" coupons overlapped with a 1" x 1" x 0.030" bond area were used.

Table 2 describes some typical characteristics of the four adhesives we tested. Only adhesive "A" required a primer, which came as a part of the package. Bonded substrates were first cleaned by blowing off dust and dry wiping with a clean paper tissue, unless stated otherwise.

After the adhesives were applied and properly cured according to recommended conditions, the lapshear was measured on an Instron machine according to ASTM D-1008 at a speed of 0.5"/min. The lapshear strength was reported on the average of at least 5 samples. The mode of failure was also recorded according to

---

* Registered trademark of DuPont

* Registered trademark of Ashland Oil, Inc.

the definitions described in Table 3. Samples with either delamination (DL) or substrate failure (SF) mode are considered to have "satisfactory" bonding.

RESULTS

EFFECT OF EXTERNAL MOLD RELEASE AGENTS AND SURFACE PREPARATION - Currently, an external mold release agent has to be applied on nearly every shot during a typical SRIM process. Either wax- or soap-based mold release agents are used. The choice of a mold release agent depends on the SRIM matrix system, mold temperature, tool material, etc.

Adhesive "A", a 2-component urethane adhesive, has bonded automotive SMC parts satisfactorily without sanding. However, as shown in Table 4, adhesive "A" did not bond ARIMAX 1150 composites well unless the substrate was sanded first to eliminate mold release. With 3M's medium 100 grit sandpaper, 30 strokes back and forth by hand was found to be sufficient.

Adhesive "B", a 2-component hybrid adhesive, is a primerless adhesive. Currently, it is used to bond SMC. Good 22°C lapshear results were obtained when ARIMAX 1150 composites were bonded with adhesive "B". As shown in Table 5, strengths ranged from 750 psi to 1160 psi and, more importantly, gave 100% DL. No sanding was required for all three mold release agents; one soap-based (A) and two wax-based (B & C).

With sanded substrates, all 82°C lapshear tests gave 100% DL. However, when the substrates were not sanded, differences were seen with the various types of mold release used. For release agent "A", greater than 1030 psi and 100% DL was obtained. For release agent "B", mixed results, from 450 psi (Adh.) to 770 psi (DL), were obtained. Results were worse for release agent "C", where only cohesive failure was obtained, although lapshear was as high as 860 psi.

The large variability found in the case of mold release agent "B" may reflect the differing amounts of mold release agents left on the substrate surface.

EFFECT OF REINFORCEMENT - Compared to SMC, significantly higher lapshear strengths were obtained when ARIMAX composites failed with composite delamination or substrate failure.

As an example, typical 22°C lapshear strengths for SMC are 650 psi with adhesive "B". However, ARIMAX 1150 composite with adhesive "B" gave strengths as high as 1160 psi. To determine the effect of reinforcement length, we tested an ARIMAX 1150 composite reinforced with 30% by weight of 2 inch chopped fiberglass mat. With adhesive "B", the 22°C lapshear strengths were identical to those of typical SMC (see Table 6). Thus, the longer fibers generally gave much higher lapshear strengths.

EFFECT OF MATRIX COMPOSITION - In many

cases, although the joints did not fail with delamination or substrate failure, the lapshear strengths were very high. For example, when sanded ARIMAX 2100 composites were bonded with adhesive "B", only light cohesive failure mode was obtained. Inspite of this, the lapshear strengths were 1090 psi, almost twice as much as the SMC (see Table 7).

We suspect this is due to the much tougher matrix of ARIMAX 2100 composites or stronger adhesion between matrix and fiberglass. In support of this, 22°C lapshear strengths for a sanded vinyl ester composite were found to be similar to those for ARIMAX 2100 composites (Table 7). Vinyl ester resins are known to be tougher than typical SMC resins.

As a result, only adhesive "C" provides satisfactory bonding with substrate sanding. Typical lapshear strengths at 22°C, 82°C, and immediately following one week soak in 54°C water, are listed in Table 8.

BONDING ARIMAX COMPOSITES WITH METALS - We have also bonded ARIMAX composites satisfactorily with substrates such as rubbers, epoxy prepreg composites, and metals. Bonding ARIMAX 2100 composites with an ELPO steel was also of interest. The results are listed in Table 9.

The ELPO steel test coupons (1" x 4" x 0.038") were obtained from ACT Corp. Hillsdale, MI. They are CRS with Chemfos®* 168 zinc phosphate and PPG's ED3158 primer.

Adhesive "C" was found to remove the primer on the ELPO steel. Adhesive "B" and "D" were found to be the best. Sanding ARIMAX 2100 composite is required. Adhesive "B" provides stronger bonding while adhesive "D" cures faster (1 minute at 116°C). Therefore, in a typical automotive assembly line, adhesive "D" would be preferred.

ASSEMBLY OF XEROX COPY MACHINE HEAT SHIELDS

Currently, the heat shields of Xerox - Technographics Products 2510 engineering copiers (see Figure 1) are made from ARIMAX 1350 resin reinforced with 24% OCF M8610. ARIMAX 1350 resin was selected because it provides cost effective performance in this demanding application. High heat resistance, low smoke emissions, and combustion modification are critical.

The final part made by bonding three pieces of ARIMAX 1350 composite with adhesive "C" has three advantages:
- Forms very strong bonds at high temperature. After 8 weeks aging in a 300°F oven, test bonds still gave delamination failure at 300°F.
- Builds handling strength very quickly. Ninety seconds in a 110°C fixture is

---

* Registered trademark of Chemfil

218

sufficient to give handling strength. Then the parts can be cured at room temperature.
- Requires minimal surface preparation.

CONCLUSIONS

We have demonstrated that current commercial structural adhesives can satisfactorily bond one SRIM composite to another or to other substrates.

Compared to SMC, external mold releases left on SRIM composite surfaces may pose a problem when bonding with structural adhesives. Therefore, sanding of the surface may be required. Another difference is that lapshear strengths of adhesively bonded SRIM composites are much higher than SMC. Typical SMC grade structural adhesives may not be suitable for some SRIM composites.

Performance of the adhesive under real use conditions is the most important factor in selecting both the adhesive and the application process. However, other factors such as curing speed and insensitivity to external mold release agents should also be considered.

As applications of SRIM become more popular and demanding, we will need to pay more attention to bonding SRIM materials properly and cost effectively. In many cases, proper bonding may determine whether or not an SRIM part can be used or is cost competitive with parts using different materials or manufacturing methods.

Therefore, more cooperation among SRIM chemical system suppliers, structural adhesive suppliers, and mold release suppliers will be needed in the future. Developments such as internal mold release modified SRIM systems, easily washable external mold releases, and one-component adhesives are some areas where benefits would be immediate.

ACKNOWLEDGEMENTS

The authors want to thank Mr. D. Perdue, Mr. B. Bowen and Mr. D. Daniel for preparing ARIMAX plaques; Mr. D. Manino, Mr. K. Williams and Ms. Judy White for teaching us how to prepare and test lapshear samples. We also want to thank the following companies for either supplying us adhesive materials or testing their adhesives with ARIMAX composites: Ashland Chemical, Company 3M, H.B. Fuller, Mobay and Ciba-Geigy.

Table 1. Typical Physical Properties of Arimax Composites

Reinforcement	Arimax 1150 30% Continuous Strand Mat	Arimax 1150 35% Carbon Fiber 10% Continuous Strand Mat	Arimax 2100 30% Continuous Strand Mat	Arimax 2100 39% Undirectional Glass Mat 25% Continuous Strand Mat
Specific Gravity	1.41	1.35	1.40	1.60
Tensile St., MPa (ksi)	114 (16.6)	255 (37.0)	132 (19.2)	476 (69.0)
Flex Modulus, MPa (ksi)	6,890 (1,000)	31,200 (4,520)	6,900 (1,000)	24,900 (3,610)
Izod, Unnotched, J/M (ft-lb/in)	779 (14.6)	822 (15.4)	801 (15.0)	3,470 (65.0)
HDT @ 264 psi, oC	>240	>240	194	217

Table 2. Description Of Structural Adhesives

Code	A	B	C	D
Chemical Composition	2-Component Urethane	2-Component Hybrid	2-Component Epoxy	2-Component Epoxy
Primer	Yes	No	No	No
Curing Process	121C fixture: 4 min. then 121C oven: 30 min. or 3 days at 25C	121C fixture: 4 min. then 121C oven: 30 min	121C fixture: 4 min. then 121C oven: 30 min. or 121C fixture: 4 min. then 25C: 2 days	116C fixture: 1 min.
Sagging	No	No	Yes	No

Table 3. Definition Of The Modes of Failure

DL (Delamination):	The Lapshear Joint Breaks Inside The Adherend(s) Such That A Layer of Adherend (Matrix And Fiber Glass) From The Joint Area Is Delaminated Away.
SF (Substrate Failure):	The Lapshear Joint Does Not Break. Instead, One Of The Adherends (Substrates) Breaks Apart Completely Outside But Near The Joint.
Coh. (Cohesive Failure):	The Lapshear Joint Breaks Inside the Adhesive Layer. The Adhesive is Left on Both Adherends. No Fiber Glass Is Exposed On The Broken Joint.
Lt. Coh. (Light Cohesive Failure):	Similiar To Cohesive Failure Except Only A Very Thin Layer/Film Of Adhesive Is Left On One Of The Adherends.
Adh. (Adhesive Failure):	The Lapshear Joint Breaks Right on The Interface Between The Adhesive And The Adherend. No Adhesive Is Left On One Of The Adherends At All.

Table 4. Lapshear Strength of Arimax 1150 Composites Bonded With Adhesive A

Composites Sanding	Arimax 1150 No	Arimax 1150 Yes	SMC No
22C Lapshear Strength, MPa (psi) Mode of Failure	2.07 (300) Adh.	7.03 (1020) 75% DL, 25% Coh.	4.41 (640) DL.
82C Lapshear Strength, MPa (psi) Mode of Failure	1.38 (200) Adh.	3.79 (550) Lt. Coh.	2.75 (400) 90% DL, 10% Coh.

Table 5. Lapshear Strength Of Arimax 1150 Composites Bonded With Adhesive B
Effect Of Mold Release Agents

Substrates	Arimax 1150	Arimax 1150	Arimax 1150	SMC
Release Agent	A	B	C	Internal
Wax/Soap	Soap	Wax	Wax	Zn Stearate
22C Lapshear Strength				
Without Sanding				
MPa (psi)	5.37 (780)	6.14 (890)	7.72 (1120)	4.48 (650)
Mode Of Failure	DL	DL	DL	DL
With Sanding				
MPa (psi)	6.14 (810)	6.55 (950)	7.99 (1160)	
Mode Of Failure	DL	DL	DL	
82C Lapshear Strength				
Without Sanding				
MPa (psi)	7.10 (1030)	3.10 to 5.30 (450 to 770)	6.69 (970)	2.62 (380)
Mode Of Failure	DL	Adh. to DL	95% Coh. & 5% DL	DL
With Sanding				
MPa (psi)	5.65 (820)	5.93 (860)	5.93 (860)	
Mode Of Failure	DL	DL/SF	DL	

Table 6. Lapshear Strength Of Composites Bonded By Structural Adhesive B -
Effect of Reinforcement

	SMC	Arimax 1150	Arimax 1150
Glass	28% Chopped Glass Roving (1-in. Long)	30% Chopped Glass Mat (2-in. Long)	30% Continuous Strand Mat
Filler	43%	0%	0%
Matrix	29%	70%	70%
Releasing Agent	Internal, Zinc Stearate	External, Wax	External, Wax
Surface Treatment	No	Sanding	Sanding
22C Lapshear, MPa (psi)	4.48 (650)	4.48 (650)	8.00 (1160)
Mode of Failure	DL	DL	DL

Table 7. Lapshear Strength Of Composites Bonded By Structural Adhesive B -
Effect of Matrix Composition

	Arimax 1150	Arimax 2100	Vinyl Ester*
Glass	30% Continuous Strand Mat	30% Continuous Strand Mat	40% Continuous Strand Mat
Filler	0%	0%	0%
Matrix	70%	70%	60%
Releasing Agent	External, Wax	External, Wax	External, Wax
Surface Treatment	Sanding	Sanding	Sanding
22C Lapshear, MPa (psi) Mode of Failure	8.00 (1160) DL	7.51 (1090) Lt. Coh.	9.31 (1350) 50% DL, 50% Coh.

* Made By Resin Transfer Molding (RTM)

Table 8. Lapshear Strength Of Arimax 2100 Composites Bonded With Various Structural Adhesives

Adhesive	A	B	C
Surface Treatment	Sanding	Sanding	Sanding
22C Lapshear, MPa (psi) Mode of Failure	11.6 (1680) Lt. Coh.	7.51 (1090) Lt. Coh.	9.58 (1390) DL/SF
82C Lapshear, MPa (psi) Mode of Failure	3.64 (480) Lt. Coh.	5.51 (800) Lt. Coh.	7.65 (1110) SF
54C Water Soak, 1 Week Pull Immediately, MPa (psi) Mode of Failure	N.A.	N.A.	7.51 (1090) SF

Table 9. Bonding Arimax 2100 Composites With ELPO Steel*

Adhesive	C	B	D
Curing Condition	4 min. @ 121C Fixture & 30Min. In 121C Oven	2 Min. @ 121C Fixture & 30 Min. In 121C Oven	1 Min. @ 116C Fixture
82C Lapshear, MPa (psi) Mode of Failure	1.38 to 6.89 (200 to 1000) Adh., or Interface Between Primer and Steel	9.10 (1320) 100% Coh.**	5.65 (820) 85% Coh.**, 15% Adh.
54C Water Soak For 1 Week Pull Immediately, MPa (psi) Mode of Failure	N.A.	N.A.	7.58 (1100) 100% @ Primer/Steel Interface

* From ACT Corp., Hillsdale, Michigan. CRS with Chemfos 168 Zinc Phosphate and PPG's ED3158 primer.
** Near Arimax 2100 Composite Surface.

222

HEAT SHIELD OF XEROX TECHNOGRAPHICS
PRODUCTS 2510 ENGINEERING COPIER

DEVELOPMENTS IN FAST CURE ADHESIVES: ENHANCEMENTS IN ADHESIVE CURE RATE AND ASSOCIATED REDUCTION IN PRODUCTION COSTS

John R. Overley
Ashland Chemical Company
Columbus, Ohio   USA

Abstract

When producing exterior body panels for automobiles by bonding inner and outer SMC parts, a major part of the production cost is the heating fixture amortization over the number of parts produced. This cost is directly related to the cycle time or the time the part remains in the fixture to achieve adequate cure. While fixture times are on the order of three minutes, the objective is to reduce this to approximately one minute to reduce part cost.

Increasing the rate at which an adhesive cures and builds strength necessary for part handling after removal from the fixture is not solely related to manipulation of the adhesive formulation. In addition to the kinetics, adhesive cure rate is also a function of heat transfer to the part during heating and cooling. Therefore, process conditions have to be considered when using a current formulation or developing new formulations to meet customer needs.

In this work, the bonding process has been modeled using the thermal energy equation, along with experimentally derived properties for the SMC and adhesive and the adhesive kinetics, to predict adhesive performance as a function of processing variations. The bonding process has been studied to determine the effects of fixture temperature and clamp time on the adhesive strength build rate to determine optimum processing conditions.

Furthermore, the modeling has permitted us to determine the importance of the heating phase, the cure temperature, and the cooling phase on the adhesive performance. This has permitted us to better modify our adhesive formulations to meet the one minute fixture time requirement.

WHEN BONDING SMC INNER and outer parts to create an automotive body panel, a major portion of the production cost per part is the amortization of the heating fixture. For example, if a heating fixture costs $200,000, is amortized over a period of five years, and parts can be bonded in this fixture with a three minute cycle time, operating 16 hours a day and 50 weeks per year, the cost of the fixture per part is $0.50. Reducing the fixture time from the three-minute cycle to a one-minute cycle would triple the production of parts per fixture per year and therefore reduce the cost per part due to the fixture to $0.167, a significant savings in part cost. The question is: how do we reduce fixture time while ensuring adequate handling strength.

The amount of time a part remains in a fixture to ensure adequate cure of the adhesive, and handling strength when the part is removed from the fixture, is a combination of the technology that goes into the adhesive chemistry and the rate at which energy can be transferred from the fixture to the part to heat the adhesive and cure it properly. Therefore, in order to determine how best to improve the cure rate of the adhesive, we need to understand the overall bonding process.

IDEAL BONDING PROCESS

Figure 1 shows the adhesive temperature versus time profile for an ideal bonding process. If we had the capability to do so, we would like to heat the adhesive instantly from room temperature to some desired curing temperature, hold the adhesive at that temperature for a sufficient time to give total cure, then instantly cool the part to room temperature prior to removal from the fixture. Current customer objectives call for a maximum of one-minute fixture time which is why the 60-seconds is used in this figure.

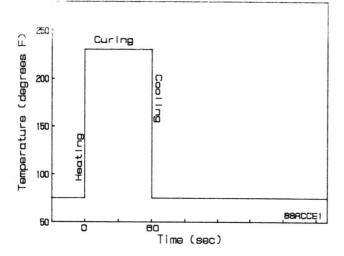

FIGURE 1. Ideal Bonding Process

In reality, we know that we cannot produce this type of temperature versus time profile. In the heating phase, assuming conduction from heat blocks is used, it is impossible to instantly conduct heat through the SMC substrate to bring the adhesive to its desired curing temperature. This heat transfer is controlled by the SMC thermal properties and dimensions, and the SMC surface temperature. The adhesive heating rate could be increased somewhat by using some type of electromagnetic heating, such as dielectric, microwave, or electromagnetic induction. This overcomes the heat transfer through the SMC, but the rate of energy input is limited by how fast the adhesive can be heated without boiling.

Before the adhesive components react and crosslink, each component has a boiling temperature range. Therefore, supplying energy too rapidly could boil the components before they react sufficiently to increase the boiling point of the system. In reality, the fastest that energy can be supplied is to heat the adhesive system to a temperature just below its boiling point as determined by the degree of crosslink. As the adhesive cures and its crosslink density increases, its boiling point increases also. As the adhesive cures, another constraint is to remain below the thermal decomposition regime. Hence, the realities of the adhesive system put a limitation on the maximum heating rate that can be used in the bonding process, regardless of the type of energy transmission to the bondline.

In the adhesive curing phase, the rate of reaction is controlled by the reactants, and the catalyst type and concentration. This is where the chemistry of the system can be manipulated to decrease the bonding cycle, but the impact of formulation changes on the adhesive service performance has to be considered.

The final phase of the bonding process, cooling the adhesive bondline and the substrate, is controlled by the substrate thermal properties and dimensions, the external heat transfer coefficient, and the temperature of the surrounding heat transfer medium which is usually air. In this phase, the heat application technique, whether it be conduction, microwave, dielectric, or electromagnetic induction, has no bearing or control over the rate at which the part cools.

In practice, the properties of the adhesive system, and heat transfer to and from the adhesive, are coupled to dictate the temperature versus time behavior in the adhesive. In commercial adhesive systems, the rate of reaction or cure increases with increasing temperature, typically being very slow at room temperature, yet reacting 10 - 20 times faster at 250°F. This behavior means that a significant amount of the adhesive cure will occur both during the heating and cooling phases of the bonding process. Therefore, in order to reduce the cycle time to produce a bond for adequate handling strength after removal from a fixture, the overall process must be understood to determine where gains can best be made to reduce cycle time. Modifications to the chemistry of the adhesive system cannot be properly set without taking into account the overall process.

PROCESS MODEL

We have modeled the bonding process using the thermal energy equation, coupled with the adhesive kinetics, to determine the effects of fixture temperature, fixture clamp time, adhesive chemistry, heat transfer coefficient during cooling, and the air temperature on cycle times. Since in most applications of commercial interest the bondline thickness is much less than the bondline width, the overall heat transfer process approaches that of single-dimension behavior, and we can therefore use the energy equation in one dimension to model the process. The energy equation for describing the unsteady-state heat transfer in one dimension is shown in Equation 1, in which the density and thermal conductivity are assumed to be constant. Our experimental measurements have shown this to be an acceptable approximation.

$$\rho C_p \frac{\partial T}{\partial t} = k \frac{\partial^2 T}{\partial x^2} \qquad (1)$$

where: ρ = Density
  C_p = Specific Heat at Constant Pressure
  T = Temperature
  t = Time
  k = Thermal Conductivity
  x = Distance Along the x-Axis

This equation has to be coupled with the reaction kinetics for the adhesive under study to predict adhesive behavior as a result of chemistry and process conditions. Therefore, in addition to Equation 1, we have to simultaneously solve the equation for the adhesion kinetics. Finally, since most adhesives are exothermic in nature and release a

significant amount of energy during cure, a term has to be added to the energy equation to take into account the magnitude and rate at which energy is evolved as the adhesive cures. Modifying the energy equation as needed and adding the kinetics results in Equations 2 and 3.

$$\rho C_p \frac{\partial T}{\partial t} = k \frac{\partial^2 T}{\partial x^2} - \Delta H \frac{dX}{dt} \qquad (2)$$

where: ΔH = Heat of Reaction (Negative for an Exothermic Reaction)
  X = Fractional Conversion (When X=1.0 the Adhesive is Fully Cured)

$$\frac{dX}{dt} = k_o \exp \frac{-\Delta E}{RT} (1-X)^n \qquad (3)$$

where: k_o = Frequency Factor
  ΔE = Activation Energy
  R = Ideal Gas Constant
  n = Reaction Order

To solve these equations, we first needed experimental data for the density, specific heat, thermal conductivity, heat of reaction, and kinetics for the adhesive being studied. Except for the thermal conductivity, these were obtained experimentally in our laboratories, with the kinetics being the most time consuming step. To determine kinetics, the differential scanning calorimeter was used to generate, at isothermal temperatures, reaction rate versus time curves. From these, the frequency factor, activation energy and reaction order were determined to give the best fit through the data.

EQUATION SOLUTION

The equation set describing the unsteady-state behavior during the bonding process was solved by putting it into finite-difference form using the Crank-Nicolson approximation for the derivatives. Since most bonding situations involve approximately 100 mil thick SMC and a 30 mil thick adhesive bondline, these are the dimensions we used in our model. The grid spacing in the the "x" direction was set at one mil, or .001 inch, and the time step-size was set at 0.2 seconds. Furthermore, to minimize the amount of calculation time, we modeled the symmetrical situation in which the same temperature is used on both sides of the part to be bonded along with the same thickness of SMC.

The final model then consists of a grid in which the first 15 points constitute a 15 mil thickness of adhesive, followed by 100 grid points which cover the 100 mil thickness of SMC. The boundary conditions are: no heat transfer through the zero grid point which is the plane of symmetry, and a wall temperature at the outer surface of the SMC set to simulate any desired fixture temperature. The program to solve the equations was also written with the option of specifying that at a specific time a heat transfer coefficient on the exterior wall of the SMC could be set. This is to simulate the type of convective cooling we desire, whether natural convection or forced convection induced by blowing air on the surface. The equation set was solved on a personal computer with the output showing at any given simulation time the grid point values for the temperature through the SMC and the adhesive, along with the fractional conversion or cure of the adhesive.

Modeling the process in this fashion provides a better understanding of what is occurs during the bonding process, but is of little value unless it agrees with the reality of the bonding situation. Therefore, the accuracy of the prediction was verified by comparing the predicted adhesive temperature versus time with measured temperatures, determined by inserting a thermocouple in the adhesive layer.

Figure 2 shows the comparison between predicted and measured temperatures and shows excellent agreement between the two. In this situation, a sample which was originally at room temperature was placed in a 300°F fixture for 40 seconds and then removed and allowed to cool by natural convection in still, ambient air. The temperature/ time profile described from 0 - 40 seconds shows excellent agreement between measured

and predicted values, and indicates that our assessment of the kinetics, the exothermic heat of reaction, and the thermal properties of the SMC and adhesive are representative of the realities. At time greater than 40 seconds, the heat transfer coefficient on the outside surface of the SMC was adjusted in the calculations to give the best fit with the cooling curve. The resulting heat transfer coefficient of four Btu/hr ft² °F agrees well with a calculated heat transfer coefficient for a surface the size of the sample tested. In effect, we gained two valuable items by modeling the process: first, we now have a better understanding of the temperature and conversion profiles in the adhesive, and second, we have a better measure of the heat transfer coefficient that exists when a part cools by natural convection in ambient air.

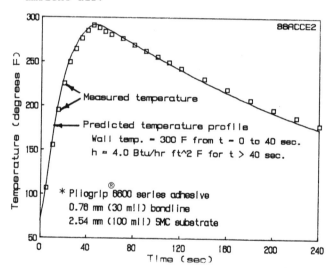

FIGURE 2. Predicted Versus Measured Temperature

BONDING PROCESS SIMULATIONS

Equipped with the equation sets and computer software to solve them, our first approach was to study the profiles for one of our adhesives when heated at different fixture temperatures and times,

---

* Pliogrip® is a Registered Trademark of Ashland Oil, Inc.

representing current practice and directions we wanted to proceed to reduce the fixture time.

Figure 3 shows the average adhesive temperature as a function of time for three different heating and cooling scenarios using a commercial urethane structural adhesive from Ashland. In many applications, the adhesive is heated in the fixture at 220°F for three minutes, then allowed to cool by natural convection in still air. The average temperature behavior versus time for this case is shown with a heat transfer coefficient on the exterior surface equal to one Btu/hr ft^2 °F. As the figure shows, the temperature rises to approximately 220° in 50 - 60 seconds then remains constant as the piece is held in a fixture. After three minutes, when the part is removed from the fixture, it will cool slowly because of the low heat transfer coefficient. This is indicated by the slow rate of temperature decrease with time after 180 seconds.

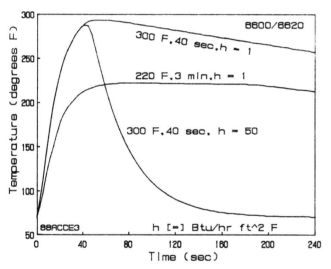

FIGURE 3. Temperature Profiles Versus Processing Conditions

Since our objective is to reduce the fixture time in a given application, and the adhesive exhibits an increasing reaction rate with higher temperatures, one approach is to use a greater fixture temperature for shorter periods of time to achieve the same adhesive conversion when the fixture is opened. Therefore,

we simulated two other situations in which the SMC and adhesive were heated at 300°F for 40 seconds then cooled using two different heat transfer coefficients; both of these curves are shown in Figure 3 with h = 1 and 50.

The heat transfer coefficient of 1 represents the very slow, natural convection cooling in ambient temperature air. After the 40 seconds of heat application the temperature is shown to drop very slowly because of the very low rate of heat transfer to the surrounding air. An h = 1 was chosen because it is representative of the heat transfer by natural convection to air when the part is very large, as is the case for a car hood, rear deck, or door. When a heat transfer coefficient of 50 Btu/hr ft^2 °F is used, representing the upper limit of forced convection with air, the resulting change in temperature with time after removal from the fixture is much faster, as would be expected. In this case, approximately two minutes after removal from the fixture the adhesive bondline is essentially at room temperature.

Because the adhesive reaction rate is a function of temperature, the different heating and cooling scenarios for a production part therefore have an impact on the time behavior of the adhesive conversion. Figure 4 shows the adhesive conversion, averaged across the bond thickness, as a function of time as it relates to the prior heating and cooling situations. For the 220°F heating case, the fractional conversion versus time continuously increases, even after removal from the heating fixture at the three minute time. This is because the adhesive temperature remains hot enough to promote reaction, even as the part cools.

Fractional conversion versus time for the 300°F heating for 40 seconds, then cooling with h = 1, also shows a continually increasing conversion after the part is removed from the fixture. This is again because the adhesive remains sufficiently hot that reaction continues to the extent that there is almost complete conversion after 4 minutes total time. In comparison, when the cooling rate after 40 seconds is rapid, as indicated by the h = 50 curve, the reaction is essentially stopped less than one minute after removal from the fixture. The net result is a lower adhesive conversion in the bonded part. If, however, this part is to be postbaked in a paint oven for proper paint cure, then the paint oven exposure will promote full adhesive cure.

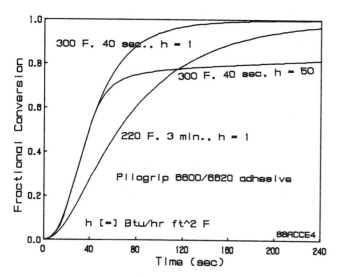

FIGURE 4. Fractional Conversion Versus Processing Conditions

Having seen the temperature and conversion behaviors as a function of fixture temperatures and time, the next question to answer is: "How does the adhesive strength build as a function of fixture temperature and time?".

ADHESIVE STRENGTH VERSUS CONVERSION AND TEMPERATURE

In its uncured state, an adhesive has negligible strength because there is no crosslinking of the reactive components. Rather, it has sufficient rheology to hold parts in place due to the Bingham yield stress which is characteristic of many polymeric materials. A crosslinking network formed by reaction of the components is necessary in order to build strength, and this strength is a function of adhesive conversion.

Figure 5 shows the cross-peel strength for one of Ashland's urethane adhesives as a function of adhesive fractional conversion. In the cross-peel test, a bead of adhesive is placed between two substrates, steel in this

case, to form a cross-shaped configuration. This sample is then placed in a device which pulls on each of the pieces normal to the bonded area to try to separate them, thereby applying a normal load to the adhesive bond.

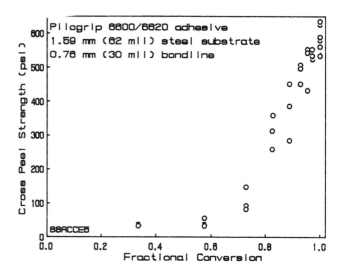

FIGURE 5. Cross Peel Strength Versus Fractional Conversion

The data in this figure were determined by taking cross-peel samples of the adhesive, heating in a fixture at 275°F for various periods of time, then quenching the samples in ice water to stop the reaction. The net result was a cross-peel strength as a function of fixture time. The energy equation and kinetics models were then used to determine the conversion at each of those times, and the information converted to the cross-peel versus fractional conversion plot shown. Figure 5 shows that up to a 70% adhesive conversion the rate of strength build with conversion is relatively low, yet drastically increases at greater than 70% conversion. However, this is only the room temperature behavior and the effect of elevated temperature on strength also needs to be considered in predicting bond strength performance.

Figure 6 gives us the final piece of information needed to predict bond strength, which is the relationship between adhesive strength and temperature for the fully cured adhesive. These data were acquired by taking cross-peel samples

with steel substrates and fully cured
adhesives, heating the samples in an oven
to the indicated temperature, then
testing them while at temperature. The
results show a strong decrease in
adhesive strength with increasing
temperature, but I should note here that
in all cases the adhesive is much
stronger than an SMC substrate. In
cross-peel testing, SMC substrates
typically fail at between 150 and 250 psi
at room temperature. Also, as the
temperature increases, the strength of
the SMC decreases, as is well known.

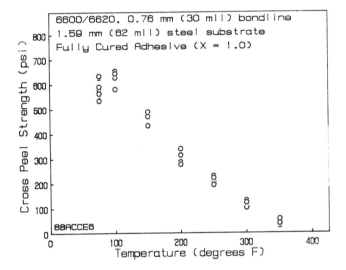

FIGURE 6. Cross Peel Strength Versus
Temperature

By coupling the data for cross-peel
strength versus conversion and strength
versus temperature, we created a model
relating cross-peel strength to these two
parameters and included it in our bonding
process software. We then had the
capability to determine the rate of
cohesive strength build as a function of
fixture temperature and clamp time.

BOND STRENGTH BUILD RATE

Computer runs were performed at
several combinations of fixture
temperature and clamp time, with an
assumed fast cooling rate after removal
from the fixture, to determine the effect
on bond strength. Figure 7 shows the
results for one of Ashland's fast cure
urethane formulations at fixture

temperatures from 250° to 325°F. For each
curve shown, the notations represent the
fixture temperature, the clamp time in the
fixture, and the heat transfer coefficient
of 50 Btu/hr ft² °F for the cooling phase.

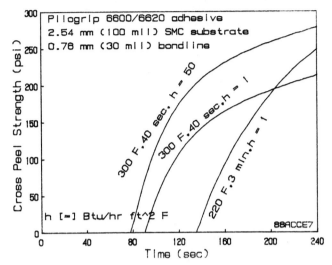

FIGURE 7. Predicted Cross Peel Strength
Versus Processing Conditions

Figure 7 indicates that the lowest
fixture temperature, in conjunction with a
50 second clamp time, gives the slowest
rate of strength build. However, note
that in this case the part is only
clamped in the fixture for 50 seconds,
well under 1 minute as required by the
automotive industry and sufficiently fast
to permit a 1 minute button-to-button
cycle time for a part. The curve shows
that at these conditions, only
approximately 25 seconds after removal
from the fixture the part will be rapidly
developing strength so that it can be
handled without damage during the
downstream assembly procedures. In each
of the cases indicated, the assumption is
that the part will be held in the fixture
for the indicated period of time, then
removed and placed in a simple fixture
which merely blows air on the part to
cool it rapidly. While the figure shows
zero handling strength at less than
approximately 75 seconds, in reality
there is some strength but we can not
measure it because the lower limit on the
cross-peel tester is approximately 10
psi. Hence, the model does not predict
strengths lower than that.

Next note in Figure 7 that for the 300° fixture temperatures, for which the clamp times are 30, 40, and 50 seconds, the fastest strength build occurs at 40 seconds. This points out the balance between clamp time to achieve adhesive conversion and cooling time to reduce the adhesive temperature to gain strength. Think of this situation as follows: assume that we have to develop a given strength 60 seconds after we have placed a part in a fixture. If we heat the part for 30 seconds in the fixture and then cool it for the remaining 30 seconds, we produce a bondline with a conversion of approximately 70% and a temperature of 220°F at the 60 seconds. In a comparative case of heating for 40 seconds and then cooling for 20, however, the conversion would be approximately 75% and the temperature around 240°F at 60 seconds. Therefore, in the longer clamp time we naturally get a greater adhesive conversion, but we also have a greater temperature at 60 seconds total time. It is a combination of these two that creates the bond strength, and our studies show that there is an optimum clamp time versus cooling time for a specified fixture temperature. This same behavior applies when there is natural convection cooling although not to the same extent, because the cooling rate is much slower in still air.

CONCLUSIONS AND RECOMMENDATIONS

As a result of these studies we have determined the sensitivity of the adhesive heating and curing process conditions to the rate of strength build. The results we have predicted by the process simulations have been verified by trials in laboratory and commercial applications.

Our current philosophy is to use the heating fixture for its original intended purpose, that is, to heat and cure the adhesive. When a part is removed from the fixture, and in many cases in commercial applications this is by means of some shuttle or truck to support the part, ultimate adhesive strength is not required at that instant. The seconds or fractions of a minute before a part is subjected to substantial handling loads is valuable time and can be used to let the part cool and the strength to build

to adequate levels. Therefore, we tend to view each of the different commercial applications as unique situations and use our process modeling and adhesive expertise to determine the optimum fixture temperature and time to achieve the necessary strength.

With our current adhesive formulations, we can achieve adequate adhesive conversion in less than 1 minute button-to-button cycle times to produce parts with dimensional integrity and sufficient handling strength for downstream processing. By minimizing the amount of time parts spend in the expensive heating fixtures, we can reduce the per part cost due to the heating fixture, thereby achieving the objectives set forth by our customers. While the development of appropriate adhesive chemistries may mean that these faster cure adhesives are more expensive than prior formulations, the combination with reduced fixture time results in a bottom line lower per part cost. The change in processing conditions to achieve the under 1 minute cure also has a bearing on the potential for causing readout in the class A side of the SMC surfaces. We have carried out development work parallel to that of the adhesives to ensure no readout, but that's another story.

THE USE OF MICROWAVE ENERGY IN
THE BONDING OF COMPOSITE PARTS

Randall C. Rains
Mobay Corporation
Pittsburgh, Pennsylvania   USA

ABSTRACT

As the use of the composites in the automotive sector increases, more and more attention is given to the relative efficiency of the production process. A significant factor in the production of composite parts is the secondary bonding operation. The current situation in secondary bonding operations will be reviewed. The use of microwave energy to improve the efficiency of the bonding operation will be discussed. Factors such as substrate response, adhesive selection, the influence of metal and fixturing issues will be reviewed. The potential for cycle times of one minute or less will be outlined.

ISSUES AFFECTING processing and production will continue to move to the forefront of development programs as the use of composite parts becomes an ever more accepted option in automotive applications. SMC is a good example of a composite which has proven satisfactory in several substantial applications but which has been burdened by a cumbersome manufacturing process. In order for SMC to compete with other technologies a continuing cycle of technological improvement has been driven by the molding industry. Great strides have been made in the areas of molding cycle time reduction and surface appearance. However, the bonding cycle time continues to be a major problem. A lengthy bonding cycle brings the obvious cost disadvantages associated with the purchase of extra fixtures. Furthermore, parts processed through a multitude of fixtures must be approved for performance and dimensional tolerances for each possible combination of mold and fixture. A drastic reduction in cycle time is the easiest way to avoid these problems.

This paper will discuss the use of microwave energy to bond composite parts quickly and efficiently while avoiding the problems outline above.

EQUIPMENT - Industrial microwave heaters come in a variety of sizes, shape, and configurations depending on the application. The unit configuration selected was a multi-mode chamber rated at 16kw with a roller conveyor bed and with vertical sliding doors at each end.

ADHESIVE RESPONSE - It was easily determined that certain types of urethane adhesive formulations respond rapidly to microwave energy. A temperature rise of 250 degrees F per minute in 0.030 in. bondline is quite normal. The response is dependent upon the formulation and one can not be sure that all commercially available adhesive respond in exactly the same way.

SUBSTRATE RESPONSE - The traditional method of increasing reactivity by increasing temperature is of only limited utility in SMC bonding. Large flat parts are susceptible to bondline readthrough and efforts are made to limit the heat history on these parts. Here the microwave process offers a decided advantage because there is little direct heating of the substrate. This desirable situation exists for a wide variety of composites, thermosets and thermoplastics.

FIXTURING - In conductive heating the fixture performs two distinct functions. It holds the inner and outer in proper register and it brings the heater elements into contact with the substrate. With microwave heating the fixture is only used for part registration. Since full bondline contact is not required it is envisioned that relatively simple fixtures will suffice. Development programs to build and test fixtures for actual production parts are underway.

SPECIAL MICROWAVE ISSUES - With multi-mode microwave applicators one must be aware that the MW field is not a constant strength at all points within the chamber. Because there are nodes a combination of wave scattering techniques and part motion is used to assure an even time-averaged MW exposure.

There are frequent questions concerning metal in microwave environments. Within broad limits metal is compatible with such environments. The standard industrial microwave heaters are metal chambers with metal rollers and chain drive mechanisms. However, fine wires and metallized fibers can cause severe arcing problems.

CYCLE TIMES - The rate of temperature increase available via microwave energy when applied to a properly formulated adhesive will yield cycle times of one minute or less. The capability of the microwave chamber to accept multiple parts offers the potential of further significant reductions in net cycle time.

35331RCR2017
GBF07

THE EFFECTS OF WATER IMMERSION ON THE SHEAR STRENGTH OF EPOXY BONDED STAINLESS STEEL

K. A. Geddes, S. P. Petrie
Plastics Engineering Department
University of Lowell
Lowell, Massachusetts 01854   USA

Abstract

In this study, the effect of water immersion time and temperature upon the shear strength of an epoxy-bonded stainless steel were investigated. Coupons of stainless steel, having six different methods of surface preparation, were bonded to form lap shear specimens using an amide-cured epoxy resin. The strength results indicated, over the range of variables investigated in this study, that the effect of temperature was of little, if any significance. An analysis of variance was used to show that water immersion time was a more significant variable than the surface treatment method. The strength-time results were fit to various models to describe the behavior of the systems.

AN EPOXY RESIN has been suggested for use in the design of a spiral, stainless steel, cylindrical shaped heat exchanger, to be used in a water storage tank for a solar energy collector system. The heated water from the solar collector flows in one side of the spiral channels and the water in the storage tank flows in the other side. These channels are alternating, letting the heat from the collector water be exchanged to the storage tank water as they both circulate around.

There are two functions that the epoxy seal must provide; one is to prevent water from leaking out of the heat exchanger cylinder, and the other is to prevent mix of the water between the alternating spiral channels. Therefore, whatever material is used must adhere well to the heat exchanger body material, providing a good seal.

When the water is pumped from the water storage tank to the heat exchanger, the water pressure can range from 20-140 psi.

Therefore, the seal must be strong enough so that it does not de-bond from the heat exchanger material. The heat exchanger is designed so that the spiral channels will not collapse on each other as the water is pumped through. However, there might be slight movement. The spiral channel in which the water from the collector circulates around is called the low pressure side. The alternating channel containing the water from the storage tank is called the high pressure side.

The environmental conditions which the heat exchanger will be subjected to, is another area to be considered. It will be in contact with water for its entire life time, therefore the effect of water, upon the sealing quality of the material must be investigated in detail. Another environmental condition to be considered is temperature. All material must be able to function properly at a maximum use temperature of $200°$ F.

The type of material under consideration for sealing the heat exchanger is a thermosetting polymer adhesive material. More specifically, it is a liquid epoxy resin which will crosslink when cured, leading to a solid material. The ends of the heat exchanger could be immersed in the liquid resin, then allowed to solidify (cure).

The study undertaken is to investigate the durability of epoxy bonded stainless steel in a hot humid environment. Based on the heat exchanger design, a lap shear joint is chosen as the type of test specimen to be used in the study. The effects of water immersion time and temperature upon the bond shear strength, of this epoxy bonded stainless steel system, will be investigated. Also the effect of steel surface preparation on the bond shear strength will be investigated.

EXPERIMENTAL AND PRELIMINARY RESULTS

STRENGTH DETERMINATIONS - ASTM Standard Test Method D1002 (1) was used to determine the lap shear strengths of the epoxy bonded stainless steel joints fabricated in this investigation. In summary, 0.01 inch thick stainless steel strips one-inch wide were cut, and lap joints, having a nominal one-half inch overlap were formed with epoxy adhesive between the two strips. The lap length, adherend width and bondline thickness were determined using a vernier caliper.

An INSTRON universal testing machine was used to load the adhesive test specimens in tension to produce failure. The jaws of the testing machine were set to give a gauge length of 5.5 inches and the jaws were moved apart at a speed of 0.05 inches/minute throughout the investigation. The load to produce failure was obtained from the chart on the testing machine. The shear strength (stress) was calculated from the ratio of the observed load to bond area.

ADHEREND MATERIAL - Type 301 (18-8) stainless steel was used throughout this investigation. The designation 18-8 stands for 18% chromium and 8% nickel content. This type of stainless (300 series) was chosen for this heat exchanger application because of its excellent corrosion resistance and its ability to be easily formed with conventional tooling. The nominal 0.01 inch thickness stainless steel sheet used in the heat exchanger and adherend specimens was obtained from Teledyne Rodney Metals of New Bedford, Massachusetts.

STAINLESS STEEL SURFACE TREATMENTS - The following surface treatments were used to treat the bond surfaces of the stainless steel strips used in this investigation:

Degreased - The stainless steel strips were immersed in 1,1,1, - trichloroethane for 15 minutes and then wiped with a lint-free tissue.

Abraded - The stainless steel strips were abraded with coarse sand paper - CARBORUNDUM ALOXITE A80 X255A.

Primed - A 0.2% solution (in iso-propanol) of a titanate adhesion promotor, KEN-REACT KR44, was applied to the stainless steel strips with a brush. The unreacted titanate was washed off with distilled water. The primed stainless steel strips were allowed to dry for 30 minutes at 150°F.

Acid Etched - The stainless steel strip was immersed 15 minutes at 63±3°C in the following solution:
100 parts by volume of Sulfuric Acid (diluted 80 parts H_2SO_4 with 50 parts H_2O)
30 parts by volume of a saturated sodium dichromate solution (75 parts by weight sodium dichromate with 30 parts by weight water).

The etched stainless steel strips were then rinsed in distilled water and allowed to dry in a 93°C oven.

These four surface treatments were used alone or in combination. A total of six different treatments were used - they are as follows:
1. Degreased
2. Degreased/Primed
3. Degreased/Abraded
4. Degreased/Abraded/Primed
5. Degreased/Etched
6. Degreased/Etched/Primed

EPOXY RESINS - Since twenty formulations of adhesive had been selected for this application, it was decided to conduct some preliminary tests to "screen" the resins to select the best candidate materials for further evaluation. (The fourteen base epoxy resins and their respective suppliers are listed on the next page in Table 1.)

Twenty adhesive formulations were used to bond degreased test coupons. The adhesives were then cured according to the manufacturers' recommendations (2). The resulting shear strengths along with the standard deviations are given in Table 2. (It is interesting to note that the observed shear strengths varied by more than an order of magnitude.)

From the previous results, EPON 828 and three other resins were selected for further evaluation. The selection was made based both on the obtained strength and the ease of application. Five candidate formulations were used to bond degreased and abraded stainless steel lap joints. The strength results can be seen in Table 3. From the results, the EPON 828/V-40 (3:1) formulation was selected for further evaluation.

CONDITIONING - Since the heat exchanger would be used at elevated temperatures for extended periods of time, it was decided to investigate the effect of these two parameters on the shear strength of test coupons bonded with the EPON 828/V-40 resin formulation.

Effect of Temperature - The shear strengths of different steel surface treated stainless steel/EPON 828 lap joint bonds, at various temperatures, were investigated. The temperatures investigated were 25, 60, and 90°C. The bonded test samples were immersed in a constant, agitated temperature bath of water for one hour (constant time). The temperature in the bath was controlled to ± 1°C. The test samples were removed one at a time, and tested immediately for required load to break using the INSTRON tensile tester.

Effect of Time - The shear strengths of different steel surface treated stainless steel/EPON 828 lap joint bonds, at various water immersion times, were investigated. The bonded test samples were immersed in a constant, agitated temperature bath (60°C) at varying times (1,4,8,25,168 hours). The samples were removed one at a time and tested immediately as previously described.

TYPE OF FAILURE - Once the bond was broken, the bond surface was examined for type of failure. Adhesive failure is when

Table 1 – Epoxy Resins and Suppliers

Resin Name	Manufacturer/Distributor
ISOCHEMREZ 402AP	Isochem Resins Co., Lincoln, RI
AR-1003	Formulated Resins, Inc., Greenville, RI
STYCAST 2651-40	Emerson & Cuming Chemical Division, W.R. Grace & Co., Canton, MA
TRA-BOND 2143D	TRA-CON, Inc., Medford, MA
TRA-BOND 2135D	TRA-CON, Inc., Medford, MA
TRA-BOND 2112	TRA-CON, Inc., Medford, MA
TRANSEPOXY 330	TRANSENE Co., Inc., Rowley, MA
LED 101-C	TRANSENE Co., Inc., Rowley, MA
1438-104A	Morton Chemical, Woodstock, IL
META-CAST 4583	Mereco Products, Cranston, RI
META-CAST 5230	Mereco Products, Cranston, RI
METRA-GRIP 321	Mereco Products, Cranston, RI
EPON 828	Shell Chemical Co., Deer Park, TX
BAKELITE ERL-4221	Union Carbide Corp., New York, NY

Table 2 – Average Shear Strengths (in psi) for Degreased Stainless Steel Lap Joint Bonded with Various Epoxy Adhesives

Adhesive Name	Average Shear Strength	Standard Deviation
ISOCHEMREZ 402AP	512	106
AR-1003	716	67
STYCAST 2651-40	470	158
TRA-BOND BB-2143D	622	69
TRA-BOND BB-2135D	820	117
TRA-BOND 2112	420	195
TRANSEPOXY 330	841	40
LED 101-C	751	90
1438-104A	277	88
4583/ #12	529	128
4583/ #14	777	13
5230/ #42	890	38
METRE-GRIP 321	986	59
EPON 828/V-40 (1:1)	897	115
EPON 828/V-40 (4:1)	732	86
EPON 828/V-40 (3:1)	1001	126
EPON 828/U	717	26
EPON 828/Y	57	13
BAKELITE ERL-4221/AP1	667	154
BAKELITE ERL-4221/AP1	711	36

Table 3 – Average Shear Strengths (in psi) for Degreased and Abraded Stainless Steel Lap Joints Bonded with Various Adhesives

Adhesive Name	Average Shear Strength	Standard Deviation
TRA-BOND BB-2143D	702	84
TRA-BOND BB-2135D	815	92
TRA-BOND 2112	521	148
EPON 828/V-40 (1:1)	1025	144
EPON 828/V-40 (3:1)	1341	125

the adhesive cleanly separates from the adherend at the interface. Cohesive failure is when the adhesive fails and part of the adhesive remains on both adherends. The percent cohesive failure was calculated by enlarging a photograph of both adherend failure surfaces and using a cut and weigh procedure.

RESULTS AND DISCUSSION

In this study, the effect of three variables upon the shear strength of an epoxy bonded stainless steel were investigated. The three variables studied were surface treatment, temperature (at constant immersion time) and water immersion time (at constant temperature).

EFFECT OF SURFACE TREATMENT - From the preliminary results, given in Table 3, it was decided to use the EPON 828/V-40 (3:1) formulation in this study to evaluate the effect of surface treatment. Table 4 gives the average shear strengths obtained for the six different surface treatments and also the type of failure which was observed in each case.

Again, there appears to be a large variation in the results. In addition to an apparent lowering of the strengths from greater than 1400 psi (maximum) there appears to be a change in the type of failure as well. In all cases, except 3, the exposure to water at elevated temperature has produced an adhesive type of failure where previously, as shown in Table 4, only the degreased stainless steel showed an adhesive type of failure.

In order to ease the interpretation of the results, an analysis of variance (ANOVA) table was constructed (3). ANOVA is a statistical technique in which the variance produced by a given variable, averaged over several levels, is compared to the error (determinded from a residual). The comparison is made by taking the ratio of the variance produced by the effect to that attributed to the error, or in statistical terms, by calculating the "F" ratio. By previously selecting a confidence level, in this case 99%, one can obtain a critical F value, Fcrit, from a statistical table. If the calculated F is greater than the critical F, the effect is statistically significant.

Table 4 - Shear Strengths of Epoxy Bonded Stainless Steel Lap Joints at Room Temperature

Stainless Steel Surface Treatment	Average Shear Strength (psi)	Type of Failure
Degreased	1001	Adhesive
Degreased, Abraded	1344	24-60% Cohesive
Degreased, Primed	1436	33-63% Cohesive
Degreased, Abraded, Primed	1222	20-38% Cohesive
Degreased, Etched	1264	37-64% Cohesive
Degreased, Etched, Primed	1082	35-52% Cohesive

From the results, it can be seen that the strength obtained varies considerably with the type of surface treatment, the highest strength being with the degreased and primed surface.

It is also interesting to note that the mode of failure changes with surface treatment. In the case of the degreased surface, the failure was adhesive. For the other treatments, the locus of failure changed from the surface to partially within the resin. Qualitatively, as the failure becomes more cohesive, the bond strengths appear to increase.

EFFECT OF TEMPERATURE - The temperature range investigated in this study was from 25°C to 95°C. The water immersion time at each temperature was held constant at one hour. Six different steel surface treatments were used.

The results of the shear strength determinations, along with the mode of failure, can be seen in Table 5 on the next page.

On the other hand, if the calculated F is less than the critical F, then the effect is statistically insignificant.

The analysis of variance table for the shear strength results presented in Table 5 is given in Table 6 on the next page.

From the table it can be determined that the value of the calculated F for temperature (471/1514) has a value of 0.31, while the critical F obtained from a statistics table has a value of 6.93. As seen from the F ratio test, Fcrit is greater than Fcalc therefore it can be concluded that temperature, in the range studied, does not have a significant effect on the bond shear strength.

A similar analysis of the results obtained for the effect of surface treatment shows a statistical significace. As can be seen from the F ratio test (Table 6), the calculated F is greater than the critical F. A further discussion of the surface treatment effect in relation to the water immersion time effect will follow.

**Table 5 – Average Shear Strengths (in psi) of Epoxy Bonded Lap Joints
Immersed in Water for One Hour at Various Temperatures**

Water Temperature in Degrees Centigrade

Surface Treatment	25 Strength	25 Type of Failure	60 Strength	60 Type of Failure	95 Strength	95 Type of Failure
Degreased	652	Adhesive	650	Adhesive	615	Adhesive
Degreased, Abraded	855	Adhesive	796	Adhesive	760	Adhesive
Degreased, Primed	827	9–52% Cohesive	815	Adh.–47% Cohesive	810	Adhesive
Degreased, Abraded, Primed	835	12–33% Cohesive	827	Adhesive	716	Adhesive
Degreased, Etched	892	Adhesive	890	Adhesive	895	Adhesive
Degreased, Etched, Primed	811	Adhesive	809	Adhesive	870	Adhesive

**Table 6 – ANOVA for the Effects of Surface Treatment and Temperature
on the Shear Strength of Epoxy Bonded Stainless Steel**

Source	Deg. of Freedom	Sum Square	Mean Square
Between Surface Treatments	5	1502132	250355
Between Temperatures	2	943	471
Error	12	18173	1514
Total	19	1521249	

F-ratio Test

	Fcalc	Fcrit
Surface Treatment	165.	4.82
Temperature	0.31	6.93

EFFECT OF TIME – Since the effect of temperature was not found to be statistically significant in the previous section, we decided to eliminate it as a variable in our further experimentation and hold it constant at 60°C while looking at the effect of time. Since the effect of surface treatment was statistically significant, it was kept as an experimental variable.

The shear strengths obtained for the epoxy-bonded stainless steels, having six different surface treatments, which were immersed in water at 60°C for times of 1, 4, 8, 25, and 168 hours are given in Table 7. It should be noted in all cases except the degreased and primed surface at one hour of immersion time that the failure was adhesive. From the results it can be seen that the effect of (immersion) time produces a drastic reduction in the shear strength of the bonded specimens. In general it can be seen that the exposure to immersion in water at elevated temperature for one week produces approximately a 50% or greater reduction in the shear stress of the adhesively bonded joints.

shear strength. The F value for the time in water is approximately 4 and 30 times that for the surface treatment and interaction respectively. A comparison of the calculated F (60.68) and the critical F (3.48) show that the effect of immersion time is highly significant.

The second greatest effect was produced by surface treatment. A comparison of the F values in Table 8 shows that the effect surface treatment is statistically significant.

The interaction between the surface treatment and time in water had the lowest calculated F value and therefore the lowest effect on the shear strength of the epoxy bonded joints. When compared to the critical F, the effect of interaction is statistically significant, but not to the level as found with the main effects of immersion time and surface treatment.

Hiroyuki Ishii and Yukisaburo Yamaguchi (4) did a study investigating the effect of immersion in water on epoxy-polyamide bond strength of a butt joint using aluminum, stainless steel and mild steel adherends.

Table 7 – Average Shear Strengths (in psi) of Epoxy Bonded Stainless Steel
Immersed in 60°C Water for Varying Amounts of Time

| | Immersion Time in Hours | | | | |
Surface Treatment	1	4	8	25	168
Degreased	650	663	645	386	332
Degreased, Abraded	796	795	742	399	360
Degreased, Primed	815	640	615	558	550
Degreased, Abraded, Primed	827	799	757	707	547
Degreased, Etched	890	867	783	571	434
Degreased, Etched, Primed	809	662	396	344	306

In order to get a better understanding of the results obtained in this set of experiments, an ANOVA table was again constructed using a 99% confidence level. The ANOVA results can be seen on the next page in Table 8.

The analysis of variance was used to determine the level of significance of each of the following effects; water immersion time, surface treatment and the interaction between the two; on bond shear strength. The error was again estimated from the residual of the above effects.

The F-ratio, which is the ratio of the variance produced by the effect to the random variance (error), is a measure of the ability of the experimental variable to produce a change in the response variable (shear strength). By comparing the magnitudes of the calculated F values, the effects can be ranked as well as tested for statistical significance.

Comparing the Fcalc values in Table 8, it can be seen the (immersion) time in water has produced the greatest effect on the adhesive

The temperature of the immersion water ranged from 20 to 100°C and the immersion time from 1 to 100 hours. The tensile strength of the butt-joined specimens was used to determine bond strength as opposed to the lap shear specimens used in this study. They also investigated treating bonding surface with silane coupling agents. In all cases the bonding surface was abraded prior to bonding. The results of their experiments showed that at water temperatures of 20 to 40°C, even up to 100 hours of immersion, the bond tensile strength was much higher, especially after 50 hours immersion time. After one hour at 100°C, the bond only retained 40% of it's original strength. In all cases, even at normal temperature the type of bond failure was adhesive at the interface. The results were basically the same for all of the adherend materials that the workers investigated.

Ishii and Yamaguchi also found that the overall bond tensile strength was twice as high when the three adherend surfaces were pretreated with a silane coupling agent. The

Table 8 – ANOVA for the Effects of Surface Treatment and Immersion Time
on the Shear Strength of Epoxy Bonded Stainless Steel

Source	Deg. of Freedom	Sum Square	Mean Square
Between Surface Treatments	5	1004258	200851
Between Times In Water	4	3049100	762275
Interaction	20	655972	32798
Error	120	1507395	12561
Total	149		

F-ratio Test

	Fcalc	Fcrit
Interaction	2.61	2.03
Surface Treatment	15.99	3.17
Time in Water	60.68	3.48

overall shape of the curve, strength versus time, at each temperature, did not change.

The results found in this study are quite similar to those reported by Ishii and Yamaguchi (4). In order to see if the general shape of our strength-immersion time results were similar to those of the other workers, we plotted our results as reduced shear stength versus time. The shear strengths at each immersion time were divided by the initial (unconditioned) shear stength, for each surface treatment, to obtain the reduced shear strength. By using a dimensionless strength, the strength-time relationships could be compared directly to each other simply by superimposing the curves. A plot of reduced strength versus immersion time for the epoxy bonded stainless steel which had been degreased and primed is given in Figure 1. (This is the surface treatment which produced the highest strength found in this investigation.) In the figure, the reduced strength values are plotted at each time along with error bars which indicate the standard deviation found at each condition. When the curves were superimposed on each other, it was found that they had similar shapes.

MATHEMATICAL MODELS - Since water immersion time had the most significant effect on bond shear strength, the strength-time results for each surface treatment were fit to two models in order to obtain a mathematical relationship to predict the effect of immersion time on the strength. The two models used were a logarithmic model

and a power law model.

Logarithmic Model - The strength-time data was fit, using the method of least squares (5), to the following equation

$$S = a + b \ln t \qquad (1)$$

where S is the reduced shear strength, t is the time in hours, and a and b are two empirically derived constants. The resultant curve derived from the model is shown as the line through the data in Figure 1. In general, the resultant curves passed through the error bars for all of the surface treatments. The curves plotted using this model had correlation coefficients between 0.814 and 0.903. The correlation coefficient indicates the "goodness of fit" of the model to the data. A correlation coefficient with a magnitude of 1 is the best, while a value of 0 indicates no correlation. This model predicted the overall strength-time behavior well. The logarithmic model predicted the strength-time behavior best for the degreased/etched surface treatment (it had the highest correlation coefficient).

Power Law Model - The strength-time data was also fit, using the method of least squares to the following equation

$$S = at^b \qquad (2)$$

where S is the reduced shear strength, t is the time in hours, and a and b are empirically derived constants. The reduced shear strength - immersion time data for the

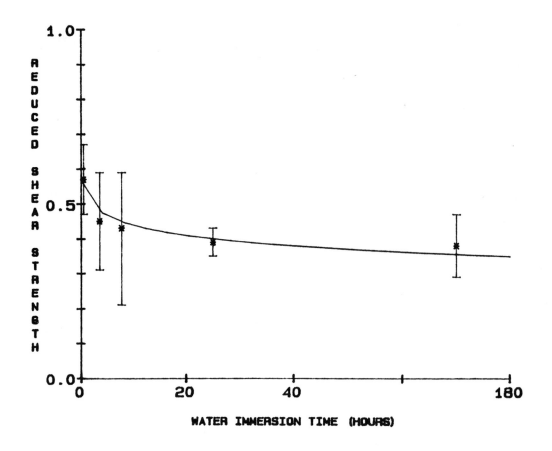

Fig. 1 – Effect of immersion time on the reduced shear strength of epoxy bonded stainless steel (degreased and primed). The line through the data represents the least squares best fit of the logarithmic model.

epoxy bonded stainless steel which had been degreased and primed are plotted in Figure 2 on the next page. Again, it can be seen that the curve, which was calculated from the model, is within the standard deviations of the data points. The power law model (Eq. 2) was fit to the strength–time data for each surface treatment. The least squares method was again used to determine the correlation coefficients for the plots. The correlation coefficients obtained were between 0.997 and 0.998. This model predicted the overall strength–time behavior very well. The power law model predicted the strength–time behavior best for both the degreased/abraded/primed and the degreased/primed surface treatment.

Comparison of Models – The power law model is a better fit to the experimental data than the logarithmic model. The correlation coefficients for the power law models are higher than for the logarithmic models.

MECHANISMS FOR BOND FAILURE – In the absence of environmental exposure, a good bond fails cohesively, however, the true test of a structural bond is cohesive failure rather than adhesive after exposure. The unexposed bond strengths in this study failed in a cohesive manner, however after even very

short exposure times, they failed adhesively.

Since the variable that produces the most significant effect on the bond shear strength is water immersion time, the mechanism for bond failure is probably diffusion of water along the interface and/or through the adhesive.

The initial failure that occured for all surface treatments, is probably caused by stress development at the interface. The stresses are probably created form the differences in the properties of the adherend and adhesive, as they both try to reach equilibrium in the constant temperature bath, at their own rates. The immediate, continued failure is probably caused by diffusion along the interface. The surface treatment methods used probably created an interfacial bond that initially resisted the diffusion of water along the interface.

CONCLUSIONS

From our experimental results, we were able to draw the following conclusions.

1. The type of epoxy used in bonding stainless steel is very important. The strengths obtained with the different resins varied by an order of magnitude.

2. Immersion in water was found to

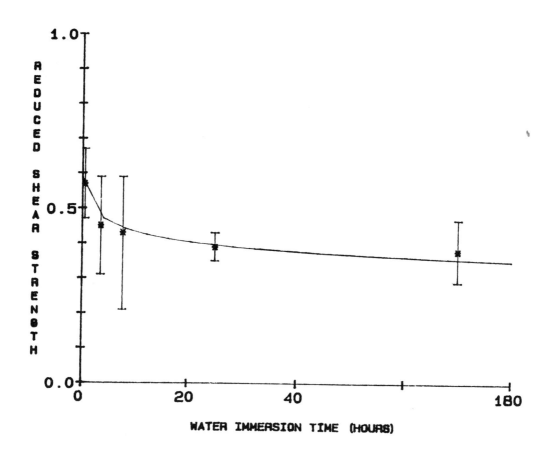

Fig. 2 - Effect of immersion time on the reduced shear strength of epoxy bonded stainless steel (degreased and primed). The line through the data represents the least squares best fit of the power law model.

produce a severe reduction in the shear strength of epoxy bonded stainless steel. In some cases, water was found to produce more than a two-thirds reduction in the observed strength.

3. An analysis of variance was used to show that the effect of immersion temperature did not have a statistically significant effect on the shear strength over the range of temperatures investigated in this study.

4. The type of pretreatment used on the stainless steel was found to have a significant effect on the bond strength obtained in the tests. Pretreatment was found to be significant in the as-bonded strength as well as the reduced strength after exposure.

5. Immersion time was found to have the greatest effect on the shear strength of the epoxy bonded stainless steel. In general, after one week of exposure at elevated temperature, the shear strength was reduced by about two-thirds.

6. The power law model was found to predict the effect of immersion time at elevated temperature on the shear strength of the epoxy bonded stainless steel in the range of times investigated in this study.

REFERENCES

1. "Annual Book of ASTM Standards", American Society for Testing and Materials, Philadelphia (1981).
2. K.A. Geddes, "The Effects of Water Immersion on the Shear Strength of an Epoxy-Bonded Stainless Steel", M.S. Thesis, Plastics Engineering Department, University of Lowell, Lowell, Massachusetts (1985).
3. E.L. Bauer, "A Statistical Manual for Chemists", Academic Press, New York (1971).
4. Ishii, H., and Y. Yamaguchi, "Effect of Immersion in Water on Adhesive Bonding Strength of Butt Joints", (Ind. Arts Lab. Saitama Prefect., Japan) Kogakuin Daigaku Kenkyu Hokoku 1978, 44, 65-72 (Japan); Translated by Dr. E. Sanz, University of Lowell, Economics Department.
5. "HP-25 Applications Programs", Hewlett Packard Company, Cupertino, California (1975).

ACKNOWLEDGMENT

The authors would like to acknowledge the generous support of Sunhouse, Incorporated, without which, this work would not have been possible.

APPLICATIONS OF THREE-DIMENSIONAL ABRASIVES IN ADVANCED COMPOSITE MANUFACTURING TECHNOLOGY

Douglas E. Earl, Beth E. Davidian
3M
St. Paul, Minnesota USA

Abstract

During the manufacturing of composite materials or assemblies, problems often arise that require light surface abrasion with little or no change in the underlying material. Many composite manufacturers are finding that three-dimensional abrasives provide a unique solution to this problem of "surface conditioning." Three-dimensional abrasives are manufactured by adhering standard abrasive materials to a base of high-loft, non-woven, synthetic web. As such, they possess a high degree of compliance which provides an even application of work pressure even when working on complex surface shapes. The surface-conditioning abilities of three-dimensional abrasives result in many applications in composite-manufacturing processes. Examples of some of these applications, in both bond site preparation and surface finishing, are presented.

THREE-DIMENSIONAL ABRASIVES are manufactured by using a resin binder to adhere abrasive particles to the fibers of a high-loft synthetic fiber web. As such, the performance of a three- dimensional abrasive can be varied by changing the nature of the abrasive (particle size and quantity), the resin binder (friability and bonding characteristics), or the fiber web (fiber size and density). Manufacturers of three-dimensional abrasives will use these variables to "tune" their products to better function an a wide range materials.

The loft of three-dimensional abrasives provides compliance which permits these products to conform to irregular surface shapes without producing excessive pressures on high spots. This characteristic permits abrasive finishing of complex shapes with minimal disruption of the underlying material. For this reason, three-dimensional abrasives are often the products of choice for surface preparation of composite materials.

BOND SITE PREPARATION

Optimum bonding of composite materials requires that the bond site be both free of surface contamination, and that it possess a surface profile of some roughness. Surface contamination can originate not only from environmental factors, but may be transferred from materials used in the fabrication of the composite itself (contamination from release liners). For this reason, even a carefully-prepared, carefully-maintained bond site cannot be considered to be contamination free. In order to achieve a clean lightly-abraded surface, most methods of bond- site preparation involve both a solvent, and some type of abrasive material.

The ability of a three-dimensional abrasive to remove surface contamination from a composite surface was investigated. A graphite/epoxy composite was fabricated using a Teflon release liner. X-ray photoelectron spectroscopy (XPS) showed that there was fluorochemical contamination originating from the release liner on the composite surface. Surface preparation was done on the contaminated composite surface using a three-dimensional abrasive and water for a solvent/lubricant. Additional XPS analysis of this sample showed that the contamination had been eliminated (figure 1). Scanning electron microscopic examination showed the three-dimensional abrasive roughens the surface without damaging the graphite fibers (figure 2).

PAINT REMOVAL FROM COMPOSITE MATERIALS

The conformable and open structure of three-dimensional abrasives make them ideally suited as media for the abrasive removal of paint from composite surfaces. The conformable nature of these products permits them to be used on irregularly shaped structures, with less chance of damaging the underlying composite structure than with conventional coated abrasives. Their open construction allows them to be used for extended periods of time, without "loading" of the abrasive with the paint residue. Further, three-dimensional abrasives may be used with a water flood which reduces the the problems associated with airborne paint residue.

The suitability of three-dimensional abrasives as a paint removal media was investigated by using these products with variety of powered tools to remove paint from composite samples. The composite samples were of graphite-epoxy construction, and were painted with a top coat of polyurethane (mil-C-832868) over an epoxy/polyamide primer (mil-P-22377D). The composite-paint system was aged seven days at room temperature and 96 hours at 210 degrees F.

The evaluation consisted of using the three-dimensional abrasive (on a power tool) to remove a measured area of paint from the composite panel. The time required to remove the paint was timed and the composite panel was visually inspected for damage. By using a conventional right-angle grinder at 4500 rpm and a 7-inch three-dimensional abrasive, it was possible to remove paint at a rate of 0.5 square feet per minute with no apparent damage to the composite.

GENERAL APPLICATIONS

In addition to the unique applications of bond site preparation and paint removal, three-dimensional abrasives have numerous general applications in composite manufacturing.

MOLD CLEANING - Three-dimensional abrasives are used to clean residue from the surfaces of forming molds. In this application, the abrasive mineral provides the cut required to remove tightly-adhered residues. The inherent conformability reduces the problem of altering critical mold dimensions.

SURFACE DEFECT REMOVAL - Three-dimensional abrasives are used when light blending of surface defects is required, either as a final finish, or prior to additional finishing steps.

GENERAL SHOP MAINTENANCE - Three-dimensional abrasives are used in many cleaning operations generally associated with steel wool, wire brushes, or conventional abrasives. In these applications, three-dimensional abrasives can often demonstrate advantages over the conventional alternatives in cost, safety, effectiveness, or productivity.

SUMMARY

Three-dimensional abrasives can be used with a water solvent/lubricant to remove surface contamination from composite surfaces. A surface properly prepared in this fashion will be free of both contaminants that are a result of the forming process and those that result from handling and storage. Further, three-dimensional abrasives provide significant advantages over traditional methods such as wire brushes, steel wool, and conventional abrasives.

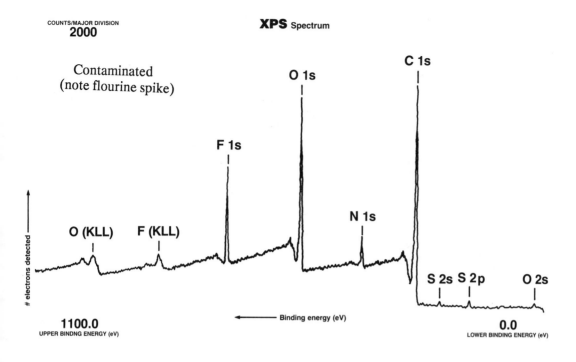

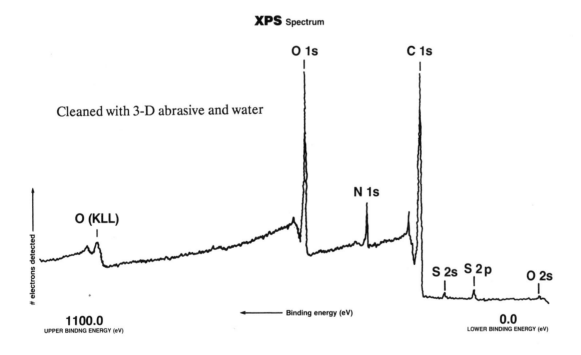

Figure 1: XPS spectrum of contaminated and cleaned surface.

100X

Figure 2: Photomicrograph of a three-dimensionally
cleaned surface.

LAMINATION OF ORDERED POLYOLEFIN SHEETS BY ULTRASONIC WELDING

T. B. Zach
Dept. of Chemical Engineering
University of Toronto
Toronto, Ontario

T. H. North
Dept. of Metallurgy and Materials Science
University of Toronto
Toronto, Ontario

R. T. Woodhams
Dept. of Chemical Engineering
University of Toronto
Toronto, Ontario

Abstract

Oriented polypropylene was ultrasonically welded using specially-depeloped tie-layer materials. A tie-layer comprising 10% by weight of amorphous polypropylene (APP) in ethylene propylene copolymer (EPC) produced the most satisfactory joints. The maximum lap shear strength values achieved were 12.5 MPa and joint failure was characterized by cohesive failure in the oriented polypropylene-materials itself. Use of smaller diameter horns produced welds that were more unifrom and much stronger than those made using larger diameter horns.

WITH ROLLING TECHNOLOGY BORROWED from the metallurgical industry, a considerable amount of work has gone into the fabrication of molecularly-oriented sheet from polyolefins such as polyethylene and polypropylene. Unfortunately the high strength and modulus offered by the uniaxially oriented polypropylene sheet in the direction of orientation are made less attractive due to the essentially unaltered strength and modulus values in the transverse direction. One way of overcoming this deficiency is biaxial orientation. An alternative to biaxial orientation is the lamination of uniaxially-oriented sheets. Lamination is a critical step in the fabrication of thick multiply laminates suitable for structural applications.

A major difficulty in the lamination of oriented polypropylene (OPP) is the formation of adequate bonds which are hindered by the non-polar, semi-crystalline nature of polypropylene (PP). This makes bonding by traditional means with adhesives and solvents unsatisfactory In an attempt to circumvent this problem, surface treatments with corona discharge and gas plasma, which can make the use of adhesives possible, have been given considerable attention in recent years. Direct heating may prove undesirable since it affects the bulk of the oriented material and may result in a loss of fibril orientation. One solution is the use of a suitable tie-layer to facilitate fusion at the joint interface of a laminate.

This investigation examined the suitability of ultrasonic welding for laminating of highly oriented PP sheets together without significant loss in their mechanical properties by the use of a suitable adhesive tie-layer.

ADHESION

Adhesion can be defined in its simplest physical terms as the joining together of two different materials. This can be broken down into two types or catagories of adhesion as presented by Mittal [1]. The first is called "basic adhesion" and approaches adhesion from a molecular point of view. It is the sum of all the internal interactions at the interface. The second type of adhesion is termed "practical adhesion" and is based on a macroscopic point of view. It measures that force or work required to effect separation of two adhering materials.

There are various theories of adhesion [2-4]. According to the wetting theory a simple way of determining the bondabiltiy of a polymer with an adhesive is to examine how well an adhesive wets a polymer surface. The condition for wetting is $\gamma_{lv} \leq \gamma_{sv}$, where γ_{lv} and γ_{sv} are the surface free energies of the liquid adhesive and solid substrate respectively. The main difficulty in wetting thermoplastics is due to their low surface free energy. Chemical modification of the surface prior to bonding is usually necessary to increase the surface free energy [5-7]. However, it should be noted that although wetting is necessary it is not a sufficient condition for high adhesive strength [8,9].

The dominant mechanism of adhesion during the welding of thermoplastics is diffusion. Voyutskii [10] has suggested that the diffusion of polymer chains across an interface determines the adhesive strength of a polymer weld. An increase in temperature results in an increase in the Brownian motion of a chain. Diffusion of polymer chains is then possible when the polymer temperature exceeds the glass transition temperature (Tg) for an amorphous polymer or the polymer melt temperature (Tm) for a semi-crystalline polymer [11]. Another criterion is that the two polymers be mutually soluble, that is, they possess similar solubility parameter values. As the solubility parameters of two polymers diverge, one polymer is highly cross linked, has bulky side groups, or the temperature is not above the Tg or Tm values then diffusion is unlikely [8,12,13].

ULTRASONIC WELDING

Ultrasonic welding is a simple method of joining thermoplastics. The technique requires the conversion of low frequency electrical power at 60 Hz to high frequency mechanical vibrations, 20 kHz in this case, which is then applied to thermoplastic components. The heat is generated in two ways. First, the interface surface asperities heat rapidly by friction to the softening point. At that time the polymer begins to flow, the rate of heating is reduced and the chain diffusion / entanglement process commences. Second, the material generates heat by hysteresis losses. When the plastic melts at the interface the components join to form a fusion bond.

When a polymer is subjected to sinusoidal deformation the stress and strain are out of phase due to polymer viscoelasticity (Fig. 1). The consequence of such a phase shift is that a portion of the mechanical energy is dissipated as heat. This is referred to as hysteresis loss (Fig. 2) and is dependent on the applied stress and the loss modulus, E", which in turn is dependant on the frequency of the applied stress and the temperature. The loss factor, or tan δ, is related to the loss modulus E" and the storage modulus E' as

$$\tan \delta = E''/E' \qquad (1)$$

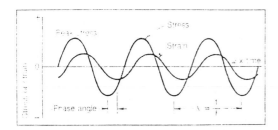

Fig. 1. Relationship between stress, strain, frequency, and phase angle in sinusoidal fatigue of a viscoelastic material. Mean stress equals zero. [14]

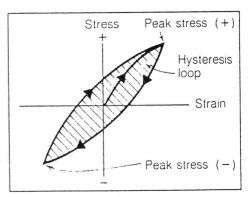

Fig. 2 Stress strain relationship in a single fatigue cycle for a viscoelastic material showing hysteresis loop. Enclosed area is the energy dissipated per cycle as heat. Mean stress equals zero.

The loss factor is thus dependent on the frequency of oscillation. Higher frequencies result in "stiffer" material behaviour as in Figure 3. According to Menges and Potente [14] the amount of heat generated, by internal friction losses can be written as

$$\varepsilon^2 E' \omega \eta t/2 = \rho C(T_e - T_a) \qquad (2)$$

where

ε: peak value of alternating strains
E': storage modulus
ω: Circular frequency ($2\pi v$)
η: tan δ
t: time
ρ: density
C: specific heat capacity
T_e: softening temperature
T_a: ambient temperature

This equation applies only until the softening temperature of a polymer after which the modulus of elasticity can no longer be meaningfully defined. The rate of heat generation can thus be written as

$$\varepsilon^2 E' \omega \eta/2 = q \qquad (3)$$

or

$$\varepsilon^2 E'' \omega/2 = q \qquad (4)$$

as given by Ferry [15].

It follows then from Equation 2 that for a constant frequency the greater the value of the loss factor, the alternating strain, or the duration of vibration time, the greater is the amount of heat generated in a polymer. Thus by controlling the amplitude or the material properties of the polymer it is possible to control the rate of heating according to Equation 4.

For the OPP sheets employed in this study no energy directors can be incorporated on the flat OPP sheets. To overcome this difficulty a tie-layer was placed between the OPP sheets to absorb and convert the vibrational energy into heat. This causes melting of the tie-layer, transferring the heat to the substrate OPP sheets, resulting in tie-layer / OPP fusion. This process takes place in a fraction of a second. To enhance the dissipation of heat in the tie-layer, the modulus of the tie-layer must be lower than the OPP modulus. The tie-layer employed was ethylene propylene random copolymer (EPC). The tie-layer modulus was further modified by the addition of impact modifiers such as EPDM or amorphous polypropylene (APP) [16,17]. The mechanical properties of the APP/EPC blend are then a function of the relative amounts of APP or EPDM in the blend.

Since the tie-layer and OPP are in series and the both materials exhibit Hookean behaviour, it can be assumed that the applied stresses produced by the ultrasonic vibrations are the same in the OPP and the tie-layer. The resultant strains are then governed by the ratio of the moduli of the OPP and the tie-layer where, $\varepsilon = \sigma/E$. The ratio of the rate of heat generated (q_{tL}) in the tie-layer compared to that in the OPP material is as follows:

$$q_{tl}/q_{opp} = (E_{opp}/E_{tl})^2(E''_{tl}/E''_{opp}) \qquad (5)$$

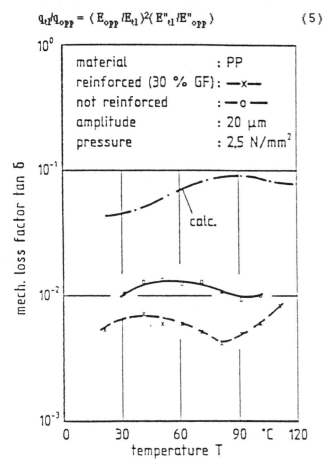

Fig. 3. Mechanical loss factor, tan δ, as a function of temperature at 20 kHz.

Experimental - The polymer used throughout this investigation was extruded PP sheet stock (Trovidur grade, Dynamit Nobel Ltd.) with a melt temperature of 165° C. The number average and weight average molecular weights were 70,900 and 377,000 respectively. The precut billets were 10 mm thick and 150 mm wide. These were then processed by roll-drawing (Fig. 4) in a Deltaplast Co. Inc. rolling mill at a temperature of 158° C [18]. This yielded 0.9 to 1.2 mm thick sheets of OPP with draw ratios between 9 and 11. The tie layer employed in these experiments was an ethylene-propylene random copolymer (EPC) (4% ethylene) supplied by the Shell Development Company [69]. Ethylene propylene diene monomer (EPDM) from Uniroyal [16] and type M-5C amorphous polypropylene (APP) from Eastman Chemical Company [17] were added as impact modifiers (zero to 20 % by weight).

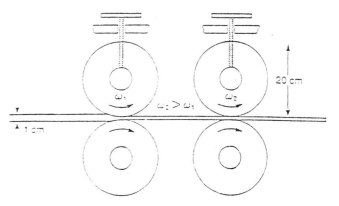

Fig. 4. The roll-drawing process.

A Gelimat K-mixer set at 3300 rpm was used to blend APP or EPDM in the EPC matrix in 150 g batches. The blend was discharged at 180° C and through a series of steps pressed to a 0.65 mm (±0.03 mm) thickness between sheets of Mylar film at 160° C. Immediately upon removal from the press the tie layer was quenched by immersion in cold water. As a result of the fast cooling rate the tie layer possessed a low degree of crystallinity. The tie layer was then cut into 25.4 mm by 25.4 mm samples and used within four hours. The tensile modulus of the 10% APP in EPC tie-layer was 0.53 ± 0.08 GPa. Samples of the pressed tie-layer were cut into rectangular sections according to notched Izod ASTM D 256-81, dipped and equilibrated in liquid nitrogen, and fractured in order to produce a smooth fracture surface for examination with a scanning electron microscope (SEM).

A 1700 watt Thermosonics microprocessor-controlled ultrasonic welder, model EO 1090, was used to laminate the 25.4 mm X 80 mm (1 in. x 3 in.) specimens of OPP as in Figure 5. Prior to welding, the OPP strips and the tie layer were rinsed in water and then placed in an ultrasonic bath of ethanol for 15 minutes after which they were air dried. The applied wattage, horn pressure, and strain amplitude were then varied individually while keeping the other two constant. The strength of adhesion in the laminate was measured by a lap shear test (ASTM D3163-73) at a constant speed of 5 mm / min. Five to seven specimens were tested for each sample. Only parallel-ply specimens were tested due to the inherent weakness of the OPP in the transverse direction. To examine the uniformity of the laminate weld zone the samples were examined using an ultrasonic C-scanning device, and an optical microscope.

Results and Discussion - The majority of lap shear specimen produced with the ultrasonic welding device operating in the constant energy mode. That is to say, a controlled amount of energy measured in watt seconds (J) was delivered to each sample as opposed to operating the unit in a controlled weld cycle time. The energy delivered to a sample is a function of the applied instantaneous pressure when all other variables are held constant. This can be expressed as

$$\Delta E = \int_0^t Pdt \qquad (6)$$

where ΔE is the applied energy, P is the instantaneous pressure and t is the time [19]. The microprocessor automatically adjusted the welding time inversely with the pressure to control the amount of energy delivered to the sample.

The operating parameters for these and subsequent samples, unless otherwise indicated, were as follows: the horn amplitude of vibration was 80 μm peak to peak; the applied static horn pressure was 0.207 MPa ; and the tie-layer thickness was 0.56 mm ± 0.03 mm.

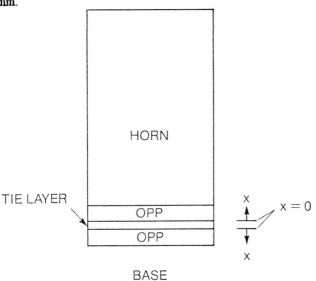

BASE

Fig. 5   Shchematic representation of the welding process showing the relative positions of the horn, OPP, and tie-layer.

The EPDM-modified EPC was compared to EPC containing 10 percent APP as shown in Figure 6. The bond strengths show a strong dependence on the energy input to the sample. For the sake of clarity the standard deviations have been omitted. The bond strength deviations were less than ± 1 MPa which was consistent with Tolunay's work [20]. Failures at the maximum bond strengths were characterized by cohesive failure of the OPP producing fracture surfaces with hair-like fibres and ribbon-like strips of OPP protruding from the fracture surface (Fig. 7). Unlike the APP-modified samples, the EPDM-modified samples exhibited localized overheating, or "hot spots", due to non-uniform dispersion of the EPDM phase in the EPC matrix.

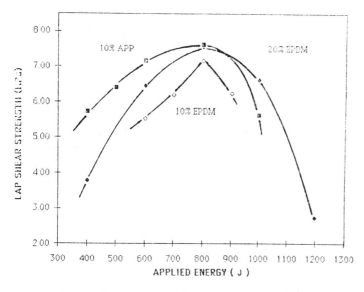

Fig.6   Lap hsear strength versus applied energy for different tie-layers compositions.
(P=0.207 MPa, ε=80 μm)

Fig. 7   Picture of cohesive OPP failure of a tested specimen.

The amount of APP added to the EPC matrix was 5, 10 and 20 percent by weight (Fig. 8). The 10 percent concentration again produced the highest bond strength. Due to increased viscoelasticity the curve of 20 percent by weight addition of APP shifted to lower energy inputs. However, all the 20 percent samples displayed some OPP surface deformation and bubbles along the bondline (except at 400 J). These features adversely affected the maximum lap shear strength values that were attained. This indicated a problem controlling the heat generated in the 20 percent APP/EPC tie-layer. The 5 percent APP/EPC system was characterized by a similar OPP failure as in the 10 percent EPDM-modified adhesive films. This suggested that an optimum blend existed under these operating conditions. Thus the 10 percent APP/EPC blend was chosen as the tie-layer for subsequent tests.

Figure 9 indicates the . strong dependence of lap shear strength on the amplitude of vibration. Since welding was performed with a constant energy input (600 J) the amplitude of vibration was inversely related to the weld cycle time [21,22]. This was confirmed experimentally as the weld time decreased with an increase in horn amplitude. The mode of failure as the horn amplitude decreased was increasingly characterized by bulk degradation of the OPP. That is, due to the increased duration of the welding cycle, the amount of heat transfer to the bulk OPP increased. Heat losses to the bulk OPP can be minimized by increasing the rate of heating i.e., by increasing the strain amplitude applied during welding This reduced the time required to transfer a prescribed amount of energy to the joint.

A smaller horn was employed to further investigate the effect of horn vibration amplitude and geometry. As a consequence of the narrower horn tip, smaller lap specimens were produced. At 80 μm amplitude the lap shear strength increased from 8.6 MPa to 10 MPa, or by 18 percent. This increase was due to the reduction in overlap length in the test samples. Wu[13] has pointed out that although the force at break will decrease with decreasing overlap length the stress will increase. Also Shonhorn has shown that the centre area of a lap shear specimen contributes little to the overall lap shear strength [23]. Another reason for the increased lap shear strength was the increased uniformity of the weld samples. Smaller samples were less likely to contain flaws and poorly welded areas and as a consequence both the load and the stress increased. When the horn amplitude was increased to 110 μm, a lap shear strength of 12.5 MPa was attained.

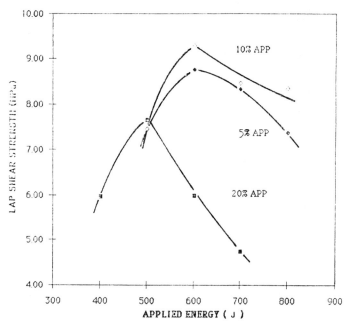

Fig. 8. Lap shear strength versus applied energy for different tie-layer modifier concentrations. (P=0.207 MPa, ε=80 μm)

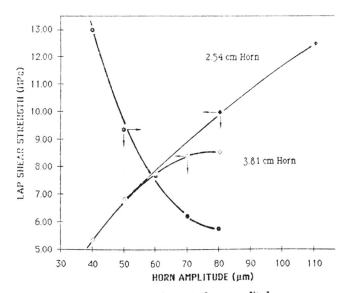

Fig.9. Lap shear strength versus horn amplitude of vibration. Two horn diameters are shown (P=0.207 MPa, Δ=600 J)

Figure 10 and 11 are plots of the lap shear strength versus applied horn pressure for three different horn amplitudes and energy inputs respectively. All curves display the same general trend: a steep positive slope followed by a plateau after reaching a critical pressure, after which the bond strength becomes independant of the applied pressure. Before the plateau, at low pressure, the horn pressure was unable to apply sufficient vibrational energy to the tie-layer resulting in incomplete fusion, wetting and chain diffusion. This is indicated by the predominantly interfacial when welding failure at this horn pressure. It is evident

then that even in the constant energy mode a minimum pressure, or critical pressure, must be applied to achieve consistent welds. It should also be noted that the final laminate thickness decreased from approximately 2.6 to 2.1 mm as the applied pressure increased. This decrease was not only a result of squeezing out the tie-layer but also loss of OPP thickness. This effect is undesireable since the laminate's ultimate mechanical properties depend upon the high degree of orientation, and loss of orientation will adversely affect mechanical properties. Thus, a minimum applied pressure is desirable to maintain the maximum bulk OPP thickness.

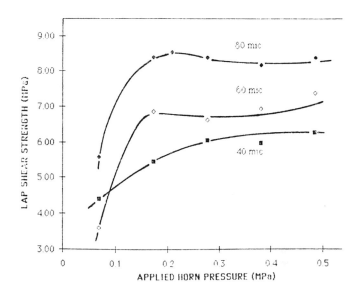

Fig. 10. Lap shear strength versus applied horn pressure for various horn amplitudes. (E=600 J)

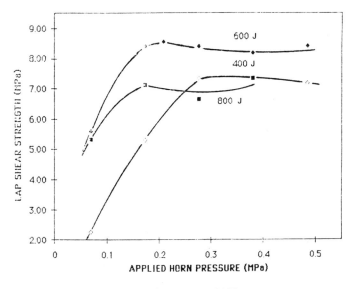

Fig. 11 Lap shear strength versus applied horn pressure for varying applied energy. (ε =80μm)

All previous samples were prepared using the constant energy mode of operation in which the weld time was automatically decreased as the pressure increased. Figure 12 shows a graph of lap shear strength versus applied horn pressure in which the weld time is controlled. It is evident that the maximum bond strength is very sensitive to the weld cycle time. For a constant cycle time the bond strength is very sensitive to the applied pressure with the mode of sample failure changing from interfacial failure to interfibrillar (cohesive) failure in OPP material. Although the maximum bond strength for the 0.68 s run exceeded that of the 0.9 s run by approximately 2 MPa, each had similar energy inputs (560

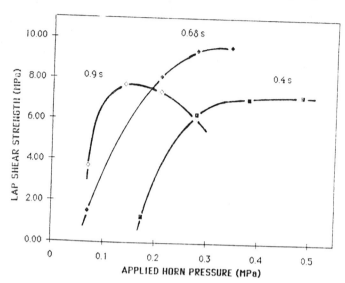

Fig. 12 Lap shear strength versus applied horn pressure for varying weld cycle times.

(ε=80μm)

and 571 J respectively). This again indicated a delicate interplay between pressure and time with respect to the rate of heating and the final joint mechanical properties produced.

A characteristic of the ultrasonically welded laminate is the non uniform clarity or transparency of the welded laminate due to variable opacity at the weld interface (Fig. 13). A sample was subjected to an ultrasonic C-scanning in an attempt to characterize the acoustic uniformity at the weld interface. The transparent and hazy regions had different acoustic properties corresponding to their observed optical characteristics. Microtomed cross sections of the laminates examined under polarized light indicated that the weld interface was non-uniform. The hazy (high acoustic impedance) regions were thicker weld zones and the clear (low acoustic impetance) regions were thin weld zones (see Fig. 14-16). The thicker weld zones contained fragmented heat affected zones that acted as stress concentrators and accounted for the failure of the hazy regions prior to that of the clear regions.

Fig. 13 Picture of ultrasonically welded sample at optimal conditions. Transparent X and haze edges displayed across sample.

Fig. 14 Optical micrograph of the bondline region (Mag. 4X) where
A = Thick (opaque) weld zone, 250 μm thick.
B = Thin (transparent) weld zone
C = OPP

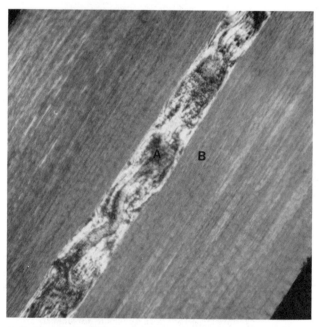

Fig. 15. Optical micrograph of the thick bondline region (Mag. 10X) where
A = Thick (opaque) weld zone, 350 μm thick.
B = OPP

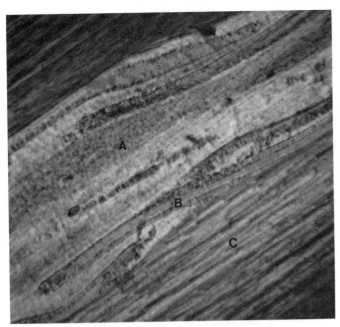

Fig. 16 Optical micrograph of the thick bondline region (Mag. 40X)

where

A = Thick (opaque) weld zone, 335 μm thick.

B = Fragmental heat affected zone, 35 μm thick.

C = OPP

Conclusions - Ultrasonic welding has been evaluated for joining oriented polypropylene. Maximum lap shear strength values of 12.5 MPa were produced in specimens distinguished by cohesive failure in the OPP material. These values compared favourably with those produced in electromagnetic induction welding.

Increased rates of energy input to the sample resulted in increased rates of heat generation at the joint interface. The amount of heat transfered from the interfacial region to the bulk OPP was thus minimized. Also sample uniformity and integrity improved as the rate of heat generation within the interfacial region increased.

In welding operations carried out in which the energy transmitted to a sample was kept constant (constant energy mode) lap shear strengths values were relatively independent of applied pressure as expected. However, it was necessary to apply a minimum pressure (critical pressure) to the laminate samples in order to effect adequate transfer of ultrasonic energy. For optimum lap shear strength this pressure was 0.173 MPa.

Welding operations carried out in which the weld cycle time was kept constant indicated a dependence of bond strength on the relationship between weld cycle time and pressure. That is, for a constant pressure an optimum weld cycle time existed. Conversely, for a constant weld cycle time an optimum pressure existed.

Weld uniformity was affected by horn geometry. The smaller diameter tip horn produced welds that were visibly more uniform and much stronger than those made with the larger horn.

The addition of amorphous polypropylene (APP) to the ethylene-propylene copolymer matrix (EPC) produced more uniform adhesive film than did the EPDM-modified film. Difficulties in uniformly dispersing the EPDM resulted in the formation of hot spots in the tie-layer and caused premature sample failure. The optimum addition of APP was 10 percent by weight.

Characteristisation of weld zone integrity using ultrasonic C-scanning (acoustic impedance measurement) revealed that melting across the joint interface was non-uniform. The thicker weld zones contained non-uniform fragmented heat affected zones. These facilitated crack propagation and accounted for failure of the thicker zones prior to that of thinner zones.

References

[1] K. L. Mittal, Surface Chemical Criteria, Polym. Eng. Sci., 17, 130 (1977).

[2] K. Herman, Z. Physik. CHEM., B10, 371 (1930).

[3] D.W.K. Krevelen, Properties of Polymers, Elsevier ISBN, 0 444-4467-3, (1976).

[4] W. Schlesinger, J. Polymer Sci. 11, 203, (1953).

[5] J. P. Jeandreau, Characteristics and Testing, (ed. R.F. Gould) p.189, Amer. Chem. Soc., Washington, (1984).

[6] H. Shonhorn et al., J. Adhesion, 2, 93, (1970).

[7] W. A. Dukes, A. J. Kinloch, Developments in Adhesives-1, (ed. W.C. Wake) 251, Appl. Sci. Publ., London, (1970).

[8] Y. Iyenger, D. E. Erikson, J. Appl. Sci., 11, 2311, (1967).

[9] J. R. Huntsberger, Treatise on Adhesion and Additives, Vol. 1, p. 199, (ed. R. L. Patrick) Marcel Decker, N.Y., (1967).

[10] S. S. Voyutskii, J. Adhesion, 3, 69, (1971).

[11] J. L. Willett, R.P. Wool, Mechanical Properties of Polymer-Polymer Welds: Time and Molecular Weight Dependance, Dept. of Metallurgy, University of Illinois, Urbana, IL 61801.

[12] A. F. M. Barton, J. Adhesion, 14, 33-62, (1982).

[13] S. Wu, J. Pol. Sci., 25, 557-566, (1987).

[14] H. Menges, H. Potente, Welding in the World, 9, 1/2, (1971).

[15] J. D. Ferry, Viscoelastic Properties of Polymers, 3rd ed., John Wiley and Sons, New York, (1980).

[16] Royalene IM 7565, Uniroyal Chem. Co., Inc., Geisman IL. 70734.

[17] M5-C, Eastbond Amorphous Polyolefins, Eastman Chemical Products, Inc., Kingsport, TN., 37662.

[18] P. E. Burke, Solid Phase Deformation of PP using the Roll-Drawing Process, M.A.Sc. Thesis, Dept. of Metallurgy and Material Science, University of Toronto, (1986).

[19] J. C. Piasecki, Plastics West, Las Vegas, RETEC (1987).

[20] C. B. Bucknall, et al., Polym. Eng. Sci., 20, (1980).

[21] H. Potente, Materials and Design, 5, (1984).

[22] G. E. Anderton, L. R. G. Treloar, Fracture and Tearing in Oriented Polyethylene, J. Mat. Sci., 6, 562-571, (1971).

[23] E. J. Frankel, K. K. Wang, Pol. Eng. Sci., April, 1980, 20, 6, (1980).

ADVANCED TECHNOLOGY FOR
REACTION INJECTION MOLDING

Jerry V. Scrivo
Micromatic TEXTRON
Holland, Michigan   USA

ADVANCED TECHNOLOGY FOR REACTION
INJECTION MOLDING

A new series of injection systems has
recently been commercialized for the
reaction injection molding process. It is
the results of a four year design and
development program and incorporates a
number of new technologies which are being
applied to the process for the first time.

This paper describes technologies,
results, and benefits for RIM molding
operations. Opportunities for future
innovations based on these technologies are
also addressed.

BACKGROUND

Micromatic began developing machinery
for reaction injection molding in 1978 with
the development of a 125 Ton press and work
on mix heads. By 1983 a decision to replace
injection molding and structural foam
machinery with RIM was made. The potential
for faster cycles, energy efficiency, and
utilization of low cost raw materials were
primary factors.(1)

In 1984, market studies and product
descriptions were completed and in late 1985
prototype assembly began. Extensive
development trials were run in 1986 and 1987
followed by field testing in 1988.

GENERAL DESCRIPTION

The Micromatic RIMCENTER is a
free-standing RRIM injection system with a

(1) Scrivo, Jerry V., "Molding without
Melting", SPE ANTE April 1986, Page 1077

dedicated hydraulic source and control panel.
The unit can accomodate up to four mix heads
and mixing pressures to 2250 psi. It is
modular in design and construction.
Production models are available for shot
sizes from 8 lbs to 200 lbs. (See Table I)

Metering is accomplished with lance
cylinders, one for each component. Additional
streams can be added to a total of four if
required. The lances are powered by
hydraulic pumps. They make it possible for
the system to handle various additive fillers
for complete RRIM capabilities. (Figure 1)

The injection systems is equipped with
two component tanks mounted on modular bases.
A third base is provided for the injection
cylinders. This construction allows for
various installation configurations to
accomodate user needs.

Low pressure pumping cylinders (Figure
2) are used for recirculation and lance
filling operations. These reciprocating
pumps utilize a single hydraulic cylinder to
power both "A" and "B" component pumps. A
special valve permits pumping action
regardless of cylinder movement direction.

Heat exchangers are supplied in addition
to jacketed tanks for temperature control of
the component streams. They are placed in
the low pressure circuit and fine tune
temperature prior to lance cylinder filling.
Separate temperature controls are provide for
each stream.

The hydraulic system consist of two
motors each driving an independent pump

system. A small third pump system is provided for auxiliary power. Main pump output is used to drive the Lance Cylinders, while the auxiliary is used to drive low pressure pumps and to power various valves. Low pressure pump speed adjustment is provided by the variable displacement auxiliary pump system.

Temperatures and pressures are monitored by sensors inserted into the component flow streams. Lance position and velocity (injection rate) are determined by integral digital sensors mounted in the lance cylinder.

OPERATING SEQUENCE

When the injection system is powered up, the low pressure pumping cylinders are started and the system is in low pressure recirculation mode. This provides constant circulation of components through the system. It is critical to maintaining constant, stable temperature and to prevent filler particles from settling. Low pressure circulation also provides controlled filling of the lance cylinders. Because a vacuum can cause nucleation gas to escape, it is desirable to provide constant pressure during lance filling.

To avoid a delay on shot request, the lance cylinders are refilled as soon as a shot is completed. Since components in filled lance cylinders are not recirculating, an automatic purge is incorporated to maintain a fresh charge at all times. If a shot is not requested in a preprogrammed time, automatic purge and refill will occure. The time interval can be adjusted by program change. During automatic purge, the lances advance forcing component through the system with low pressure circulation being diverted to tank through a bypass line. When a shot request is received, the interval timer is reset and an automatic purge does not occure.

Several functions occure when a shot is requested. Lance cylinders accelerate to required velocity, a series of valve shifts ensure that the entire output is directed to the appropriate mix head and critical parameters are monitored.

When all parameters are within their acceptable windows, the shot commences. It is terminated when the lance cylinder rams reach their predetermined "end point" where sufficient volume has been displace to fullfil the shot parameters requested. The mix head closes, and the lances continue to advance (high pressure recircultion mode)

until end of stroke. Retract and refill then occure immediately.

The shot can occur only when the machine is in "Auto-Mode". In the alternate "Hand Mode" operation is nearly identical except that the shot is inhibited to allow for adjustment of the mix head needle valves.

NEW TECHNOLOGY FOR RIM

This new injection system brings a number of unique technologies to the field of RIM processing. These are digital servo lance cylinders, mass flow monitors, bulk nucleation control, "no shock" mix head, multiple mix head operation, and a CRT operator interface. In this section, each will be presented in detail including its benefits for the molder.

DIGITAL SERVO LANCE CYLINDERS

At the heart of the RIMCENTER injection system is the new DSLC (Digital Servo Lance Cylinder) The system is equipped with two of these cylinders. They are slurry rated for RRIM operations. They provide for complete separation of hydraulic actuator and component displacement sections. Digital hydraulic servo valves, component valves, rupture disc and velocity position sensors are all integral parts of these assemblies. (Figure 3) Removing four (4) mounting nuts between component and hydraulic portions of the assembly allows separation. Once separated, the component and hydraulic seals are exposed for easy replacement.

Digital servo lance cylinders provide the ultimate in injection control. That is the ability to set parameters from memory and to accurately repeat them until they are updated with new parameters.

The DSLC assembly is comprised of two sub systems; a hydraulic drive and a lance ejector.

DRIVER FEATURES

The hydraulic driver is an integrated unit consisting of hydraulic actuator, feedback sensor and closed loop servo valve. (Figure 4) The processor, feed back and valve drive electronics are all implanted in the servo valve iteself.(Figure 5) Since all control functions are digital, this system eliminates external D/A conversions, potentiometer adjustments and 80% of the separately mounted electrical devices in a conventional system.

It eliminates a highly regulated DC power supply, is insensitive to electrical noise and is free of drift caused by temperature variations.

Each lance cylinder is therefore an intelligent perpheral which responds only to commands addressed to it. Its movement can be selected, sequenced or synchronized relative to any other cylinder by means of software. Internal electronics (Figure 4) provide a complete closed loop. This simplifies software, and operation. Internal algorithms eliminate inaccuracies from externally generated error signals by providing more frequent signal updates. In the digital lance cylinder, a pulse width modulation technique is used to drive the servo valve torque motor. It responds to the RMS valve of the pulse train which is interpreted as a specific DC voltage. High speed communications with immunity to industrial noise allow operations at distances to 4000 feet. The on board electronics also provide for readout of diagnostic data. Position, velocity, acceleration, deceleration, velocity ratio, zero, span, gain, fault detection, baud rate, timing and ranges are all programmable functions.

EJECTOR FEATURES

The lance displacement ejector is close coupled to its driver. (Figure 3). A 4" open space provides visual confirmation of seal condition and prevents contamination should leakage occur. The design allows a small bore/long stroke configuration for metering accuracy while maintaining acceptable overall length. Special seals have provision for external lubrication.

The ejector has an integral component valve built into its end cap. Component is pumped into the cap during refill and low pressure recirculation. When a shot is requested, the component valve shifts to allow high pressure recirculation and shot. This diverts low pressure pumping cylinder output to tank; isolating the high and low pressure circuits. The component valve body also incorporates a built-in rupture disc which simplifies installation and piping. Should an "overpressure" condition occur, the rupture disc will break, relieving pressure by venting it to the component tank.

MASS FLOW MONITORS

The purpose of a RIM injection system is to meter reactive components, mix them and inject the mixture into a mold to form a part. Since the finished part must be a specified weight, the system must deliver pre-determined mass flows of several components. A lance cylinder, however, meters on a volumetric basis. Temperature, pressure, aeration, solids, and viscosity all effect its delivery of specific mass flows. For this reason, mass flowmeters (one for each component) are incorporated to monitor actual mass flow delivery. The meters, based on Corialis Forces, feature a non-intrusive sensor, in-line installation, and have no moving parts. They are located in the high pressure circuit as close as possible to the mix head. Remote electronics, good tolerance for vibration, and a 3000 psi pressure rating allow their use in this location. Since mass flowmeter accuracy of .4% is unaffected by changes in viscosity, temperature and density, they are ideally suited for RIM monitoring.

The benefits of mass flow monitors are that they measure independent of fluid properties, they delivery superior accuracy, are maintenance free and provide a universal means of on-board calibration.

BULK NUCLEATION CONTROL

Modern RIM injection systems almost always provide a means for component nucleation. Generally the system consists of an aeration device for the "B" component and a system to monitor density and control gas injection. Density monitoring technologies employed to date have included compressibility factor, gamma ray attenuation, vibrating U tube, and weight-volume measurement types. Problems encountered have included difficult to calibrate, lack of sensitivity, unstable, transcient effects, and licenses for handling nuclear materials. The RIMCENTER introduces a system based on differential hydrostatic pressures. It is made possible by on-board electronics which combine high sensitivity, fast responce and high accuracy. The system includes an in-tank sparger, a high shear mixer, dry gas injection control, and two in-tank hydrostatic transducers. It provides a simple accurate means to determine overall specific gravity of day tank material on a continuous basis. Two high sensitivity pressure transducers are spaced a known distance apart vertically in the day tank. By comparing the difference in readings from the two sensors and comparing to a known value, the specific gravity is calculated directly.

Using differential pressure sensors and venting to the dry gas blanket, eliminates any effect of blanket pressure on sensor readings. Thus, specific gravity of tank contents can be calculated directly and can be changed by modulating dry gas flow into the tank. Since the specific gravity calculated is an overall valve for the total contents, local variations due to non-uniform distribution are averaged out. This eliminates instability which occurs with point sensors.

Because high sensitivity sensors are used, they are effected by transcients from agitation, low pressure pumps and recirculating material. These transcient effects are screened out by proprietary software to provide stable accurate, component nucleation control. Benefits are rapid response, continuous reading, specific gravity of entire contents, stability, accuracy, non-nuclear, (no permits required), and no external pipe and pump circuit required.

"NO SHOCK" MIX HEAD

The key component of any RIM system is its mixhead; especially on a system rated for reinforced RIM operation. In the Micromatic mix head, all critical parts such as orifices, needles and clean out plunger are constructed of special wear resistant alloys. (Figure 6) The head features high pressure recirculation through the primary pour orifices. This prevents entrapment of filler materials and allows easy pressure balancing. The mix head clean out plunger does not incorporate recirculation grooves which makes it stronger and more wear resistant. Recirculation passages are machined into the mix head sleeve. A concentric sleeve valve provides control of high pressure recirculation. Because a sleeve valve controls all four streams, their opening and closing are always synchronized. Shifting is provided by a fluidic circuit that synchronizes operation of the valve and main plunger. (Figure 7) The result is elimination of pressure spikes which normally occur when a RIM mix head opens and closes. (Figure 8) Caused by momentarily dead heading the component streams, these spikes are indicative of a flow disruption which adversly effects mixing. They are also unnecessary and undesireable machine loads which may cause reliability problems. Their elimination provides continuous high pressure flow with no disruption when opening and closing the mix head.

By incorporating interchangable clean out plungers and sleeves, this design offer several advantages. A mix head can be easily repaired by removing and replacing the sleeve and plunger set. It is also possible to change mix chamber size by fitting a plunger and sleeve set with a larger or smaller plunger diameter.

Repairability is enhanced by use of U.S. standard seals, a pre-sealing technique which eliminates "break-in" on new mix heads, simplified actuator assembly removable as a unit, and replacable orifices in the needle valve assemblies. These features provide the user with the benefits of high reliability and low maintenance costs. They also provide a high degree of operational flexability so that fewer mix heads can cover a wider range of applications. Micromatic mix heads come in sizes from 5mm to 40mm, injection ratings from 1 lb/sec to 50 lb/sec, a choice of 2, 3 or 4 streams, with fixed orifice or needle valves and with or without high pressure recirculation. Special heads for specific applications have also been custom engineered to meet specific needs when necessary.

MULTIPLE MIX HEAD OPERATION

The RIMCENTER answers industry's need for a range of integrated, pre-assembled units designed to work together as a complete system. It can be configured to specific requirements in a turn-key production center. The basic system is a complete single station molding center incorporating all functions required for high volume RIM manufacturing.

With only the addition of mix heads, control valves, and field piping, the injection system can be expanded to accomodate up to four injection stations. These stations can be utilized on multi cavity tooling or on separate tools in multiple presses. With digital servo lance cylinders, shot size and ratio can be different for each station. The shot parameters are changed "on the fly" with new instructions for each station being down loaded to the on-board microprocessor immediately prior to the shot command. Thus, operating parameters can be changed without the need for machine shut down. Where multi point sequential injection is needed on a single tool (Figure 9) only minor software modifications are necessary.

Digital servo lance cylinders can be operated in multiples (up to sixteen) from a single control. With sizes from 2" to 12" diameter, it is possible to provide a machine (Figure 10) to operate multiple heads in a simultaneous injection mode.(2) In this concept, multiple mix heads would be injected simultaneously to accomodate fast gellation times. Small injection cylinders being provided for each head. The main control would down load shot parameters to each cylinder ahead of time. The main control then gives a "go" signal, and the injection sequence is internally controlled within the lance cylinders themselves. This avoids overloading the main control and a resultant loss in accuracy.

Other possible configurations are multi cylinder machines to minimize recharge time and multi cylinder machines for molding multiple materials.

CRT BASED CONTROLS

A CRT based control system was designed from the beginning. (Figure 11) It incorporates the latest in programmable controllers and includes full mathematical calculation capability. The central controller also provides "BASIC" language capability. This combination provides a number of benefits. These include friendly controls, internal calculations, set up "on the fly", diagnostic, and on board SPC.

The operator selects the appropriate screen from the main menu (Figure 12) and enters shot parameters using the CRT and key pad. Some parameters need to be entered for each station. These are: Shot weight, weight ratio, shot time, and component shot pressures.(Figure 13) Other parameters apply to all stations and are entered on the screen labeled "Global" parameters.(Figure 14) They are specific gravities, temperatures and nucleation.

After shot parameters are entered, the calculations to control injection are performed internally. A sample follows:

- shot weight   18 lbs
- weight ratio (B/A)   2:1
- secific gravity A   1.2 g/cc
- specific gravity B   1.6 g/cc
- shot time   3 seconds

18 lb shot at 2:1 (B/A) ratio = 12 lb (B) and 6 lb (A). At a 3 second shot:

(2)Scrivo, Jerry V., "Composites - Laboratory to Commercial Reality", ASM/ESD advanced composites conference Sept. 1987

B injection rate = 4 lb/sec or 1816 g/sec
A injection rate = 2 lb/sec or 908 g/sec

1816 g/sec at 1.6 g/cc = 1135 cc/sec

908 g/sec at 1.2 g/cc = 756.7 cc/sec

Once the volume rate is determined; the factor for volume to lance displacement is applied to determine lance velocity during the shot. Controlling this velocity yields the total shot weight.

Doing these calculations with on board software yields consistent and accurate results, while minimizing the need for technically trained personnel.

Practical limits have been incorporated into the machine's software to prevent entry of invalid data (Figure 13 & 14). A number of other operational parameters are not immediately accessible from the keyboard. They can only be accessed through programming changes. These include lance fill speed, lance purge speed, nucleation speed and the duration between purge cycles.

When a shot is requested, the injection system shifts to high pressure operation and the cylinders ramp up to injection velocity and pressure. When these are achieved, the mix head opens providing sufficient lance stroke remains to make the shot. The shot end point is established when the head opens and displacement is monitored until this point is reached. At that time the head closes. This avoids variations caused by minor velocity errors.

Machine accuracy and repeatability determined with a calibration head showed excellent linearity over the shot range and outstanding tracking of A & B component flows. At a three sigma confidence level, the standard error for the total shot was .069, (Figure 15) with standard errors for A and B respectively at .038 and .031 (See Figure 16 & 17). Evaluation showed the mix head closing signal to be the critical parameter. Electronic update time within the control system iteself is the source of the variation observed. Maximum effect occurs at maximum throughput, with accuracy and repeatability being even better than reported when operating at lower injection rates.

A substantial advantage of this control is that it is common to all Micromatic products. Thus, software developed for any Micromatic product can be adapted for RIM. Specifically, this allows Microstat SPC software to be applied to RIM. Microstat is an analysis program which provides statistical information on the manufacturing process. It can be used to monitor any variable available through the machine's control system.

Data is logged for the parameter being monitored and stored within the controller. Observations are broken into sub groups each consisting of five consecutive observations. A measureable period of time separates each sub group. Upon request, the data table is processed and appropriate reports produced. These include:

SUM, X Bar and range values for all sub groups, and four charts. The charts are Range, X Bar, Median and Histogram.

This information can be displayed on the CRT or down loaded to a factory computer.

COMING ATTRACTIONS

The RIMCENTER injection system provides a substantial base for future development. While much of this development cannot be presented at this time, the following items are illustrative:

High Throughput - accumulator based injection systems with rates to 50 lb/sec for polyurea processing.

Low Throughput - accumulator based injection systems for RTM and SRIM with shot sizes to 200 lbs.

Horizontal RIM Machines - Similar to injection molding machines for fast cycle (45-60 sec) polyurea molding. Injection rates to 50 lb/sec and clamp tonnage to 2500 Ton.

Mini RIM - Small high volume production machine to replace P.U. Casting. Available in 1 to 4 lb shot sizes with up to 1 lb/sec injection rate.

Bar Code System - Optional addition to the CRT operating system. Bar Codes and prints labels with information on manufacturing and/or SPC parameters for each part produced.

SUMMARY

Micromatic's RIMCENTER injection system is the culmination of a multi year design and development effort based on extensive operational experience with a wide variety of chemical and machinery systems. The resulting product is designed for simplicity, accuracy, operational

flexibility, high productivity and high reliability. State-of-the Art concepts and components are incorporated wherever possible. This high level of new technology makes molding the future with RIM more productive than ever before.

Bibliography

1. Scrivo, Jerry V., "Molding Without Melting", SPE ANTE April 1986.

2. Scrivo, Jerry V., "Composites - Laboratory to Commercial Reality", ASM/ESD advanced Composites Conference, September 1987.

RIM METERING SYSTEMS
SPECIFICATIONS

Item	Model 8-3	Model 15-5	Model 25-9	Model 50-18	Model 100-34 Model 200-34
Nominal Shot Size	8.0 lb.	15.0 lb.	25.0 lb.	50.0 lb.	100.0/200.0 lb.
Maximum Delivery	3 lb./sec. 180 lb./min.	5 lb./sec. 300 lb./min.	9 lb./sec. 540 lb./min.	18 lb./sec. 1080 lb./min.	34 lb./sec. 2040 lb./min.
Tank Size	80 gal.	150 gal.	150 gal.	150 gal.	300/600 gal.
Running Time/Tank	160 shots	160 shots	100 shots	50 shots	50 shots
Metering Type	Cylinder	Cylinder	Cylinder	Cylinder	Cylinder
Cylinder Size	4″ diam.	6″ diam.	6″ diam.	8″ diam.	10″/12″ diam.
Multi Shot/Fill	No	No	No	Optional	Optional
Pump Capacity (High Pressure)	14 gpm	23 gpm	41 gpm	81 gpm	153 gpm
Pump Capacity (Low Pressure)	4 gpm	6 gpm	12 gpm	22 gpm	46/90 gpm
Fillers	Yes	Yes	Yes	Yes	Yes
Nuculation	Automatic	Automatic	Automatic	Automatic	Automatic
Programable Control	Yes	Yes	Yes	Yes	Yes
Flowmeters	Standard	Standard	Standard	Standard	Standard
Mix Head	4 Stream	4 Stream	4 Stream	4 Stream	4 Stream
Temp. Control	Automatic	Automatic	Automatic	Automatic	Automatic
Diagnostics	Optional	Optional	Optional	Optional	Optional
Multi Station	4 Optional	4 Optional	4 Optional	4 Optional	4 Optional

NOTE: All machine ratings are based on a nominal 1:1 mix ratio of materials having a specific gravity of 1.0. Consult factory for specific machine ratings based on other operating parameters.

TABLE I

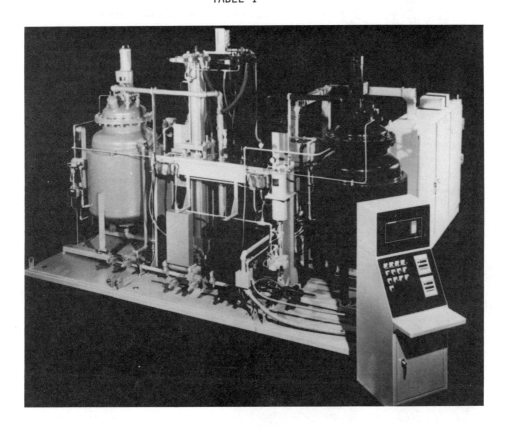

FIGURE 1

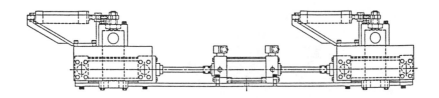

FIGURE 2

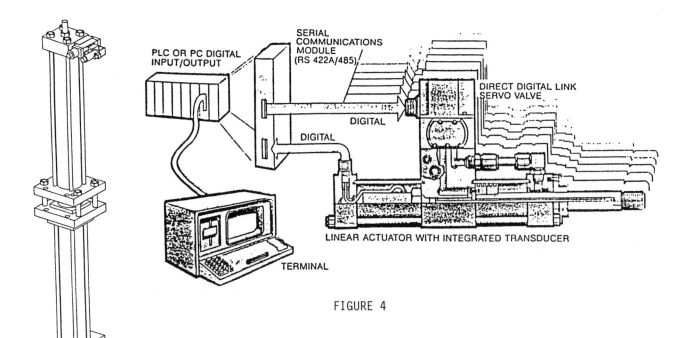

FIGURE 4

FIGURE 3

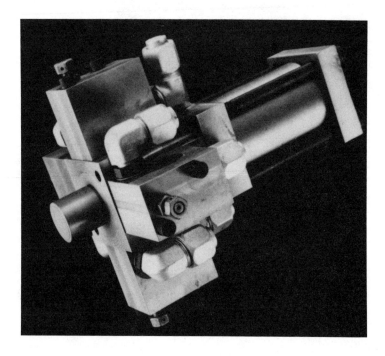

FIGURE 6

SERVO VALVE BLOCK DIAGRAM

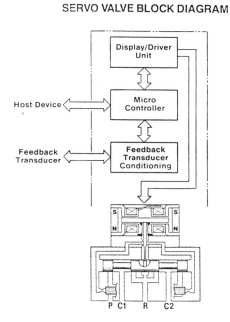

FIGURE 5

264

RRIM Mix Head Section

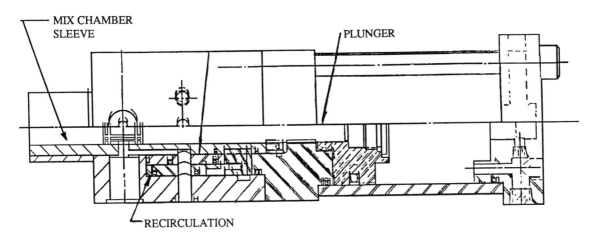

FIGURE 7

SHOT PROFILE

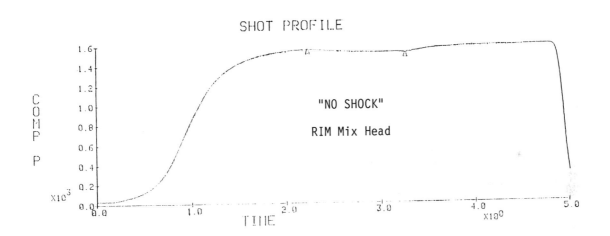

"NO SHOCK"

RIM Mix Head

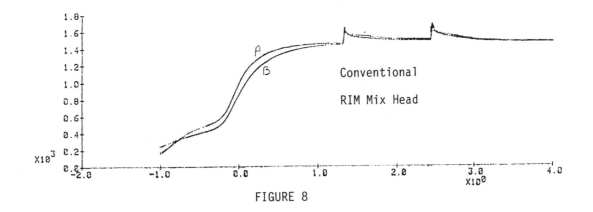

Conventional

RIM Mix Head

FIGURE 8

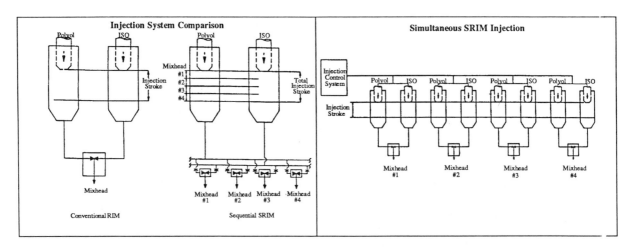

FIGURE 9

FIGURE 10

RIM screens for monitoring and changing machine parameters

FIGURE 11

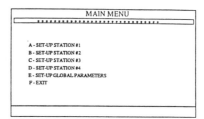

FIGURE 12

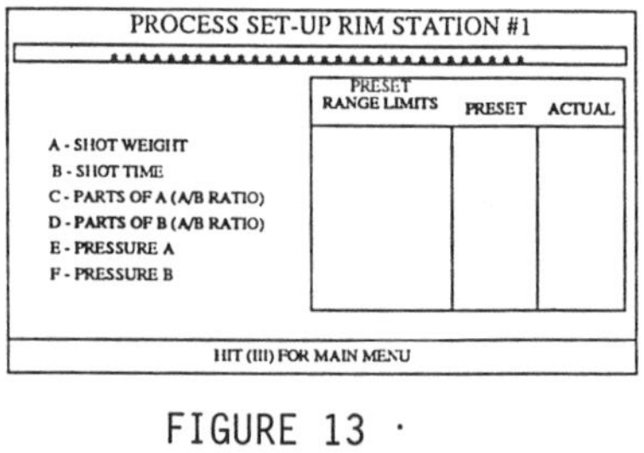

FIGURE 13 ·

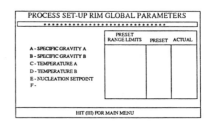

FIGURE 14

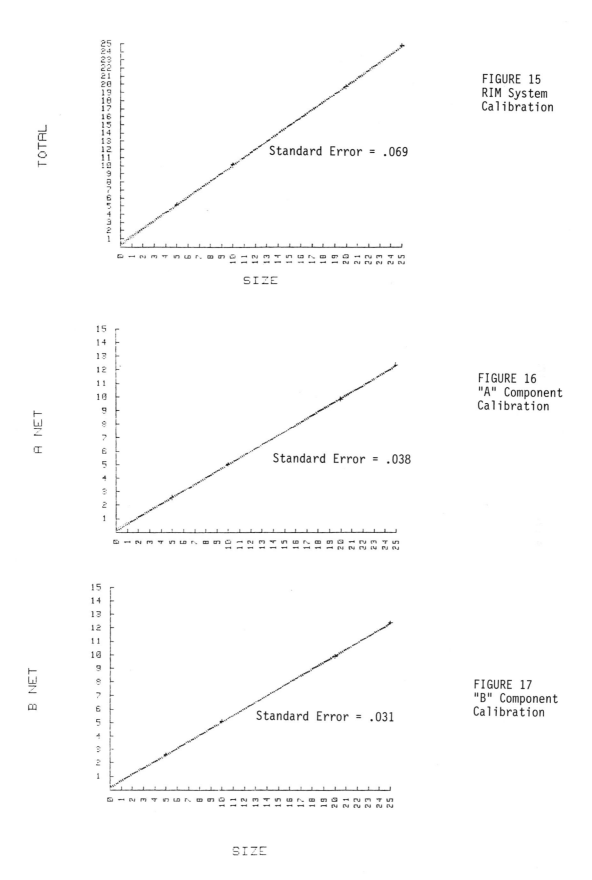

FIGURE 15
RIM System
Calibration

Standard Error = .069

FIGURE 16
"A" Component
Calibration

Standard Error = .038

FIGURE 17
"B" Component
Calibration

Standard Error = .031

EVOLUTION OF RIM WITH
VINYL ESTER RESINS

D. Babbington, J. Barron , J. Enos, R. Ramsey, W. Preuss
Dow Chemical Company
Freeport, Texas 77541   USA

ABSTRACT

Vinyl ester resins are changing the shape of the Resin Transfer Molding (RTM) Process as we know it. Although RTM has been around for some 20 years, it is only recently that the process is receiving attention in fast cycle, medium-high volume structural applications. This paper will review resin, catalysis, process and equipment advancements required to make fast cycle times a reality.

The paper will begin by discussing resin system requirements for RTM. Next, catalysis options for "fast" cycle times at a variety of molding temperatures will be reviewed. And last, process and equipment parameters such as the effects of fill nozzles, vent ports, pumping equipment and vacuum will be discussed. Touching upon each of these topics, the feasibility of RTM with vinyl ester resins for medium-high volume structural applications would have been demonstrated.

INTRODUCTION

As composites production becomes more commonplace, it becomes increasingly important to develop methods of production which are efficient and inexpensive. There are many methods to produce high quality composites and each has advantages for particular applications. Resin transfer molding (RTM) has evolved into one of the most versatile and efficient methods of producing high quality structural composite parts. The RTM process, illustrated in Figure 1, can be described as the introduction of a liquid thermoset resin into a mold which contains fibrous reinforcement. The resin is then cured in the mold, producing a composite part. Typical advantages of RTM include low pumping pressures, low equipment costs, and short lead times for tooling. The process provides parts of close tolerance dimensions with good surface quality to all sides. Additionally, reinforcement location, content, and direction can be closely controlled and varied throughout the part.

Due to its early history and uses, the RTM process carries with it a reputation for long cycle times. In the past, only a handful of parts were molded per day with simple epoxy tools, pressure pots, and room temperature cure systems. Cycle time was far from optimized. This mode of operation was, however, often economical compared to other methods of fabrication such as hand lay-up or spray-up and worked well within its limitations. Today, the RTM process is being expanded into high volume automotive and industrial applications such as floor pans and structural crossmembers (1) where strength requirements are high, part geometry is complex, and cycle time must be held to a minimum.

Vinyl ester resins are ideal for use in the RTM process whether prototyping a few or mass producing several thousand parts. This paper discusses the evolution of the RTM process with vinyl ester resins, the process versatility, and some recent advances which are allowing vinyl ester resins to enter new areas of composite applications. The issues of resin selection, kinetics, mold release agents, process equipment, tool design, reinforcements, and economics are discussed.

RESIN SELECTION

When thermal, fatigue, and chemical resistance are required, vinyl ester resins are often specified for a composite application. DERAKANE* vinyl ester resins are an established family of resins offering a variety of thermal, mechanical, chemical resistant, and fire retardant properties.

The glass transition temperature (Tg) of a thermoset resin is a good indication of the degree of crosslinking in the polymer and the ability to withstand elevated service temperatures. Typical glass transition temperatures for several vinyl ester resins are summarized in Table 1. Note that the DERAKANE* 470, which contains a novolac backbone, has the highest Tg while DERAKANE* 8084, the elastomer modified vinyl ester resin, exhibits the lowest Tg of the family of resins. Depending on the service temperature requirement of a specific application, a range of vinyl ester resins are available.

A second important characteristic of a resin is its resistance to fatigue crack propagation. Conventional fatigue testing requires an extended period of time, numerous samples, and often yielding a high degree of scatter in the results. A new method of fatigue testing, utilizing fracture mechanics (2) principles, measures the material's resistance to fatigue crack propagation (FCP). The FCP studies are time efficient and very useful in differentiating between resin systems.

Referring to Figure 2, an improvement in fatigue resistance is associated with a decrease in slope or a shift of the curve to the right. A decrease in slope implies a decrease in crack growth rate at the same stress intensity. A shift to the right implies that a higher stress is required to grow the crack at a given rate. As would be expected, the elastomer modified resin was found to give significantly better fatigue resistance than the highly crosslinked novolac vinyl ester resin. Also, the polyester resin in this study exhibited lower fatigue resistance compared to the vinyl ester resins.

KINETICS

Vinyl ester resins offer some of the most versatile curing kinetics of any resin type. Using a wide variety of well known peroxide catalysts and amine accelerators, the gel time can be controlled from 10 seconds to over one hour for cure temperatures ranging from room temperature to 300 degrees F. Generally, three options are available to control the cure kinetics: catalyst formulations can be adjusted for room temperature cure, the mold can be heated to decrease the gel time of these formulations, or heat activated catalysts can be utilized for a very fast gel time and long pot life.

Typical room temperature catalyst formulations available for vinyl esters are summarized in Figure 3. Gel times were obtained with a Shyodu gel tester. Depending on the catalyst and accelerator type and level, gel times ranging from several minutes to over an hour were obtained. In general, increasing the catalyst and accelerator levels increases the rate of cure. However, if an excessive amount of accelerator is used, gellation proceeds rapidly but the part does not cure completely. An example of this is shown in the thermal analysis in Figure 4. Two BPO formulations are being compared containing differing levels of the accelerator DMT. When 0.5% DMT was used, the Tg was found to be only 63.8 C (147°F) whereas cutting back to 0.2% accelerator increased the Tg of the material to 107.4 C (225°F) which approaches the ultimate value for this resin.

To achieve a faster gel, the molding temperature can be increased while using similar catalyst formulations as for the room temperature molding. Gel data for catalyst systems to be used between 100- 175°F are summarized in Figure 5. A platen gel test was used for this analysis since this test most closely simulates process conditions. Again, depending on the catalyst type and level, a wide range of gel times are available. However, as with concentration, the temperature range of the use of accelerators is limited. Beyond 200°F, rapid gel times may not translate into a fully cured part.

If further reductions in gel time are desired, heat activated catalysts must be used. These systems have an extremely long pot life at room temperature but cure very rapidly at temperatures above 200 degrees F. A high degree of cure is quickly developed in the tool during molding allowing for rapid demold times while often eliminating the postcure step. Figure 6 shows the gel data of several of these systems. A platen gel test was again used for analysis. The gel time can be controlled over a wide range by varying temperature and catalyst concentration at elevated molding temperatures. A gel time as low as 18 seconds was obtained using 2% of the USP 245 catalyst at 250°F. Volatiles began to boil from the resin on samples run at platen temperatures approaching 300°F causing gas entrapment to occur and suggesting an upper limit on molding temperatures. Experience has shown that the demold time will run at three to five times the gel time to obtain a full cure for the heat activated catalyst systems. Therefore, using the USP 245 catalyst at a 2% concentration and a mold temperature of 250°F, one could expect a demold time of approximately 54-90 seconds.

To demonstrate the rapid cycle time, a heat activated catalyst was processed on RTM equipment to produce a composite panel for evaluation. The process conditions used in the demonstration are summarized in Table 2. Physical and thermal properties, given in Table 3, were measured on this composite part both with and without postcure. As the glass transition temperature indicates, a high degree of cure was accomplished during molding without postcure. The physical properties of the sample without postcure were found to approach those of the the sample which had been postcured.

MOLD RELEASE AGENTS

The area of mold release has also received much attention over the years. Waxes have been used for years and release well but they are difficult to apply, especially to hot molds. Spray and wipe-on releases make application easier and last for repeated moldings. However, eventual buildup on the mold can occur and the application of the release agent adds to the cycle time. The optimum release agent is internal to the resin providing for multiple part releases without the use of an external spray.

Several commercial internal mold release agents for vinyl ester resins have been screened yielding some promising candidates. Consistent mold release has been good when evaluated on a flat plaque tool without excessive buildup of release agent. The effects of two such internal release agents on gel time and physical properties are summarized in Table 4. Results are compared to a control which used an external spray to provide for part release during molding. At low molding temperatures, the release agents had a greater effect on cure rate than at elevated temperatures (>200°F). External waxes or release sprays are a better choice for room temperature and moderate temperature molding to assure the cure of the part is not affected. However, at elevated temperatures, the internal release agents tested were not found to adversely affect the gel rate nor the physical properties at the levels tested.

The effectiveness of an internal release agent in production will depend on part design and complexity, tool material, surface finish, and resin shrinkage.

PROCESS EQUIPMENT

As the number of parts required increases and the cycle times decrease, the process becomes more complex and expensive. Simultaneously, the kinetics of the chemistry become faster and more difficult to control. Automation is necessary for the accelerated cycle times as machines are required to perform more of the functions previously handled by man. Equipment must be designed to resist wear and tear from repeated operation. Safety becomes a greater issue as chemical inventories increase and as equipment gets heavier, more powerful, and automated. The process must be made foolproof to prevent costly down-time.

The sophistication of resin delivery systems can vary drastically from open pours or pressure pots to two-component pumping systems or high-speed automatic metering systems depending on the needs of the process. Vinyl ester resins are adaptable to all of these. Using the various catalyst systems described above, the pot-life and gel time of the system can be adjusted to meet the process requirements.

The room temperature and moderate temperature cure systems can be poured, pressured, or pumped as long as the process equipment containing catalyzed resin is flushed with solvent before gel. The peroxide catalyst can be stored separate from the resin and accelerator and combined only as needed in a two-component pumping system to ease handling.

The high temperature activated catalyst systems have pot-life's of several days at room temperature and are ideal to treat as one-component systems with fast processing speeds in heated tools. Care must be taken that the catalyzed resin storage and delivery systems do not contain hot spots, or incompatible materials of construction which could accelerate the resin advancement.

These heat activated catalyst systems can also be treated as two-component systems in high speed metering equipment as shown in Figure 7. The hydraulically operated piston pumps provide high output and precise metering and ratio control. The equipment configuration and processing speeds begin to approach RIM in nature but with some processing simplifications. Since the catalyst is not added in stoichiometric amounts, ratio control is less critical. The long pot-life of the resin system at room temperature allows a static mixer to be used without fear of gellation. The pressure needed for static mixing is much lower than that necessary for the impingement mixing of the RIM process.

TOOLING

Mold technology is another area of varying sophistication. For example, if only a limited number of parts are to be produced, epoxy tooling with tubular fill and vent lines could be employed. For a high volume production operation, however, elevated temperature tooling with self cleaning vents and inlet ports would be used to hold cycle time to a minimum.

Epoxy tooling is inexpensive, requires little lead time, and is lightweight. However, it withstands little pressure, has a very limited life, and can only be heated to moderate temperatures. Electroformed molds offer a harder surface and better heat conduction. To allow faster cycling of the mold temperature, heating wires can be placed close to the mold surface with insulation between the surface material and the mold reinforcing material (3). Aluminum and steel tooling are constructed with hot water, steam, or oil heating. They can withstand high pressures and repeated moldings.

Injection and vent ports in the epoxy tool typically consist of plastic tubes run through the epoxy to the part surface. The vents are sealed off successively during fill as air is bled from the part. Once the resin is cured, these tubes are either cleaned out or discarded. Disposable static mixers are often used for injection to eliminate solvent flushing.

If fast cycle times are required for high volume part production, the operator cannot be bothered with cumbersome inlet ports and vents. Isolated self-cleaning ports can prevent solvent flushing during steady state operation. The venting can be accomplished by placing small grooves along the parting line (4). Instead of venting, a vacuum can be pulled on the tool prior to fill as a means of eliminating voids. With the very low viscosity of the vinyl ester resins, some type of sealing gasket is typically required to restrict resin leakage out of the tool. Very little success has been seen at Dow by sealing a tool with a metal-to-metal fit.

REINFORCEMENT

The placement and orientation of reinforcement give the designer freedom to take advantage of a composite's unique properties. A variety of reinforcements can be used in RTM. The continuous strand mat is by far the most commonly used reinforcement. It is often combined with various directional reinforcements (stitched or woven mats) or specialty mats (polyester, graphite or Kevlar fiber) in high stress applications.

Besides affecting the resultant composite physical properties, the level and type of reinforcement will also affect the fill behavior through the part. Care must be taken in selecting the reinforcement layup such that it will be compatible with the molding process. The reinforcement must be held in place or combined in such a way as to prevent it from distorting during fill. Consistent reinforcement content is desirable throughout the part to provide even flow distribution and eliminate voids and dry areas.

One of the most time consuming steps in the RTM process is the loading of reinforcement into the tool. Several options are presently available for obtaining a "preform" to simplify this operation. Preformable continuous strand mat(5) and spray-up chopped preforms(6) allow the reinforcement to be shaped in a separate operation prior to being loaded into the tool. A thermoset binder material is available which can be sprayed onto the glass to make almost any mat preformable (7). Custom braided preforms can be built if a part has high stress requirements where directional reinforcement is needed. Any of the above preforming methods can be combined within a part to fully utilize the advantages of each method.

ECONOMICS

Throughout this paper the versatility of the RTM process was discussed. Options available in equipment and catalysis for controlling cycle time were evaluated. Using an economic cost model developed at Dow (8), this section will discuss the effects of these process alternatives on part cost.

As one would expect, the number of parts being produced will economically dictate the type of equipment and cycle time required. One would not want to spend excessive amounts of money on a tool from which only a small number of parts would be produced, nor could one tolerate a very long cycle time if a large volume of parts was required. This is demonstrated in Figure 8 for two different cases modeled; fast RTM with steel tooling and slow RTM with epoxy tooling. As expected the faster RTM process would result in reduced part costs for the higher part volumes with the slow RTM process and epoxy tooling being a more attractive alternative for the lower part volumes.

To understand the key variables affecting cost, the overall part cost was segmented into costs associated with materials, tooling, capital, labor and energy as shown in Figure 9. In the first case of the fast RTM cycle time, for the lower part volumes, the burden of under-utilized pumping equipment and expensive steel tooling becomes apparent, whereas at the higher part volumes, materials costs dominate. For the second case using the slow RTM cycle time and the epoxy tooling, both labor and capital costs begin to enter into the part costs due to the excessive presses and pumping equipment along with the manpower support required to accommodate the long cycle times. The use of such modeling enables one to effectively specify equipment and cycle times to minimize part costs for any given application. The significance of the versatility of the RTM process becomes apparent enabling one to optimize their part costs by tailoring the process to meet their specific needs.

CONCLUSIONS

Resin Transfer Molding is a versatile process with respect to choices in resin systems, catalysis and process equipment. A wide variety of vinyl ester resins are available to meet the service temperature and fatigue requirements for specific applications. Vinyl ester resins can be catalyzed for either room temperature or elevated temperature molding to meet both prototyping and fast cycle time requirements. Fabrication equipment can also be selected to accomodate these varied cycle time requirements. Proper selection of each of these parameters through discussions with knowledgeable material suppliers and molders is necessary to capitalize on this process versatility and to economically meet the needs of a specific application.

*Trademark of The Dow Chemical Company

REFERENCES

1. C.F. Johnson, N.G. Chavka, R.A. Jeryan, C.J. Morris, D.A. Babbington, "Design and Fabrication of a HSRTM Crossmember Model." Advanced Composites Conference. Detroit, September 15-17, 1987.

2. D.L. Steinbrunner, "Fatigue Crack Propagation of Composites." Advanced Composites Conference. Detroit, September 15-17, 1987.

3. D.A. Babbington, J.M. Cox, J.H. Enos, "High Speed Transfer Molding of Vinyl Ester Resins." Autocom '87 Conference. Detroit, June 1-4, 1987.

4. A. Harper, "Resin Transfer Molding - The Low Pressure Alternative." SPI Conference. Cincinnati, February 1-5, 1988.

5. S.G. Dunbar, "Preformable Continuous Strand Mat." SPI Conference, Cincinnati, February 1-5, 1988.

6. P. Emrich, "Directed Fiber Preforms." Materials Week '87 Technical Presentation. Cincinnati, October 10-15, 1987.

7. K.A. Seroogy, W.N. Reed, S.L. Voeks, "Thermoformable Preforms for Advanced Composites." Advanced Composites Conference, Detroit, September 15-17, 1987.

8. G. Ellerbe, "RIM Mat Molding - Economic Evaluation of a New Structural Composite." SPI Conference, Cincinnati, February 2-6, 1987.

TABLE 1

DOW VINYL ESTER RESINS FOR RESIN TRANSFER MOLDING

RESIN SYSTEM	DESCRIPTION	VISCOSITY	Tg	HDT**	ELONG.
DERAKANE* 411C-50	BIS A EPOXY VER	125 CPS	240°F	215°F	5-7 %
DERAKANE 530	BROMINATED BIS A EPOXY VER	450 CPS	270°F	230°F	4-5 %
DERAKANE 510A-40	BROMINATED BIS A EPOXY VER	250 CPS	260°F	230°F	4-5 %
DERAKANE 470-36	EPOXY NOVOLAC VER	200 CPS	305°F	300°F	3-4 %
DERAKANE 8084	FLEXIBILIZED BIS A EPOXY VER	375 CPS	230°F	180°F	8-10 %
EXPERIMENTAL PRODUCT XU 71835.01L	FLEXIBILIZED HIGHLY CROSSLINKED VER	350 CPS	250°F	200°F	6-8 %

*Trademark of The Dow Chemical Company
**Heat Distortion Temperature at 264 psi, unreinforced

TABLE 2

PROCESS CONDITIONS FOR DEMONSTRATION
OF RAPID CYCLE TIME

DERAKANE* 411-C-50 / 2 phr USP-245**

(5 layers) 1-1/2 oz./ft^2 OCF M8608 Continuous Strand Mat

Liquid Control Pumping Equipment

15" X 20" X 1/8" Part Size

250°F Mold Temperature

18 Seconds Fill Time

90 Seconds In-Mold Time (Including Fill Time)

*Trademark of The Dow Chemical Company
**Product of The Witco Corporation

TABLE 3

THE EFFECT OF POSTCURE ON PHYSICAL PROPERTIES

RAPID CYCLE TIME

	NO POST CURE	POST CURE[1]
GLASS CONTENT, WT. %	47.0	47.0
GLASS TRANSITION TEMPERATURE, °F	178	240
TENSILE STRENGTH, PSI	22,960	24,885
TENSILE MODULUS, 10^6 PSI	1.23	1.24
ELONGATION, %	2.43	2.44
FLEXURAL STRENGTH, PSI	33,885	35,666
FLEXURAL MODULUS, 10^6 PSI	1.23	1.27
COMPRESSIVE STRENGTH, PSI	21,081	24,743
SHORT BEAM SHEAR STRENGTH, PSI	3891	4245

[1]PANEL WAS POSTCURED AT 300°F FOR 2 HOURS.

DERAKANE* 411C-50 VINYL ESTER RESIN WAS USED IN THIS STUDY
SAMPLE CONTAINED OCF M8608 CONTINUOUS STRAND MAT

*Trademark of The Dow Chemical Company

TABLE 4

EFFECT OF INTERNAL MOLD RELEASE AGENTS ON GEL TIME AND PHYSICAL PROPERTIES

	CONTROL	1% INT EQ-6[1]	1% KANTSTIK FX-9[2]
PLATEN GEL TIME, 225°F 1% BPO (SEC)	70	61	58
GLASS CONTENT, WT.%	54.2	52.4	56.4
TENSILE STRENGTH, PSI	37,140	32,631	37,420
TENSILE MODULUS, 10^6 PSI	2.08	1.73	2.07
ELONGATION, %	2.5	2.5	2.6
FLEXURAL STRENGTH, PSI	54,958	53,638	54,711
FLEXURAL MODULUS, 10^6 PSI	1.98	1.89	2.00
COMPRESSIVE STRENGTH, PSI	36,432	34,153	33,877
UNNOTCHED IZOD STRENGTH FT-LB/IN	31.9	34.8	33.0
NOTCHED IZOD STRENGTH FT-LB/IN	21.5	22.0	21.6

- EXPERIMENTAL PRODUCT, XU 71835.01L VINYL ESTER RESIN WAS USED
- SAMPLE CONTAINED VETROTEX U750 CONTINUOUS STRAND MAT MEASURED IN THE TRANSVERSE DIRECTION

(1)INT EQ-6 IS A PRODUCT OF AXEL PRODUCTS
(2)KANSTIK FX-9 IS A PRODUCT OF SPECIALTY PRODUCTS COMPANY

FIGURE 1

THE RTM PROCESS

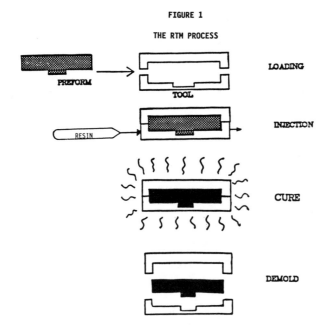

FIGURE 2

FATIGUE RESISTANCE

LOG da/dn vs LOG delta K

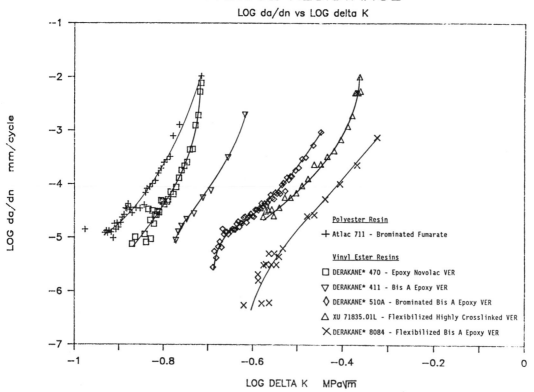

∗TRADEMARK OF THE DOW CHEMICAL COMPANY

FIGURE 3
EFFECT OF CATALYST FORMULATION ON GEL TIME
ROOM TEMPERATURE CATALYST SYSTEMS

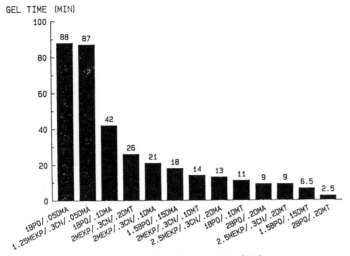

GEL TIME (MIN)

CATALYST FORMULATION (PHR)

<u>MATERIALS USED:</u>

DERAKANE* 411-C-50 VINYL ESTER RESIN, 60% METHYL ETHYL KETONE PEROXIDE (MEKP), BENZOYL PEROXIDE (BPO), 6% COBALT NAPHTHENATE (CN), DIMETHYLANILINE (DMA), DIMETHYL TOLUIDINE (DMT)

*TRADEMARK OF THE DOW CHEMICAL COMPANY

FIGURE 4

EFFECT OF CATALYST FORMULATION ON GLASS TRANSITION TEMPERATURE
DSC ANALYSIS

CATALYST FORMULATION: 1.4% BPO(ACTIVE)/0.5% DMT

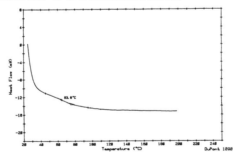

CATALYST FORMUALTION: 1.4% BPO(ACTIVE)/0.2% DMT

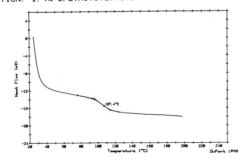

<u>MATERIALS USED:</u>

DERAKANE* 411-C-50 VINYL ESTER RESIN, BENZOYL PEROXIDE (BPO),
DIMETHYL TOLUIDINE (DMT)

*TRADEMARK OF THE DOW CHEMICAL COMPANY

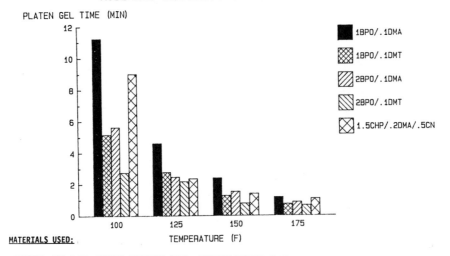

FIGURE 5
EFFECT OF TEMPERATURE ON PLATEN GEL TIMES
MODERATE TEMPERATURE CATALYST SYSTEMS

MATERIALS USED:

DERAKANE* 411-C-50, BENZOYL PEROXIDE (BPO), DIMETHYLANILINE (DMA), DIMETHYL TOLUIDINE (DMT), CUMENE
HYDROPEROXIDE (CHP), COBALT NAPHTHENATE (CN)

*TRADEMARK OF THE DOW CHEMICAL COMPANY

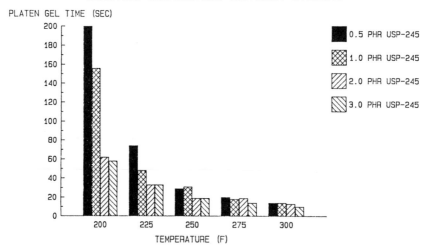

FIGURE 6
EFFECT OF TEMPERATURE ON PLATEN GEL TIME
ELEVATED TEMPERATURE CATALYST SYSTEMS

DERAKANE* 411-C-50/ USP-245
*TRADEMARK OF THE DOW CHEMICAL COMPANY

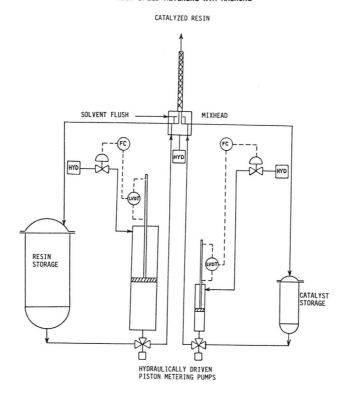

FIGURE 7
HIGH SPEED METERING RTM MACHINE

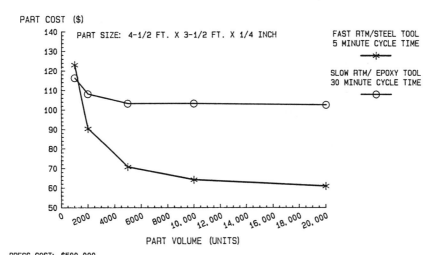

FIGURE 8
EFFECT OF PART VOLUME ON PART COST

PRESS COST: $500,000
RTM MACHINE: FAST $100,000, SLOW $25,000
TOOLING: STEEL $200,000, EPOXY $50,000

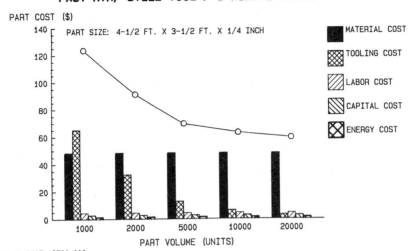

FIGURE 9
EFFECT OF PART VOLUME ON PART COST
FAST RTM/ STEEL TOOL : 5 MINUTE CYCLE TIME

PART SIZE: 4-1/2 FT. X 3-1/2 FT. X 1/4 INCH

PRESS COST: $500,000
RTM MACHINE : $100,000
TOOL COST: $200,000

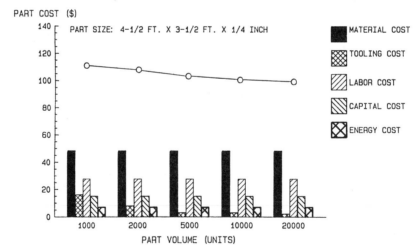

EFFECT OF PART VOLUME ON PART COST
SLOW RTM/ EPOXY TOOL : 30 MINUTE CYCLE TIME

PART SIZE: 4-1/2 FT. X 3-1/2 FT. X 1/4 INCH

PRESS COST: $500,000
RTM MACHINE: $25,000
TOOL COST : $50,000

PROCESS OPTIONS FOR THERMOSET COMPOSITE MOLDING

Richard A. Gastaldo
Ashland Chemical Company
Dublin, Ohio   USA

ABSTRACT

Thermoset composite molding covers a broad
base of processing techniques. In many cases,
the materials used in composites require
specific processing steps. Thus, by specifying
the composite material, one may be limited by
the recommended manufacturing process. Ashland
Chemical's ARIMAX® resins, primarily of
thermoset acrylamate chemistries, can be molded
in several manners. These different molding
procedures were studied to determine the
economical influence of process variables such
as production volume and capital investment.
Costs evaluations will explain these process
alternatives that, in certain cases, can be
more economically advantageous.

THE TRANSFORMATION OF THE "PROTOTYPE INTO
PRODUCTION" scenario can be hampered by the
strict processing conditions of some polymer
composites. These sometimes expensive
transformations are compounded by other
factors. One factor is a proposed production
volume which is either too high or too low for
economically producing composites with the
recommended molding process. A material which
is capable of being molded by various
processing conditions has the advantage of
permitting the selection of the most economical
molding process for a given situation.

Process economics play a key role in
determining which materials are used to
manufacture structural components. Ashland
Chemical's ARIMAX materials, which were
originally developed as pioneering resins in
Structural Reaction Injection Molding (SRIM),
are based on hybrid chemistries. These hybrid
chemistries expand the processing window of
composites made from ARIMAX resin systems
beyond that of other conventional composite
materials and add to their range of composite

molding techniques. In addition to SRIM
processing, acrylamate chemistries have been
developed for prepreg applications(resins
PREimPREGnated into glass mat), for HMC
applications (High glass content sheet Molding
Compound), and for newly developed SRTM
(Structural Resin Transfer Molding)
applications.

SRIM processing combines the high physical
properties of fiber reinforced polymers with
the high speed production rates of reaction
injection molding. The ARIMAX SRIM resin
systems are rapidly injected into a mold in
which fiber reinforcement had previously been
placed. Reaction rates can be catalytically
controlled for high speed demolding of large or
small parts. Conventional urethane RIM
equipment is used for the SRIM process. The
moderately priced capital equipment combines
with low clamping/molding pressures and high
production rates to make up the economically
attractive production scheme of the SRIM
process using ARIMAX resin systems.

The SRTM set-up is similar to the SRIM
scenario with the major process difference
being in the mix/metering of resin into the
mold cavity. Unlike SRIM, SRTM uses low
pressure injection of resins into the mold
which, like SRIM, has preplaced fiber
reinforcement. Viscosity altering additives
and catalyst modifications allow the ARIMAX
acrylamate resin systems to fill the mold under
these SRTM resin pumping conditions. ARIMAX
SRTM acrylamate systems can be pumped in ratios
which are standard for polyesters (100:1, resin
to catalyst) by making use of a
pre-coupling reaction designed for this
process. The SRTM system can also be injected
in standard SRIM ratios (1:2, A side to B side)

® Registered trademark of Ashland Chemical
Co.

providing the resin transfer machine has adequate pumping facilities. The low capital investment of the SRTM process includes relatively inexpensive mix/meter machines and low pressure, alternative tooling. The longer cure rates obtained in SRTM from the chemistry modifications necessary for mold filling and from lower tooling temperatures can suggest that this process would be ideal for the low volume production of structural reinforced composites.

The HMC and prepreg applications are included in the processing of acrylamate chemistries. The HMC systems, marketed as AROTECH® resins, can be processed in a conventional compression molding process. Unlike typical SMC materials, a coupling reaction also occurs as the A and B sides of the resin system are mixed and delivered to the doctor box. Once the HMC is manufactured, the compound is compression molded. Elevated temperatures during molding trigger the final crosslinking reaction. HMC process economics will encompass only the compression molding operation. The HMC material costs will include the possible offsetting costs of inexpensive glass reinforcement and SMC material preparation. Thus, for this example, HMC processing of AROTECH resins combines familiar SMC processing with even higher reinforcement levels to provide another alternative processing method for acrylamate resins.

Like the HMC resins, ARIMAX prepreg resins are compression molded into their final state. Acrylamate prepreg formulations are precoupled in a solution and this solution is then impregnated on fiber reinforcement. Cure rates and molding conditions of this staged material are comparable to SMC molding. The comparative capital costs, the moderate material costs and cycle times, along with the familiarity of SMC processing, make the acrylamate prepreg and HMC materials ideal for prototype and production applications respectively.

DETERMINATION OF PROCESSING ECONOMICS

A range of manufacturing variables will be set for each of the four molding processes which utilize ARIMAX acrylamate resin systems. Because of weighted factors (e.g. "button to button" cycle times) only the final molding process is analyzed. The typical molding parameters of ARIMAX resin systems are given in Table 1. As one can imagine, these variables have a direct effect on the molded part costs. Possibly the most influential variable in manufacturing is the cycle time for fabricating these reinforced parts. Processing related issues such as mold carrier capabilities and reinforcement preforms can drastically alter cycle times. Chemistry related issues such

as resin cure time and part/mold release, are also important to cycle determination. Cycle related processing issues,while not specifically discussed, contribute to the wide range of cycle times attained in the industry.

CAPITAL INVESTMENT - One of the initial questions asked by molders is what equipment is required for processing a material. For some manufacturers, these immediate expenditures control their interests in the process and ultimately in the composite material. Having a choice of alternative processing techniques opens the door to selection of an optimal economic process by possibly matching or modifying existing equipment as necessary. The major capital expenditures for the four processing schemes of these thermoset resins are listed in Table 2. This table shows that the SRTM process involves lower capital investment while the other processing schemes required similar significant increases.

PROCESS ENERGY USE - The estimated range of energy consumption of this capital equipment is shown in Table 3. Although manufacturing energy costs are the lowest contributor to molded part costs (as will be shown later), they will be included in this analysis. The increased energy consumption of the mold carriers for the compression molding processes, as well as the increased costs of these clamps, is directly related to the higher molding pressures required during the forming of the composite. As far as equipment and energy are concerned the SRTM process shows, at least on initial investment, a 50% cost savings over other acrylamate processes considered.

TOOLING CAPABILITIES - The tooling costs for SRTM can also be more economical as shown in Table 4. A major reason for this is that there is a longer list of alternatives for SRTM tool composition. Selecting a tooling material is based upon the process maximum in-mold pressures, molding temperatures, and production volumes. However, reinforcement levels and part design should be noted to ensure the tool's longevity.

MANPOWER REQUIREMENTS - Although molding operations are trying to become more automated, there is still a major necessity for quality personnel to run molding equipment. Estimated manpower requirements for each ARIMAX resin system processing procedure are listed in Table 5. Manpower percentages are used in this analysis as a percentage of time spent at that particular work station in one work shift.

MATERIALS - The materials used for ARIMAX resin system processing include thermoset resins and fiber reinforcement. For this analysis, the SRIM and SRTM resin prices are equal to the system prices which include both A and B sides, catalysts, and additives such as fire retardants and colorants. The HMC and prepreg material costs include raw material costs and a molders manufacturing cost to produce the HMC resins into its final moldable state. For this discussion, it is approximated that the SMC manufacturing costs add to the resin costs, while lower priced additive costs such as glass reinforcement and fillers deflate the cost of the HMC product. These factors make the approximate price per pound of HMC material to be equal to that of the SMC resin cost.

ARIMAX resin systems used for SRIM and SRTM range in price from $1.35 - $2.00 per pound, depending on previously mentioned variables. Reinforcement costs for most applications can entail using fiberglass mats which range from $1.30 - $1.75 per pound. However, more elaborate reinforcements such as carbon fibers or surface veils can cost $5.00 - $15.00 per pound, which drastically affect final part costs. HMC resin systems use a chopped glass reinforcement and are composite priced at $2.50 - $3.00 per pound. Prepreg systems, which are maintained for "prototype applications only", are delivered in their final state with costs ranging from $18.00 - $22.00 per pound, depending on reinforcement levels.

CASE STUDY

A computer program was set up to directly compare the three production viable acrylamate processing schemes of ARIMAX SRIM and SRTM resins and AROTECH HMC resin. All processing parameters which contribute to final part costs were taken as the median values from the ranges given in Table 1-5. The part design was chosen as a C-section beam. The projected area of this beam was 1200 sq. in. and it contained approximately 40 weight % fiberglass reinforcement. The simple design of this beam did not necessitate the use of fiber preforms although this will be considered later. System processing costs to manufacture these beams were calculated over a spread of production volumes ranging from 1,000 to 600,000 units per year.

RESULTS

Many variables are involved in each of the acrylamate processing systems which produce the same final structural part. Processing costs of the proposed beam for the ARIMAX SRIM and SRTM resin systems, and AROTECH HMC resins are graphed versus production volume. Graph 1 shows processing costs versus an extended production volume while Graph 2 shows a more detailed view of each processing system at lower production volumes. As one would expect, each process's capital and tooling investment significantly contributes to part cost at low volumes. The ARIMAX SRTM system seems to offer the most economical procedure for producing these structural beams at low volumes. The SRTM

process is most economical up to a volume of approximately 12,000 - 13,000 units per year. At this point, the ARIMAX SRIM resin system becomes more economical for producing these beams. The SRTM processing costs cross the HMC processing cost line at production volumes of approximately 20,000 units per year at which HMC processing becomes more economical. The economic confrontation between ARIMAX SRIM systems and AROTECH HMC resin processing begins around a production volume of approximately 40,000 units per year. At this point and up to infinite production levels, SRIM and HMC molding costs are nearly equal with each system incurring various staggered capital and tooling steps. The lower capital investment of SRIM is evident in the processing cost breakdown (Graph 3). However, HMC processing levels these costs by having slightly lower labor requirements.

Production volume is definitely not the only consideration given to selecting a production method for making structural components. As mentioned previously, the composite molding cycle times are probably the single most important factor in determining production costs, especially at high production volumes. An example of this is shown in Graph 4 which shows a range of ARIMAX SRIM resin system processing costs at different cycle times. This translates to either a savings or debit of approximately $0.50/part for a ±10% change of cycle times based at 160 sec. Similar results for HMC processing costs are ±$0.47/part for ±10% alteration in cycle time based on 140 sec. SRTM processing cost change ±$0.68/part for same ±10% change in cycle times based at 720 sec.

While the chemistry and compression molding technique of the HMC process limit major cycle time improvement, SRIM and even SRTM cycle times could be improved by greater than 10% by improving molding techniques, possibly with automation. By referring back to Table 1, one notices that cycle times of the SRIM and SRTM system are controlled less by the chemistry (cure time) and more by the process mechanics than are the cycle times of HMC processing. Thus, clamp closing speeds, resin injection speeds, part demold time and other cycle altering conditions can greatly influence final part cost.

One way to possibly reduce cycle times for ARIMAX SRIM and SRTM resin system processing (reduced part cost) is by the use of reinforcement preforms. However, with the current preforming market prices, the possible 20-30% cycle time improvement may not justify the 500-600% cost increase of using preformed reinforcements instead of mat reinforcements. This is especially true at higher production volumes where material costs for any reinforced molding operation can reach 50% of the part's total processing costs. Resin costs can also greatly contribute to final part costs. For the beam analysis, a change in the resin price of ±10% will change the final processing costs

±3%. This is valid for SRIM and HMC processing conditions with an annual production volume of 100,000 beams per year. The sensitivity of SRTM processing costs to resin costs is slightly lower. It should also be mentioned that processing costs become less sensitive to material costs as production volumes decrease below 50,000 parts per year.

Processing costs per part, however, are not listed as the sales price per part. Items such as profit margins, sales commissions, administration costs, and inventory/interest all add their certain percentage. This can total up to 20-35% of the processing cost and is added to the final sales cost.

CLOSING

When considering the manufacture of structural composites, two major decisions are needed. The first is material selection. The ARIMAX SRIM and SRTM resin system and the AROTECH HMC resins bring similar high temperature, high strength properties to the design of lightweight structural composites. The second is the process. The versatility of the acrylamate chemistry allows it to be processed in several manners. This paper focused on the SRIM, SRTM and HMC molding processes using Ashland Chemical materials. Guidelines such as the ones presented can help molders select a composite molding process that would allow economical production of structural composite parts.

REFERENCES

Ellerbe, Gilbert, "RIM Mat Molding - Economic Evaluation of a New Structural Composite," SPI Reinforced Plastics/Composites Proceedings, 3-C (1987).

Stevens, M.G., G.M. Gynn, W.C. Howes, "Designing Prototype Structural Parts with a Squeeze Molding System," 42nd Annual Conference, Reinforced Plastics/Composites Proceedings, SPI, 17-B (1987).

Eckler, J.H., D.A. Rust, "Development of Flow Models for SRIM Process Design," ASM/ESD 3rd Annual Conference on Advanced Composites Proceedings, p. 109 (1987).

ACKNOWLEDGEMENTS

This writer would like to thank those in the Composite Polymers Group who helped put this paper together. Thanks to those who helped gather the information, to those who reviewed the content, and to B. Smyers who formatted the structure.

Table 1 - ARIMAX Resin System Processing Conditions

	SRIM	SRTM	HMC	PREPREG
Resin Cure Time (min.)	0.2 - 1.0	3.0 - 6.0	1.5 - 2.0	1.5 - 2.0
Processing Cycle Time (min.)	1.0 - 4.0	6.0 - 15.0	2.0 - 3.0	2.0 - 4.0
Molding Pressures MPa (psi)	0.14 - 2.41 (20 - 350)	0.03 - 1.72 (5 - 250)	4.83 - 10.34 (700 - 1500)	5.51 - 11.72 (800 - 1700)
Clamp Tonnage (for 1000 in^2 projected area)	10 - 175	3 - 125	350 - 750	400 - 850
Mold Temperature °C (°F)	71 - 105 (160 - 220)	55 - 71 (130 - 160)	143 - 160 (290 - 320)	143 - 160 (290 - 320)
Glass Levels (wt. %)	20 - 60	20 - 70	40 - 75	40 - 60

Table 2 - Capital Equipment for Composite Processing Using ARIMAX Resins

	SRIM	SRTM	HMC	PREPREG
Mold Pressures MPa (psi)	0.14 - 2.41 (20 - 350)	0.03 - 1.72 (5 - 250)	4.83 - 10.34 (700 - 1500)	5.51 - 11.72 (800 - 1700)
Clamp Tonnage (for 1000 in² projected area)	10 - 175	3 - 125	350 - 750	400 - 850
Mold Carrier Costs (x 1000 $)	50 - 220	45 - 170	280 - 440	300 - 480
Mix/Meter Machine (x 1000 $)	60 - 160	20 -45	N/A	N/A
Mold Temperature Control Unit (x 1000 $)	2 - 3	1 - 2	3 - 4	3 - 4
Trim Equipment (x 1000 $)	2 - 3	2 - 3	2 - 3	2 - 3
Material Handling (x 1000 $)	2 - 3	2 - 3	2 -3	2 - 3
Total Capital (x 1000 $)	116 - 389	60 - 213	287 - 450	307 - 490

Table 3 - Estimated Energy Requirements for Composite Processing Using ARIMAX Resins

	SRIM	SRTM	HMC	PREPREG
Mold Carrier (Amps)	60 - 100	30 - 80	140 - 200	140 - 200
(Volts)	440	440	440	440
Mix/Meter Unit (Amps)	90 - 170	5 - 10	N/A	N/A
(Volts)	440	220	N/A	N/A
Mold Temperature Control (Amps)	30 - 60	15 - 30	40 - 90	40 - 90
(Volts)	220	220	220	220
Trim Equipment (Amps)	15 - 30	15 - 30	15 - 30	15 - 30
(Volts)	220	220	220	220

Table 4 - Tool Material Selection for ARIMAX Resin System Processing

Tool Material	% of Injection Molding Tool Costs	SRIM	SRTM	HMC
Machined Steel	100 - 80%	Production	Production	Production
Machined Aluminum	70 - 50%	Production	Production	Production
Cast Kirksite	65 - 45%	Limited Production Prototype	Production	Prototype
Cast Aluminum	55 - 35%	Limited Production Prototype	Prodcution	Prototype
Sprayed Metal	45 - 25%	Prototype	Limited Production	N/A
Nickel Sheel	35 - 20%	Prototype	Limited Production	N/A
Epoxy	30 - 15%	N/A	Prototype	N/A

Description	SRIM	SRTM	HMC
Material Preparation	1 - 25%	1 - 25%	1 - 50%
Press Molder	1 - 100%	1 - 100%	1 - 50%
Press Demolder	1 - 50%	1 - 50%	1 - 50%
Trimmers	1 - 50%	1 - 50%	1 - 50%
Inspectors	1 - 10%	1 - 10%	1 - 10%

Note: Manpower based on percentage of time spent at the specific job station

GRAPH 1

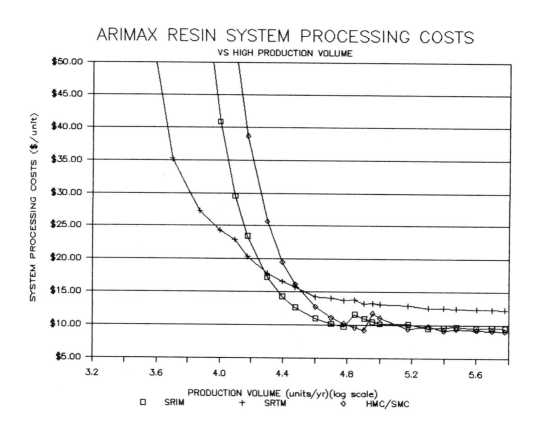

ARIMAX RESIN SYSTEM PROCESSING COSTS
VS HIGH PRODUCTION VOLUME

GRAPH 2

ARIMAX RESIN SYSTEM PROCESSING COSTS
VS LOW VOLUME PRODUCTION

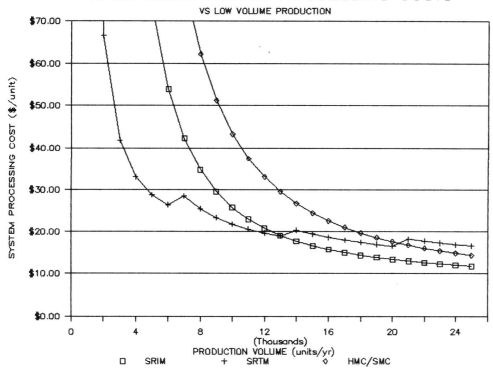

GRAPH 3

ARIMAX RESIN SYSTEMS COSTS BREAKDOWN
(for production volume = 100,000 units)

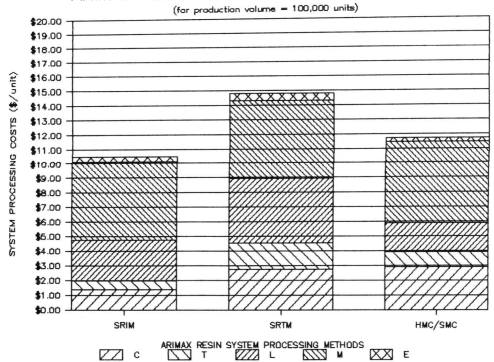

GRAPH 4

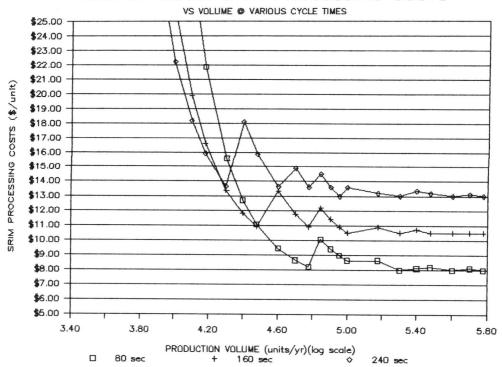

ARIMAX RESINS SRIM PROCESSING COSTS
VS VOLUME @ VARIOUS CYCLE TIMES

□   80 sec   +   160 sec   ◊   240 sec

PRODUCTION USING LIQUID COMPOSITE MOLDING

Dan A. Kleymeer
Ardyne Inc.
Grand Haven, Michigan   USA

ABSTRACT

Production of composite parts using Liquid
Composite Molding polymers requires identifying
and controlling process variables. There are
unique characteristics to be considered when
designing, tooling, molding and finishing LCM
parts. From a molder's viewpoint, some of these
variables and characteristics will be discussed
with regards to design and processing issues.
Structural RIM start-up will be presented,
including variables found in the chemical and
mechanical systems. Methods of control from
actual practice will be shown and supported so
the experience base will be broadened. Composites are the forefront of materials to improve
our quality of life; LCM processes are proving
to be significant players.

LIQUID COMPOSITE MOLDING (LCM) using structural
reaction injection molding (SRIM) and resin
transfer molding (RTM) polymers to make parts
requires the integration of design, material,
and manufacturing parameters during product
development. Such integration is necessary to
assure meeting performance criteria and smooth
production start-up. Despite comments to the
contrary, LCM is successful only when the part
is carefully designed for this process. Design
principles from other materials should be put
aside. LCM materials selected to match design
criteria have specific processing characteristics that must be made known for production to
occur. Manufacturing is successful only when
these unique characteristics are treated as
process variables needing control. LCM is not
an off-the-shelf technology to be place into
production without paying careful attention to
the variables found in part design, materials
selection, and manufacturing procedures.

The purpose of this paper is to present
some of the variables encountered during

product development using the LCM process.
Before such development can occur, a clear understanding of design parameters, material characteristics, and manufacturing variables is
required. These can be integrated by a project
management program once the broad subjects are
incremented into segments that can be easily
handled. Bringing a part to production with
LCM can be successful once process variables are
understood and controlled. This paper can serve
as a beginning point.

DESIGNING FOR LCM

The real key to part performance resides
in the reinforcement package designed into a
composite part. The package involves reinforcement type, size (diameter and length), orientation, lay-up combination, binders, and loading. With this wide range of reinforcement
variables, it is no wonder that part design can
be optimized by selective use of orthotropic or
anisotropic properties as required. Of course,
having all these choices makes the design exercise much more difficult. Even more difficulty
arises when part geometry choices are considered. Often the geometry is fixed, but slight
changes can have dramatic effect on performance.
If a part is ever to be brought to production,
these design choices must be narrowed.

Designers are best advised to think of the
design process as a funnel (see Fig. 1). The
large end is a first generation design defining
geometry, stress paths, reinforcement package,
and polymer matrix. The next level down is the
use of finite-element analysis to determine
stresses, deflections, and strengths of the
composite structure. Design changes are made,
thereby narrowing the funnel toward the final
design. Because of such a wide variety of options available to the designer, the next level
of the design funnel is prototyping. Fortunately, LCM can use low-cost, easily-made epoxy
composite molds for prototyping. Thicknesses,

reinforcement packages, and polymer matrices can be confirmed prior to arriving at the narrowest end of the funnel, the final production design.

POLYMER SELECTION AND FLOW - The polymer matrix is analogous to "glue" that holds the reinforcement package together. Consequently, its importance is somewhat downplayed. The ability to capture the strength of the reinforcement loosely identifies the role of the polymer. However, the designer will count on the reinforcement package to determine the composite part's physical performance. Other performance requirements influencing polymer choice may include finish, impact resistance, ease of processing, cost, gel time, etc. These design parameters as dictated by the application can be thrown into the design funnel to help select the best polymer matrix.

A part can be made of LCM composites only when polymer flow has been considered during the design process. Flow characteristics are a function of the reinforcement package, polymer matrix, and part geometry. Too high a reinforcement volume fraction will prevent the polymer from flowing through the fibers. The strength of the reinforcement is captured only when the matrix surrounds individual fibers and polymerizes to hold it in the desired shape. Directional fibers will give higher fractions, but may route polymer to the mold's parting line before filling the rest of the cavity. Having all random fibers may cause wrinkling or packing that increases the fraction in certain places. The polymer will follow the path of least resistance and go around areas of high fraction.

Often the geometry affects the volume fraction. A radius that is less than three times the thickness increases the risk of restricting the polymer flow. A tight radius also can damage the fibers as the mold is closed, thereby weakening the reinforcement. As one can imagine, perpendicular ribs are difficult to make with LCM because the reinforcement does not conform well to sharp angles; and polymer does not flow well into blind areas. The need for ribs can be handled by providing greater local stiffness through modifying the reinforcement package; or by designing a flowing rib in the shape of a U. Such considerations of design are critical to achieve good polymer flow for maximum wet out. Another reason to design with smooth geometry is the ease with which reinforcement preforms can be made.

DESIGNING FOR PRODUCTION - Once the desired reinforcement package has been selected, manufacturing methods should be determined. The annual volume of parts to be made decides the degree of automation to be implemented. Requirements of, say, 50,000 per year may be produced via manual or mildly semi-automatic means. The complexity of the reinforcement package influences the nature of the preforming operation. Several layers of different styles of fiberglass to be used in a part having a very uneven contour, will be handled much differently than a relatively flat part needing little reinforcement. After the reinforcement is cut to size, it is either placed into the mold cavity, assembled with other cut pieces into a layup, or formed to shape via heat and pressure. Combinations of these procedures can be used to achieve the desired preform. The subject of preforming is beyond the scope of this paper, but is an extremely important element of the LCM process. The method of preparing the reinforcement package for the molding operation is a classic example of using good manufacturing engineering principles. Quite often, the design can be altered to simplify preform production. Simultaneous engineering efforts have obvious advantages during the design and manufacturing engineering steps when the goal is to bring a part to production using the LCM process.

These advantages again appear when determining post-molding, or secondary, operations to finish the part. The design can greatly affect finishing procedures. For example, one method of making sure reinforcement material extends all the way to the edge of the part, is to make the part larger and then trim it to size. Another method is to make the preform the exact size of the part so it fits precisely into the mold cavity. As one can see, these two designs pose different challenges for the manufacturing engineer. The single most difficult design problem involves fastening. A vast menu of mechanical and chemical fasteners exists for composite materials, including LCM. One obscure method is the use of inserts. As part of the preform, other materials can be sandwiched between fabric layers to provide a means of solid attachment with fasteners. Steel, wood, and aluminum are often used to carry localized fastener loads and help to spread stress outward into the composite. Inserts can simplify secondary operations by making them a drill or drill-and-tap only step. The designer may consider molding holes in place, but as polymer flows around any obstruction, teardrop air pockets can be formed. Reinforcement would be parted by a pin being used to form the hole, leaving the hole with polymer-rich edges. Such a situation is not desirous because most polymers used in LCM are quite weak by themselves. Therefore, holes should be drilled and chamfered as a secondary operation. The designer can supply hole location marks and depressions during mold design to make this step easier. Trimming and drilling are two finishing steps that can be greatly influenced by design; consequently, it is important to know what can be done economically and practically during the design stage.

Designing for production with LCM involves controlling the variables found in materials selection, part geometry, reinforcement preforming, and secondary operations. Before these variables can be isolated, the design intent needs to be understood as a function of performance. Some of the variables and how they can be controlled were presented for your consideration. Each application will have its own unique optimum design, including reinforcement, geometry, polymer matrix, and manufacturing method. Successful designs match these variables to performance criteria; LCM offers very wide choices with some limitations. Such a wide range affords greater design success because optimization can be greater. It also requires more effort because of so many variables. Once the design has been finalized, analyzed, and prototyped, the part can be brought to production. The discussion that follows is how this might be done for a somewhat simple design.

MANUFACTURING WITH LCM

When it comes time to actually manufacture a part using LCM, the process will involve preforming, chemical handling, molding, and finishing. Each of these subjects have been touched on during the design discussion above. These items will again be presented, but this time in relation to an actual production experience. The first commercial application of structural reaction injection molding (SRIM) began in 1984 when Ardyne Inc. was contracted to manufacture spare tire covers for General Motors. Reference to this beginning has been made in other writings, but not from someone on the inside. The account that follows will deal with what should be done rather than dwell on the effort it required to arrive there.

The spare tire cover is a somewhat contoured panel that provides a false floor in the trunk of the larger GM cars. It must withstand the impact of a dropped weight at 82° C (180° F) and -29° C (-20° F). It has only a textured surface finish specification and must have a label adhered to the topside. To satisfy these requirements, the tire cover has a thickness of 1.6 mm (0.063 in.) and a reinforcement content of about twenty weight percent. The reinforcement package is a single 66 kg/m^2 (1.5 oz./ft^2) continuous strand fiberglass mat. The polymer matrix is polyisocyanurate blended with proprietary amounts of catalyst and pigment. After being in production over two years, Ardyne has ran over two million pounds of polymer and one million pounds of reinforcement.

PREFORMING - The reinforcement preform is made by cutting the shape from rolls of continuous strand fiberglass mat. Several rolls are loaded into a rack so they can be dispensed simultaneously in a stack. A wringer-washer style feeder directs several layers of CSM onto a lower platen area of a steel rule-die press. A single operator guides the mats during the feed, disposes of the trim after the cut, and stacks the preforms onto a skid (see Fig. 2). The entire operations takes place in an enclosed room having dry ventilation. It is imperative to minimize moisture and foreign materials from being exposed to the preforms. Contaminants can cause poor reactions to occur during polymerization. The tire cover contour does not

necessitate forming the mat into a shape of matching contour. The mat stretches enough while the mold closes to fill the cavity without wrinkling. From the die cutting operation, the preforms are staged near the production molding presses.

MOLDING - The first systems acquired for molding spare tire covers were from the conventional soft RIM industry. It was quickly realized that conventional equipment was not quite suitable for the easily reacted, very stiff, but brittle SRIM polymer matrix. Temperatures within the two delivery tanks require controlling spans within 5.5° C (10° F). The chemicals in the tanks are agitated very slowly with a customer-made impeller to reduce gradients without entrapping gases. The space above the liquid level was originally blanketed with extremely dry air, and now by nitrogen at a pressure of 0.14 MPa (20 psi). The pressure and liquid level are important because together they provide the head pressure to force the chemicals through a bag filter and into a non-priming rotary piston pump. The pumps and mix heads are the most critical components of the system. They are kept free of contamination by the filters which are changed when the pressure differential exceeds 0.03 MPa (5 psi). The pumps are driven by two-speed motors so throughput ranges can be doubled without changing pumps. All the components in the system are installed with valves, unions, quick-disconnects, and plug/socket connectors to simplify and speed replacement. Each chemical system is situated in its own diked area so mixing will not occur in the event of a spill (see Fig. 3). The chemicals are pumped to a ring line that delivers them to the mix heads situated on each mold. Six presses, each holding two molds, are fed from the ring line. At this point each mold has its own chemical handling system.

Very fast acting ball valves dead-head the ring line from its normal recirculation of the two chemicals when a shot is called for; that is, when an operator wants a part to be made. Pressure is allowed to build to a tightly controlled, predetermined level. Once reached, ball valves route both chemicals to a mix head mounted to a mold (see Fig. 4). Inside a two-stream mix head, the polyol and isocyanurate pass through individual metered orifices and into separate grooves in the side of the piston blocking the mix chamber. The chemicals are routed back to the supply tanks until the pressure at the mix head reaches a stable predetermined level. At this time, a hydraulic cylinder on the mix head retracts the piston. The head hydraulic system is small in capacity, has a heat exchanger for cooling, and is isolated from any other system. Contamination to the mix head hydraulic system has occurred from the chemical reaction that takes place in the mix chamber. With the piston retracted, impingement mixing occurs and the liquid polymer flows into the mold cavity. The shot duration is strictly a function of time and the flow characteristic through the reinforcement and cavity. At the end of the shot, valves quickly return the chemicals to the previous recirculation route. The mix head piston simultaneously closes and wipes residual chemicals from the mixing chamber. If properly combined, the polyol and isocyanurate will polymerize within a few seconds in the presence of heat and catalyst. In-mold cure times are part dependent; usually fifteen to twenty seconds per pound of polymer. Exact cycle times for molding parts from any LCM process is absolutely part dependent. The entire process is successful only when the variables are controlled.

Within the chemical handling portion of LCM, the following variables appear: supply tank blanket pressure, supply tank temperature, filter size, supply pump pressure, supply pump volume, diverter valve operating time, pipe size, mix head operating speed, mix ratio, catalyst level, shot time, contamination amounts and mold temperature. Most of these variables are monitored and controlled by a high-level programmable controller. A pair of desktop computers provide system operation and SPC monitoring. Although it seems like overkill, this degree of control has provided a trouble-free environment for over a year of continuous production.

Another portion of the process having many variables is the design of the mold. Some of these include the following: part surface area and geometry, surface finish, injection gate location and shape, parting line seal integrity, vent or vacuum locations, temperature level and uniformity, and construction material. Again, starting from conventional RIM, mold design variables can easily be handled. The part area and geometry along with the reinforcement package largely control polymer flow. Injection gates can be essentially straight without static mixing devices. The parting line seal should be one of matched metal for production applications. Temperatures are chemistry-dependent but are usually 93-110° C (200-230° F). Uniformity is a function of media throughput, and heat source line quantity, size, and location. Since 90% of the polymerization cure occurs in the mold, cooling is not required. Molds for the spare tire cover are made of steel, and have a 12 mm tube serpentined 30 mm below the surface. Water heated to 110° C (230° F) under a pressure of 0.10 MPa (15 psi) is continually circulated through all the production molds from a central boiler and heat exchanger system. Control of the mold temperature as well as maintaining the molds to original designs is all that is required to keep them in production.

During the molding operation, the adhesive nature of the polyurethane-based polymer causes the part to adhere to mold surface. A thin coating of a conventional mold release sprayed from a paint-style air spray gun is all that is required to prevent sticking. Of course, this introduces another variable into the process; one that is largely operator dependent. Auto-

mating this step is possible but not economically attractive. Internal mold release is a better solution, but is not yet commercially available. At this time, operator training, equipment maintenance procedures and mold release chemistry control are the best means of handling variables involving mold release.

FINISHING - The spare tire cover is molded larger than need be and then trimmed to size in a secondary operation. The centrally located gate is removed via a carbide hole saw mounted in a pneumatic drill press. LCM materials can be machined by drilling, sawing, punching, abrasive grinding, routering, and water-jet cutting. Of these, abrasive grinding is perhaps the easiest and quickest. Water-jet cutting, with multiple-axis control, probably provides the greatest accuracy and holds the greatest potential for low to medium production volumes. As stated before, the degree of automation depends on the number and complexity of parts to be made. The spare tire cover is finished with a two-head, six station CNC router adapted from the woodworking industry (see Fig. 5). The carbide router bits had their life greatly extended by showering them in a water spray. The trim and dust are carried by the water and separated in a filter. Residual glass fibers are removed via light sanding. Dust particles are cleaned off in a car-wash style device prior to labeling and packaging for shipment. The secondary operations are neither capital nor labor intensive for this application.

Production with the spare tire cover gives confidence in the future of LCM as a high-volume process for composite parts. This manufacturing experience has opened doors to many applications regimented to traditional processes involving multiple assemblies with isotropic materials. Now that LCM has broken the "hand lay-up" mindset, new applications will be conceived in many commercial areas.

PRODUCT DEVELOPMENT USING LCM

Basic project management elements exist in LCM product development projects as they do with most every project undertaken. The four major steps are concept, development, implementation, and audit. Project management practices are important so tasks are completed on time and within expected resource allotment. The people who created the project will have expectations to see their vision accomplished. A project having structure and order is likely to succeed whereas one conducted in chaos will tend to fade away, especially when many resource people are involved. Developing a product so it is manufacturable with the LCM process is again a matter of controlling variables. This time, however, the variables are found in the people working on the project, from design through production.

CONCEPT - During the concept portion of a product development project, the single most important task is to clearly identify the exact scope of the program. For example, the scope of the spare tire cover project was not to make a cover with contour that could pass low temperature impact tests. Certainly this was one of the visible goals, but not the main thrust of the program. The scope instead was to develop and prove LCM technology for use in the automotive industry. Once the scope is clear, tasks can be listed and resources assigned. The concept phase should conclude with a basic design that has a 75% chance of satisfying performance criteria. An economic analysis and shallow market survey will determine if the program should proceed.

DEVELOPMENT - The development phase is where most of the difficult work is done. Plans are made incrementally and checked against a global overview. Drawings are completed and finite element analyses conducted. In short, the development phase undergoes the aforementioned design funnel steps to arrive at a final part. Plans for production are in this phase as well. Methods, equipment, staffing, facilities, etc. are all considered and identified. Layouts and process flow diagrams are generated. Market studies are conducted and sales strategies planned because a product without a buyer is merely a research project. Until someone wants or needs what was developed, the program is often regarded as technology advancement. The development phase concludes with hard piece prices, tooling costs, capital requirements, manufacturing plans, tested prototypes, etc. Again, a check is made to be sure the scope is being met and that the project is still economically viable.

IMPLEMENTATION - The implementation phase involves carrying out the plans made during development. Capital monies are obtained, task schedules are generated, and mechanical work is begun. The process variables uncovered during the early days of the spare tire cover, became controlled during the implementation of the new production environment outlined before. Budgets are established and monitored. Maintenance procedures are generated. Raw materials and purchased items are procured. Acceptance trials and final product approvals are given. Purchase orders, patents, contracts, and other paperwork take place. Of course the list of tasks and accomplishments continues until the first order is shipped and paid for. Projects can continue indefinitely if allowed to do so. It is important at the beginning of this phase that some ending point is clearly defined and understood by all, especially those responsible for maintaining production. Early involvement through simultaneous engineering really makes this last point happen. The implementation phase has the most visibility but is only as good as the plans made during development.

AUDIT - The last phase is important because people's paychecks depend on a favorable outcome. The audit phase assesses the economic virtues of the project and verifies that the scope has been met. If the results are not acceptable, a new project containing all four phases must be created to define and rectify problems. The audit phase does not have to end; it should continue as a means of tracking production performance through accounting procedures. For the spare tire cover, the chart in Figure 6 shows the dramatic impact of implementing Ardyne's project to control process variables by building a new production environment. The audit phase gives final accountability to the previous phases and can provide a definite end to a project.

Although this discussion on project management is quite generic, it draws together previous design and manufacturing discussions. The design begins as a concept and is developed to a final part. Manufacturing develops from the design and is implemented into part production. Because of the vast array of variables found in LCM, a systematic, structured approach is required to install controls successfully.

SUMMARY

Liquid composite molding has variables in the areas of design, materials processing, and manufacturing environments. Material options were discussed but not in regards to performance. An analogy was created by depicting the design process funneling ideas and experience toward the output: a production-worthy part design. Polymer flow is a function of the reinforcement package, polymer, and part geometry. Suggestions were given on how to handle these variables. A relationship between design and manufacturing was established as a responsibilty of the designer. A case study of manufacturing spare tire covers with one LCM process, SRIM, was presented. Some of the process variables and how they are controlled is important to any LCM production program. Finally, a general project management discussion showed ways to bring a part to production by following four phases: concept, development, implementation, and audit.

CONCLUSIONS

Liquid composite molding is now a proven process suitable for manufacturing structural components for a variety of consumer and industrial products. It has been accomplished by controlling variables found in design methods, material choices, and production processes. These variables were identified and controlled by using project management procedures. A part targeted for the LCM process must be designed with consideration for unique requirements involving geometry, reinforcement packaging, and polymer matrix. Funneling design ideas through a four step process will result in a part that can be made by LCM.

Production variables will exist with every application brought to LCM and each will have uniqueness. This means that LCM is not an off-the-shelf commodity that can produce parts with standard technology. To dispell concerns that LCM may not be worthwhile, Ardyne's experiences of placing a spare tire cover into production were recounted. With proper design and manufacturing knowledge, project management principles can guide potential producers toward successful outcomes.

SELECTED REFERENCES

Carleton, P.S., D.P. Waszeciak and L.M. Alberino, "A RIM Process for Reinforced Plastics", Society of the Plastics Industry Conference (January, 1986).

Eckler, J.H. and T.C. Wilkerson, "Composite Manufacture Using the Reaction Injection Molding Process", Society of the Plastics Industry Conference (January, 1986).

Johnson, C.F., "Resin Transfer Molding", p. 564-568, International Engineered Materials Handbook Volume I (November, 1987).

Nelson, D.L., "Structural Composites from RIM Mat Molding", Society of Automotive Engineers International Congress (February, 1987).

Rouse, N.E., "Optimizing Composite Design", Machine Design (February 25, 1988).

Shirrell, C.D., "Liquid Composite Molding - The Coming Revolution", AUTOCOM '88 (May, 1988).

APPENDIX

Fig. 2 - Reinforcement preforms are made by cutting them from rolls of continuous strand mat using a steel rule die press.

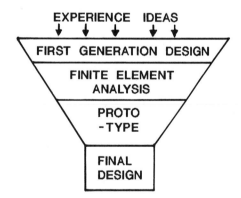

Fig. 1 - Arriving at a final part design using the "Design Funnel" analogy.

Fig. 3 - The chemicals are dispensed from equipment housed in an isolated room and designed for expedient component change out.

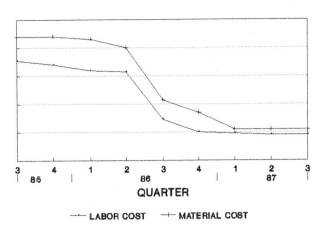

Fig. 4 – Mix heads are mounted atop molds mounted in a press that is situated in a paint spray booth.

Fig. 6 – The dramatic drop in labor and material costs was a result of controlling process variables.

Fig. 5 – Spare tire covers are trimmed on a two-head, three-axis router.

DESIGN AND PROCESS CONSIDERATION FOR STRUCTURAL RIM

G. H. Slocum, D. W. Schumacher, M. F. Hurley

Mobay Corporation
Pittsburgh, Pennsylvania, USA

ABSTRACT

A brief description of the Structural RIM process is given and compared to other composite production techniques. Typical physical properties of SRIM composites are given for comparison, and discussed with respect to suitability for various applications. Factors which affect the costs of SRIM composites compared to other methods of production are discussed, and the importance of resin flow through the reinforcement is noted. Techniques for measuring the flow of resins through reinforcement filled molds are described, and results of some of these measurements are given. Models are discussed and conclusions are drawn about processing SRIM systems.

COMPOSITES OF VARIOUS ORGANIC RESINS reinforced with numerous organic and inorganic fibers and flakes have been used in thousands of applications over the years. The sort of physical properties necessary for "structural" applications were achieved decades ago, but the growth of composites usage in this type of application has been modest. This has been due mostly to the limitations on properties attainable in most high volume production processes. Certain of these processes have made possible the production of significant volumes of parts in the automotive and other industries, but these have been mainly of a semi-structural or cosmetic nature.

The reaction injection molding (RIM) process was invented approximately twenty years ago, and has been used extensively for the production of various parts in the automotive industry, including fascias and exterior body panels. This process, while adapted to and developed for high volume applications, has also shared the limitation that the possible physical properties achieved are not suitable for structural applications. In the past several years, though, as the general interest in composites has grown, there have been developments in RIM materials which provided properties that are suitable to structural applications. Some of these developments, though, have had costs in terms of the process and its suitability for high volume applications.

The nature of these developments point to several key issues as a guide to further development. These issues include the material costs, of course, but most important is the total cost of the process, in terms of cycle time, scrap rates, production equipment costs and lifetimes, and environmental and worker safety concerns. In this paper we will discuss some of the recent developments in structural reaction injection molding (SRIM), and how these and anticipated developments will affect these overall costs.

HISTORICAL

Structural RIM is defined by the nature of the resin matrix, the form of the fiber reinforcement, and the process by which the composite is produced. The resin system is a multicomponent thermosetting polymer mixed via high pressure impingement. A number of different approaches to resin formulation have been evaluated and are being offered in the market, most of which are variations of polyurethane chemistry. The composite is produced by the direct injection of the resin into a closed mold containing typically 20 to 50 percent by volume long fiber reinforcement, meaning the mixed resin viscosity must necessarily be very low. In fact, as will be seen, the viscosity of the resin system is critical to the processability of SRIM. The reaction profile of the thermoset reaction must be controlled in order to allow as a long flow time as the part size and complexity would require, yet allow for the composite part to be demolded after reasonable interval. In addition, the resin matrix must itself possess high stiffness and impact properties as well as sufficient thermal properties to be suited to the demands of automotive applications. These properties should be obtainable without requiring a secondary postcure operation.

The specific processing conditions for SRIM vary according to the basic chemistry and formulation chosen for the matrix resin. Recommended processing parameters for Mobay's STR400, which is based on polyurethane chemistry, are listed in Table I. It may be observed that these parameters parallel those of conventional RIM (such as the processing of Bayflex 110 Systems) with the exception of mix ratio and lower pressure required for satisfactory mix quality. Studies have shown that conventional mold temperatures (140- 150°F) give optimized processing and property characteristics, a particular advantage of STR400 chemistry. The injection rate, as will be discussed, is determined by the part size and geometry and the amount and type of reinforcement. Depending on the catalyst level of the specific formulation, the demold time will generally be about three times the allowable flow time.

Structural RIM is a relatively recent development as a method of producing composites. Its utilization for specific automotive applications is only beginning. Historically, a number of composite materials have been developed and made available in the last 35 years for use in automotive applications. However, Structural RIM, as a new generation, will allow greater use of composite structures in automotive designs due to its versatility and cost-effectiveness.

Resin transfer molding (RTM) is closely related to SRIM in that a thermosetting resin is injected into a mold containing a preplaced fiber mat reinforcement. Because a stationary reinforcement is used, control is

maintained over physical property uniformity and variation, as needed, within the part. Also, very large parts can be produced with reasonable mold and metering equipment costs. The characteristically slow reactivity, high viscosity (controlled to some degree by dilution with reactive monomer), and consequential high flow resistance through the reinforcement allow for very simple metering equipment and relatively slow injection of the resin into the cavity. The curing of the resin is accelerated by heat and therefore cycle times are a function not only of catalyst level but cross-sectional thickness as well. The cycle times inherent in the RTM chemistry and process do not provide a very efficient production cycle, precluding application to high volume (greater than 50000 pieces per year) production.

SMC also has a thermoset matrix with similar chemistry to typical RTM resins and has some of the same processing restrictions inherent in free radical initiated thermosets. But other factors are also to be considered which are characteristic of compression molded composites. SMC has gained considerable attention for composite body panel parts such as horizontal body panels where exceptional stiffness is required. Compounds have been developed which give exceptional surface qualities appropriate for class A applications. Certain secondary structural parts such as bumper beams are being evaluated with SMC because of its apparent low cost. However, the compression molding process and the nature of SMC compounding itself limit its adaptability to certain design constraints. In a relatively flat part, such as those for which SMC is usually specified, the resin and the fibers flow together throughout the tool as the compound is compressed. Physical properties remain adequately uniform throughout the part. With a more complex geometry, though, which requires the compound to flow a significant distance in directions at angles to the compression plane, the flow characteristics of the resin and the reinforcement vary. In particular, at these angles, the resin tends to flow without carrying the fiber reinforcement. The result is an inconsistent distribution of reinforcement and hence a variation in physical properties. Measurements from an actual prototype part with perpendicular walls demonstrated a 40% decrease in flexural and 75% decrease in tensile properties in the side walls as compared to the wall that was in direct compression. Since most structural applications involve complex shapes such as C-sections, modified box sections and long draws governed by styling considerations, such a dramatic variation in properties becomes detrimental, especially when considering primary structural components. Variations in wall thicknesses are more viable with compression molded composites than with stationary reinforcement, yet the typical chemistry of the matrix in the case of SMC may lead to other processing restrictions. A 2.5 mm section may require only a 1 minute cure, while a 6 mm section could require as much as 4 minutes to achieve a uniform cure. These limitations are worthy of consideration as well as the high capital and operating expenses associated with high molding pressures (1000 psi) and temperatures.

A variety of composite applications have been developed as the result of the introduction of glass mat reinforced thermoplastics. Although a number of thermoplastic matrices have been utilized, polypropylene remains the most commonly used. The composite blanks are processed by preheating in an oven and then charging into the mold in a similar manner to SMC. The mold compresses the material forcing it to flow to fill the mold, which serves as a heat sink to cool the composite. Because no cure is involved, an attractive cycle time can be achieved, provided the placement of the blanks in the tool is simplified. The moderate stiffness of these materials is in some cases offset by the low specific gravity in achieving the performance specified. Although directional fibers have been incorporated with some success, for the most part, reinforcement selection to obtain specific local properties is limited. Also, the maximum fiber content is considerably less that for RTM or SRIM. As with SMC, the physical property variations inherent in compression molding composites are manifest again as a function of the ability of the glass and resin to flow in concert through complicated geometries. High molding pressures required to cause the flow of the softened thermoplastic severely limit the practical size of the composite part to be molded. This type of composite seems well suited for certain small secondary structural applications, but is constrained by economic and processing factors.

PHYSICAL PROPERTIES

The physical properties of an SRIM composite depend primarily on the glass type and content in the composite. The data in Table 2 illustrate the variation in physical properties achievable through changing the level and orientation of the glass with one particular resin system. The properties in column A are those of a polyurethane matrix reinforced with a random, continuous strand glass mat. By replacing some of the random mat with a directional glass mat, the properties in the direction of the glass are increased even for lower glass levels, as shown in column B. The properties in the perpendicular direction do suffer somewhat, but at the same level of glass (comparison of columns A and C) composites with some amount of directional glass offer exceptional properties in one direction while maintaining reasonable properties in the other.

An additional advantage of SRIM is the capability for the engineer to adjust the composite properties to meet the needs of the application. While the properties of SRIM composites compare favorably with those of mat reinforced thermoplastics at their available levels of glass (typically 40% random glass), SRIM is not wedded to either this glass level or configuration. As shown in column D, the properties with this particular level and configuration compare favorably with both mat reinforced thermoplastics and typical SMC, even perpendicular to the glass orientation. The properties parallel to the glass orientation exceed even high glass SMC, especially in terms of impact strength.

The properties of the neat, or unreinforced, resin may be important for some applications. The properties of one resin are listed in Table 3. The flexural modulus of most SRIM resins available today is in the range of 3-500,000 psi. The modulus of the composite is not greatly affected by differences in this range, because the main contribution to the modulus is from the glass. Although the tensile properties of the resin can have some affect on the composite properties, they tend to follow more closely the flexural properties, with little variation from one resin formulation to another, at least among the polyurethanes. The two areas which can affect the composite properties are the neat resin impact strength and thermal stability, which affect long term composite properties most importantly. The neat impact strength may be critical to maintaining part integrity in applications where there is considerable exposure to small impacts, such as a floor pan or underbody which may see millions of impacts from gravel and larger rocks during the life of the part. The thermal stability of the part will be important in design when considering use temperature and shielding.

COST CONSIDERATIONS

If the properties of a material are completely suited to an application, whether that material will be used will depend on how it competes economically with alternative methods of production. This statement is true even if the alternate production material is in some way physically inferior, as long as the requirements of the application can be met through the use of more of the alternate material or other still economically competitive means. This competition must be viewed in terms of the cost to the ultimate consumer of the application. Therefore, while material costs, capital costs, and labor costs are all important, every significant cost must be considered.

With the few applications of SRIM materials actually in production, there may still be considerable question about what significant costs exist. The answers to these questions are what will determine whether SRIM materials or one of the alternates will be used. Given that there will still be questions, a detailed look at the process, from the beginning, is the best way to delve for answers.

The process begins long before the part is actually made with the design of the part. A perfectly suitable material with a perfectly efficient process is not likely to fit a part designed for a completely different material and process. This consideration is important for several reasons. Most structural parts to date have been made from steel or other metals. Even here design is critical, in that a design optimized for steel will not be necessarily optimum for aluminum. Designs optimized for steel are often assemblies of several steel parts joined by welding, fasteners or other means. A one for one replacement of a steel part by a composite part from whatever production method almost never makes good economic sense. Parts consolidation, though, is not the answer to everything. Volume has a significant effect on the impact of material, capital and labor costs for a part of any particular design. Making 1,000,000 of a part per year may well be less expensive in steel while 100,000 per year of the same part may be less expensive in a composite. Design of the part to fulfill the same function may be possible for either, but ultimate costs will depend not only on volume but on which material the design of the part was optimized. The implications of this consideration are that a certain amount of design expertise in alternate materials is required along with a realistically complete cost accounting system in order to do the rigorous analysis necessary to choose the design leading not only to the best part but to the best profit for the producer and the lowest costs for the consumer.

The next step in the process, at least parts of which are often ignored in an overall analysis, is the tooling. While the tooling costs and timing are discussed in comparing various plastics technologies and metals with respect to volume, the specific design related aspects of tooling are sometimes left for discussion until after decisions are made which those aspects should impact. The most obvious example of this is in the area of gating and venting injection molded parts (whether thermoplastic or RIM). Trial and error procedures are used more than occasionally, and the costs involved are seldom anticipated or even counted after the fact.

Once the part is designed and the tooling made and "debugged," the conventional cost comparisons for processes can be done. Here material costs, energy costs, equipment costs, and labor costs all interact with cycle time, reliability, scrap rates and capacities to affect overall piece cost. Each of these considerations affect the competitiveness of SRIM materials.

If the design and tooling inputs are made with the SRIM process in mind, the parts coming out the end of the process will have a lower overall cost. If one SRIM part can perform the same function as an assembly of several parts in another material, some of which take multiple forming operations to make, it is likely that the SRIM process will be more cost effective. This, of course, depends on volume. If the equipment necessary to do the multiple forming operations, and assemble the several parts can be operated at a significantly faster rate than the SRIM production, and the number of parts required, both annually and over the life of the operation, is large, then it is conceivable that cost per part will be lower for the alternate process. The energy costs tend to favor the SRIM process, particularly if there is no necessity for post curing operations. The cycle times and labor costs of the process at its present stage of development, though, tend to favor the competition. For SRIM to become a really competitive process, a great deal of development must be done in the area of overall cycle time reduction and in the automation of the cycle.

It is in these areas where our main development efforts have been concentrated. The overall cycle can be reduced in a number of ways, but one of the simplest is by reducing the demolding time. With the dependence of the demold time on the flow time, the faster the material can be put in the mold, the faster the part can be demolded. The speed with which the material can fill the mold depends on several factors, the exact nature of which have important implications on the whole process as discussed above. How we have characterized these factors and some of those implication is discussed below.

EXPERIMENTAL APPARATUS

Our main tool for this characterization, which is pictured in Figure 1, is our large, instrumented plaque tool. The plaque size is nominally 1 meter by 2 meters, however it is typically blocked to a 0.25 meter strip down the center of the plaque. The tool has possibilities for either center or end gating. Down the center are placed nine pressure transducers ports and nine thermocouples. If the tool is end gated, there are ports for two more transducers in the aftermixer area. Although the standard width for mold filling characterization is 0.25 meter, there are inserts for the tool which allow for adjustment in the width of the plaque from 0.1 meter to 1.1 meter. The transducers and thermocouples in the tool are connected to a data acquisition system allowing for digitizing analog signals, counting of pulse trains, and storage and processing of these data. Until recently this tool was run in a 125 ton press. Now the tool is run in a new development machine which has a 600 ton press and a metering unit capable of running from 225 g/s to 6000g/s.

RESULTS AND DISCUSSION

What this tool has given us is the ability to monitor what is happening throughout the part during the injection and cure of the resin. We can now see what the pressure and temperature are at the mixhead and every twenty centimeters removed, from the time the injection

begins until the part is removed from the tool. Some of the typical data from this tool are shown in Figure 2.

This Figure shows the data for an injection from the center of the tool. Each trace in this graph represents the pressure at one transducer. All of the transducers start at zero pressure. As the shot begins, the transducer at the center of the mold begins to show pressure first. As the material reaches each transducer, it begins to register the pressure at that point. The pressure at a transducer at any particular time is the pressure necessary to continue pushing material past it as the flow front extends further beyond that point. Thus, the pressure in the first trace rises from zero to something in excess of 400 psi as the material front moves out of the mixhead to the vent. As the material reaches the vent the pressure plateaus until the end of the shot, when it re turns to around zero. As the material cures, a slight rise in pressure is observed, consistent with the expansion of the material by the exotherm of the curing reaction.

While a number of conclusions can be drawn from this type of data, a bit more is needed in order to make this type of data really useful. The pressure drop that is measured between two of the transducers is proportional to several factors, as given in the following equation:

$$\frac{\triangle P}{\triangle L} = \frac{Q}{A} . \eta . \frac{1}{k},$$

where P is the pressure differential necessary to push an amount Q of a material with a viscosity through an area A in a unit time over a length L. "k" is a constant related to the permeability of the reinforcing mat through which the material must flow. It can be thought of as the fraction of the volume of the mold which is effective open for resin flow. This constant can be treated theoretically, or, with the proper experiments, it can be treated empirically with convenience. The determination of this permeability constant involves pushing a material of known viscosity through the mold loaded with various reinforcement lay ups. Figure 3 shows these empirically determined constants for a random continuous strand glass mat as a function of the number of layers of mat. The technique for this determination is being refined and the true amount of curvature to this line will be defined with this refinement.

Even without the refinement, though, it is possible to use these constants and data of the type shown in Figure 2 to determine the actual viscosity of the mixed resin as it flows through the reinforcement in the mold. Figure 4 shows the viscosity of Mobay's STR400 as a function of the number of layers of glass as determined in some of our first experiments of this sort. The scatter resulting from the need for a refined technique is evident, but the magnitude of the number within the scatter is significant nonetheless. Refinements of the techniques we are presently making will help us define more exactly where within the range of 10 to 20 cps the mixed resin viscosity actually lies.

Characterization of the permeability of various reinforcement arrangements, along with the assignment of exact viscosities of the reaction mixture as a function of the distance from the mixhead, the velocity of the material and the temperatures of the mold and material can be used to model the filling of a mold. This type of information has already been used to accurately predict the flow of thermoplastic melts through real mold geometries and, with proper development, can be used in the same way for structural RIM molds. Compilation of the data base necessary to do this sort of calculation and modelling, both for our various systems and for the different reinforcement combinations, is one of our primary goals in the near future.

Perhaps more important is the comparison of processing Structural RIM systems with alternative composite part production method, which even our as yet unrefined data allow. The data shown in Figure 2 are for flow of one of our systems through 60 wt.% glass at about 10 cm/sec flow velocity. In order to push the material one meter through the mold takes approximately 400 psi at the mixhead. This does not mean, though, that the clamp is required to hold 400 psi over the entire mold, but the integrated pressure represented in the graph. Since the pressure drop between the transducers is nearly linear, this calculation is simply the pressure at the middle transducer times the projected area of the mold, in this case 200 psi times 80 inches by 10 inches, or 80 tons. This particular mold configuration was used to simulate filling a bumper beam. By properly designing the part and the tool even this can be reduced, perhaps to the point where the press is used mainly for opening the tool rather than holding it closed during the shot. This type of design could also allow for filling the part in 1-3 seconds allowing for demold in well under one minute. Development of this modelling capability and its application to the design of parts and tools will allow the production of SRIM parts with equipment that is very small compared to alternate composite processes and with cycle times that are competitive with the best current technology has to offer.

ACKNOWLEDGEMENTS

There are a number of people who have contributed to our development efforts. Among those who are particularly significant are Dr. Neil Nodelman, Dr. Henry Mueller, Dr. David Sounik, David D'Errico, Kenneth Naleppa, Carl Holsinger, and Albert Magnotta. We would also like to acknowledge the extremely strong support of our management for this development, particularly Dr. Ronald Taylor, Kieth Spitler, Dr. Richard Keegan, Maurice Courtney, Dr. Robert Volland and H. Lee Noble.

Table 1

Processing Parameters for STR-400
Structural RIM System

Mix Ratio (pbw)	194/100
NCO/OH Index	1.05
Material Temperatures	90 ^{o}F
Mold Temperatures	140 - 150 ^{o}F
Injection Rate	0.3 - 2.0 lb./sec.
Mold Release	Wax
Gel Time	5 - 180 sec.
Demold Time	15 - 540 sec.

Table 2

Properties of Mobay Structural RIM System Composites

Glass Configuration U= unidirectional mat 8 oz/yd2 R= random mat 2 oz/yd^2	5XR	1XU 3XR 1XU	1XU 4XR 1XU	2XU 4XR 2XU
Total wt % glass	46	40	46	49
Specific Gravity (Theoretical)	1.60	1.54	1.60	1.63
Tensile Strength (psi)	3300	33600/18500	37200/22200	44600/24200
Tensile Modulus (ksi)	1830	2100/1260	2230/1640	2480/1380
Flexural Strength (psi)	41300	40600/24500	48000/31600	43700/29200
Flexural Modulus (ksi)	1610	2100/836	2450/998	2650/903
Izod Impact, notched (ft-lb/in)	25.5	35.5/10.5	41.8/13.0	43.8/14.4

Table 3

Neat Resin Properties of One Mobay Structural RIM System

Tensile Strength (psi)	13200
Flexural Modulus (psi)	505000
Izod Impact, notched (ft-lb/in)	1.2
T_g by DMA (^{o}C)	102
Specific Gravity	1.21

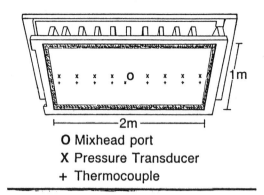

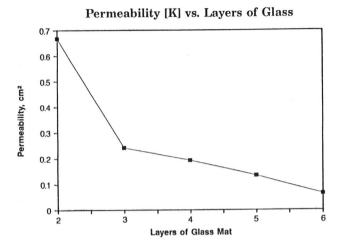

Figure 1 - Schematic of Instrumented Plaque
         Tool

Figure 3 - Graph of Calculated Permeability
         (K) versus Layers of Glass Mat

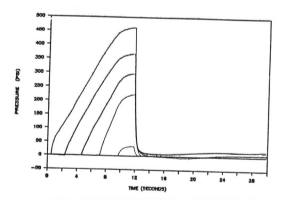

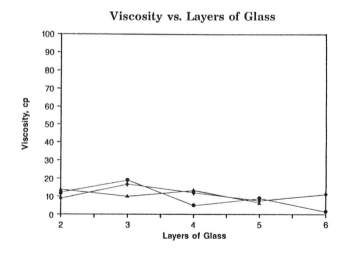

Figure 2 - Pressure Curves from Instrumented
         Plaque Tool for Reaction Injection
         Through Pre-placed Long Glass
         Reinforcement

Figure 4 - Graph of Calculated Viscosity
         (N) versus Layers of Glass

COMPOSITE INTEGRATION AND OPERATION OF DISTRIBUTED OPTICAL FIBER SENSORS FOR VEHICULAR APPLICATION

Richard W. Griffiths
G2 SYSTEMS CORPORATION
P. O. Box 666
Pacific Palisades, California 90272-0666   USA

ABSTRACT

A patented electro-optic structural monitoring system has been developed and demonstrated for application to conventional structures, initially for strain-change measurements. Using a proprietary, continuously-distributed and structurally-attached optical fiber sensor, a structural profile is produced at a terminal which is connected to one end of the sensor. Optical Time-Domain Reflectometry (OTDR) is employed, producing data on both the location and magnitude of structural strain changes.

The system has now been expanded for measurements of pressure and temperature changes as well as strain, separately or in combination, and for application to composite-material structures. Three potential vehicular applications are presented.

Strain tests with an aluminum structure are referenced and considerations outlined for composite integration, including the results of an embeddment in a graphite-epoxy pultrusion. The progress of composite integration, including current developments, is summarized.

The combination of a proprietary optical-fiber sensor and terminal, including structurally-programmed electronic-data processing, has led to the selection of the term "FIBER-TRONIC_{TM}" to describe the structural monitoring system.

A PROTOTYPE "STRUCTURAL MONITORING SYSTEMS USING OPTICAL FIBERS" has been developed and demonstrated in a strain-monitoring configuration, based on recently-issued U.S. Patent No. 4,654,520, which bears the above-quoted title. References (a) and (b) provide background information for marine application of the system and demonstration data for the deflection of a 16-foot aluminum I-beam. The method of sensor-structure coupling in the demonstration involved the secure external attachment of a cable-type, optical-fiber, strain sensor. This procedure is acceptable for many conventional civil structures such as bridges, dams, pipelines, power transmission lines or tunnels. The major installation concerns are strain sensitivity and range, thermal insensitivity, attachment integrity, placement for maximum strain coupling (usually tensile), attachment integrity and protection from the environment (including vandalism).

For composite structures, installation of a distributed sensor can be approached on either an external attachment basis, as above, or an embedded basis. For the former, the procedure is very similar to that referenced for conventional structures. For the latter, consideration must be given to several factors, including sensing for pressure and temperature as well as for strain:

COMPOSITE METHOD - Wet layup/curing, filament winding/curing, pultrusion;

COMPOSITE PROCESSING - Materials compatibility, sensor survivability and process control;

SENSOR INTEGRATION - Placement, placement control and connection access;

SENSOR APPLICATION - Pressure sensitivity and range, temperature sensitivity and range, strain sensitivity and range (tension and compression).

Thus, the integration of distributed optical fiber sensors with composites by embeddment is much more complex than for external attachment. Reference (c) provided an assessment for sensor adaption to aerospace structures. This paper addresses composite integration with operation of the FIBERTRONIC$_{TM}$ SYSTEM for vehicular application. The results of an embeddment experiment with a prototype pressure sensor and conclusions are provided. A summary of composite integration progress is included.

POTENTIAL VEHICULAR APPLICATIONS

Three composite-integration applications have been identified, in which FIBERTRONICS$_{TM}$ offers unique processing, testing, NDE and operational features for advanced-design vehicles.

DRIVE-SHAFT FABRICATION, QUALITY CONTROL, TORQUE TESTS AND HEALTH HISTORY MONITORING:
1. Pressure sensing by embedding a FIBERTRONIC$_{TM}$ pressure sensor with filament winding for periodic measurement of density uniformity throughout the entire fabrication process (sensor connection by a Fiber Optic Rotary Joint, or FORJ);
2. Utilize the same pressure sensor for completed parts NDE by measurement of distributed pressure profile (sensor connection by fusion splicing;
3. Strain sensing by embedding a FIBERTRONIC$_{TM}$ strain sensor in final layers of filament winding, at 45 degrees, for torque machine loading, typically 5 to 50,000 pounds (sensor connection by fusion splicing);
4. Utilize the same strain sensor for periodic static checks of increased strain with life (pre-fatigue set), or for dynamic measurements in essentially real time (with

advanced OTDR and use of the FORJ).

BUMPER-BAR FABRICATION, QUALITY CONTROL AND TEST CERTIFICATION:
1. Pressure sensing for process monitoring and control by embedding a FIBERTRONIC$_{TM}$ pressure sensor, as above for filament winding or by embeddment with lay-up or pultrusion;
2. With the same sensor being utilized for quality control/-NDE of finished parts, both before and after crash tests;
3. Strain sensing by embeddment of a FIBERTRONIC$_{TM}$ strain sensor in outer layers of composite, for all methods of fabrication, and measurement of strain changes (tension, or compressing with pre-tensioning) between pre- and post-crash tests with the data being used for certification;
4. With advanced OTDR, similar but essentially real-time testing for certification printout (sensor connection by FORJ or by fusion splicing).

SHELL-TO-FRAME BONDING FOR QUALITY CONTROL AND HEALTH HISTORY:
1. Pressure sensing with FIBERTRONIC$_{TM}$ pressure sensors in joint and measurement of distributed pressure for joint-profile uniformity, during and after bonding;
2. Measurement periodically thereafter for health history (sensor connection by fusion splicing).

SYSTEM AND SENSOR CONSIDERATIONS

A block diagram of the prototype FIBERTRONIC$_{TM}$ Strain-Sensing System is outlined in the accompanying figure. The cable-type sensor is shown coupled to a previously undeflected structure with the OTDR instrument measuring the attenuation change in the sensor resulting from structural deflection. The processor, programmed with sensor, structure and sensor-orientation parameters, converts the attenuation change to the corresponding deflection profile. A similar system can be utilized for pressure and temperature, in which the coupled sensor responds to one of these factors as experienced by the structure. With the three types of sensors in a single cable run, various locations in the structure can be selectively monitored for these factors.

The basic sensor, as described by the referenced patent, consists of an optical-fiber type of light waveguide with a proprietary coating to achieve light loss (attenuation change) proportional to pressure, through refractive index change (microbending) in a reversible manner. For monitoring pressure, temperature or strain, a finite length of sensor must receive a sufficient pressure change over that length to cause a measurable light loss (at the OTDR terminal). This length is known as the "gauge length". At the present time, pressure measurements have been made over a 0.3 (one-foot meter) gauge length with an accuracy of 10 centimeters (4 inches) for locating the mid-point of that length. This appears to be adequate for many vehicular structures. For aerospace, the current objective is to achieve a gauge length of 0.3 meter (one-foot) with an accuracy of 2.5 centimeters (one-inch).

The pressure sensor can be made essentially temperature insensitive, and the strain sensor can be made essentially temperature insensitive but must be isolated from pressure. Similarly, the temperature and pressure sensors can be made essentially insensitive to strain, but the temperature sensor must be isolated from high or non-uniform pressure.

For installations involving external attachment to either conventional or composite structures, these isolations can be incorporated into jacket or sheathing elements of the sensor cable. For embeddment, involving composite structures, these isolations must be considered as a part of the composite materials design and formation process.

For installation on conventional structures or composites by external means, the basic sensor is jacketed or sheathed for secure attachment (clamping or adhesion) under continuous distribution compatible with the desired gauge length. The jacket/sheath design is also made selectively responsive to the structural pressure, temperature or strain to be sensed. For embeddment with composites, however, integration of the jacket/sheath function with the composite materials is desired for structural

simplicity and minimum effect on integrity. Consequently, the embeddment of the sensor involves the following composite design requirements:

SENSOR FUNCTION ISOLATION- Temperature sensor, from high or non-uniform pressure; strain sensor, from high or non-uniform pressure.

SENSOR FUNCTION INCORPORATION- Pressure sensor, pressure proportional to pressure; strain sensor, pressure proportional to strain; temperature sensor, pressure proportional to temperature.

The following sections deal with composite designs for strain and pressure sensing as initial approaches to the embeddment of distributed optical fiber sensors.

STRAIN SENSING

As noted above, FIBERTRONIC$_{TM}$ sensors can be integrated into composite structures fabricated by a variety of different process ranging from hand layup to pultrusion. The difficulty in placing the sensor and having it survive the processing environment increase when more sophisticated processing methods are used.

For strain sensing, the basic pressure sensor must receive pressure proportional to strain. For conventional structures, this is achieved by adding a braided jacket. For composites, this means a relatively large sensor diameter and requires that the braided jacket be free to move as the structure is strained. Consequently, techniques must be developed to strategically place the sensor within the structure at the time of manufacture without bonding the braided jacket to the structure. A second consideration is to isolate the pressure developed during autoclave or die cure and prevent damaging the sensor.

Such a strain sensor can be integrated into a hand-layed-up structure by covering the sensor with a teflon shrink jacket to prevent bonding to the matrix thus allowing the braided jacket freedom of axial movement. This technique will be successful as long as the pressures developed during cure do not collapse the teflon jacket, or other methods or materials must be employed to allow braided jacket movement.

The pultrusion processing of composite parts presents a severe environment for the sensor. Die pressures at peak exothermic reaction can approach several hundred pounds per square inch and exothermic temperatures exceed 400°F. One method of overcoming the risk of sensor failure due to these processing conditions is to introduce the sensor into the structure on a continuous basis after the structure has exited the die.

As an example, consider the case of a pultruded graphite/epoxy `I' Beam into which the sensor is to be placed at the intersection of the web and the flange, as shown by Figure 1 - Strain Sensor Placement. If a small hollow mandrel is placed in the die cavity at the intersection point of the flange and the web, a cavity in the structure will be formed. This mandrel can be extended beyond the downstream or exit end of the die. The sensor can then be introduced into the hollow mandrel upstream of the die entrance, with the jacket adhered to the composite. The mandrel protects the sensor as it passes through the die. As the sensor exits the mandrel, it is positioned in the cavity formed by the mandrel. Since the structure is cured inside the die, the sensor's braid will be free to move axially within the beam to sense strains due to external loadings.

A future alternative to the use of the braided jacket (to achieve pressure on the sensor proportional to strain) is to integrate a radial compression action on the sensor through design of the composite structure. This also involves both placement and placement control of the sensor. This would reduce the sensor diameter and reduce sensor-induced voids, thereby minimizing any effect on structural integrity.

PRESSURE SENSING

The basic sensor, as previously noted is a pressure sensor. This can be used to sense variations in compaction within a composite structure. Voids or delaminations would appear as low pressure areas while areas which were highly densified would exert higher pressures on the sensor, the magnitude of which would be dependent on processing conditions during

cure.

In order to assess the feasibility of such a device to withstand process conditions the following project was selected as an NDE Experiment:

PRESSURE SENSOR EMBEDDMENT-Prototype, distributed-pressure sensor in a 5 x .25 (inch) graphite-epoxy panel formed by pultrusion, and testing with a commercially-available, short-haul-type OTDR; inclusion of a simulated void on a blind-fold basis for location identification.

For the composite, collaboration was established with Goldsworthy Engineering, Torrance, California for the embeddment of a 800-micron diameter, distributed pressure sensor in a graphite epoxy. Also, with Buehler Ltd., Lake Bluff, Illinois, for composite-sample evaluation and for polishing/connection. For the terminal, contractual use of a short-haul OTDR was arranged. The instrument selected was the STANTEL OFR7, made in England for telecommunications purposes (Stantel Components, Schaumburg, Illinois, and Fibernetics, Inc., Barker, Texas).

The cooperation of the above-named firms is acknowledged with appreciation.

NDE EXPERIMENT

The following outlines the experiment performed on the 5-inch panel, the test results and conclusions:

EXPERIMENT

Specimen
Quadaxial layup ($\pm 45^{\circ}$ and 90°) of multiple-ply graphite cloth, 0.027 inches thick, formed by pultrusion, approximately 5 inches wide by 15 feet long, 1/4 inch thick.

Process Environment
Temperature, to 400°F; pressure, to 1,000 PSI.

Embedded
Two lengths of proprietary, distributed-pressure sensors, approximately 3 meters long, centered (spaced about 5 cm) longitudinally.

Void Stimulation
Nylon tape applied by Goldsworthy over a portion of one sensor to reduce pressure, location and length

not identified before tests.

Reference Sensor
Sensor taken from same spool, length 3.5 meters; placed in straight line.

Test
Per Figure 2---OVERVIEW DIAGRAM, same pigtail used in each test (3 dB/Km).

LAYOUT

Instrument Utilized
Stantel OFR7 Optical Time Domain Reflectometer, for attenuation change published specifications:
Dynamic Range (Backscatter) 12dB
Attenuation Resolution   0.05 dB
Location Resolution   +0.1 m
Laser Pulse Widths   2.5 and 10 ns

TEST RESULTS, 10 NS LASER PULSE WIDTH (SEE FIGURE 3, A AND B, 4A)

Reference Sensor (#2002)
Length, 3.0 meters (cursor to cursor) Attenuation, 0.31 dB.

Embedded Sensor (#0001)
Length, 3.0 meters (cursor to cursor) Attenuation, 0.45 dB, with positive slope over 0.4 meters.

Embedded Sensor (#0002)
Length, 3.0 meters (cursor to cursor) Attenuation, 0.85 dB, with uniform slope (negative) over entire length.

TEST RESULTS, 2.5 NS LASER PULSE WIDTH (SEE FIGURE 4B)

Reference Sensor (#2002)
Length, 3.0 meters (cursor to cursor) Attenuation, 0.38 dB.

Embedded Sensor (#0001)
Length, 3.1 meters (cursor to cursor) Attenuation, 0.56 dB; corrected to 3.0 meters; 0.54 dB; with positive slope over 0.6 meters. (See Figure 4B).

MEASUREMENT CONCLUSIONS

Embeddment Effect
Increase in attenuation, over Reference Sensor:
At 10 ns, #0001, +0.14 dB (+50%)
#0002, +0.54 dB (+177%);
At 2.5 ns, #0001, +0.23 dB (+80%)
#0002, +0.70 dB (+235%).

Identification of Void
Because of marked difference in comparative loss between embedded of sensors, with #0001 showing the least increase in attenuation, and because of positive slope over a portion of sensor as noted above for #0001, sensor #0001 was cited for void location/length as noted by the OTDR printout figures. This has been verified by Goldsworthy.

Utility of Results
A marked relative difference in attenuation increase, in this case of the order of 127% to 155%, accompanied by a region of reduced light loss (positive slope) can be used to detect and locate the presence of voids.

SPECIMEN ANALYSIS (PHOTOMICROGRAPHS)

Referenced Pressure Sensor
A short section of the sensor was hot glued in a plastic ferrule and the tip epoxied for holding in place for polishing. The uniform relation of the fiber and coating is depicted in the two photos, 100 and 400 magnification respectively.

Embedded Sensor
A short length of the sensor-embedded composite was similarly placed in a fixture and photographed. The void introduction caused by the relatively-large sensor, and rather severe ovalization of the coating caused by the pultrusion process, but shape integrity of the fiber, are depicted by the two photos, 100 and 400 magnification respectively.

EMBEDDMENT CONCLUSIONS

1. In spite of the high temperature and pressure (to 400°F and 1,000 PSI) experienced by the two sensors, they survived the pultrusion process and the electro-optic system correctly identified the location of the void-simulated density variation in the composite.

2. For layup and pultruded composites, the electro-optic system could be readily adapted, by the use of one or more continuously-distributed (embedded) pressure sensors, to provide comparative material density measurements during the

forming process. For filament winding, the addition of a low-loss Field Optic Rotary Joint (FORJ) would be necessary.

3. For composite parts, with fusion-splice connection to the pressure sensor (to eliminate connector reflection), NDE-density quality could be determined by the FIBERTRONIC$_{TM}$ system.

4. For cut-off composite parts with no pressure sensor leads, development of a uniform optical connection will be necessary to perform reliable NDE of density variations; this is particularly applicable to pultruded parts.

5. Reduction in the diameter of the basic sensor and improved coating adhesion (under pressure) are necessary.

SUMMARY OF COMPOSITE INTEGRATION PROGRESS

While budget restraints have limited the pace of composite integration FIBERTRONICS$_{TM}$, the present status is as follows:

1. The FIBERTRONIC$_{TM}$ System has successfully demonstrated the capability for distributed strain in an attached-sensor mode and distributed pressure in an embedded-sensor mode, for structures comparable in size to those typical of composites in advanced-design vehicles.

2. These accomplishments have established confidence that a similar capability can be similarly achieved for distributed-temperature monitoring.

3. Reduction of the basic pressure sensor diameter to less than 200 microns (0.007 inch), with a stronger adhesion high-temperature coating, has now been accomplished; this facilitates improved composite-materials compatibility.

4. Using this small-diameter FIBERTRONIC$_{TM}$ Sensor, preliminary pressure and attenuation tests have been made, and embeddment by filament winding

is in process. Repeatable test results, using a Laser Precision TD-9960 OTDR as the FIBERTRONICS Terminal, indicate the complete absence of any hysteresis.

5. The temperature rating of the new FIBERTRONIC$_{TM}$ Sensor has been improved to 300°C+ by the use of a polyimide coating to assure composite-processing survival.

6. Development of a low-loss rotation-calibrated Fiber Optic Rotary Joint has been undertaken; the preliminary test data indicates that the FORJ attenuation can be subtracted, permitting sensor measurements for static and dynamic (advanced OTRD) conditions.

7. With further FIBERTRONIC$_{TM}$ Terminal development for essentially real-time operation and Sensor refinement for embedded strain, pressure and temperature, together with positioning control in composite formation, the goal of "Intelligent Materials" appears realizable for:

* Process control in composite formation;

* NDE and testing of composite parts, including the detection of density variations indicative of delamination (voids);

* Distributed structural-integrity (strain and pressure) monitoring on a periodic check basis, or in essentially real time, for forewarning of significant operational changes and the monitoring of health history.*

Inquiries regarding vehicular applications of the patented FIBERTRONIC$_{TM}$ System are welcomed.

*Note: For these purposes, an assessment for system adaption to aerospace structures is outlined in Reference (c).

310

REFERENCES

Previous papers on FIBERTRONIC^{TM} System development and application, by the same author/coauthor

(a)   Offshore Technology Conference paper #5119, 1986, Houston, TX.

(b)   Offshore Technology Conference paper #5564, 1987, Houston, TX.

(c)   SPIE, E/O FIBERS '87, paper #838-54, San Diego, CA

(d)   SME, paper #MS87-834 Composites in Manufacturing 7, Long Beach, CA.

(e)   SNAME, Marine Technology Journal, July 1988.

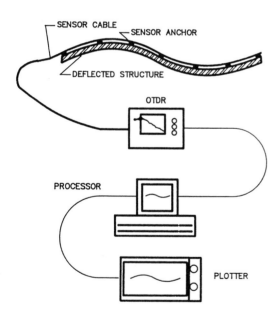

FIBERTRONIC_{TM}
Strain Sensing System

Composite Integration

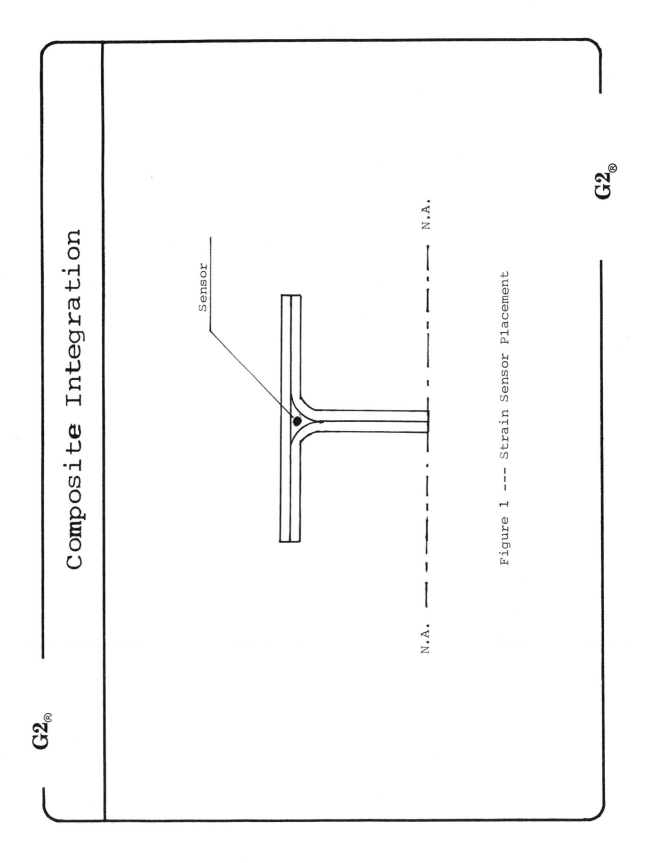

Sensor

N.A. — · — · — · — · — · — N.A.

Figure 1 --- Strain Sensor Placement

Typical Test Output

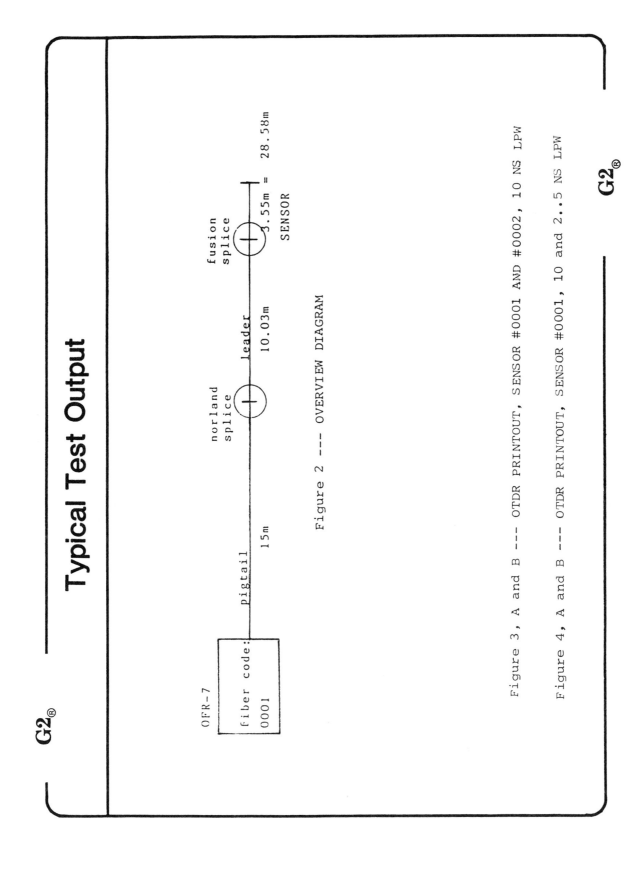

```
OFR-7
┌──────────┐
│fiber code:│  pigtail    norland      leader    fusion
│  0001     │              splice                splice
└──────────┘
            15m          10.03m      3.55m =
                                            28.58m
                                     SENSOR
```

Figure 2 --- OVERVIEW DIAGRAM

Figure 3, A and B --- OTDR PRINTOUT, SENSOR #0001 AND #0002, 10 NS LPW

Figure 4, A and B --- OTDR PRINTOUT, SENSOR #0001, 10 and 2..5 NS LPW

Typical Test Output

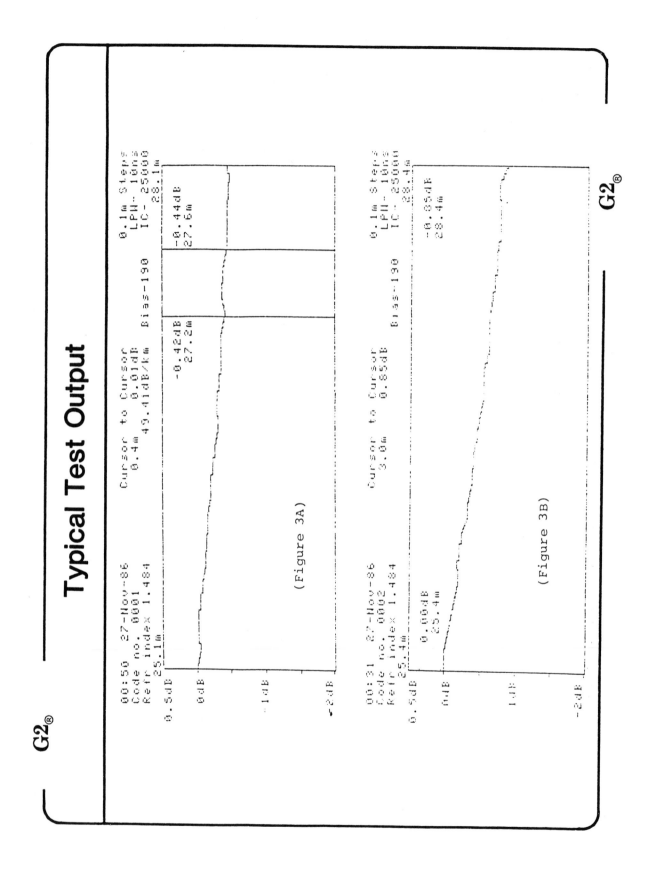

(Figure 3A)

(Figure 3B)

Typical Test Output

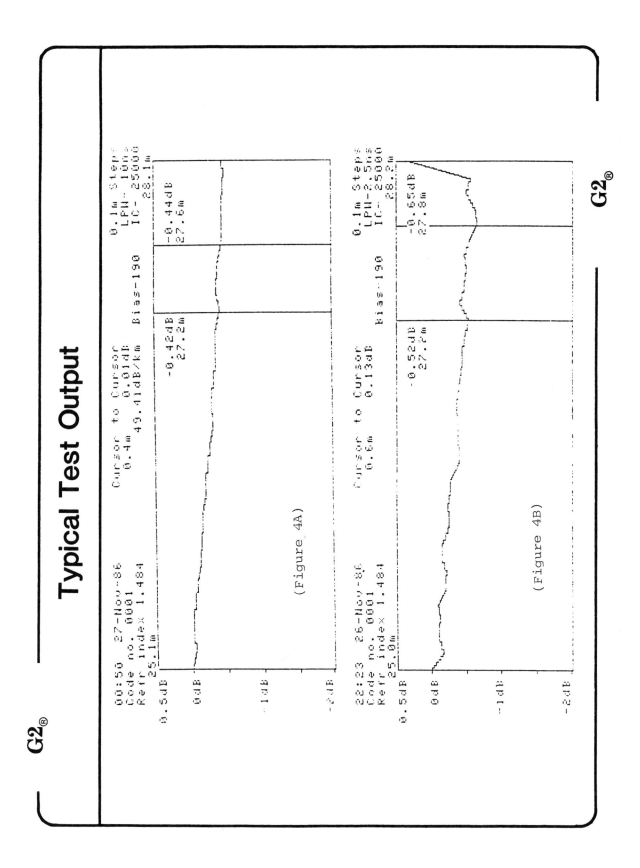

(Figure 4A)

(Figure 4B)

Embedded Sensor, Cross Section – Not Supported

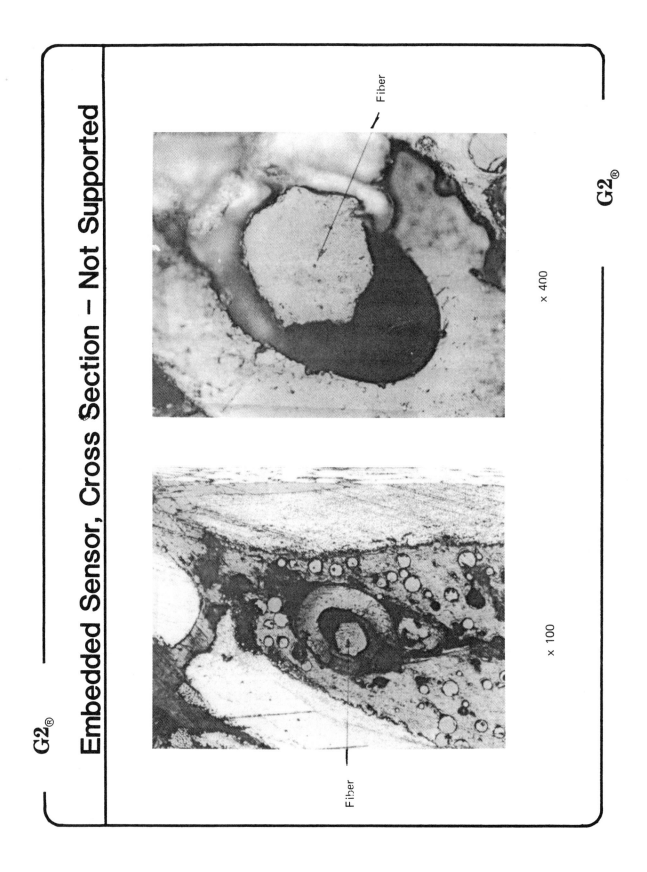

x 400

x 100

G2®

Reference Pressure Sensor – Cross Section

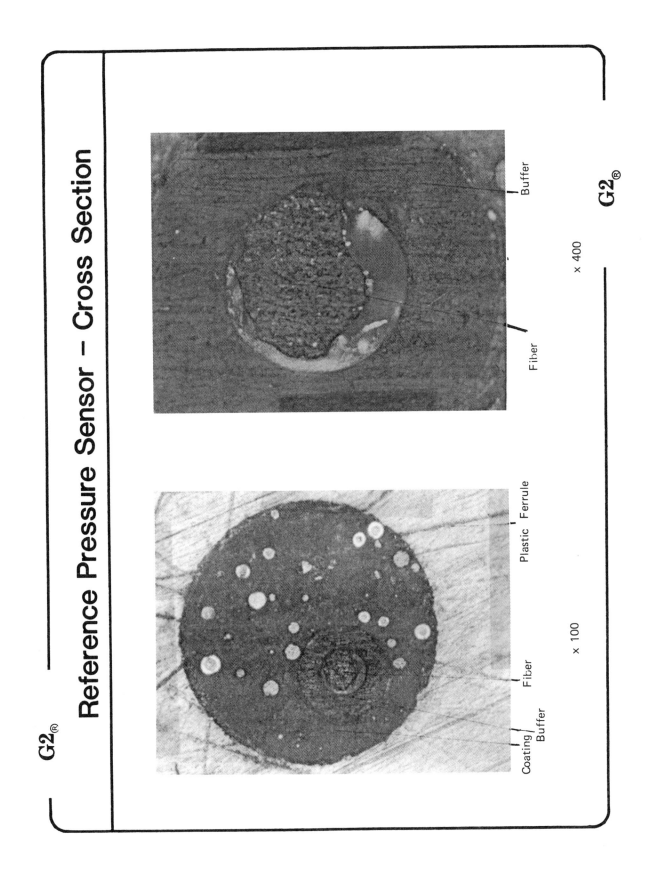

G2®

MODE II AND MODE III INTERLAMINAR
FRACTURE OF LAMINATED COMPOSITES

Herzl Chai
Polymers Division
National Bureau of Standards
Gaithersburg, Maryland 20899   USA

Abstract

The energetic aspects of interlaminar shear fracture were studied for a number of advanced composites using DCB test configurations and scanning electron microscopy. The mode III fracture was characterized by a resistance behavior similar to that observed for mode I. The initial fracture work in mode III and the fracture work in mode II were indistinguishable, the nominal value is believed to be controlled by plastic deformation processes in the interlaminar resin layer ahead of the crack tip.

Introduction

Most research to date on interlaminar fracture have focused on Mode I, which proved to be the critical component in first-generation, brittle-matrix composites. Use of tough matrix systems was shown to greatly improve the composites damage tolerance [1], but unfortunately the accompanying reduction in G_{IIC}/G_{IC} ratio for such materials necessitates the additional consideration of shearing modes in the fracture analysis. Although research on mode II interlaminar fracture has grown considerably in the last few years [2-7], a number of controversial issues remained, including a meaningful interpretation for the observed nonlinearity in the load-deflection curve, and the considerable data scatter (i.e., for T300/BP-907: 1250-1950 [7]; for AS4/PEEK: 1765 [2], 2700 [7], 1800-2500 [6])[*].

The case of mode III seems more cumbersome due to a lack of valid testing methodology and large disparity in the limited data available. Utilizing a double-crack flexure test, the mode III interlaminar fracture work of a number of unidirectional composites was shown to display a "resistance" behavior attributable to a fiber-bridging effect [8]. An initiation and a plateau values for G_{IIIC} were identified corresponding, respectively, to onset and maturity of the damage tied zone at the crack tip. The initiation value was as little as the neat resin G_{IC} while the plateau value exceeded the initiation by a factor of 10 to 30, being about twice as large as G_{IIC}. A similar G_{IIIC}/G_{IIC} ratio (~2) was also reported for wood [9] or AS4/3502 composite [10], though no "resistance" behavior was noted in these works. The validity of such findings is questionable, however, in light of recent tests on adhesive joints [11] showing G_{IIC} and G_{IIIC} to be identical, on the one hand, and the excellent composite-joint toughness correlation found for mode I [12], on the other hand.

As an extension to [12], this work has the dual goal of establishing a valid characterization of interlaminar shear fracture and determining quantitatively the role of matrix constituent on it. The test variables were resin ductility and crack extension.

Experimental

a. Fabrication

As shown in Fig. 1, the mode II and mode III specimens used are variance of the popular mode I DCB configuration, the particular fracture mode is affected by the loading manner. These tests, which provide nearly pure fracture modes (note that due to beam width finiteness, some mode II component should be present at the edges for the mode III specimen) as well as a continuous type crack growth, have been originally designed for mode II [13] or mode III [9] testings of wood. The particular choices of a relatively large beam height, h, and a nearly quasi-isotropic layup (Fig. 1c) were made to prevent adherends large deformation and/or plastic deformation in applications to tough composites, and to minimize the considerable transverse shear deformation present in the commonly-employed unidirectional layup since its effect on G_{II} is controversial.

As for mode I [12], the three composites tested were AS4/3502, AS4/BP-907 and AS4/PEEK (APC2). The first two matrix constituents are amorphous thermosetting resins while PEEK is a semi-crystalline thermoplastic. Also, the 3502 resin is brittle while BP-907 and PEEK exhibit a nonlinear strain-stress behavior. The test specimens were sized as shown in Fig. 1 from panels that were fabricated from commercially available prepreg tapes per manufacturer recommendations. The layup (Fig. 1c) consisted of 62 plies, the first 30 in each split arm are quasi-isotropic. An initial crack was formed prior to curing by inserting a 25 μm thick Kapton film (AS4/3502, AS4/BP-907) or aluminum foil (AS4/PEEK) between the central 0 deg plies.

The initial crack length was approximately 40 and 100 mm in mode II and III, respectively. The corresponding crack length to half-span ratio, a/L, of over 0.6 insured that G_{II} was a decreasing function of a. To facilitate natural crack, prior to testing the debond was extended approximately 1 mm from the insert border by means of careful wedging [12]. Crack tip visibility was enhanced by painting the crack line white and attaching a prescaled paper strip.

b. Testing

The specimens were loaded at a rate of 0.51 mm/min (mode II) or 2.5 (mode III) in a closed-loop servohydraulic testing machine (MTS) operated in a stroke control mode. Load P and deflection δ were traced on a chart recorder. The mode II deflection was monitored from an extensometer (Instron) mounted under the loading pin, thus circumventing the non-negligible effect of machine deformation. Figure 2 exemplifies loading histories for several cases; the markings on the chart indicate instantaneous crack length established visually (mode III) or using an X20 traveling microscope (mode II).

All loading traces in Fig. 2 exhibit a nonlinear behavior similar to that found for adhesive joints [11]. This effect, which is more pronounced for the ductile matrix systems, was due to irreversible resin deformation at the crack tip, not a gross plastic deformation of the test specimen. After completion of a mode II test, the

specimen was shifted horizontally, and a new test carried out. In this way as many as three tests could be done on a single specimen.

For each individual test a compliance plot was generated, see Fig. 3. In this and all subsequent figures, filled and open circles identify mode II and mode III fracture, respectively. The data for these plots were either taken from the fracture test (mode III) or generated separately prior to testing (mode II). In the latter case, the applied load was only a fraction of the peak load. Figure 3 shows that the compliance, $C(=\delta/P)$, in each mode is well fitted by a straight line such that

$$C_{II} = c_0 + c_2 \, (a/L)^3 \quad , \quad C_{III} = c_3 a^n \qquad (1)$$

where c_0, c_2, c_3 and n are experimentally determinable constants. The energy release rates were calculated from the compliance method [11]:

$$G_{II} = \frac{1.5 \, c_2 \, (aP)^2}{bL^3} \quad , \quad G_{III} = \frac{n}{2b} \, \frac{P\delta}{a} \qquad (2)$$

where b is the beam width. The experimental values for c_2/c_0 and n were typically 1.3-1.7 and 2.5-3.0, respectively. The corresponding beam theory predictions are 1.5 and 3.0.

Calculation of fracture works necessitates elucidation of the critical conditions from Eq. 2. It has been found in the analogous study of adhesive joints that these conditions prevail once the peak load on the P-δ curve is reached; thereafter the fracture works were independent of crack length [11]. Evidently, that load signify the maturity of plastic deformation zone and the start of a self-similar crack growth. Consequently, the peak load and its corresponding crack length, a_1 were used in (2) to calculate G_{IIC} or G_{IIIC}. This is in contrast to common practice employing the initial crack length, a_o, in conjunction with peak load for calculating G_{IIC} [2-7]. Figure 2a shows that the relative difference between these two values becomes very significant for the highly ductile PEEK. As shown from Fig. 2b, for the present mode III test the onset of peak load is obscured by a continuous increase in P that was consequenced by a development of an additional damage tied zone characterized by fiber-bridging. Nevertheless, the P-δ traces from Fig. 2b or alike do show a distinguishable slope change starting at a=99 mm (AS4/3502) or a=105 mm (AS4/PEEK). Such an event was taken to signify the maturity of plastic deformation zone, and thus the critical conditions for crack growth. The corresponding G_{IIIC} is termed "initial".

Results and Discussion

Following testings, all specimens were split open and the fractured surface classified using a stereo-optical microscope as either interlaminar (i.e. planar, no loose fibers) or otherwise intralaminar. Table I summarizes the fracture data pertaining to a truly interlaminar failure. The mode II values were averaged from a number of tests per specimen. The G_{IIC} values for AS4/3502 (590 N/m) compare favorably with published value of 570 N/m [5] but for the ductile PEEK (2315) it is substantially larger than the value of 1765 given in [2]. This may partially be due to use in (2) of the initial crack length, a_o, as opposed to the concurrent value pertaining to peak load that was employed in this work. As indicated from Fig. 2a, the ratio a_1/a_0 for AS4/PEEK is substantially greater than 1 (i.e. 38/30), which would amount to over 60% increase in the reported G_{IIC}. Similar arguments holds for AS4/BP-907 which value (2325) exceeds the range of values in [7] (1250-1950). It is interesting to note that the present G_{IIC} value for AS4/BP-907 is over twice that obtained most recently by the author in testing of ASTM Round-Robin specimens. However, in view of the great sensitivity of G_{IIC} to adhesive bond thickness, Fig. 5, this variation is likely a consequence of different (thicker) interlaminar resin layer.

The fracture behavior for mode III is exemplified in Fig. 4 (open symbols). For all materials, the fracture work exhibits an R-curve which is the result of a damage tied zone characterized by detached fibers bridging the delaminating interfaces. The initial rise is followed by a leveling off, the corresponding crack extension which greatly exceed the extent of the initial plastic zone. As is apparent from Fig. 4a, however, the R-curve characteristics are highly probablistic, the disparity includes the extent of damage tied zone, plateau level and crack growth manner within the plateau range. Judging from the mode I case [12], the R-curve should also depend on specimen geometry, which further diminish its significance. Conversely, the initial G_{IIIC} (i.e. first data point) as well as the asymptotic value (at $\Delta a=0$) are quite reproducible. This, together with a planar morphology make the initial value the true measure of mode III toughness.

Fig. 4 shows that the value of G_{IIIC} thus defined agree well with its mode II counterpart (solid circles) for all materials, a trend consistent with that found for adhesive joints [11]. The apparent exception to that role found for wood [9] or composites [10] are likely due to incorrect association of G_{IIC} with the plateau level.

Indeed, the reported value for AS4/3502 (1200 N/m [10]) is in the ball park for large crack extension from Fig. 4c. Conversely, the initiation value in [4,6,9] which is only a fraction of the plateau value [8], is also misleading because it was derived from the first appearance of damage at the crack tip. As discussed earlier, that event occurs well before the plastic deformation zone becomes fully-developed.

Summary and Conclusions

The failure process in mode II and III started with the development of plastic deformation in the interlaminar resin layer ahead of the crack tip and continued with a self-similar, stable type crack growth upon maturity of the plastic zone. For mode III, the latter process was masked by the development of an additional damage tied zone characterized by fiber-bridging which dimension well exceeded that of the plastic zone. This gave rise to a resistance type fracture behavior which highly probablistic nature ruled it insignificant. However, the initial value, evaluated at the instant of plastic zone maturity, was reproducible, associated with an interlaminar morphology, and thus considered the true measure of G_{IIIC}. Extrapolation of the R-curve to $\Delta a=0$ produced an asymptotic value differing from the initial one only by the energy lost to fiber bridging during the development of plastic zone, the amount that is expected to decrease with increasing specimen width or with the approach to a true anti-plane conditions. The so identified G_{IIIC} value coincided with its G_{IIC} counterpart, thereby reinforcing earlier proposition [11] that fracture is a two-parameter phenomenon.

The great sensitivity of shear fracture work to adhesive bond thickness implies a great sensitivity of G_{IIC} or G_{IIIC} to material processing variables affecting the interlaminar resin layer. This as well as inadequate treatment for nonlinearity in the load-deflection curve may well account for the large scatter in G_{IIC} value reported in the literature.

References

1. Toughened Composites, ASTM STP 937 (1987), S. W. Johnson, Ed.

2. A. J. Russell and K. N. Street, in Toughened Composites, ASTM STP 937 (1987), 275-294.

3. Vanderkley, P. S. "Mode I-Mode II Delamination Fracture Toughness of a Unidirectional Graphite/Epoxy Composite," Master's Thesis, Texas A&M University, December 1981.

4. E. A. Armanios, L. W. Rehfield and A. D. Reddy, in Composite Materials: Testing and Design, ASTM STP 893 (1986), 232-255.

5. W. M. Jordan and W. L. Bradley, in Toughened Composites, ASTM STP 937 (1987), 95-114.

6. L. A. Carlsson, J. W. Gillespie and B. R. Trethewey, Journal of Reinforced Plastics and Composites 5 (1986), 170-187.

7. T. K. O'Brien, G. B. Murri and S. A. Salpekar, "Interlaminar Shear Fracture toughness and Fatigue Thresholds for Composite Materials," presented at ASTM Conference on Composite Materials: Fatigue and Fracture, Cincinnati, OH, April 28-30, 1987.

8. G. R. Sidey and F. J. Bradshaw, in Carbon Fibers, Their Composites and Applications, Proceedings of the International Conference of the Plastics Institute, London (1971), 208-213.

9. G. R. DeBaise, Mechanics and Morphology of Wood Shear Fracture, Ph.D. thesis, State University College of Forestry, Syracuse University (1970).

10. S. L. Donaldson, Mode III Interlaminar Fracture of Composites, Master Thesis, University of Dayton (1987).

11. H. Chai, International Journal of Fracture (in press).

12. H. Chai, Engineering Fracture Mechanics 24 (1986), 413-431.

13. J. D. Barrett and R. O. Foschi, Engineering Fracture Mechanics 9 (1977), 371-378.

*Certain commercial materials and equipment are identified in this paper in order to specify adequately the experimental procedure. In no case does such identification imply recommendation or endorsement by the National Bureau of Standards, nor does it imply necessarily the best available for the purpose.

Table I Interlaminar Fracture Results*

Material	Mode II G_{IIC} (N/m)	Mode III G_{IIIC} (N/m) Asymptotic	Initiation
AS4/3502	590(50)	410(130)	620(100)
AS4/BP-907	2325(240)	2065(255)	2485(55)
AS4/PEEK	2315(390)	1735(75)	2695(55)

*parenthesis enclose standard deviation

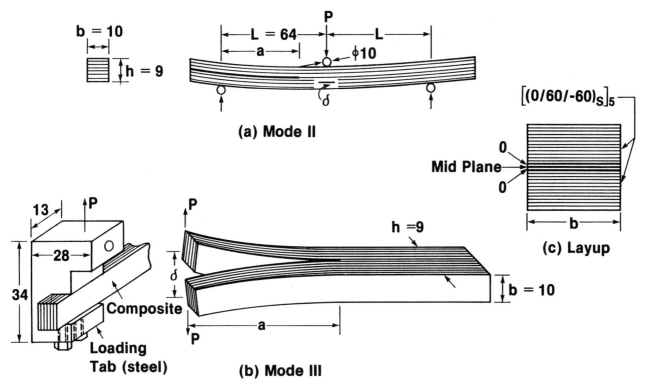

Fig. 1   Interlaminar shear fracture test
specimens. All dimensions are in mm.

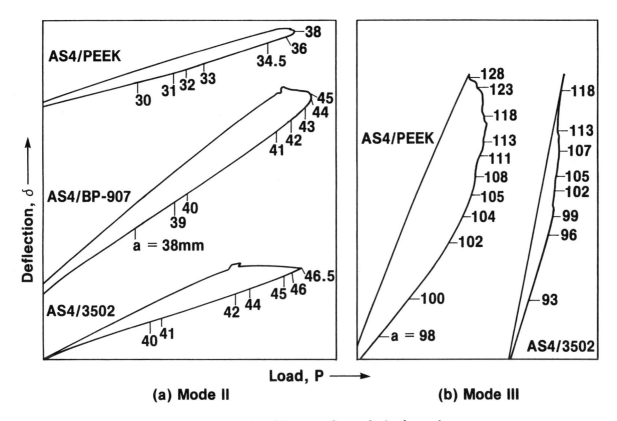

Fig. 2   Loading history for a brittle and
ductile-matrix composites. Tic mark
indicates crack length.

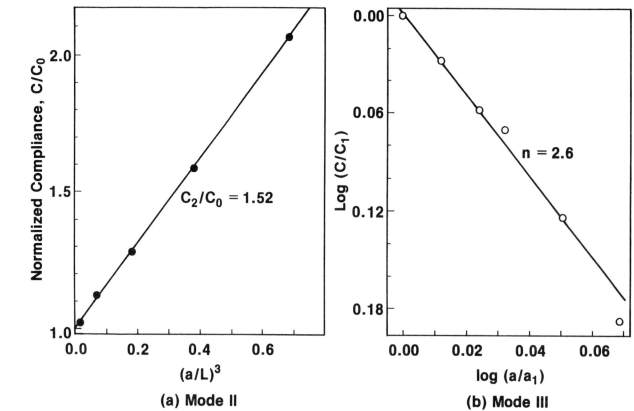

(a) Mode II   **(b) Mode III**

Fig. 3   Compliance plot for the AS4/3502 (a)
          and AS4/PEEK (b) composites detailed in
          Fig. 2. a_1 is the first crack length
          at which data were first reduced.

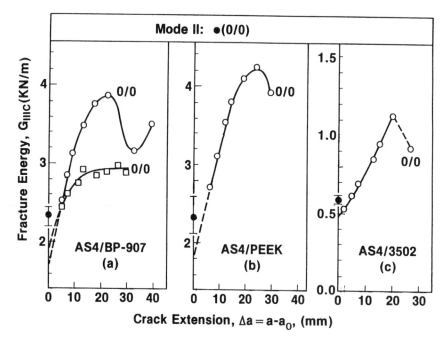

Fig. 4   Mode III fracture energy (open symbols)
          as a function of crack extension (a_0 is
          initial crack length) for three
          materials. Solid symbols denote mode
          II fracture values. Straight line
          between data points indicate a slip-
          stick type growth.

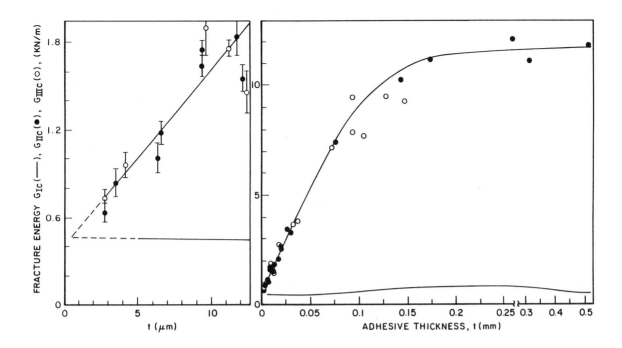

Fig. 5   Effect of adhesive bond thickness on
adhesive fracture work [11].   Adhesive
is BP-907.

SIGNAL PROCESSING TECHNIQUES FOR IMAGING IMPACT DAMAGE IN COMPOSITES

B. G. Frock, R. W. Martin
The University of Dayton
Research Institute
Dayton, Ohio   USA

T. J. Moran
AFWAL/Materials Laboratory
Wright-Patterson AFB, Ohio   USA

ABSTRACT

Three different signal processing techniques
have been applied to the ultrasonic
interrogations of impact damaged quasi-
isotropic graphite/epoxy composites. The
capabilities of the techniques for imaging the
extent of impact damage near the entry surface
in composite materials are discussed and
compared. The first technique uses multiple
gates to acquire data from the rectified and
filtered versions of the ultrasonic pulse echo
A-scans. The second technique uses multiple
software gates to digitize and store data
directly from the Rf A-scans while the sample is
being raster scanned. The third technique uses
multiple software gates to digitize and store
data from the results of Wiener deconvolutions
of the Rf A-scans. Application of the third
technique requires that the Rf waveforms be
digitized and stored in the computer.

Significantly less blurring is present in images
generated from gated Rf A-scans than in images
generated from rectified and filtered A-scans.
Even less blurring is present in images
generated from Wiener deconvolved A-scans than
in images generated from gated Rf A-scans. The
Wiener deconvolution process also reduces the
background noise level. These reductions in
depth-wise blurring and background noise level
lead to significant improvements in the ability
to define the extent of damage near the entry
surface of a graphite/epoxy composite panel.

ULTRASONIC C-SCAN IMAGING TECHNIQUES can be used
to determine the boundaries of impact damage at
interfaces in graphite/epoxy composites. If
conventional methods are used, the C-scan images
are generated from data acquired at specific
"gated" locations along rectified and low-pass
filtered A-scans. Placing the gates at temporal
locations which correspond with interfaces
should allow the integrity of the interfaces to
be examined. However, if the interfaces are
closely spaced, as is the case for quasi-
isotropic graphite/epoxy composites, the
information from adjacent layers will be blurred
together. There are two reasons for this
blurring. First, the duration of the Rf pulse
from the transducer is finite and may extend
(ring) through several interfaces. Second, the
process of rectifying and low-pass filtering the
A-scans causes additional temporal or depth-wise
blurring and destroys phase information. The
combination of these effects results in blurring
at and below the first interface. This blurring
makes precise definition of the boundaries of
damage at specific interfaces very difficult.

These blurring effects can be reduced by
acquiring data directly from the Rf waveform.
Although the echoes from adjacent interfaces may
still overlap, the C-scan images which are
generated from Rf signals that are gated at the
interface locations are less affected by depth-
wise blurring than are the C-scan images
generated from the rectified and low-pass
filtered A-scans[1,2]. Significant depth-wise
blurring may still exist at lower frequencies,
however, leading to imprecision in the
definitions of the boundaries of damaged regions.

The application of the one-dimensional
Wiener filter deconvolution process to Rf A-scans
decreases the finite time duration of the Rf
echoes and, thus, decreases the depth-wise
blurring. Collecting data from gates placed at
the appropriate locations along the axially
deconvolved A-scans should allow more precise
definitions of the boundaries of the damage at
specific ply interfaces. The boundary definition
should be more precise than that which is
achievable with either the gated rectified and
filtered A-scans or the gated Rf A-scans.

WIENER FILTER DECONVOLUTION

The deconvolution technique utilized for
this work is known as constrained deconvolution

327

or Wiener filtering[3-6]. It assumes that the measured signal from the reflector, y(t), is a convolution of the overall system response, h(t), and the flaw response, s(t), or

$$y(t) = h(t) * s(t) \qquad (1)$$

where h(t) is obtained by recording a reference waveform from a planar reflector. The measured signals, y(t) and h(t), are transformed into the frequency domain and the flaw spectral response is calculated from

$$S(\omega) = \frac{Y(\omega)H^*(\omega)}{|H(\omega)|^2 + K^2} \qquad (2)$$

where $H^*(\omega)$ is the complex conjugate of $H(\omega)$ and K is a constant. Noise in $S(\omega)$ is minimized when $H(\omega)$ is zero or very small by setting K to a predetermined percentage of the maximum amplitude of $H(\omega)$.

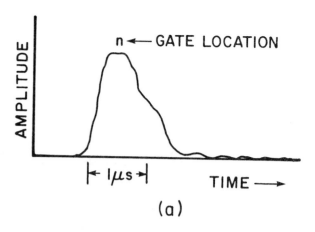

(a)

SAMPLE

The sample used in this study is a 32-ply thick, quasi-isotropic graphite/epoxy composite. Prior to ultrasonic inspection, it had been intentionally damaged by a 5.4 joule impact from a 12.7 mm diameter stainless steel ball on a pendulum impacter.

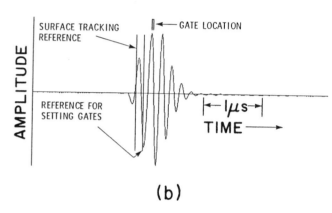

(b)

DATA COLLECTION

A computer controlled ultrasonic immersion C-scanning system was used to collect data at each of 6400 discrete points in a 40 mm by 40 mm square area surrounding the impact damage site. The separation between discrete data collection points was 0.5 mm in each planar direction. A 5 MHz center frequency, 12 mm diameter, 76 mm focal length transducer was used for all data collection, with excitation for the transducer provided by a broadband spike pulse.

GATED RECTIFIED/FILTERED A-SCANS - Data from a gate on the rectified and low-pass filtered A-scans were digitized and stored using 8-bit resolution at each of the 6400 discrete sampling points around the damage site. The gate was positioned at the approximate location of the interface between the second and third plies (about 0.25 mm below the entry surface), and the average value in the gate was digitized and stored at each of the discrete locations. A rectified and low-pass filtered A-scan with the gate superimposed is shown in Fig. 1.

RF A-SCANS - Rf A-scans (512 points long) were digitized and stored using 8-bit resolution at all 6400 discrete points around the damage site. These digitized Rf A-scans were used for direct generation of C-scan images from the Rf data and for axial Wiener deconvolutions[6] followed by C-scan image generation. A reference A-scan (Rf echo from the front surface of a flat plexiglas plate) was collected and stored for use during the Wiener deconvolution process. The reference echo and its Fourier amplitude spectra are shown in Fig. 2.

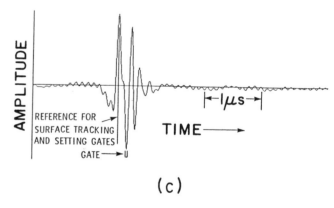

(c)

Fig. 1. Gating for C-scan image generation:
(a) rectified and filtered A-scan;
(b) rf A-scan; (c) deconvolved A-scan.

328

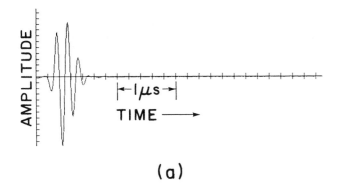

(a)

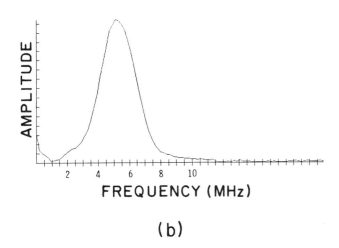

(b)

Fig. 2. Reference A-scan: (a) rf A-scan;
(b) Fourier amplitude spectra of "a".

C-SCAN IMAGE GENERATION

GATED RECTIFIED AND FILTERED A-SCANS - C-scan
images were generated in the standard manner from
the gated data after completion of the scan.

GATED RF A-SCAN METHOD - This technique is
analogous to gating the raw Rf A-scan during
scanning, except that in the present case the Rf
A-scan had already been digitized and stored in
the computer. Peak signal levels were collected
and stored from a gate located over a 40
nanosecond-wide region of the A-scan (see
Fig. 1a). This location corresponded with the
location of the interface between the second and
third plies. The C-scan image generated from
that data provided information concerning the
structural integrity of the interface.

GATED WIENER DECONVOLVED A-SCAN METHOD - All
6400 of the Rf A-scans were deconvolved using a
Wiener filter of the form given in Eq. (2). The
amplitude cut-off, K, was set at 0.5% of the
maximum amplitude of $H(\omega)$ because of the very low
noise level of the Rf A-scans. Phase reversal
of the echoes from the delaminated interfaces
yielded a negative-going spike in the deconvolved

A-scan at the echo return locations. Thus,
the minimum value of the portion of the
deconvolved A-scan which was within an
appropriately placed 40 nanosecond-wide gate
(see Fig. 1b) was collected from each of the
6400 deconvolved A-scans and stored in the
computer. A C-scan image of the interface
between the second and third plies was then
generated from the data[7].

RESULTS

C-scan images of the interface between the
second and third plies (measured from the entry
surface) are shown in Fig. 3. The images in

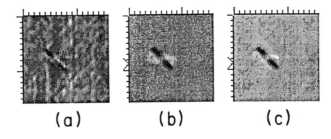

(a)   **(b)**   **(c)**

Fig. 3. C-scan images: (a) from rectified and
filtered A-scan; (b) from rf A-scan;
(c) from deconvolved A-scan.

Fig. 3a, 3b and 3c were generated from the
rectified and filtered A-scans, the gated Rf
A-scans and the Wiener deconvolved A-scans,
respectively. The delamination between the
second and third plies is difficult to resolve
from the background in the C-scan image
generated from the rectified and filtered
A-scans (Fig. 3a). However, that delamination
is easily resolved in the C-scan images
generated from the gated Rf A-scans and from
the Wiener deconvolved A-scans as the dark
somewhat peanut-shaped object. The lighter
regions on either side of the second-to-third
ply interface delamination in the gated Rf and
Wiener deconvolved images are "precursors" of
the delaminations at the interface between the
third and fourth plies. These "precursors" are
more evident in the C-scan image generated
directly from the Rf A-scans and are a result
of the "ringing" of the transducer. This
"ringing" is partially removed by the
deconvolution process.

Plots of the individual pixel amplitudes
along selected horizontal lines in the C-scan
images are shown in Fig. 4. Figure 4
demonstrates that the general signal-to-
background level is significantly higher for
the C-scan image generated from gated Rf
A-scans than is the signal-to-background level
for the C-scan image generated from gated
rectified and filtered A-scans. The signal-to-
background level is even higher for the C-scan
image generated from the gated Wiener

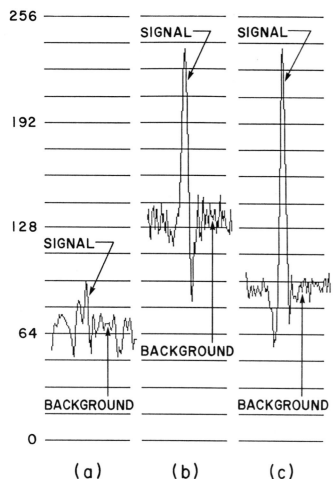

256

192

128

64

0

SIGNAL

SIGNAL

SIGNAL

BACKGROUND

BACKGROUND

BACKGROUND

(a)   (b)   (c)

Fig. 4. Pixel amplitude plots from line 45 of
         C-scan images: (a) from rectified and
         filtered A-scan; (b) from rf A-scan;
         (c) from deconvolved A-scan.

deconvolved A-scans than is the signal-to-background level for the C-scan image generated from the gated Rf A-scans. In these cases, the "signal" is defined as the pixel amplitude in the delaminated region and the "background" is defined as the pixel amplitude everywhere else. The higher signal-to-background level greatly improves the ability to precisely define the boundaries of the damaged regions at the interrogated interface.

The rectified and filtered A-scans, the Rf A-scans, and the axially deconvolved A-scans from two locations (one "signal" and one "background") along line 35 of the C-scan images are shown in Figs. 5 and 6. The locations of the "gates" are superimposed on each of the A-scans. The dynamic range of the portion of the Rf A-scan which lies within the gated region is significantly larger than the dynamic range of the portion of the rectified and filtered A-scan which lies within the gated region. The dynamic range of the portion of the deconvolved A-scan which lies within the gated region is greater than the dynamic range of the portion of the Rf A-scan which lies within the gated region. It is these differences in the

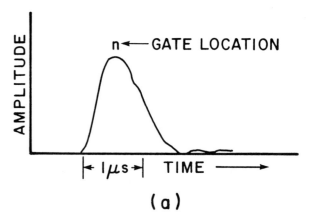

(a)

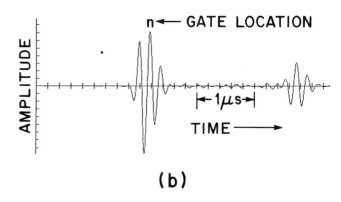

(b)

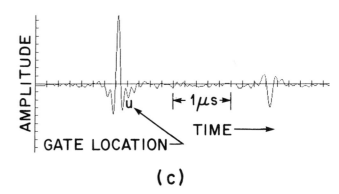

(c)

Fig. 5. A-scans from background region of
         C-scans: (a) rectified and filtered
         A-scan; (b) rf A-scan; (c) deconvolved
         A-scan.

dynamic ranges which cause the different signal-to-background levels in the C-scan images generated from the three different types of A-scans.

A second advantage of the C-scans generated from Wiener deconvolved A-scans relative to the C-scans generated from the gated Rf A-scans is the lowered "noise" level which is superimposed on the "signal" and "background" regions of the images. This can be seen in the pixel amplitude plots in Fig. 4 and the Fourier amplitude spectra in Fig. 7. The Fourier amplitude spectra are from line number ten (a "background"

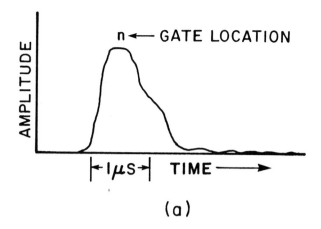

(a)

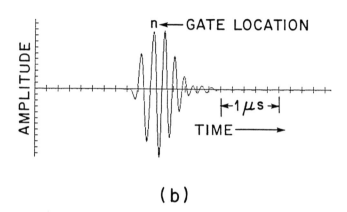

(b)

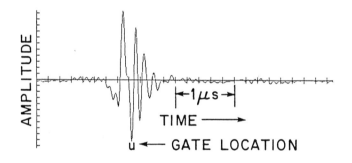

(c)

Fig. 6. A-scans from signal region of C-scans:
(a) rectified and filtered A-scan;
(b) rf A-scan; (c) deconvolved A-scan.

region) of the C-scan images in Fig. 3. The
Fourier amplitude spectra of line 10 of the
C-scan data generated from the Rf A-scans has
more energy in the high spatial frequency region
than does the spectra of line ten of the
C-scan data generated from the deconvolved
A-scans. The lower noise level allows for a
more precise definition of the delaminated
regions and for better visual definition of
other features such as the fiber tows.

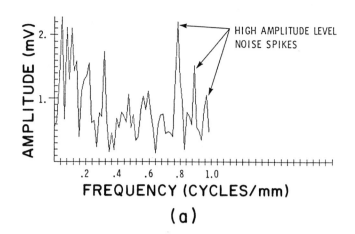

(a)

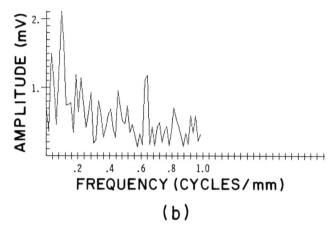

(b)

Fig. 7. Fourier amplitude spectra of line 10
from C-scans in Fig. 3: (a) from image
in Fig. 3b; (b) from image in Fig. 3c.

CONCLUSIONS

The signal-to-background level ratio in the
C-scan image generated from the gated Rf A-scans
is significantly greater than is the signal-to-
background level ratio in the C-scan image
generated from the gated rectified and filtered
A-scans. This is primarily due to the temporal
blurring and loss of phase information which
results from rectification and low-pass
filtering. The higher signal-to-background
level ratio in the C-scans generated from gating
the Rf A-scans allows for a more precise
definition of the boundaries of the delamination
in those C-scan images.

Deconvolution removes some of the "ringing"
in bandwidth-limited ultrasonic pulses. This
decreased "ringing" in the deconvolved waveforms
leads to a greater dynamic range between the
"signal" regions and the "background" regions of
C-scan images generated from these deconvolved
waveforms than for C-scan images generated
directly from the Rf waveforms. The
deconvolution process also performs some
smoothing which results in less "noise" being
superimposed on the "signal" and "background"

regions of the C-scan images generated from the deconvolved waveforms. The greater dynamic range and the noise suppression which result from the deconvolution process allow more precise visual and machine definition of the boundaries of damaged regions in impacted graphite/epoxy composites.

ACKNOWLEDGMENTS

Research sponsored by the AFWAL/Materials Laboratory under Contract No. F33615-86-C-5016.
The authors thank Mr. Mark Ruddell for his data collection efforts.

REFERENCES

1. Buynak, C. F. and T. J. Moran, "Review of Progress in Quantitative NDE," 6B, 1203-1211, edited by D. O. Thompson and D. E. Chimenti, Plenum Press, New York (1987).

2. Buynak, C. F., Moran, T. J. and S. Donaldson, Sampe Journal 24 (2), 35-39 (March/April 1988).

3. Neal, S. and D. O. Thompson, "Review of Progress in Quantitative NDE," 5A, 737-745, edited by D. O. Thompson and D. E. Chimenti, Plenum Press, New York (1986).

4. Furgason, E. S., Twyman, R. E., and V. L. Newhouse, "Proceedings DARPA/AFML Review of Progress in Quantitative NDE," AFML-TR-78-55, 312 (1978).

5. Murakami, Y., Kuri-Yakub, B. T., Kino, G. S., Richardson, J. M. and A. G. Evans, Applied Physics Letters 33 (8), 685-687 (October 1978).

6. Liu, C. N., Mostafa, F. and R. C. Waag, IEEE Transactions on Medical Imaging MI-2, 66-75 (June 1983).

7. Smith, B. T., Heyman, J. S., Moore, J. G. and S. J. Cucura, "Review of Progress in Quantitative NDE," 5B, 1239-1244, edited by D. O. Thompson and D. E. Chimenti, Plenum Press, New York (1986).

PREFORMABLE CONTINUOUS STRAND MAT

S. G. Dunbar
Owens-Corning Fiberglas
Technical Center
Granville, Ohio 43032   USA

ABSTRACT

There is increasing interest in SRIM and RTM processing methods for large, complex automotive shapes. The reinforcements for these processes are placed in the mold and then resin or reactive materials are injected into the closed mold. SRIM and RTM processes have the advantage of much reduced initial capital expense and lower operating costs when compared to traditional SMC techniques. In addition, parts consolidation opportunities go far beyond those available for SMC, wet compression molding, or stampable thermoplastics.

One problem area common to both SRIM and RTM is how to place reinforcements in the mold without the time consuming problems of mat springing out of the mold or sliding and changing its position during mold closure. There is general agreement that what is needed is for the reinforcements to be preformed to shape. Preforms are needed.

This paper discusses preformable continuous strand mat and how it may be processed to make complex shapes. The mat is thermoformable and requires a holding frame, heating chamber, and a shaping die to make preforms. Also discussed is the concept of blanket reinforcements, which are combinations of layers of random continuous strand mat and unidirectional or bidirectional knitted rovings. These multiple layers may be placed in a specified sequence and then sewn together so as to be handled as a single blanket. The blanket is then preformed to shape for use in structural applications.

INTRODUCTION

There is increasing interest in SRIM (structural reaction injection molding) and RTM (resin transfer molding) processing methods for large, complex automotive shapes.

The fiberglass reinforcements for these processes are placed in the mold and then resin or reactive materials are injected into the closed mold. Currently, most parts being considered are structural or semi-structural parts without critical surface appearance requirements. By structural, we mean that the parts have the ability to carry a load.

SRIM and RTM processes have the advantage of much reduced initial capital expense and lower operating costs when compared to traditional SMC (Sheet Molding Compound) techniques. In addition, parts consolidation opportunities go far beyond those available for SMC, wet compression molding, or stampable thermoplastic.

One problem area common to both SRIM and RTM is placement of the reinforcements in the mold without the time consuming problems of mat springing out of the mold or sliding and changing its position during mold closure. There is general agreement that the reinforcements need to be preformed to shape. For this reason, OCF has been aggressively working to develop a continuous strand mat that can be readily shaped into a preform.

Previously, if one wanted to mold a deep draw part such as a tote box, one had limited choices. One had to make a chopped strand preform, or use SMC or use a time-consuming method of cutting out a mat pattern and laboriously fitting it to the mold. With the advent of preformable mat, a preform can be made in less than two minutes from a flat piece of mat. The mat is heated and stretched over the mold shape and then cooled to hold the shape. The corners show excellent consistency with no overlaps or thin spots. The equipment needed to preform a shape like this consists of a mat holding frame, a heater, and a preform mold, which in the case of the tote box was a matched steel die set. Mat preforming equipment will be discussed in detail later on in this presentation.

Another example of a mat preform is an experimental engine compartment side panel made by a major automotive company. The battery tray area extends into the shock absorber tower surround and ends in the firewall. This part consists of four plies of 1½ oz/sq ft preformable mat. The flat mat was anchored in a holding frame, heated, then compressed to shape where it was cooled to hold its shape. The preform mold in this case was made of glass fiber reinforced plastic.

The finished part has uniform glass content and, interestingly, in the corner where the plane of the battery tray intersects with the two side walls, there is no overlapping of mat layers or thin spots. The preformable mat is able to be drawn uniformly over the corner.

I am going to discuss some details of molding the Ford Aerostar cross-member preform, but before I do, I would like to take a couple of minutes to explain some of the features of using preformable continuous strand mat.

Product Features

One feature is precise fiber placement. Unlike other molding compounds, where glass flow occurs, or there are knit lines or other unwanted fiber orientation, with continuous strand mat, you place it where you want it and it stays there.

Another feature is superior mechanical properties. Continuous strand mat, because of its continuous strand nature, is stronger than the chopped strands in either a chopped strand preform or sheet molding compound.

Another important feature is that preformable continuous strand mat is wash resistant and, in most applications, will not wash.

Any resin formulation which contains styrene will generally cause wash during resin impregnation due to the chemical action of the styrene on the mat binder. Most RTM resins and some RIM resins contain styrene. However, due to the unique nature of the proprietary binder used on preformable continuous strand mat, it does not wash and yet still retains its thermoformable properties.

Preformable mat can also be combined with other reinforcements (unidirectional Cofab or bidirectional reinforcements by Knytex and Hexcel, for example) and then formed to a shape. When unidirectionals are used, stretching is very limited, but bending to shape is quite possible. In some applications, we have molded up to ten plies of Cofab and mat to make a structural part.

Preformable continuous strand mat is quite different from other grades of continuous strand mat. While all other mats are bonded with thermosetting binder which does not soften during heating, preformable

mat is made so that it will soften when heated. When hot, the mat can be easily drawn to shape. Upon cooling, the soft binder regains rigidity so as to hold the preformed shape during subsequent handling operations.

We see this type of product as a key to expanding the use of the RTM and SRIM processes. Both processes require placement of glass into the mold. Use of a preform is quick, eliminating the aggravation of pushing mat into tight corners and having it spring out or slip from position.

Preformable mat is ideal for load-bearing parts. When used with unidirectional fabrics in specific combinations, shapes can be made that provide for handleable preforms at high glass content while maintaining easy impregnation.

The mat is typically available in 1½ oz/sq ft weight and 50 inch width.

Ford Aerostar Cross-Member

I want to spend the rest of my time describing the work we have done on the Aerostar cross-member preform. I will describe the materials, equipment and process we used to make cross-member preforms which were later molded in another facility by Ford. Important here are the concepts that were used. While specific techniques will change as we gain more experience, the basic concept will likely remain the same.

The Ford Aerostar cross-member preform mold that is discussed here is about four feet long and two feet wide, with twin cavities each about eleven inches deep. What we propose to do is to lay down four plies of flat mat in a holding frame, heat the mat to about 300°F and then form it between two cold FRP dies.

A steel holding frame is used, the bottom half of which is placed on a suitably sized table. An open press gives an idea of the contours to be formed in this very difficult shape. The press we used was selected not because of its 860 ton capacity, but rather for its large platen area (85" x 78") which will allow us to preform the Aerostar cross-member as well as considerably larger automotive shapes.

Each ply of mat is laid down on the frame for a total of four plies of 1½ oz/sq ft preformable mat.

Other parts may require more or fewer plies of mat depending upon part thickness and the desired glass content. Parts have been molded with up to ten plies of glass with plans underway to form parts with many more plies of reinforcement.

The Aerostar glass preform cavity has four mold alignment posts which have to be taken into consideration during forming of the preform. When the mold is closed, mold-to-mold contact is made only at the alignment posts.

If four plies of mat are under the posts, then preform thickness and stiffness will be affected adversely. So to compensate for this, cutouts were made where the alignment posts come in contact with the mold. The areas to be cut out are placed so that as the mold closes and mat stretches, the cutout areas fall into the right position, i.e., where the alignment posts make contact. Common heavy-duty scissors are used for the cutouts but other equipment might be more efficient such as electrical cloth cutting equipment or even a steel rule die.

Next, the top of the frame is placed over the bottom half with the glass fiber sandwiched in between. The frame is bolted down on all four sides. For the concept we are using for this part, the mat is stationary on all four sides and made to stretch to the final shape. Eventually, I expect that in addition to stretching, the mat will be also allowed to slip in a controlled fashion to form the preformed shape.

The framed mat is now ready to be heated, then formed to shape and cooled so as to hold its shape. The preform mold is inspected to remove any glass buildup that may occur at sharp corners. Mold release may also be applied to facilitate removal of the finished preform.

Immediately in back of the preform press is an oven used for heating the mat. The particular oven we used had Calrod heating units in the bottom of the oven and resistance heaters in the top. An oven can use resistance heaters, Calrod heaters, quartz tubes, heat lamps, or can be a gas fired convection oven. If radiant heaters are used, it is necessary to heat from both sides of the mat. If a convection oven is used, baffles must be used so as to force heated air through the mat and not allow the air to go around the mat. The oven is located directly behind the preform press so that, upon completion of the heating cycle, the mat can be removed from the oven, placed in position and the mold closed--all within 10 seconds.

The mat and frame are lifted from the lay-up table onto a track that extends through the preform mold area and into the oven. The mat and frame are pushed over the mold and into the heating oven where the heating cycle is soon to begin. Once the mat is in the oven, a frame handle is attached in position that will be used to pull the mat out of the oven and positioned over the preform mold.

There is a door at the rear of the oven which remains closed during heating to make more efficient use of heat energy. The door can be opened for manual positioning of the mat and adding thermocouples so that the heating cycle may be monitored. We used two thermocouples, one at the center of the mat pack, and the other buried just beneath the upper surface. The oven door is closed and the heating cycle begun. Temperature is monitored on the thermocouple graph.

Heating zones may be cycled on and off to obtain uniform heating throughout the mat pack. When radiant heaters are used as we did, the surfaces of the mat will heat quicker than the center. If the surfaces get too hot, they will smoke and cause discoloring. By cycling the heaters on and off, we can prevent smoke while getting the center sufficiently hot to make a good preform. We want to heat the mat up to about 300°F-350°F for good performance. Hot mat molds best and is the reason why the time to remove mat from the oven and complete forming of the part is kept down to about 10 seconds. The time to heat up the mat was about two minutes in this case, but can vary depending upon how many plies need to be heated and the capacity of the heat source.

There are some important points to remember. Heating for a long period of time does not help in molding. Actually, mat held at 350°F for 10 to 15 minutes will begin to lose its effectiveness as a preform. Also, carrying the mat to a distant mold, or slow mold closure that might take a minute or longer, allows the mat to cool so that it resolidifies before it is shaped. Thus, it will not hold a preform shape.

The best sequence is to rapidly heat the mat to desired temperature, move it to the mold, and close the mold -- fast. Once the mat is at the desired temperature, it is ready to be taken out and placed in the preforming mold. As the mat comes out of the oven, sag or bagging in the mat is normal.

The mat is hand placed in the correct position followed by mold closing. As the mold reaches half closure, mat stretching begins. The press continues to close and when fully closed, the mat has formed to shape and begins to cool to build up rigidity. We used a cooling period of up to 1½ minutes. Shorter periods are possible by using air- or water-cooled molds. For example, when making the tote box preform described earlier, a cooling time of 20 seconds was sufficient.

When the press opens, the mat frame is still in position on the guide rails and the depression of the preformed shape can be observed.

The finished preform is cut away from the frame and is raised out of the mold which allows one to see the full complexity of the shape.

During a trimming operation, one gets even a better idea of the part complexity. While trimming was done by hand, more automated techniques would be used in production such as steel rule die stamping.

The basic concepts for making a preform that have been discussed here are:

a. mat holding frame
b. mat is clamped on all sides in the holding frame
c. a heater is used to heat the mat to 300°F
d. a cold preform die is used to shape the mat

Remember to rapidly heat the mat to temperature, then quickly place it into a cold preforming mold and close the mold quickly.

As we gain more experience, the techniques will be refined. Automated mat frame clamps may be used and many changes are envisioned that would speed up the preforming process.

While the Aerostar cross-member is a large part, we have yet to determine the maximum size that can be preformed. I expect the only limitation is the limit of one's imagination.

Handling several plies of reinforcements can be cumbersome. Some structural parts already contain ten layers of reinforcements while others under consideration may contain up to twenty-four plies of random and unidirectional layers. For this purpose, we have found that reinforcement blankets can be very helpful.

Reinforcement blankets are layers of random mat and oriented fiber placed in a predetermined sequence and then sewed together, for example, with a Malimo machine. When heated and formed to a shape, the preformable mat is stiff enough to hold the blanket to shape even though the unidirectional rovings may not have any binder of their own. Further, when two or more plies of unidirectional roving are placed together without any added binder, preform delamination could occur if it were not for the fact that the blanket is sewed together which prevents delamination.

We also found that the action of sewing the blanket formed small needle holes through the structure which acted like mini-channels for improved resin penetration. The blankets, when impregnated, showed much better wetout compared to individually plied laminates.

SUMMARY

Increased interest in molding large, complex automotive shapes has resulted in development of a preformable continuous strand mat. Preforming techniques have also been developed which aid in making complex deep draw shapes that previously could only be made by chopped strand preforms, SMC or time consuming mat pattern cutouts.

Fabrication of Aerostar cross-member preforms is described as an example of the steps that must be taken to form a complex shape.

Preformable mat is held in a frame and heated to about 300°F in an oven which is placed directly in back of the preforming press. Upon reaching the desired temperature, the frame holding the mat is removed from the oven, placed in position and formed to shape by a cold preform mold, all in about ten seconds. After cooling, the mold is opened and the finished preform is removed for subsequent molding into a finished part.

If many plies of reinforcements are to be used, such as combinations of preformable mat and unidirectional mats, then reinforcement blankets made by sewing several layers together may be used. Blankets facilitate handling of large numbers of reinforcement plies.

EPOXY/POLYESTER COMPATIBLE GLASS FIBER YARNS IN STRUCTURAL BRAIDING

T. P. Hager

Owens-Corning Fiberglas Corporation
Technical Center
Granville, Ohio 43023   USA

ABSTRACT

Over the past several years, the braiding process to produce structural reinforcements for a variety of composite products has been gaining in importance. A structural braid combines yarns in an oriented pattern over a mandrel or form which, when impregnated with resin and cured, results in a rigid composite that is lightweight yet displays high strength and modulus.

Although glass fiber yarns have long been used in braiding for sleeving, wire jacketing, etc., these yarns are normally starch-sized, and thus are not resin-compatible. Coatings designed for resin compatibility are available on glass fiber rovings, but are not usually available on textile glass yarn, because they have not been designed to protect the glass fiber during rigorous textile processes such as braiding.

The objective of this paper is threefold:

(1) To discuss the characteristics of glass fiber yarn sized with a new, epoxy and polyester resin-compatible coating that also allows for efficient processing during braiding, relative to both conventional glass yarns finished with starch sizings and rovings coated with resin-compatible binders.

(2) To discuss the strength characteristics of various epoxy- and polyester- cured resins reinforced with this new yarn versus conventional starch-sized yarns and resin-sized rovings.

(3) To discuss the design and performance characteristics of a triaxially braided snow ski using the new yarn compared to a conventionally produced, hand lay-up ski using conventional rovings.

STRUCTURAL BRAIDING FOR REINFORCING COMPOSITES offers manufacturers unique processing and product advantages relative to conventionally manufactured composites such as by hand lay–up methods. Braiding permits planned cross–directional fiber orientation to control tensile strength. The braid is conformable, following the shape of most cylindrical or cylindrical-like forms. Braiding is more predictable than hand lay-up techniques, ensuring repeatable processing from one part to the next. The braid configuration inherently resists twisting and thus displays torque stability. The braid, due to its 3-dimensional character, improves stress translation along and through the form being reinforced. The braid configuration can be varied over the length of a product, producing variable fiber angle and, thus, controlled performance. Composite structural braids have been shown to exhibit outstanding impact characteristics. Finally, relative to hand lay-up, braiding is a more cost-effective process.

Glass fiber yarns have long been used in constructing braids for sleeving and wire jacketing. To process efficiently, glass yarns of small fiber diameter (<10 micrometer) are used. The fine diameter is more resistant to damage than the large diameter fiber (>13 micrometer) due to radius of curvature considerations. Also to improve textile processing, the fine fiber diameter glass yarns are sized with a starch-based finish. However, this finish is not resin compatible. The glass can be desized and refinished with silane coupling agents to gain resin compatibility, but tensile strength is greatly reduced.

Continuous glass strand with resin-compatible sizings are normally available only on large filament diameter strands (rovings). These products do not process well in braiding due to both the heavy filament diameter and the high coefficient of friction coating.

As will be described in this paper, Owens-Corning Fiberglas Corporation has developed a continuous filament yarn with a specially formulated sizing termed Composite Yarn[TM] 603-0 that displays good epoxy and polyester resin wetout and cured resin compatibility. Furthermore, the sizing also allows the glass yarn to be processed on conventional winding and braiding equipment as efficiently as conventional starch-based (nonresin-compatible) sizings.

MATERIALS

The materials evaluated in this work were: yarns ECG-75 1/2 2.8s 603-0 (plied yarn, epoxy/polyester compatible sizing), ECG-75 1/2 2.8s 620 (plied yarn, starch/oil sizing), ECG-75 1/0 1.0Z 620 (singles yarn, starch/oil sizing), ECG-75 1/2 2.8s 620 HC (starch sizing heat cleaned off and replaced with a mixture of silanes to confer epoxy/polyester resin compatibility) ECG-75 1/0 1.0z 366 (singles yarn, resin-compatible roving sizing); 366-AC-250 (Type 30[R] roving, epoxy/polyester compatible sizing); Dow Chemical epoxy resin DER 331 with Lindau Chemical anhydride curing agent Lindride 6; OCF polyester resin E-701 (isophthalate/maleate polypropylene condensate in styrene) with Caddox methyl ethyl ketone peroxide M-50 and 6% Cobalt.

EXPERIMENTAL

The yarns and roving were evaluated for abrasion resistance, relative resin wet-out, resin compatibility, and impregnated strand tensile. To simulate and quantify abrasion damage during braiding of the various glass fiber samples, an abrasion method was devised that consisted of an enclosed box equipped with steel pins of constant diameter (1/32 inch) and configuration. See Figure 1. The box was attached to a vacuum source, and the outlet was covered with a fine mesh metal screen to collect the broken filaments and powder (sizing that had shed off.) The previously tared screen was then weighed, burned at 1250°F for 20 minutes, and reweighed to determine the amount of inorganic glass and organic sizing.

Applied strand tension (T_1) and resulting strand tension (T_2) were also recorded as grams of force. This allowed for a measure of the relative frictional characteristics of each coating as well as for a measure of abrasion damage (fuzz and powder) as a function of applied tension. Fuzz and powder were recorded as milligrams per kilogram of yarn run.

The speed with which either the epoxy or the polyester resin wet each of the glass samples was measured by placing approximately 0.4 gm of 2 cm chopped fiber lengths in a loose mat on a 2 x 3 cm square target, pouring excess neat resin at room temperature on top

of the chopped glass, and timing in seconds the point at which the target could be discerned (minimum translucency).

Resin compatibility is the ability of the resin to adhere to the glass fiber surface, usually after an accelerated weathering treatment. NOL ring shear tests were used to quantify this parameter (ASTM D2344). Samples were prepared by passing the glass strand through a bath of either the epoxy or polyester resin mixed with the appropriate curing agent, winding the impregnated strand onto a narrow doughnut-like mold to form a flat ring, and curing the assembly in an oven. The resulting rings were cut to 1 inch lengths and tested in an Instron Model TTC for binder/resin matrix shear strength. The Instron was set up for 0.5 inch span, 0.05 inch/minute crosshead speed, and 0.5 inch/minute chart speed. The ring samples were tested as received or after 2 or 24 hours immersion in boiling distilled water. Shear strengths were then recorded as a function of water exposure.

Tensile measurements were performed on impregnated strands using an Instron Model 1125. The glass strands were prepared by passing the sample into a bath of epoxy or polyester resin, through a die to control pick up, and onto a 16-inch I-shaped wooden form. This form was then placed in an oven to cure. Strands of appropriate lengths were then cut and tested. The Instron was set up for 10 inch gauge length, 0.2 inch/minute crosshead speed, and 2.0 inch/minute chart speed. Tensile values were recorded in pounds per square inch of reinforcement to normalize for the differences in the amount of glass from sample to sample.

RESULTS AND DISCUSSION

ABRASION RESISTANCE - The second and third columns of Table 1 lists the frictional characteristics of the yarns and roving (illustrated in Figure 2) which show resulting running tension (T_2) of the glass samples as a function of applied tension (T_1) before and after the strands pass over the abrading pins.

The G75 620 starch-sized yarns, singles and plied, were used as control. As can be seen, the G75 1/2 603-0 epoxy/polyester compatible yarn has virtually the same strand running characteristics as the starch yarn. The G75 1/0 366 yarn ran at a significantly higher tension. A plied yarn construction was not available with the 366 sizing at the time of these tests. However, G75 1/0 1.0z 620 singles yarn ran with even lower tensions than the plied construction, perhaps due to less glass and thus less surface contact. Therefore, the large increase in running tension of the G75 366 is due to the binder, not differences in yarn construction. Strand running tensions affect how hard a braider would be made to work, how fast braider parts would wear out, and how well the braid lay is controlled. This latter point would be expected to become very important as the complexity of the braided shape increases.

The 366 250 roving could not be tested. The abrasion equipment was damaged while trying to measure this heavy strand. Even at very low applied tensions, running tensions were extremely high (>500 gf). This was due to: running about 15 times the amount of glass per unit length relative to the plied yarns, the nature of the 366 sizing on the roving, and the fact that rovings are untwisted, and thus offer the highest surface contact area.

Values are also not available for the G75 1/2 620 HC heat cleaned yarn. Heat cleaning (burning the organic starch sizing off the glass) and recoating the glass strand with a solution of silanes damages the strength of the glass, making it very unworkable and too fragile for the abrasion test. The large drop in tensile strength from heat cleaning is discussed later.

Running tensions also affect the amount of glass fiber damage that occurs. Damaged fibers at best cause housecleaning problems, and at worst, can affect machine processing and tensile strengths. Table 1, columns 4 and 5, list the amounts of broken filaments and powder collected from each sample at low (25 gf) and high (75 gf) applied tensions. These results are illustrated in Figures 3 and 4. Taking the area under each curve of applied tension versus fuzz or powder offers a single number or "index" to compare the glass strands to one another. This integration takes into account the different slopes and intercepts of the lines, reflecting the different rates at which the various strands abrade. The G75 366 has an index 30% greater than the G75 1/0 620 control yarn. The G75 1/2 603-0 has an index of only 25% that of the G75 1/2 620 control. 603-0 sizing protects the glass fibers from abrasion damage better than the starch, and much better than the roving binder.

From a practical standpoint, these abrasion values correlate with actual field experience from our major braiders. The 603-0 produces a braid with fewer broken filaments than the 620 starch sizing. Rovings cannot be braided, or are braided only with extreme difficulty.

Starch sizes usually produce some degree of powder during processing. The starch film formers in the formulation are brittle and do not have the adhesion characteristics to the glass surface of resinous coatings. Also, the starch sizings are present in levels greater than 1% by weight of the glass. The 603-0 and 366 coatings account for only about 0.5-0.6% of the weight of the glass. Thus, the powder indices for these two binders are much lower than that of the starch, with the coating on the broken filaments probably accounting for most of the "powder".

RESIN WET-OUT - Values for resin wet-out, or better, wet-through are listed in Table 2. The numbers are in seconds to reach minimum translucency. The results for both epoxy and polyester resins are listed. In general, the epoxy DER 331 wet-out times are longer due to the higher viscosity at room temperature of this resin (1100-1400 cps) compared to the polyester E-701 (600-650 cps).

The wet-outs of each strand can be ranked from fastest to slowest. For epoxy resin the ranking is: G75 366 > G75 603-0 > G75 620 HC > 366 250 > G75 620. The slow response of the M250 roving may be due to the higher glass mass per filament, since the same binder on a G fiber has the fastest wet time. As expected the 620 starch has the slowest response: about 400% longer than the G75 366. 603-0 takes 30% longer than the G75 366. Heat cleaned and silanized yarn takes 90% longer than the G75 366.

For the polyester resin the wet-out rankings are: G75 366 > G75 603-0 > G75 620 HC > 366 250 > G75 620. The starch 620 again takes over 400% longer than the G75 366. The 603-0 and the 620 HC took 13% and 27% longer to wet, respectively.

RESIN COMPATIBILITY - NOL ring shear results of the various glass strand samples are listed in Table 3 (epoxy resin) and Table 4 (polyester resin). Figures 5 and 6 illustrate these results. Strengths were generally higher for the the epoxy resin than for the polyester. The 366 250 roving had the best strength retention in either resin (100%) after 24 hours exposure to boiling distilled water. The G75 620 starch had the worst retention (20-30%). 603-0 had fair to good strength retention (60-75%). These values illustrate the necessary compromise between processability and resin compatibility. While the resin compatibility of 603-0 is much better than a starch sizing, the sizing does not have the compatibility of a good roving binder designed principally for resin reinforcement. However, the roving binder performs very poorly in textile processing operations.

TENSILE STRENGTH - Table 5 lists the tensile measurements of the various strands. All strands were impregnated and cured prior to testing. The results are listed in pounds per square inch. The tensile calculation to normalize the results takes into account the resin and the variable amounts of glass of each sample due to differences in yields.

The 603-0 and 620 had slightly higher normalized tensile values than the 366 roving. This is probably due to the fact that the yarns are of smaller filament diameter. Since glass fiber tensile is controlled by surface flaws, the larger fiber would be expected to have a higher ratio of surface flaws per unit mass of glass compared to finer filament yarns, and thus give slightly lower measured tensile values.

The G75 1/2 620 HC heat cleaned yarn gave the lowest tensile values. Heat cleaning is very detrimental to glass strength. Besides the necessity for additional handling, the high temperatures reached during the burn-off step (700°-1200°F) may cause mechanical and chemical corrosion of the glass fiber, introducing flaws into the fiber surface, and thus lowering tensile strength. A certain amount of densification also occurs at elevated temperatures, which may or may not introduce stress into the fiber.

PERFORMANCE CHARACTERISTICS OF A BRAIDED SNOW SKI

K2 Corporation has developed a new ski that incorporates the 603-0 epoxy/polyester compatible yarn in a triaxially configured reinforcing braid. Table 6 lists the physical characteristics of a conventional hand lay-up ski to the braided ski. (Data courtesy of K2 Corporation.) The conventionally-made ski was manually "wet wrapped" around a foam core. The triaxial ski was machine braided around a wooden core. Cores affect damping and the vibration a skier feels, but do not appreciably affect strengths. Both skis were the same length (204 cm).

Figure 7 shows the horizontal, 64-carrier New England Butt braider braiding the Fiberglas[R] yarn around a wooden core. Figure 8 shows a close-up of the glass fiber strands as they are laid down in the braid around the core. Note how well the braid conforms to the underlying wooden core. Figure 9 shows the braided cores before they are placed in the resin transfer mold.

The braided ski shows significant increases in strength, impact resistance, and permanent deformation resistance compared to the wet wrapped ski. Minimum break strength in work-energy units is 4,000 inch-pounds applied vertically to the ski tip and tail. The braided ski exceeds the standard by 20%. The impact test consists of alternately slapping the tip and tail against a steel plate with an energy of 8,000 foot-pounds. 30 cycles before failure is minimum acceptance. The braided ski far exceeds this minimum. Permanent deformation resistance is measured by deflecting the ski at a given load, releasing the load, and measuring the recovery after 60 seconds. Permanent deformation is the load at which 100% recovery is not attained. The braided ski offers significant improvements in this parameter also.

The above improvements of the braided ski over the wet wrapped ski were achieved with 30% less glass relative to the hand-wrapped ski. K2 found from finite elemental analysis, that the longitudinal and torsional forces exerted on a turning ski couple to form diagonal stress lines along the ski. The braided structure was designed to parallel these stress lines. Furthermore, by braiding the glass onto the core, rather than applying a pre-braided sock around it, the glass is laid down under tension. The high twist yarns also help to dampen vibration and improve the efficiency by which the stress on the ski is translated to the reinforcing strands. K2 found that this design maximized the energy transmitted from the skier to the ski's edge with little loss of mechanical energy, and subsequent less exertion to the skier.

CONCLUSIONS

It has been shown that the epoxy/polyester compatible glass fiber yarn ECG-75 1/2 2.8s 603-0 has superior abrasion characteristics relative to starch-based and resin-compatible coated glass yarns and rovings. The 603-0 sizing displays good epoxy and polyester resin wet-out and compatibility. Although the resin compatibility of the 603-0, as measured by NOL ring shears, is not as high as a "good" roving binder, the 603-0 gives much better values than a starch sizing. Tensile measurements of impregnated strands of this binder are typical of twisted, fine filament yarns, and are much greater than the comparable, heat cleaned and silane treated sample. The processability and resin reinforcing characteristics have been further corroborated in actual field experience. One customer, K2 Corporation, has designed and is now producing a line of snow skis incorporating this yarn in a structural braid for the ski. Their tests have shown the combination of this yarn with the braided structure to result in a ski with much improved strength, impact, and deformation resistance relative to a conventional hand lay-up ski. Moreover, K2 has achieved these increases with 30% less glass in the composite.

TABLE 1
ABRASION RESISTANCE OF THE VARIOUS GLASS FIBER STRANDS AS A FUNCTION OF APPLIED TENSION (T_1) AT CONSTANT PIN DIAMETER (0.79mm)

Glass Fiber Input	T1 (gf)	T2 (gf)	Fuzz (mg/kg)	Powder (mg/kg)	Abrasion Index Fuzz	Powder
G75 1/2 2.8s	25	108	0.161	0.185	11.90	7.65
603-0	75	283	0.317	0.119		
G75 1/0 1.0z	25	153	1.825	0.428	202.18	20.78
366	75	357	6.258	0.403		
G75 1/2 2.8s	25	110	0.638	3.895	45.38	208.55
620	75	280	1.179	4.443		
G75 1/0 1.Oz	25	87	1.187	1.166	154.10	43.90
620	75	237	4.977	1.427		

TABLE 2
RESIN WETOUT TIMES (SECONDS)
OF THE VARIOUS GLASS FIBER SAMPLES

Glass Fiber Input	Epoxy (sec.)	Polyester (sec.)
G75 603-0	106	17
G75 620	313	69
G75 620 HC	150	19
G75 366	80	15
366 250 Roving	303	32

TABLE 3
NOL RING SHEAR RESULTS OF VARIOUS GLASS INPUTS IN EPOXY RESIN

Glass Fiber Input	Shear Strength (PSI $\times 10^3$) After Exposure to Boiling H2O (hours) 0	2	24	Percent Retention	Percent Water Absorption	Percent Glass
G75 1/2 2.8s 603-0	8.9	7.8	6.6	74	1.088	84.4
G75 1/2 2.8s 620	7.6	4.4	2.1	28	0.722	82.9
366 250	8.7	9.2	8.9	102	0.083	82.0

TABLE 4
NOL RING SHEAR RESULTS OF VARIOUS GLASS INPUTS IN POLYESTER RESIN

Glass Fiber Input	Shear Strength (PSI × 10³) After Exposure to Boiling H2O (hours)			Percent Retention	Percent Water Absorption	Percent Glass Content
	0	2	24			
G75 1/2 2.8s 603-0	5.6	5.0	3.5	63	0.392	87.8
G75 1/2 2.8s 620	2.6	0.8	0.6	23	2.387	84.8
366 250	8.4	9.2	8.6	102	0.233	74.8

TABLE 5
IMPREGNATED STRAND TENSILE STRENGTHS OF VARIOUS GLASS FIBER INPUTS

Resin	Glass Fiber Input	Tensile Strength (psi × 10³)	Percent Glass
Epoxy	G75 603-0	374	61.8
	G75 620	358	61.0
	G75 620 HC	125	59.1
	366 250	342	66.6
Polyester	G75 603-0	367	64.5
	G75 620	377	55.2
	G75 620 HC	143	61.6
	366 250	360	68.7

TABLE 6
COMPARISON OF THE PHYSICAL CHARACTERISTICS OF A CONVENTIONAL HAND LAY-UP SKI TO A BRAIDED SKI

	K2 Slalom 66 (204 cm)	K2 TRC (204 cm)
Construction	Wet Wrap Foam Core	Triaxial Braided Wood Core
Weight	2,100 grams	1,950 grams
Swing Weight	5.2 HZ	5.8 HZ
Break Strength	At Standard	20% Over Standard
Impact Resistance	40 Impacts	70 + Impacts
Permanent Deformation Resistance	820 lb.	850 lb.

Data courtesy of K2 Corporation, Vashon, Washington.

FIGURE 1
METHOD TO QUANTIFY
ABRASION RESISTANCE
OF GLASS FIBER STRANDS

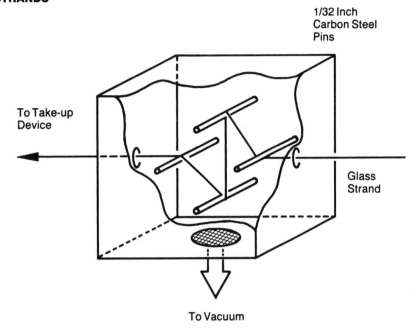

1/32 Inch
Carbon Steel
Pins

To Take-up
Device

Glass
Strand

To Vacuum

FIGURE 2
FRICTIONAL
CHARACTERISTICS OF
YARNS COATED WITH
VARIOUS SIZINGS—
RESULTING TENSION VS.
APPLIED TENSION

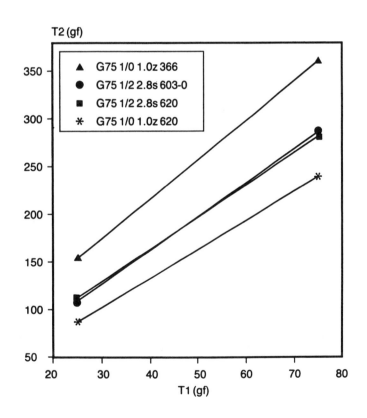

T2 (gf)

▲ G75 1/0 1.0z 366
● G75 1/2 2.8s 603-0
■ G75 1/2 2.8s 620
✳ G75 1/0 1.0z 620

T1 (gf)

FIGURE 3
ABRASION RESISTANCE OF GLASS FIBER YARNS COATED WITH VARIOUS SIZINGS—BROKEN FILAMENTS (FUZZ) VS. T1

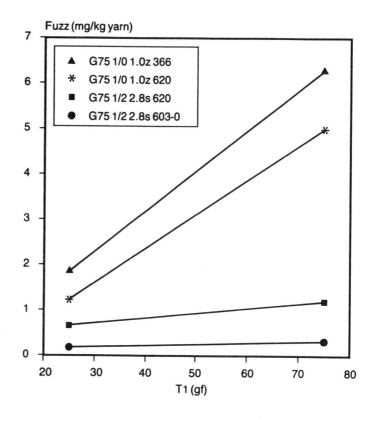

FIGURE 4
ABRASION RESISTANCE OF GLASS FIBER YARNS COATED WITH VARIOUS SIZINGS—SHED COATING (POWDER) VS. T1

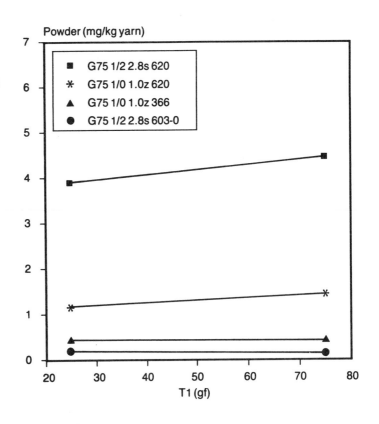

FIGURE 5
NOL RING SHEAR DATA OF
VARIOUS GLASS INPUTS
EPOXY/ANHYDRIDE RESIN

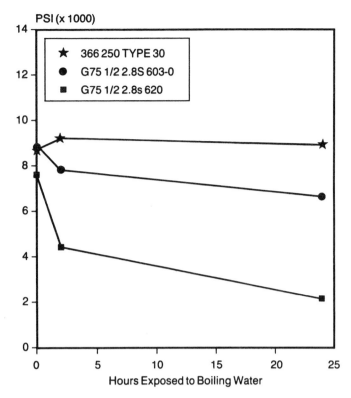

FIGURE 6
NOL RING SHEAR DATA OF
VARIOUS GLASS INPUTS
POLYESTER/PEROXIDE
RESIN

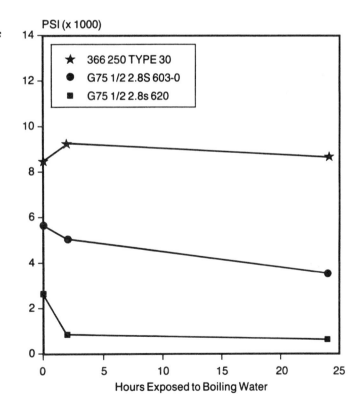

A NEW APPROACH TO
BULKED ROVING

J. V. Gauchel, H. I. Glaser
Owens-Corning Fiberglas Corporation
Granville, Ohio 43023   USA

ABSTRACT

Owens-Corning Fiberglas has developed a bulked roving designed to increase the transverse property performance of pultruded parts. This product is an outgrowth of Owens-Corning's spun roving programs in Europe, Canada, and the U.S. The material format is designed to utilize the best qualities of traditional bulked roving products without the processing disadvantages.

The roving is produced on a process which allows the placement of loops in the axial and transverse direction. The physical properties of the strand, such as yield, bulkiness, and extendability, are controllable within a wide range. Currently, Owens-Corning Fiberglas produces products within the yield range of 10-70 yards/pound and with loop ratios designed to meet the needs for specific processes and applications.

This paper will compare the performance of several versions of the new bulked roving to Fiberglas Canada 852 HC 4035 spun roving and to Type 30® roving hybrid structures. Preliminary test data for these systems will be given.

Although akin to spun roving, this is a new product form with different and exciting characteristics. Early testing indicates that this new bulked roving has the ability to expand and to contract within the available space and still mold to a solid section. This opens product design potential which goes beyond currently available roving products.

OWENS-CORNING FIBERGLAS has developed a new approach to the manufacture and format of a bulked roving product. This process creates a family of products designed to impart multiaxial reinforcement with uniaxial input. The program is an outgrowth of Owens-Corning's spun roving programs in Europe, Canada, and the U.S. Originally designed to increase the transverse property performance and improve die fillage in pultruded parts, these material systems and benefits have been translated into other processes including filament winding, and preform molding.

This new product form with different and exciting characteristics has the ability to distribute itself uniformly through a cavity over a wide range of glass contents. This allows it to fill a mold of variable cross-section -- expanding and contracting within the available space to mold a solid section. It also can be formed to have controlled axial extensibility ranging from high for preforming to low or zero for pultrusion and filament winding.

Since the product is new, only preliminary information is available on any given process or application. This paper will outline the process used to form the product, discuss the format and concept of the product, define selected performance aspects, and show comparisons to existing products such as Fiberglas Canada 852 HC 4035 spun roving and OCF 424BA Type 30® roving. Information on the performance of Type 30®/bulked roving hybrids will also be presented.

DISCUSSION--PROCESS

The process used to form the new bulked roving is designed to take advantage of recent developments and process control which help maintain product uniformity. The process is based on the controlled placement of longitudinal and transverse loops of glass at a given ratio resulting in low yield variation and product performance uniformity.

Because the process is highly flexible, multiple products may be produced from many input fiber systems. Currently, yields of 10-70 yards per pound are being experimentally produced. Work on expanding the yield range is continuing.

DISCUSSION--PRODUCT

Figure 1 schematically depicts the format of the new bulked roving. In describing the product, emphasis must be placed on the concept of control. The process is designed to independently control the ratio of transverse to longitudinal loops formed, the strand yield produced, the degree of twist induced, and the type and amount of sizing applied for each product.

The process is sufficiently flexible so that a wide variety of products potentially can be produced with functional performance matched to the needs of the end-use process and application. Initially, only products designed for the pultrusion, filament winding, and preform markets will be available for evaluation.

When comparing the new bulked roving product to traditional spun and texturized roving products, there are some similarities and many differences which can be noticed. The new product is designed to handle similarly to other spun and texturized roving products. It is designed to process through standard creels at line speeds and tension associated with current FRP processes. It will be compatible with existing resin systems and capable of being impregnated by transfer, immersion or injection techniques. Two of its stated performance attributes are that it maintains uniform fiber dispersion at lower glass levels than nonbulked roving products and offers multiaxial reinforcement with a uniaxial input format.

The differences associated with the new product are that it is formed from a selectable ratio of unbroken multiaxial loops. Unlike some of the traditional spun rovings, it has no core strand and therefore will not strip back during processing. Its bulking effect is also permanent. That is, unlike the current texturized or air jet systems, the product will remain in the bulked state even when impregnated and under tension. The product is a low yield product with fine fiber diameter. The low yield allows reduced creeling time and space. The fine fiber enhances wetting speed and stress transfer efficiency. The process allows for controllable in-process quality assurance and product reliability.

DISCUSSION--PERFORMANCE

During the course of developing the new bulked roving product, a series of trials were performed to evaluate its mechanical and physical performance. Since pultrusion was the primary target process, most trials were performed using this manufacturing technique. A 0.5-inch diameter rod specimen was chosen to represent a typical thick section pultruded product. This selection was also made because of the ability to directly manufacture ASTM

Standard test specimens for both flexural and shear performance. Tensile, compression and compression-shear specimens could also be machined from a 0.5-inch diameter input specimen. Evaluation and manufacture of single end roving/bulked roving hybrids were also easily accomplished with the symmetrical cylindrical shape.

The first series of trials were designed to evaluate the relative shear performance and die fillage characteristics of the new systems versus a spun roving (Fiberglas Canada 852 HC 4035) and a single end roving (OCF 424BA). The flexural shear strength (ASTM D4475) of each system versus glass content is shown in Table 1. Where data is absent, the product would not make acceptable rod in the manufacturing process. Two things are evident from the table: the new systems have a wider range of glass contents at which products can be produced, and the shear performance of rods produced with the new bulked roving are higher than those produced from either single end roving or spun roving at equal glass contents. The relative shear performance of these systems can be seen graphically in Figure 2. These results are for a 0.5-inch diameter rod made with polyester resin reinforced with a glass volume equivalent to 40 ends of 113 yards per pound yield single end roving (approximately 53 volume percent). In all cases, acceptable crack free rod was manufactured, all reinforcement was sized with polyester compatible systems, and process variables were held constant among the systems. The new bulked roving systems show a 14% improvement over the single end roving system and 50% higher performance than spun roving system.

The higher shear performance of the new roving can also be seen when the product is hybridized with single end roving. Figure 3 depicts the flexural shear strength of single end roving/bulked roving hybrids versus equivalent ends of 113 yield glass in the axial direction. The equivalent yield in the axial direction of the bulk roving product was estimated by

$$N_{ax} = (1 - \frac{LFR}{1+LFR}) N_T \qquad \text{Eq. (1)}$$

where

N_{ax} = equivalent ends of 113 glass in axial direction for the bulked roving products

N_T = total equivalent ends of 113 glass for the bulk roving product in hybrids

LFR = loop formation ratio =

$\frac{\text{amount of glass in transverse direction}}{\text{amount of axial glass}}$

The LFR is an attribute of the new product that can be broadly viewed. The LFR for spun roving was estimated to be equal to .2. The LFR for a single end roving product equals zero.

Interpreting Figure 3 requires a basic understanding of the flexural shear test and the failure mechanisms available during that test. The flexural shear strength is composed of two components: a flexural component and a shear component. Increases or decreases in the flexural shear strength are related to the net effect of the input materials on these two components. In Figure 3, the interpretation has been simplified by keeping the flexural component of the hybrid systems equal. This was accomplished by placing the bulked products in the center of the rod as a core and keeping the total glass content of the rod and the ratio of bulked fiber to single end roving constant. Failure mode was also observed to remain constant. The use of 3:1 span to depth ratio for the test as specified in ASTM D4475 helped ensure this. The all single end roving specimens did have changes in the flexural component. However, the flexural shear results seem to fit within the scatter of the hybrid trend line.

The basic conclusion from Figure 3 is that as one increases the effective amount of transverse glass in the core of the hybrid sample, the effective shear performance of the sample increases predictably. It also indicates that because the new product more effectively places glass in the transverse direction, hybrids made with it out-perform those made with a spun roving roving core.

The flexural modulus performance of the new bulked roving product can be seen in Figures 4 and 5. In Figure 4, the flexural modulus (ASTM D790) of 0.5-inch diameter rods containing 77% glass by weight is plotted versus loop formation ratio for various input slivers. Two important observations may be made for this graph: axial flexural modulus is a function of both loop ratio and input sliver and the flexural modulus for the new product projected to LFR=0 is equivalent to the flexural modulus expected for an all single end roving product. The importance of these observations is that there appears to be no change in the axial effectiveness of the reinforcement caused by the looping of the product (i.e. no stress concentration or loss in coupling efficiency is seen -- just a glass volume effect). The change in flexural modulus with sliver is associated with the distribution of the bundles near the surface of the part. The larger the input sliver, the less glass is distributed near the outer surface of the rod. Since flexural modulus is highly dependent on the skin stiffness, this change in glass content near the surface shows up as a loss of flexural modulus with sliver increase at constant loop ratio.

In Figure 5, the effect of glass content at constant sliver input for a 22.6 yards/pound yield input sliver is shown. The flexural modulus is a function of both loop ratio and total glass content. For both glass contents, the flexural modulus projected to a loop ratio of zero is equivalent to the flexural modulus predicted for a single end roving product of that glass content.

If one accepts the hypotheses that the effects shown in Figures 4 and 5 indicate that axial flexural properties are directly related to the effective amount of glass oriented in the direction of test, one can calculate the effective volume fraction of bulked fiber oriented transverse to the axial direction. Calculating this effective transverse volume fraction glass and inserting it into the simple rule of mixtures for predicting modulus allows one to predict the transverse modulus for the rod systems tested.

$$E_{cf} = E_f \, \Phi_{ef} + E_m \, (1-\Phi_{ef}) \qquad \text{Eq. (2)}$$

where

E_{cf} = the predicted modulus

E_f = the modulus of the reinforcement

Φ_{ef} = the effect volume fraction fiber in the direction of the test

E_m = the modulus of the matrix

NOTE: The modulus of the matrix is not necessarily the modulus of the unfilled resin used to make the rod. But for simplicity sake, it is considered to be that modulus and the equation is used as a lower bound.

Figure 6 graphically projects how this predicted transverse performance would vary with LFR. Owens-Corning Fiberglas is currently developing this transverse modulus data on flat plate systems in order to assist the effectiveness of this prediction and to develop a data base on the transverse performance of the new bulked roving systems.

SUMMARY

This paper has presented a discussion of the process, product, and performance of a new bulked roving product. Preliminary indications are that this system represents an improvement in flexibility, die fillage and mechanical performance over existing bulked systems. Owens-Corning Fiberglas is continuing efforts to characterize this system's performance over a range of applications and processes to help establish the product formats necessary for our customer's needs.

TABLE 1
FLEXURAL SHEAR STRENGTH VERSUS INPUT GLASS WEIGHT[1] FOR BULKED ROVING PRODUCTS

Equivalent Ends of 113 Yield Roving	424 BA[A]	852 HC[B]	Roving Input		
			LR1	LR.6	LR.3[2]
25	N.M.	N.M.	6.4	N.M.	N.M.
30	N.M.	5.77	7.5	7.1	7.4
35	7.6*	6.1	8.6	8.15	8.16
40	7.2	5.47	9.0	8.4	8.3
45	5.66	*	9.0	8.3	8.5
50	N.M.	N.M.	N.M.	8.07	8.3

(A) Manufactured by Owens-Corning Fiberglas Corporation.
(B) Manufactured by Fiberglas Canada, Incorporated.
N.M. = not manufacturable.
* Unstable process produces intermittant cracks.
(1) Die Diameter = 0.5"
45 ends of 113 yield nominally gives 78% glass content by weight.
(2) Loop formation ratio of the new bulked product.

FIGURE 1
STRAND CONSTRUCTION

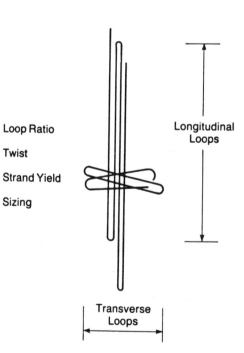

Loop Ratio

Twist

Strand Yield

Sizing

Longitudinal Loops

Transverse Loops

FIGURE 2
FLEXURAL SHEAR STRENGTH vs. INPUT
FOR VARIOUS ROVING SYSTEMS

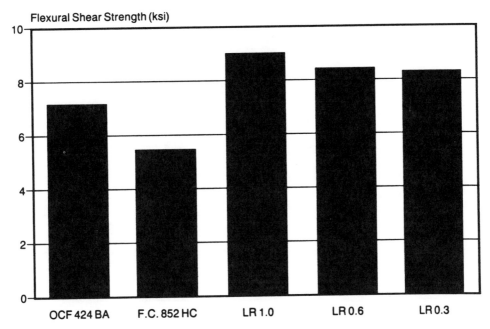

Flexural Shear Strength (ksi)

Glass Input Equiv. 40 Ends 113 Yield
In 0.5 In. Dia. Rod

FIGURE 3
SHEAR STRENGTH vs. EQUIV. ENDS OF 113
FOR VARIOUS BULKED ROVING/SINGLE END
ROVING HYBRID SYSTEMS

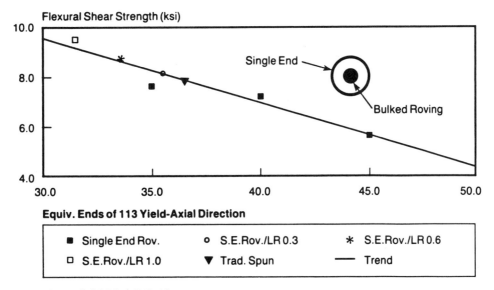

Flexural Shear Strength (ksi)

Equiv. Ends of 113 Yield-Axial Direction

| ■ Single End Rov. | ○ S.E.Rov./LR 0.3 | * S.E.Rov./LR 0.6 |
| □ S.E.Rov./LR 1.0 | ▼ Trad. Spun | — Trend |

Hydrid = 24 ends D.R./15 Ends Bulked Rov.
Glass Contents = Hybrid 53 Vol %
Single End Rov. = 48-56 Vol %

FIGURE 4
FLEXURAL MODULUS VERSUS LOOP RATIO FOR
NEW BULKED ROVING SYSTEMS—22.6 yd./lb.

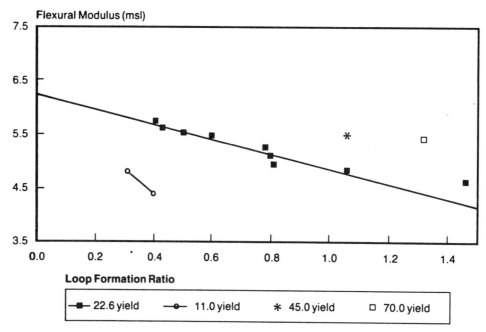

Glass Content = 77%

PREFORM DEVELOPMENT FOR A STRUCTURAL COMPOSITE CROSSMEMBER

C. F. Johnson, N. G. Chavka,
Ford Motor Company
Research Staff
Dearborn, Michigan 48121   USA

Introduction

The cost-effective production of derivative vehicles presents a new challenge to automotive material and design engineers. There is an increasing demand in the market place for a number of vehicles which are in some respect distinctive from other similar offerings. Currently there are some 600[1] nameplates in the American automotive market place, with that number likely to grow over the next decade. To be viable for production, derivative vehicles must return adequate profit at low volumes. The use of composite structures appears to be one alternative material driven strategy which can, through the use of large-scale part integration and high specific physical properties, provide truly unique design potential at the low investment levels required for economic feasibility at low volumes.

The crossmember module concept described in this paper enables a standard two-wheel-drive van to be differentiated as a four-wheel-drive vehicle with the addition of a single multifunctional crossmember module. While it is feasible to design such a structure as a multipiece steel component, the exceptional strength and durability characteristics of composite materials allow its fabrication as a single complex piece. To date, a weight savings of 25 lbs. has been achieved and a durability greater than that of a steel component demonstrated in limited testing.[2] There are, however, open issues which need to be addressed prior to adaptation of this technology to an actual production component. Improved preforming of reinforcement materials and a demonstration of rapid process cycle times projected are necessary elements in the development and demonstration of the crossmembers feasibility.

From a research standpoint, the crossmember had some key features which made it an ideal demonstration component. The part was complex in geometry, included foam cores, molded in metal inserts, and required highly stressed attachment points. Installation to a vehicle could be accomplished by bolting in the assembly making the component essentially self-contained. The complex geometry also provided realistic challenges in preforming, mold design and mold resin filling. Most importantly, the size and complexity combined with the high performance requirements were judged to provide sufficient difficulty to provide a credible demonstration of both the preforming process and ultimately the HSRTM process. An additional benefit not perceptible at the onset of the program was the coincidence of the crossmember program with one of the likely senarios for the implementation of composites in vehicles, i.e., low volume, unique performance modules with minimal surface finish and crash energy management implications.

The crossmember program was, therefore, designed to address two major open issues, the demonstration of automotive production rates and to prove feasibility of complex preforming. Investigation of both issues are ongoing. This paper will focus on the progress to date in preforming.

Desired Attributes in Preforming

Experience gained to date in a number of automotive structural composite programs[3,4] has resulted in a generalized set of attributes likely to be required of a preform process. Of primary importance is the capability to fabricate preforms to net shape. This must include a minimum of excess bulk in preform mass as well as a minimum of free reinforcement strands both of which tend to interfere with loading and closing of a tool.

Any preform to be production viable must be easily handled manually or robotically. It should remain stable during storage at normal production plant conditions, including elevated temperature and humidity. Preforms currently available generally have a low tolerance to abrasion during handling which results in a loosening of the preform structure and subsequent difficulty in loading the preform into the tool.

Economic considerations are critical in any envisioned composite application. Low labor content, rapid cycle times and low material waste are all key features of an economical preform process. Ultimately there must be an optimization of cost versus performance of the composite structure which results from the preform. Such an optimization dictates that a preform process must be capable of utilizing the anisotropy of the reinforcement as well as varying thickness where appropriate. It is likely that any process will need the ability to use both random and directional reinforcement orientations to achieve maximum performance at minimum cost. While the optimization of cost versus performance is ultimately going to be a key requirement, there are near-term applications where some of the advantages of composite materials can be applied using less than optimal preforms.

The crossmember which is the subject of this paper uses a stamping process which could likely be further optimized; but it is important to get some production experience at this time. Future optimization may further improve the benefits of a composite structure, although some benefits such as reduced weight are available now. Several additional preform process attributes are likely to be needed if future preforms are to be fully optimized.

Predictive tools for the design of preforms and the relationship of the preform structure to performance do not yet exist. Computer codes which go well beyond classical laminate analysis will be needed and must be developed to interface with other design codes used in part design. Outputs from such codes must help define fiber orientation in three planes, performance, preform shape or in the case of stamping, blank shape, and binder design.

Additional efforts are also needed to address connectivity between separate preform sections. Today a laminate approach, i.e., patches over seams, is used to give the preform structurally sound joints. Future preforms would ideally be applied as a unified form to the core or directly into the mold. If multiple pieces are used, a faster more reliable joining technique will be required.

Cut and Sew Glass Preform Fabrication

Cut and sew techniques were used to fabricate preforms during the preliminary phase of the crossmember program. Patterns for orienting and cutting the glass materials to conform to the shape of the part are developed using trial and error hand-shaping methods. The preform laminate construction consists of several different types of glass materials. A complete set of precut to size glass material panels and patches (55 individual pieces) are shown in Figure 1. The prototype preform includes seven strategically placed patches consisting of two layers of (+-+ 45) balanced double bias fabric in the highly stressed areas of the component. Local patches applied to the foam core are shown in Figure 2. The entire center section of the preform is covered with four layers of (+-+ 45) balanced double bias material. Sandwiched between these layers are seven layers of 1.5 ounce random chopped mat material covering the entire part with the exception of the shock towers and frame rail pockets. These areas are covered with eight layers of (+-+ 45) balanced double bias fabric.

Build-up of the glass preform begins with a light sanding of the foam core to remove any sharp edges and to roughen the core surface for improved adhesion to the laminate. Loose particles are blown away and glass application to the core is begun. Glass material is pulled tightly over the foam core and stapled in place. A complete layer of (+-+ 45) balanced double bias material is applied to the center section of the foam core using overlap and butt joints as required. Next seven reinforcement patches are installed, followed by another complete layer of (+-+ 45) material in the center area of the part. Two layers of the (+-+45) are applied to the shock towers. This procedure is repeated to apply additional layers of glass until the preform is complete. A typical arrangement of glass laminate material with butt joints is shown in Figure 3.

The lower suspension control arm attachment areas or "eyes", are fabricated by wrapping Knytex Promat 0/90/Rc fabric onto aluminum mandrels and attaching them to the preform using tabs.

Reinforcement unidirectional glass "straps" are applied to the preform to provide additional reinforcement in the eye attachment area as shown in Figure 4. Following the attachment of the lower control arm eyes, the glass preform is then ready for molding. The Cut and Sew preform assembly method is labor intensive and requires several hours to complete. This method is effective for producing prototype parts for demonstration and testing programs, but does not lend itself to the high volume production levels ultimately required for automotive production. A more automated preforming method will be needed.

Automated Glass Preform Fabrication

Stamping is a mature, highly automated production technique used in producing high volume parts for the auto industry. Ford Motor Company Research has been working with Owens/Corning Fiberglas to developed a technique for stamping thermoformable continuous strand mat (OCF M8608-X5) into a complex shape for use in the c/m glass preform.

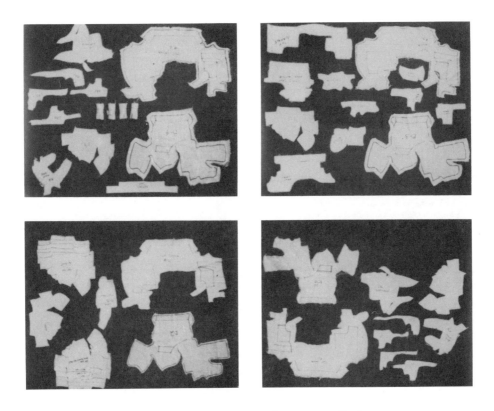

Figure 1 - Cut and sew complete set of glass
panels and patches.

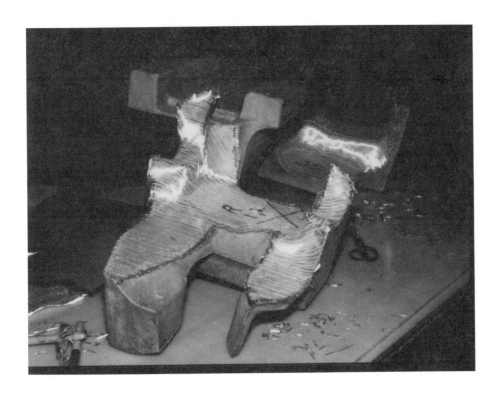

Figure 2 - Reinforcement patches on foam core.

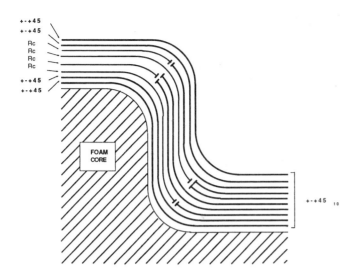

Figure 3 - Cut and sew typical arrangement of glass laminate materials with butt joints.

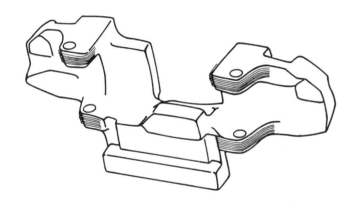

Figure 4 - Reinforcement unidirectional glass "straps" for eye hole attachment area.

Using this material combined with other materials, such as engineered continuous fabric and braided materials, an automated approach was developed for fabricating the glass preforms.

As in the cut and sew method the structural glass preform laminate surrounds a molded polyurethane foam core. The core is used as a tooling aid to achieve the parts three dimensional shape, as well as, a mechanism to hold metal attachment plates in their design location. The thickness of the laminate structure varies with respect to the anticipated stress levels in the particular locations of the part. Three thicknesses were selected; 7.5, 9.0, and 10.5 millimeters. The shaped preform clam shells are designed to have a laminate thickness of 7.5 mm. The thicker sections of the laminate are achieved by

attaching glass reinforcement patches to the foam core, as previously shown in Figure 1.

The matched Hi-Temp epoxy shaping tools, Figure 5, were designed to form four layers of 1.5 oz/sq ft continuous strand thermoformable mat (R_sf) to a 7.5 mm thickness. One additional arrangement of glass materials containing three layers of (R_sf) and one layer of 2.91 oz/sq ft balanced double bias (+-+ 45) fabric, was also successfully preformed. In future programs we plan to form the entire laminate arrangement in one stamping, which consist of eight layers of R_sf and two layers of +-+ 45 or a similar engineered fabric.

In the automated preforming technique, the sequence of events is as follows. The shaping tool for the preform top clam shell is placed in a press having a large platen area. Behind the press is a large Calrod oven for heating the stacked layers of material. Four layers of (Rsf) are placed in the metal holding frame and clamped in place. Holes are cut out in each of the four corners of the assembled layers of material, Figure 6, to provide clearance for the tool's alignment posts. Cutting clearance holes is time consuming and the holes act as defects during the forming process. Future tools will have removable alignment posts to eliminate the need for cut outs.

The frame holding the mats is lifted onto a track that runs between the opened tool halves and extends into the Calrod heating oven. The thermoplastic binder on the (R_sf) mat becomes molten in the ovens at approximately 300 F. The frame holding the heated mat is quickly pulled out of the ovens and positioned between the forming tools. The press rapidly lowers the upper tool onto the mat then slowly draws and stretches the mat into the cavity. The mat is held under pressure as the binder solidifies in the cooled tools. The tool is opened and the shaped mat is cut out of the holding frame, as shown in Figure 7. The same procedure is used to produce the bottom clam shell preform. Production forming tools may require steel rule cutting blades to cut the formed shell to net size. This would speed up the entire forming and trimming process.

Crossmember suspension attachment points "eye" hole preforms will be discussed next. The initial eye preforms were produced by hand wrapping 0/90/Rc multi layer engineered material onto an aluminum mandrel. Attachment tabs were designed into the cutting pattern to provide for attachment of the eye to the preform body, as shown in Figure 8. Wrapping could be accomplished using automated winding devices for production.

A second approach to attachment eye preform design utilizes a braided glass configuration and a bobbin as seen in Figure 9. Figure 10 shows a schematic of the braiding paths followed in fabricating the eye. Excess material is allowed

Figure 5 - Crossmember glass preform Hi-Temp
epoxy shaping tools.

Figure 6 - Assembled layers of glass mat with
clearance hole cut outs being loaded
on transfer track.

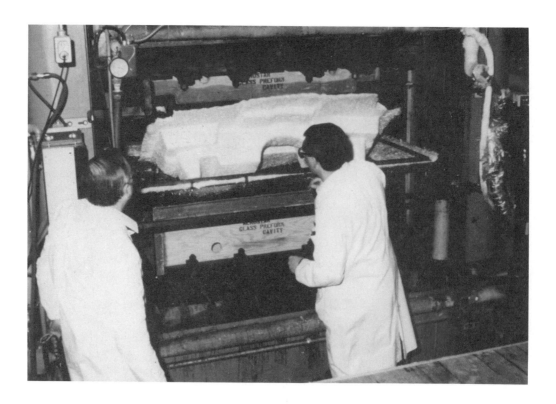

Figure 7 - Shaped mat is cut out of holding frame.

Figure 8 - Wrapping an "eye" preform with Promat 0/90/0 fabric.

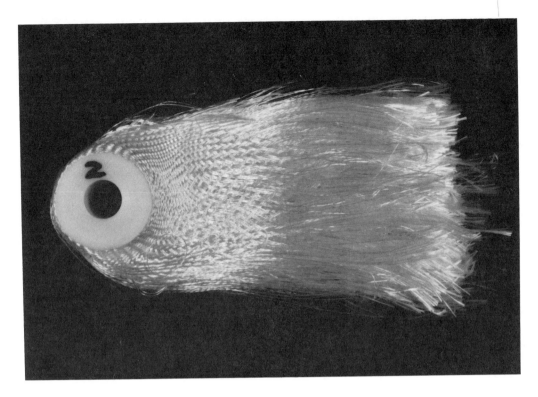

Figure 9 - Braided glass attachment "eye"
preform.

to extend over the end of the eye to serve as the
attachment to the main body of the preform
similar to the function of the tabs in the
previous design. Several bobbins could be set in
place on the same braiding machine and be braided
over simultaneously to enhance the speed of the
operation for production. This concept is shown
in Figure 11. Ford Motor Co. and Drexel
University have an ongoing research program which
is aimed at optimizing this process.

Stamped Preform Assembly

The preform assembly starts as it did in the cut
and sew method with a light sanding of the foam
core for better laminate adhesion. Patches are
applied to the foam core in their design
locations with the use of metal staples. The top
and bottom stamped preform shells containing four
layers of R/$_s$f material each are fitted to the
foam core and trimmed to butt join to each other
Figure 12. A second set of patches are applied to
the preform. A layer of balanced (+-+ 45) is
hand shaped and trimmed to cover the entire part.
A second set of stamped preform shells are
installed. Once again the entire preform surface
is covered with a layer of balanced (+-+ 45)
fabric. The 7.5 mm thick portion of the
crossmember laminate structure is shown in Figure
13. The eye preforms are fitted to the preform
and attached by stapling the attachment tabs to
the preform. Reinforcement straps are applied
around the eye attachment areas. The preform is
now completed and ready for molding.

Stamped Preform Optimization

Further optimization of the stamped preform
process was attempted in subsequent prototype
preform trials. An additional reduction in
preforming labor was achieved through
incorporation of the balanced (+-+ 45) fabric
with the R$_s$f material in a single stamping. This
eliminated the hand fitting and shaping fabric to
the preform. Since this material is not coated
with the thermoplastic binder it is placed
between layers of the R$_s$f to help in retaining
the stamped shape. In this design, to achieve
the proper strength level in the part, two upper
and lower shells are assembled plus an additional
two ply upper and lower shell of the Rsf
material. The stamped 7.5 mm laminate layer for
this design level is shown in Figure 14. Since
the (+-+ 45) material in the stamped version of
the preform laminate structure is not continuous
over the butt joint, a two inch reinforcement
strip of(+- 45) is installed over the entire
length of the butt joints of two upper and lower
clam shells. As shown in Figure 15, there are
considerably fewer parts (20 individual pieces)
to assemble in this preform build than in our
earlier builds. Figure 15,16,17 show the planned
evolution of reducing the part count to 6
individual pieces in the assembly of the
composite crossmember preform using a stamping
approach.

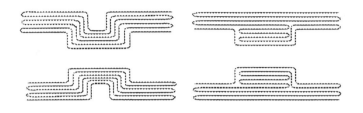

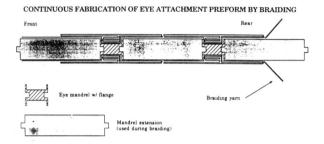

CONTINUOUS FABRICATION OF EYE ATTACHMENT PREFORM BY BRAIDING

Figure 10 - Cross-sectional views of possible braiding paths.

Figure 11 - Continuous fabrication of eye attachment preforms by braiding.

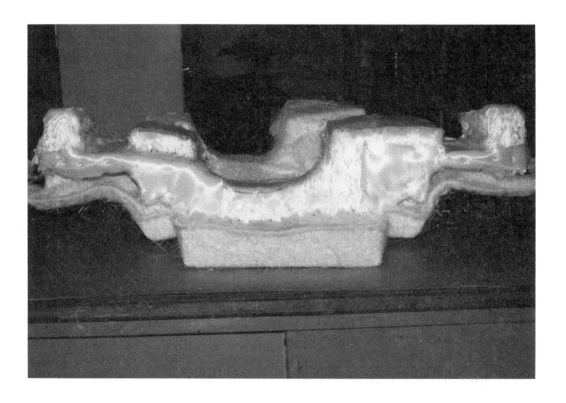

Figure 12 - Stamped preform clam shells being assembled.

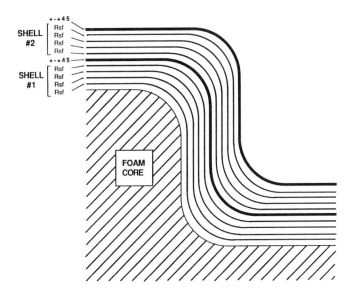

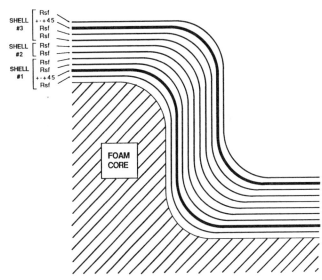

Figure 13 - Stamped preform clam shell laminate structure.

Figure 14 - Consolidated (+-+ 45) fabric in stamped clam shell laminate structure.

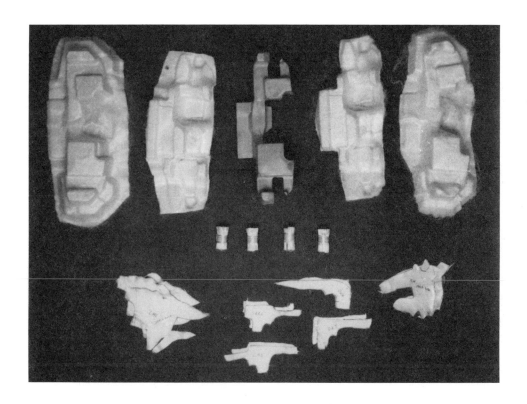

Figure 15 - Phase I complete set stamped preform material.

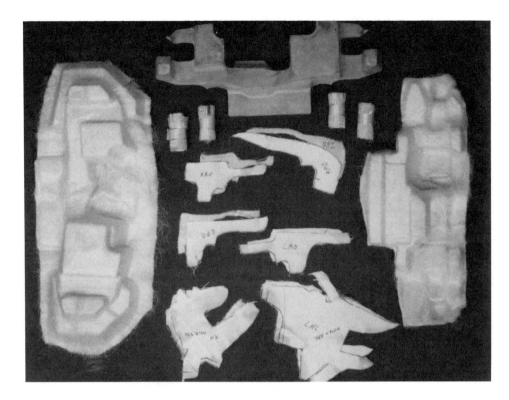

Figure 16 - Phase II complete set stamped preform
material.

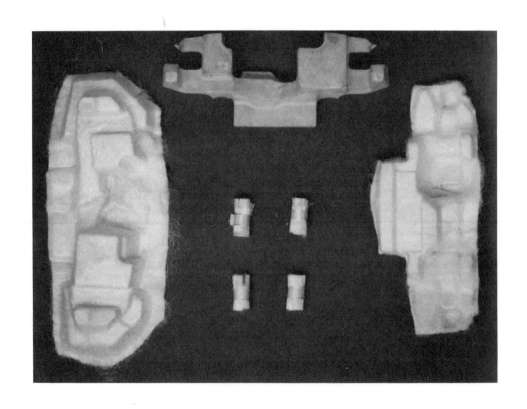

Figure 17 - Phase III complete set stamped
preform material.

Figure 18 - Assembled preform with braid
material.

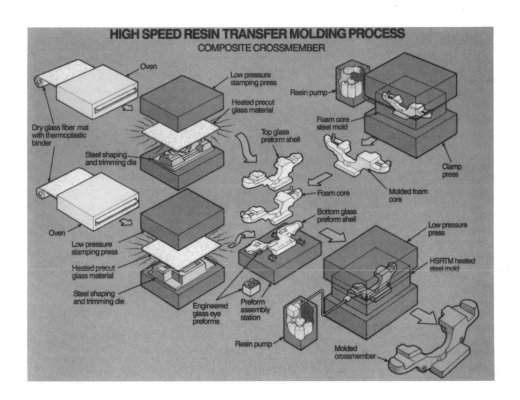

Figure 19 - High speed resin transfer molding
process for the composite
crossmember.

Braided Crossmember Preform

As part of the braiding program at Drexel University, a more unified approach to preforming the cross member through the use of braiding was also evaluated. "E" glass yarn was braided over a modified crossmember shaped mandrel in an effort to replace the engineered glass fabric used in previous preform builds. A fundamental problem which required solution was the inability to braid the glass yarn into the concave surfaces of the part. Using Lundy computer design techniques a convex surface design of the part was generated for the braiding mandrel. Collapsible wire mesh convex panels were installed over the concave areas of the mandrel. After the braiding process, the braided preform was treated with a binder material to hold it in its formed state, a cut along one edge of the braid was made for removal from the braiding mandrel.

The installation of the braid in the preform build was accomplished with a great deal of difficulty. In the preform build using the braided material, the assembly sequence was the same as shown in figure 13, except the(+-+ 45) fabric was replaced with the braid. The assembled preform with braid is shown in figure 18.

HSRTM for the Aerostar Crossmember

The high speed resin transfer process (HSRTM) has been previously described by Johnson, et. al.[Ref.] The process drawing presented in Figure 19 shows the HSRTM process as it is applied specifically to the Aerostar crossmember. Reinforcement material in pre-oriented multi-layer sheet form, pretreated with a thermoplastic binder system (4% by weight is typical) enters the process. A blank is cut from a roll of preform material and clamped in a binder frame. The blank is then preheated to soften the binder after which it is rapidly transferred to a set of stamping or forming tools and pressed to shape. The binder solidifies upon cooling and acts to retain the desired preform shape. Trimming is ideally accomplished in the stamping operation and the preform section is then transferred to the assembly area. Other stamping operations form additional features of the preform which are likewise transferred to the assembly area.

Attachments for highly loaded areas of the component will typically require a more fully engineered attachment than is attainable with a process such as stamping. In this case form attachment "eyes" are prebraided or prewound from fabric and come to the process as subassemblies.

Foam cores, which often are not required for their structural contribution, provide location for molded in attachments and serve as a tool core for the attachment of preform sections. In the schematic a separate foam production line produces the crossmember foam and delivers it to the assembly station.

All components of the preform are next assembled and the completed preform is transferred to the HSRTM molding tool. Once the preform is in place, the tool is closed and resin is injected rapidly but at relatively low pressures (50-200 psi). The tool is heated to accelerate the resin cure. Following cure the component is demolded, trimmed if required, and sent on for inspection and subsequent machining.

While certain portions of this process have been proven out in previous molding and preforming trials, the process as a whole is still largely conceptual and there are not as yet demonstratable cycle times for the entire process. It is critical for the acceptance of this technology by the auto industry that cycle time be realistically demonstrated and that a capable supply base be developed for the technology.

Conclusions

During the 10 molding trials conducted to date, preforming of the crossmember has been optimized a great deal. Stamping of a preform may not be the optimum long-term approach to preform fabrication but near-term it appears to be the best available technology.

Blank optimization to reduce scrap and further development of more conformable continuous materials is key to the success of this approach. Minimization of scrap and the reuse of scrap reinforcement in another product such as BMC (bulk molding compound) will improve the economics of the stamped preform approach.

Components such as the Aerostar crossmember which are modules with high functionality could hold the key to the penetration of the HSRTM process into automotive applications.

References

1. "Picking The Niches", J. Lowell, Wards
    Auto World, Feb. 1988, p.32-36.

2. Johnson, C.F., R.A. Jeryan, N.G. Chavka,
    C.J. Morris, and D.A. Babbington, "Design
    and Fabrication of a HSRTM Crossmember
    Module", in Proceedings of the Advanced
    Composite III Conference of ASM
    International, Paper 8707-
    014,(1987).

3. Johnson, C.F., N.G. Chavka and R.A.
    Jeryan, "Resin Transfer Molding of Complex
    Automotive Structures", in Proceedings of
    the 41st Annual Conference of the Society
    of the Plastics Industry, Paper 12-A
    (1986).

4. Chavka, N.G. and C.F. Johnson,"A Composite
    Rear Floor Pan",in Proceedings of the 40th
    Annual Conference of the Society of the
    Plastics Industry, Paper 14-D (1985).

5. Johnson, C.F. "Rapid Manufacturing
    Processes for Integrated Automotive
    Structures", in Proceedings of the First
    Advanced Composites Conference, ASM
    International,pp.95-99 (1985).

3-D KNITTED PREFORMS FOR STRUCTURAL REACTION INJECTION MOULDING (S.R.I.M.)

G. T. Hickman, D. J. Williams
Courtaulds Research
Coventry, England

ABSTRACT

Using advanced computer controlled knitting techniques pioneered by Courtaulds, 3-D fibre preforms suitable for Structural Reaction Injection Moulding (S.R.I.M.) can be produced in an automated and highly reproducible manner. Development of these techniques over the last two years has allowed glass, aramid and carbon fibres to be processed successfully. The technique is very versatile and a wide variety of 3-D shapes is possible.

For example, a flanged 'T'-junction preform in glass fibre has been evaluated as a concept component in a Courtaulds S.R.I.M. research programme. Specific preforms have been developed for applications in the automotive, aerospace and general engineering industries.

The advantages of these knitted 3-D preforms are reproducibility, high production rates, fibre continuity throughout the structure, zero waste and the ability to mix fibres easily.

THE RESIN INJECTION MOULDING PROCESS has considerable potential for the low cost volume production of a diverse range of composite structures. The process offers short demould times for composite components from relatively low cost raw materials and cooling. However, the technique has made limited progress for the volume production of structural composite components, partly because of resin system limitations, but mainly because the mould assembly procedures for the reinforcing fibres are time consuming, wasteful and do not allow the realisation of the full design requirements for structural components.

Significant progress has been made by resin manufacturers in developing resin systems to enable the injection process to produce, in extremely fast demould times, composite components with optimum resin matrix properties. By comparison, a limited amount of development has been undertaken in the reinforcing fibre area where there is a need to develop 3-D fibre preforms to give structurally acceptable composite components.

Such 3-D fibre preforms will almost certainly include the use of woven, random mat, unidirectional and knitted materials, or hybrids of them. The design of acceptable load-bearing composite structures from RIM depends on the development of such preforms. This paper concerns the development by Courtaulds of knitted 3-D fibre preforms for SRIM.

THE KNITTING TECHNOLOGY

Advanced computer-controlled presser foot knitting techniques, pioneered and patented by Courtaulds, have formed the basis of the development of 3-D preforms in glass, Kevlar and carbon fibres. Presser foot technology enables the fibre tensions developed during knitting to be maintained uniformly across all the needles on the bed of the knitting machine. This innovation gave, for the first time on powered knitting machines, the individual needle control required for the manufacture of complex 3-dimensional shapes. With the advent of modern computer controlled machines, the full potential of presser foot technology was realised and complex 3-D forms could be produced directly at the knitting machine.

The technique now developed with glass fibres is very versatile and a wide variety of 3-D shapes is possible. The basic surface shapes which can be produced include:

* plane, flat surfaces
* cylindrical surfaces
* spherical surfaces
* conical surfaces

Several of these basic surface shapes can be integrated into one structure with the technique, highlighting the potential for significant parts integration in 3-D preforms for SRIM. For example, the integration of a plane surface with a cylindrical surface produces an integrally flanged tube. Similarly, other 3-D shapes can be produced such as tube junctions, integrally flanged tube junctions, large fully integrated flanged tube preforms and integrally flanged T-junctions.

USE OF THE KNITTING TECHNOLOGY

CONCEPT RTM PARTS

T-Joint - This flanged T-junction preform in glass fibre has been evaluated as a concept part in an internal Courtaulds 3-D preform moulding programme. Using an RTM technique with low cost GRP tooling, good quality flanged T-junction composites have been produced in a variety of resin systems (e.g. polyester, epoxy, phenolic and methacrylate materials) at fibre volumes up to 35%. Evaluation of structural properties has centred on the comparison of random mat and knitted T-joint components. Preliminary test data indicates that the knitted components are superior to random mat materials. The test rig is shown in Figure 1. The loads are applied via aluminium mandrels. A typical stress/strain curve resulting from this test is shown in Figure 2. It is typical of a random mat composite showing classical brittle fracture. However, the knitted preform composite does not show conventional composite fracture behaviour in that the component is capable of carrying loads after a clearly defined yield point which is absent from the random mat curves (Figure 3). The integrity of the knitted structure is maintained following test. The results are summarised in Tables 1 and 2 which show the better reproducibility and higher energy of destruction characteristic of the knitted preform. This "fail-safe" behaviour has important implications for the automotive industries.

Flanged Box - A flanged box part is also under evaluation. The knitted glass preform is fully fashioned and has no discontinuities either at the corners or at the changes of section. In particular, the flange itself is continuous around the box perimeter. A composite part from this preform, which has been partially rigidised prior to resin injection, has been produced successfully by RTM techniques.

DEVELOPMENT APPLICATIONS

K-Joint - An application showing the jointing capability of integrally knitted preforms is the K-joint. This joint is to be used in antenna mast constructions where pultruded tubular sections are connected by appropriate angled K-joints. The capability of producing an all composite mast structure will thus be demonstrated.

Structural Beam - Complex tubular preforms have been produced, e.g. a support beam. The beam changes cross section throughout its length; starting as a rectangular section, it changes progressively into a triangular section, an elliptical section and finally returns to a rectangular section. No fibre discontinuities are present in the structure. Such continuity eliminates any problems of delamination in the composite.

Others - There is an expanding applications programme for knitted 3-D preforms in SRIM developments in the automotive, aerospace and general engineering industries. Automotive applications for which prototype preforms have been produced include pick-up box structures, hoods, bumpers and tailgates.

ADVANTAGES OF KNITTED PREFORMS

Such components have highlighted a number of advantages of knitted 3-D preforms for the SRIM technique:

*   Reproducibility - There is no variation from preform to preform when produced on a computer-controlled knitting machine and, therefore, no variation in distribution of the reinforcing fibres when placed in the mould, i.e. component consistency is assured.
*   High Production Rates - A fully automatic knitting machine is capable of producing large numbers of complex 3-D preforms cost effectively in realistic time scales. Thus, high volume production of composites by SRIM would not be restricted by time consuming fibre assembly techniques.
*   Fibre Continuity - With knitted 3-D preforms, complex integral structures can be produced without fibre discontinuities at important changes of section - the fibre "flows" throughout the structure.
*   No Waste - Preforms are produced at shape so that assembly directly into the mould without cutting or trimming (even with complex shapes) could be achieved. Additionally, little or no post-trimming operation should be required.
*   Ease of Handling - The fully shaped nature of the preforms allows quick and easy handling by operatives with consequent improvements in overall demould times. In the future, robotic handling of the preforms appears feasible.
*   Special Features - A variety of special features can be built into the preforms, e.g. "button holes" to provide integral ready-made bolt or fixing holes, with subsequent reduction in composite machining operations. (This could be particularly advantageous for Kevlar composites.)
*   Mixed Fibres - Hybrid preforms containing a mixture of reinforcing fibres are possible. Positioning of the various fibres can be specific, including the introduction of unidirectional fibres, allowing structural engineering requirements to be rationalised with fibre costs.

CONCLUSIONS

Exploitation of SRIM technology is still hindered by the absence of a comprehensive preform technology. Knitting technology goes a considerable way towards solving this problem and in many cases it is sufficient in its own right. It is accepted that it may need to be combined with some other preform technologies so as to meet all structural engineering requirements.

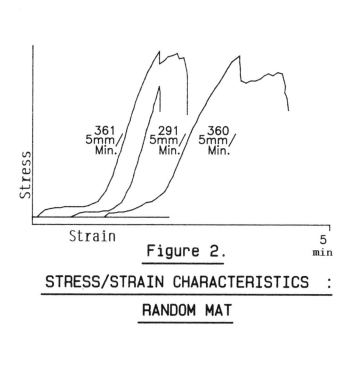

Figure 2.

STRESS/STRAIN CHARACTERISTICS :

RANDOM MAT

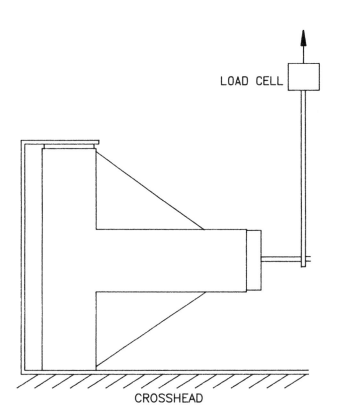

Figure 1 : T-JOINT TEST ASSEMBLY

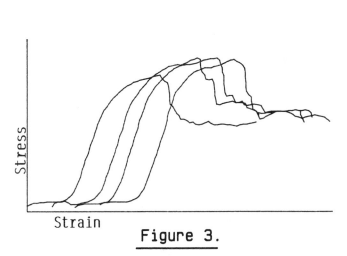

Figure 3.

STRESS/STRAIN CHARACTERISTICS :

KNITTED PREFORM

Table 1 : BREAKING LOAD T-JOINTS

Reinforcement Type	Sample Ref.	Breaking Load/N	Statistical Data
Knitted Preform	362	724.0	Standard Deviation
	363	645.5	σn = 38.1 N.
	367	717.0	Coefficient of variation
	369	723.5	= 5.4 %
	Mean = 702.5 N		
Random Mat	291	434.0	Standard Deviation
	358	680.5	σn = 101.4 N.
	360	532.0	Coefficient of variation
	361	542.5	= 18.5 %
	Mean = 547.25 N		

Table 2 : ENERGY OF DESTRUCTION T-JOINTS

Reinforcement Type	Sample Ref.	Energy of Destruction J	Statistical Data
Knitted Preform	362	3.03	Standard Deviation
	363	2.46	σn = 0.24 J
	367	2.42	Coefficient of variation
	369	2.72	= 9 %
	Mean = 2.67		
Random Mat	291	0.58	Standard Deviation
	292	1.68	σn = 0.44 J
	360	1.60	Coefficient of variation
	361	1.14	= 35 %
	Mean = 1.25		

370

A PROCESSING SCIENCE MODEL
FOR THREE DIMENSIONAL BRAIDING

Christopher M. Pastore, Frank K. Ko
Drexel University
Dept. of Materials Engineering
Philadelphia, Pennsylvania USA

ABSTRACT

This paper demonstrates a unified approach for the design, analysis, and manufacturing of three dimensional braids for structural composites. Combining a topological model with geometric and mechanical models a process-structure-properties relationship has been established for 3-D braid reinforced composites to form a communication link between structural design engineers and textile materials engineers.

ADVANCED STRUCTURAL TEXTILES for composites are finding their way from the aerospace industry to the high volume, low cost industrial marketplace due to the excellent performance, increased productivity and availability, and reduced labor compared with traditional hand lay-up laminates. In order to become fully accepted and utilized by the industry it is necessary to establish well founded scientific bases for these materials backed by a good understanding of the processing of the preform. This type of scientific modelling of the textile material provides the necessary analytical tools to construct a communications link between the structural designer and the preform fabricator.

The application of textile structures for industrial applications is certainly not new. It has been demonstrated time and again in recent history that the identification of industrial demands backed by new analytical tools, new approaches, and new materials tend to stimulate the development of new technologies and in the textile community. The pioneering work of Peirce [1] in the 30's and 40's on the geometric modelling of woven cloth was a timely response providing the foundation for a smooth transition from a natural fiber based industry to a man-made fiber based industry. The work by Hearle, Grosberg, and Backer [2], Hamburger [3], and Platt [4], on textile structural mechanics

provided an analytical basis for the analysis and use of textiles as engineering structures for military applications. The entrance into the space age in the 60's and 70's stimulated the rediscovery of textile structures and fostered the development of machines to construct multiaxial structures (such as the triaxial weave) as well as recognizing the importance of 3-D fabrics for aerospace applications. The birth of the computer age in the 70's and 80's promises the necessary tools, as demonstrated by Konopasek [5], for the design and analysis of complex structures such as textiles. The explosive growth of high technology in the 80's, supported by the development of new fibrous materials and powerful computers has brought about the possibility of computer integrated manufacturing and is pushing toward the development of artificial intelligence and expert systems for the 1990's and beyond.

In this paper, a textile system for advanced composite structures, the three dimensional braid, is presented as an example of the processing science of advanced textile materials. The development of the processing science base for the 3-D braid is outlined in terms of the structural geometry, mechanical properties, and processing. Using these elements, the groundwork is laid upon which to build computer integrated manufacturing (CIM) systems and later expert systems to translate structural design requirements into textile processing control.

THREE DIMENSIONAL BRAIDING

Three dimensional braiding is an extension of two dimensional braiding technology in which the fabric is constructed by the intertwining or orthogonal interlacing of two or more yarns to form an integral structure. Well-known examples of three dimensional braids are the diagonal, or packing, braids that are produced by the intertwining of three or more groups of yarns in a square arrangement of horn gears as shown in Figure 1. Instead of moving in a continuous maypole fashion, as does the square braider, these 3-

D braiding methods move the carrier in a sequential, discrete manner, which is quite suitable for adaptation to computer control. The basic loom set up for 3-D braiding is shown in Figure 2.

As detailed by Ko [6] the 3-D braiding system can produce thin and thick structures in a wide variety of complex structural shapes such as I-beams and hat sections. By proper selection of the yarn bundle sizes, the dimension of these structures can be as thick as desired. Fiber orientations can be chosen, and 0° longitudinal reinforcements can be added as desired. Extensive analytical research has been done in this area, and comprehensive models have been developed relating final shape to manufacturing processes.

STRUCTURAL GEOMETRY OF 3-D BRAID FABRICS

The development of a processing science base for 3-D braided fabrics for composites consists of two basic components: quantification of fabric geometry, and the determination of fiber volume fraction. With these components, and a knowledge of fiber and matrix properties, a composite preform can be formed to meet structural design specifications. The mechanical analysis of the composite depends upon the fabric properties that can be quantified using the properties, architecture, orientation, and volume fraction of the fiber. A description of the development of constitutive equations to relate these parameters to the actual fabrication of the fibrous network follows.

To establish a geometric model and method for analyzing the properties of the 3-D braid, it is necessary to identify the orientation of the yarns in the structure. This is accomplished by identifying a macroscopic unit cell.

The height, width, and thickness of the unit cell can be represented by the parameters w, v, and u, respectively. From trigonometric relationships of the braid, w can be given as:

$$w = (u^2 + v^2)^{1/2} / \tan(\theta) \qquad (1)$$

where θ = angle of inclination of the yarn,
and w = height of the unit cell.

From the above relationship, the pick spacing necessary to produce a desired fiber orientation, θ, can be determined.

Because composite structures are usually made according to a predetermined fiber orientation and volume fraction, the ability to predict the fiber volume fraction in a 3-D braided composite structures is necessary. In making a 3-D braid with a given fiber volume fraction, the volume fraction of fiber can be defined as:

$$V_f = V_y / V_c \qquad (2)$$

where
V_y is the volume of the yarn
and V_c is the volume of the composite.

This can be rewritten as

$$V_f = N_y L_y A_y / (L_c A_c) \qquad (3)$$

where
N_y = the total number of yarns in the fabric,
L_y = the length of each yarn
A_y = the cross-sectional area of a yarn
L_c = the length of the composite
and A_c = the cross-sectional area of the composite.

From geometry, the following relationship can be established:

$$w = v / \tan(\theta') \qquad (4)$$

where
θ' = surface angle of the braiding yarn.

Combining this with the previous definition of θ, suitable manipulation of the equation gives

$$\theta = \tan^{-1}((1 + k^2)\tan(\theta')/k) \qquad (5)$$

where
k = ratio of track to column movement = u / v

Incorporating these identities into our original relationship, one can determine the total number of yarns required to make a 3-D fabric with a given fiber volume fraction:

$$N_y = V_f A_c \rho\, 9 \times 10^5 \cos(\theta)/D_y \qquad (6)$$

Using this equation one can easily determine the total number of yarns required to make a fabric with a given fiber volume fraction and cross-sectional area if the parameters of fiber density, yarn linear density, and yarn surface angle are known.

The volume fraction attainable with a given construction has a maximum that is dependent on the fiber architecture. With fibers of circular cross-section, the maximum attainable value of V_f in a uniaxially aligned fibrous network is $\pi / 2\sqrt{3}$, ($\sim$ 0.906). It can be shown that the maximum fiber volume fraction of a 3-D braid unit cell with circular fiber cross-sections is $\pi \sqrt{3/8}$ ($\sim$ 0.6801). This means that the maximum fiber volume fraction attainable without distortion is about 68%.

MECHANICS OF 3-D BRAID COMPOSITES

The mechanical properties of the 3-D braid fabric reinforced composites can be predicted with a knowledge of the fiber properties, matrix properties and fiber architecture through a modified laminate theory approach. A geometric unit cell defining the fabric structure (or fiber architecture) can be identified as shown in Figure 3. For a general 3-D braid fabric there are three basic yarn components to the unit cell, defined according to the yarn orientation: 0° (longitudinal), 90° (transverse), and $\pm\theta$, the braiding yarns. The fractional volume of fiber in each of the directions can be calculated geometrically based upon yarn size, braid angle and braid pattern.

With this information a 6 x 6 stiffness matrix can be formed for each system of yarns using the stiffness matrix of a comparable unidirectional composite and transforming it by the appropriate fiber orientation [7]. This stiffness matrix relates applied strains to the corresponding stresses. For each system of yarns, this stiffness matrix is expressed as:

$$[C_i] = [T_{\varepsilon,i}] \, [C] \, [T_{\varepsilon,i}]^{-1} \qquad (7)$$

where

$[C_i]$ = stiffness for the ith system of yarns

$[T_{\varepsilon,i}]$ = geometric strain transformation for the ith system

of yarns

and $[C]$ = stiffness matrix for a comparable unidirectional composite.

The stiffness matrix for the unidirectional composite is based upon fiber and matrix mechanical properties and fiber volume fraction using superimposition according to the rule of mixture:

$$[C] = V_f \, [C_f] + (1-V_f) \, [C_m] \qquad (8)$$

where '

$[C]$ = stiffness of unidirectional composite

V_f = fiber volume fraction

$[C_f]$ = stiffness of fiber

and $[C_m]$ = stiffness of matrix.

Then the stiffness matrices for each system of yarns are superimposed proportionately according to contributing volume:

$$[C_s] = \sum k_i [C_i] \qquad (9)$$

where $[C_s]$ = total stiffness matrix

and k_i = fractional volume of the ith system of yarns.

Thus, the stress-strain behavior of the composite can be determined as

$$\{\Delta\sigma\} = [C_s] \, \{\Delta\varepsilon\} \qquad (10)$$

where $\{\Delta\sigma\}$ = incremental stress vector (6x1)
and $\{\Delta\varepsilon\}$ = incremental strain vector (6x1)

From this, the stress vector can be determined as

$$\{\sigma\} = \{\sigma\} + \{\Delta\sigma\} \qquad (11)$$

where $\{\sigma\}$ = stress vector (6x1)

To accommodate the potential nonlinear contribution of the matrix, fiber/matrix interface and/or fabric geometry the system stiffness matrix is determined anew at each strain level. Theoretical predictions can be made from this analysis by incrementing the strain on the composite and plotting the corresponding stresses.

A failure point for the composite is determined for each system of yarns by a maximum strain energy criterion. If the strain energy on the fiber exceeds the maximum allowable, that system of yarns has failed. Mathematically, if the following expression is true, the system has failed:

$$U_{c,i} \geq U_{max} \qquad (12)$$

where $U_{max} = ||V_f(\sigma_{f,u}(\varepsilon_{f,u})\varepsilon_{f,u})/2)$

$\qquad + (1-V_f) \int_0^{\varepsilon_{f,u}} (\sigma_m(\varepsilon) \, d\varepsilon)||$

$\sigma_{f,u}$ = fiber ultimate strength

$\varepsilon_{f,u}$ = fiber ultimate strain

$U_{c,i} = \int_0^\varepsilon \sigma_{c,i}(\varepsilon_k) \, d\varepsilon$

$\qquad$ = strain energy of the ith composite system

and $\sigma_{c,i}$ = stress on the ith composite system.

Using this maximum energy criteria, a failure point for each system of yarns can be found. When a system of yarns fails, its contribution to the total system stiffness is removed. When all systems have failed the composite is said to have failed. In this way, the entire stress-strain curve for the composite can be predicted up to the point of composite failure.

COMPUTER AIDED ENGINEERING OF 3-D BRAIDS

Computer modelling has provided an efficient means to simulate the orbits associated with given braiding machines [8]. The modelling is based upon the group notation of the machine and generates a mathematical simulation of the machine process.

The software takes as input the number of tracks and columns associated with the machine under consideration. The track and column motion vectors, u_i and v_j are input graphically. The model then generates the Ψ operator for this machine. Using this operator the user can identify the orbit of any desired element. Figure 4 shows a flowchart describing the basic program flow. The program was written in Macintosh Pascal®.

When the appropriate size machine has been identified, the next screen is a movement input screen. In this mode the user identifies the desired displacements associated with each track and column. The input is graphical in nature. A thin arrow represents a displacement value of 1, a thicker arrow represents 2, thicker yet represents 3, etc. The orientation of the arrow represents the direction the displacement should take.

Since $\Psi = \prod \psi_i$, it is necessary to define the input parameters for each ψ_i. This is done by cycling through the move numbers. Each move number corresponds with the subscript i. There is an option, "Default", which generates the nth movement value based upon the previous $n-1$. This forces a "conservative" loom. A conservative loom is a loom wherein $\sum u_{i,j} = 0$ for all j (j=1, 2, 3, … ,N) summed over i=1 to M, and $\sum v_{i,j} = 0$ for all j (j=1, 2, 3, …, M) summed over i=1 to N. The restriction of a conservative loom is necessary for Ψ to be a permutation mapping of G.

When the input phase is completed the program develops the full Ψ operator. Using the track and column displacements, two matrices is generated which maps 1-1 with the machine. These matrices, X and Y have dimensions [N, M, k]. The application of the operator to the elements takes the form

$$\psi_i(\mathbf{y}_j) = \psi_i(a,b)$$
$$= (a+X_i(a,b),\ b+Y_i(a+X_i(a,b),b)) \qquad (13)$$

This sequence of points is a description of the trace of the operator Ψ. From this interpretation, the operators A_i and B_i can be defined as

$$A_i(c,d) = (c + X_i(c,d),\ d) \qquad (14)$$
and $\quad B_i(c,d) = (c,\ Y_i(c,d) + d) \qquad (15)$

Then, in order to identify the orbit of the element $\mathbf{y}_j$, the Ψ operator is employed as

$$\mathbf{y}_{j+1} = \Psi(\mathbf{y}_j) = \psi_k\ \psi_{k-1}\ \cdots\ \psi_1\ (\mathbf{y}_j) \qquad (16)$$

Thus the sequence of points $\{\mathbf{y}_i\}$ describes the orbit of the element. Figure 5 shows a typical display in the process of generating the orbit of the element (3,3) in the $\{6 \times 14;\ (1)-(2,2,-2,-2)\}$ machine. The black line indicates the trace of the orbit.

CONCLUSIONS

A processing science model has been introduced in this paper in terms of the 3-D braid for advanced composite structural applications. Providing a communications link between composites structural designers and textile materials engineers, this processing science model lays the foundation for the creation of a CAD/CAM system for 3-D braided complex structural shape composites. Specifically the linkage can be achieved through the following steps:

- structural design is translated to the necessary mechanical properties of the composite;

- mechanical properties are translated to geometric properties such as fiber volume fraction and fiber spatial orientation;

- geometric properties of the fabric as well as the geometric properties of the component are related to the process (i.e. the braiding machine) through a topological model.

Correspondingly, the three basic components of the CAD/CAM system cited above are developed in this processing science model: Topological, Geometric, and Mechanical. The topological model relates the fundamental aspects of the braiding system as a mathematical entity to the mechanism of braiding. In this way, a model for the formation of complex shaped fabrics is developed. The geometric model employs the physical properties of yarns and fabrics and correlates these with the topological model to produce a physically realistic model of the fabric, which includes physical dimensions of the fabric, fiber volume fraction and spatial orientation. The mechanical model relates the geometric properties of the fabric to the mechanical response of the composite system.

REFERENCES

[1] Peirce, F. T. (1937), *Journal of the Textile Institute*, **28**, T45.

[2] Hearle, J.W., Grosberg, P., and Backer, S., Structural Mechanics of Fibers, Yarns, and Fabrics, John Wiley and Sons, New York, 1969.

[3] Hamburger, W.J., A Technology for the Analysis, Design, and Use of Textile Structures as Engineering Materials, Edgar Marburg Lecture, ASTM, Philadelphia, PA . 1955

[4] Platt, M. M. (1950), *Textile Research Journal*, **20**, 1.

[5] Konopasek, M., "Program Package for Large Deflection Analysis of Thin Rods and

Their Assemblies", Proceedings, CAD 74: International Conference on Computers in Engineering and Building Design, London, 1974.

[6] Ko, F. K. "Braiding", p. 519, in Engineered Materials Handbook, Editors: Reinhart, T. J. et al., ASM International, 1987

[7] Whyte, D ,"On the Structure and Properties of 3-D Braided Composites", Ph. D. Thesis, Drexel University, Philadelphia, PA, June, 1986.

[8] Pastore, C.M., "A Processing Science Base for Three Dimensional Braids", Ph. D. Thesis, Drexel University, Philadelphia, PA, March, 1988.

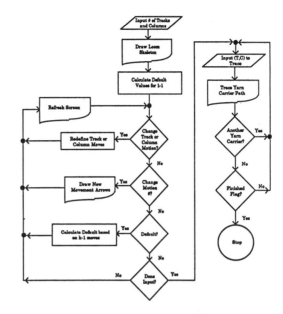

Figure 3. Program Flow Chart for 3-D Braid Simulation Based on Braid Group.

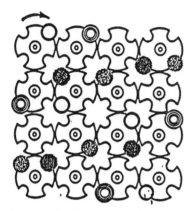

Figure 1. Schematic Illustration of Square Braider Machine.

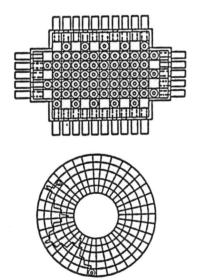

Figure 2. Schematic Illustration of 3-D Braiding Machines.

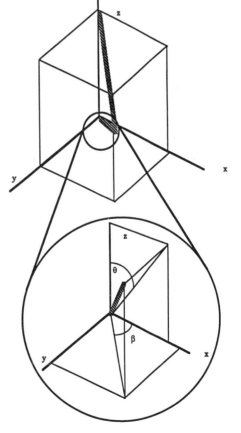

Figure 4 Spatial Orientation of a Yarn in a Three Dimensional Braid

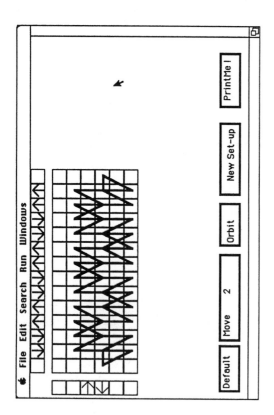

Figure 5. Computer Display of 3-D Braid Simulator.

MULTIAXIAL WARP KNIT FOR
ADVANCED COMPOSITES

Frank K. Ko, John Kutz
Drexel University
Fibrous Materials Research Lab
Philadelphia, Pennsylvania   USA

ABSTRACT

This paper introduces a new family of textile structural preforms - the Multiaxial Warp Knits (MWK) - for structural composites. To assess the process-structure relationship of the MWK fabrics, an omnidirectional tensile testing method is introduced. The tensile and flexural properties of the MWK glass/vinylester composites are also characterized and compared to woven composites.

IN THE THIRD ANNUAL Advanced Composites Conference, the need for mass production of structural composites was clearly identified by Mr. Jardon [1]. In order to make the use of structural composites in automobiles a reality, as illustrated by Johnson et al. [2], the ability to produce large integrated structures is quite necessary. It is anticipated that, through part consolidation and judicious selection of composite fabrication processes, the systems cost of composites should be competitive with steel through savings in tooling and assembly cost, providing a far superior specific performance. The key to the successful implementation of the composite fabrication processes is an understanding of the dynamics of materials-process-structure interaction. While the fibers and matrices are of basic importance in the performance of a composite, the fiber architecture of the preform plays a crucial role in the translation of the fiber properties to the composite and it dictates the fabricability of the composite. As introduced by Ko [3], there is a large family of textile preforms available for advanced composites ranging from 3-D integrated net shape structures to thin and medium gage multilayer fabrics. In addition to properties translation efficiency, structural integrity and formability, a key requirement for advanced preforms for automobiles is their availability at a reasonable fabrication cost.

One emerging family of preforms which has the potential to meet the demand for automotive structural composites is the multiaxial warp knit (MWK) fabrics. Although a comprehensive data base is not yet available for the MWK composites, several studies have been carried out to assess the potential of MWK fabrics for structural composites. The design flexibility of MWK fabrics was demonstrated by Ko et al. [4,5]. In another series of studies the structure, properties, and design of MWK fabric preforms and composites were presented [6,7]. In this study the conformability of several commercially available fiberglass MWK fabrics and the tensile and flexural properties of their vinylester composites were characterized. An attempt was made to relate the fiber architecture of these preforms to their composite properties to account for the multidirectional nature of the deformation which the fabrics experience during stamping process. A simulated omnidirectional tensile test was employed to assess the conformability or moldability of these MWK preforms.

THE MWK PREFORMS

The mechanical and physical responses of a fabric depend on the structural arrangement (or fiber architecture) of the component fibers. Fiber architecture plays a primary role in dictating the mechanical behavior of composites through its influence on fiber orientation; fiber volume fraction and fiber linearity. From the structural geometry point of view, the MWK fabric systems consist of warp (0°), weft (90°) and bias (±θ°) yarns held together by a chain or tricot stitch through the thickness of the fabric as illustrated in Figure 1. The basic distinctions of these fabrics are the linearity of the bias yarns; the number of yarn axis; and the precision of the stitching process. Examples of nonlinear (or zig-zag) MWK fabrics are the Multibar Weft Insertion Warp Knit (WIWK) such as the 4 DIR fabric produced by the J.B. Martin Company and the Malimo system, with the WIWK fabric being a more

precise structure. Another way of introducing the bias yarns is by laying in a system of linear yarns at an angle. Examples of these fabrics are the Spanply fabric produced by the Hi-Tech Composites Inc. and Kyntex's DBM fabric, with the Spanply fabric being able to maintain a more precise and linear continuous geometry in the diagonal directions. Depending on the number of guidebars available and the yarn insertion mechanism, the warp knit fabric can consist of predominantly uniaxial, biaxial, triaxial or quadraxial yarns. The latest commercial MWK fabric is produced by the Mayer Textile Machine Corporation utilizing a multiaxial magazine weft insertion mechanism. The attractive feature of this system is the precision of yarn placement with linear or nonlinear bias yarns arranged in a wide range of orientations. Furthermore, the formation of stitches is done without piercing through the reinforcing yarns at a production rate of over one meter per minute. Quite similar to the Mayer method, another West Germany development is the LIBA Copcentra machine which can produce fabrics up to six yarn layers plus one nonwoven mat (Figure 2). Unlike the MWK fabrics produced by the Mayer machine, some impalement of the linear yarns occur during the stitching process.

For this study, E-glass, MWK fabrics were obtained from the following companies:

- Advanced Textiles, Inc., Seguin, TX
- J.B. Martin Company, Inc., Leesville, SC
- King Fiberglass Corp., Arlington, VA
- Kyntex, Inc., Seguin, TX
- Mayer Textile Machine Corp., Clifton, NJ

Since the objective of this paper is not to compare the relative performance of commercial MWK fabrics, the physical and mechanical properties of the fabrics and composites were not identified with the manufacturer. The candidate MWK fabrics for this study are quite representative of the fiber architecture and manufacturing method for the state of the art MWK preforms. Common to all the fabrics are the linear reinforcing yarns and the chain stitch employed to hold the reinforcing yarn together. Varying from two yarn layers to four yarn layers, these fabrics have various stacking sequences including (0/90), (0/90/±45), (±45), and (±45,-45,+45). The (0/90) and (0/90/±45) fabrics were made by unimpaled stitching method wherein the stitching yarns wrap around the reinforcing yarns. The (±45) and (±45,-45,+45) were made by impaled stitching method wherein the stitching yarns pierce through the reinforcing yarns. Figures 3 and 4 illustrate the MWK fabrics produced by the impaled and unimpaled methods. A summary of the physical properties of the various MWK fabrics is provided in Table I. For comparison purposes a 90/0/90 woven fabric was also included in this study. Figure 5 illustrates the construction of the woven fabric.

BEHAVIOR OF MWK FABRICS

The behavior of the MWK fabrics were characterized in terms of their inherent bending stiffness and their response to multidirectional loading. The ability of the fabrics to bend under its own weight is a measure of the interaction of material property and geometric contribution to conformability. The resistance of the fabric to multidirectional loading simulates the dynamic response of the fabric assembly to the forming process such as stamping or molding.

BENDING STIFFNESS - The bending stiffness test is based on a standard textile testing method ASTM D1388 for the stiffness of fabrics. This cantilever type bend test measures the ability of fabrics to bend under its own weight. Accordingly it provides an indication of the conformability of a fabric in composite formation processes such as open mold lay-up wherein the loading on the preform is relatively small. Five, 1" wide x 10" long fabric strips were cut at +45°, -45°, 0° and 90°

directions. The fabric specimen was placed on the cantilever testing device, as shown in Table II, and slid along the length direction. The length overhang '0', when the specimen is bent to 41.5° as indicated in Table II, is recorded. The flexural rigidity or bending stiffness, G, of the fabric was calculated based on the following equation:

$$G = W \times (O/2)^3$$

where W = fabric weight unit area, mg/cm^2
O = length of fabric overhang, cm

The same procedure was repeated for the opposite side of the fabrics for each of the specimens cut in the various directions. Five replications were made for each specimen. Table II provides a summary of the bending length and the average bending stiffness of the fabrics. Obviously, the longer the fabric's overhang, the stiffer a fabric would be. It can be seen that the bending characteristics of the fabrics are directional and fabric geometry dependent. It is generally true that the direction where the fibers are oriented tend to have high resistance to bending deformation. A quasi-isotropic structure such as No. 3 has uniformly distributed but higher bending stiffness. Fabric thickness plays a significant role in bending stiffness. In general, as illustrated by fabrics No. 1 and No. 3, thicker fabrics have higher bending stiffness.

The freedom of fiber mobility also plays a significant role in the bending resistance of the fabrics. For example, fabric No. 2 is much heavier than fabrics No. 1 and No. 4, its bending stiffness is much greater than these two fabrics due to the high density of stitch yarns which tend to restrict the freedom of fiber mobility. Comparing to the woven fabric, No. 5, fabrics with similar orientation should have similar bending characteristics.

OMNIDIRECTIONAL TENSILE RESPONSE - In order to simulate the multidirectional nature of fabric deformation during stamping and forming processes such as in high speed resin transfer molding (HSRTM), the fabrics were subjected to the CBR Puncture Strength Test [8]. Developed for the measurement of the puncture strength of geotextiles, this method provides a means to measure the load- deformation characteristics of fabrics under multidirectional loading. With a well defined plunger geometry, one can create various loading conditions which the fabrics encounter in the molding processes.

The unidirectional tensile test was carried out by placing a 10" x 10" fabric in the test fixture as shown in Table III. Five specimens were tested for each fabric with a plunger speed of 50 mm/min (~2 in/m). Taking the areal density into consideration the load on the fabrics was normalized by the following equation:

Specific Stress $S = (P/A \bullet L)$ (g/tex)

where P = load, g
A = areal density, g/m^2
L = plunger perimeter, mm

The specific stress of 1 g/tex for the glass fabrics is equivalent to approximately 35.55 ksi.

The strain on the fabrics is defined as follows:

$$\varepsilon = ((\sqrt{d^2 + a^2}) - a^2)/a^2$$

where d = plunger displacement
a = gage length or distance between plunger,
boundary and jaw = 2 inches

Accordingly, the stress-strain response of the fabrics can be plotted as shown in Figure 6. For discussion purposes, the ratio of the rupture strain and stress (C.I.) were calculated and summarized in Table III. The ratio C.I. is a measure of the compliance which reflects the conformability of the fabrics and the ability of the fabric to deform without failure. From Table III, it can be seen that multilayer, multidirectional reinforcement and structural integrity are necessary conditions for conformability and processability. Although fabric No. 1 has high specific strength, there is a lack of uniform fiber coverage which resulted in low puncture elongation and therefore lowest in conformability index. Contrarily, with three layers of bias yarns and high level of stitch density, fabric No. 2 provides a well balanced performance resulting in the highest in conformability index. Likewise, the woven fabric (No. 5) has a high conformability index due to the dense fiber coverage and structural integrity provided by the large number of yarn interlacing. Fabrics No. 3 and 4, although quite different in weight and thickness, their stitch density is similar resulting in similar conformability index.

BEHAVIOR OF MWK COMPOSITES

In order to establish an objective evaluation of the fabric preforms, composite specimens were made to the same fiber volume fraction of approximately 37%. To avoid using stacking sequence favorable to some fabrics, all the composites were stacked in a quasi-isotropic manner. Obviously, a thin ply fabric such as No. 1 would require more layers (6 layers) for a final composite thickness of 150 mil. Whereas a fabric such as No. 3, which has four yarn orientations per ply, requires only three layers to make up the laminate. The preforms were impregnated with a room temperature cured vinylester resin - Deva Kane 411-45, and compression molded to 15" x 1" x 10" composites. Tensile testing was performed according to ASTM D-638. Since the fiber volume fraction and fiber orientation distribution are similar for all the composite systems, as shown in Table IV, the tensile properties of the composites are quite similar, with strength ranging from 28.9 to 36.3 ksi and modulus ranging from 2.87 to 3.42 Msi. In order to assess the strength translation efficiency from the fabric to the composite the ratio of composite strength to fabric strength was computed. As shown in Table V, the MWK composites tend to have lower strength translation efficiency than the woven composites. This can perhaps be attributed to the large linear yarn bundle size along the loading direction.

The bundle size effect is more obvious in the three point flexural test, as shown in Table VI, wherein the specimen experiences both tensile and compressive loading. The flexural modulus of fiber No. 5 was 2.43 Msi, comparing to an average of 1.7 Msi for the rest of the MWK composites.

CONCLUSIONS

As a preliminary study into the dynamic interaction of process-structure relationship, representative MWK fabrics were obtained from the industry. The wide variety of structures examined in this study illustrated the preform design options available for tailoring automotive structural composites. With the flexibility in fiber placement and high productivity (1 meter to 2.5 meters wide by 1 meter/min) the MWK fabric is an attractive preform for structural composites.

To assess the processability of the MWK preforms, an omnidirectional tensile test method was introduced along with a more traditional method for the evaluation of the bending stiffness of the preforms. The test results suggested that these methods are sufficiently sensitive to the fiber architecture and provide guidance for preform design. Although fabric thickness has a strong influence on the conformability of fabrics, the uniform fiber distribution (coverage) and sufficient level of structural integrity are necessary for moldability.

This study also demonstrated that regardless of the per ply orientation of linear yarns, similar tensile properties can be achieved with the MWK preforms. At 37% fiber volume fraction, the glass MWK composite has tensile strength and modulus of approximately 30 ksi and 3 Msi, respectively. The distinction between these composites lies only in the number of plies required for composite formation. With more yarn orientations per ply, there is an obvious saving in the labor or time required for ply lay-up. Where tension and compression are included in the deformation process, this study suggested that larger linear yarn bundles can be beneficial. Since this study is preliminary in nature, further refinement of test methods are planned. For fiber conformability assessment, the resistance of fabrics to shear deformation will be included. A more thorough characterization of the elastic constants of the composites will also be required to provide engineering design data for structural analysis.

ACKNOWLEDGMENTS

The authors gratefully acknowledge the unrestricted financial support provided by the Ford Motor Company through the Textile Structural Composites Consortium and the Drexel Industrial Internship Program. The interest that Carl Johnson and Norm Chavka of the Ford Scientific Research Laboratory have shown in this work is also greatly appreciated.

REFERENCES

[1] Jardon, A. and Costes, M., ASM Advanced Composites Conference, Sept 1987 Detroit, MI.
[2] Johnson, C.F., Chavka, N.G. and Jeryan, R.A., "Resin Transfer Molding of Complex Automotive Structures", Proceedings of the SPI/RP-CI 41st Annual Conference, 1986.

[3] Ko, F.K., "Recent Advances in Textile Structural Composites", Proceedings of Advanced Composites Conference, ASM, 1985, p. 83.

[4] Ko, F., Bruner, J., Pastore, A., Scardino, F., "Development of Multi-Bar Weft Insertion Warp Knit Fabric for Industrial Applications", ASME Paper No. 90-TEXT- 7, October 1980.

[5] Ko, F., Krauland, K. and Scardino, F., "Weft Insertion Warp Knit for Hybrid Composites", Proceedings of the 4th International Conference on Composites, 1982.

[6] Ko, F., Fang, P., and Pastore, C., "Multilayer Multidirectional Warp Knit Fabrics for Industrial Applications", Journal of Industrial Fabrics, Vol. 4, No. 2, 1985.

[7] Ko, F.K., Pastore, C.M., Yang, J.M. and Chou, T.W., "Structure and Properties of Multi-Layer, Multi-dimensional Warp Knit Fabric Reinforced Composites", Proceedings of the Third U.S.-Japan Conference on Composites, Tokyo, 1986.

[8] Drexel University GRI Test Method GS1-86.

MULTIAXIAL WARP KNIT

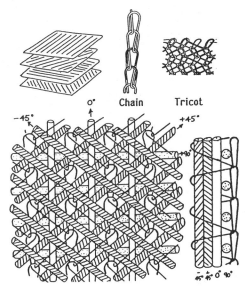

Figure 1. Structural Geometry of MWK Fabrics.

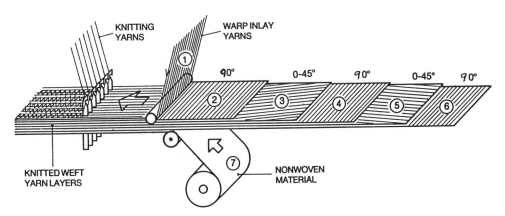

FIGURE 2. **Principle of The LIBA Multi-Axial Magazine Weft Insertion Warp Knitting Machine. Up to 6 Yarn Layers and 1 Fleece Layer are possible.**

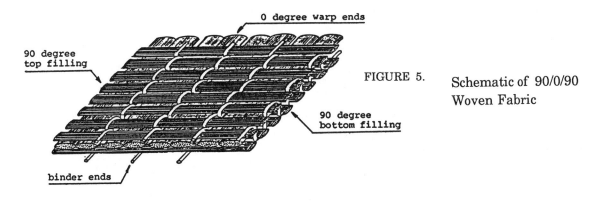

FIGURE 5. Schematic of 90/0/90 Woven Fabric

0/90 Non-impaled MMWK (20X)

0/90/±45 Impaled MMWK (20X)

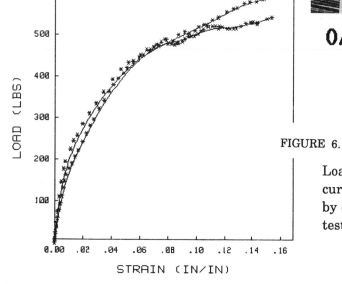

FIGURE 6.

Load-deformation curve of MWK fabric by omnidirectional test method.

382

TABLE I

FABRIC CHARACTERIZATION							
Fabric type	Construction	Thickness (in)	Weight (oz/sq.yd)	# of stitch yarns per	Yarn orientation	Yarn Linear linear density (tex)	# of lay-in yarns per in.
1. +- 45	knit	0.02	19.2	9	45	685	9
					-45	685	9
2. +45-45+45	knit	0.029	27.7	18	45	1196	7
					-45	420 (ea.)	7 (x2)
3. 0/90/+-45	knit	0.045	35.2	6	0/90	1228 (ea.)	8
					+-45	1228 (ea.)	6
4. 0/90	knit	0.025	17	7	0	1133	7
5. 90/0/90	woven	0.036	30.1	4	0	820	8
					90	730 (ea.)	10 (x2)

TABLE II

STIFFNESS OF FABRICS							
Fabric type	Fabric overhang at various orientations(cm)				ave. length of overhang(cm)	fabric wt. (gm/cm)	flex. rigidity (mg-cm)
	+45	-45	0	90			
1. +- 45	9.4	9.2	7.2	3.8	7.4	65.1	3297
2. +45-45+45	14.4	13.1	8.9	8.7	11.3	93.9	16930
3. 0/90/+-45	10.8	10.3	11.6	11	11.1	119.3	20339
4. 0/90	7.3	7.8	17.5	13.2	11.4	57.6	10672
5. 90/0/90	6.9	6.5	14.1	10.5	9.5	101.9	10921

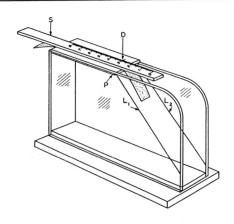

TABLE III

	Fabric type	Load (lbs)	σ_u stress(g/tex)	deflection(in)	ϵ_u strain (%)	C.I. strain/stress
	OMNIDIRECTIONAL FABRIC TEST RESULTS					
1.	+- 45	3101	31.8	0.645	5.1	0.16
2.	+45-45+45	3560	25.3	1.2	16.5	0.65
3.	0/90/+-45	4567	25.6	0.802	7.7	0.3
4.	0/90	2114	24.5	0.774	7.2	0.29
5.	90/0/90	3622	23.7	1.004	11.9	0.5

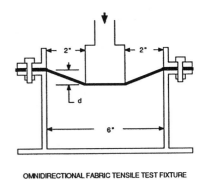

OMNIDIRECTIONAL FABRIC TENSILE TEST FIXTURE
CROSSECTION

TABLE IV

COMPOSITE TENSILE TEST RESULTS

	Fabric type	% Fiber (wt)	Stress (ksi)	E (msi)
1.	+- 45	61.4	33.8	3.21
2.	+45-45+45	60.3	30.5	3.19
3.	0/90/+-45	59.9	28.9	2.93
4.	0/90	61.6	30.6	2.87
5.	90/0/90	62.2	36.3	3.42

TABLE V

STRENGTH TRANSLATION EFFICIENCY

	Fabric type	Construction	Fabric stress (ksi)	Composite stress (ksi)	Efficiency (%)
1.	+- 45	knit	103.4	33.8	32.7
2.	+45-45+45	knit	75.8	32.7	43.1
3.	0/90/+-45	knit	81.3	28.9	35.5
4.	0/90	knit	69.6	30.55	43.9
5.	90/0/90	woven	77.1	36.3	47.1

TABLE VI

COMPOSITE FLEXURAL TEST RESULTS

	Fabric type	% Fiber (wt)	Stress (ksi)	Deflection(in)	E (msi)
1.	+- 45	61.4	37.6	1.27	1.76
2.	+45-45+45	60.3	34.7	1.29	1.66
3.	0/90/+-45	61.6	22.5	0.62	1.74
4.	0/90	59.9	28.9	1.37	1.82
5.	90/0/90	62.2	36.3	1.01	2.43

1"

.150" (nom)

6"

THREE POINT BEND FIXTURE

LOCAL BUCKLING AND MAXIMUM STRENGTHS
OF PLATE TYPE COMPOSITE COMPONENTS

H. F. Mahmood, Jian Hua Zhou
Ford Motor Company
Dearborn, Michigan   USA

ABSTRACT

Local buckling strength and material ultimate strength of composite component play an important role in the crush analysis and design of composite structures. In the crush process the relationship between the elastic buckling and the material strength dictates the crush characteristics of deep collapse and the amount of energy absorbed by the structure. The elastic buckling is mainly a function of the geometry and the stiffness of the component. The maximum strength is shown to be strongly related to the component geometry for thin walled components and directly associated with ultimate strength for thick walled components. The thin walled components exhibit a folding type mode while the thick walled components crushing type mode. In this paper a simple analytical approach to evaluate the local buckling strength and maximum strength of composite components is presented. Also various failure criterion are discussed to accurately predict the ultimate strength of the material. A computer code, COMP-CRUSH, is developed, which has the capability of calculating the above various strengths, and the elastic buckling wave length in compression.

INTRODUCTION

The importance of material selection to improve the impact performance of automotive structures is evident through increased design requirements that govern crashworthiness behavior. With the recent increased emphasis on lightweight vehicle structures, the use of composite materials in automotive structures has created the need to analyze and understand the behaviors of composite structures subjected to crash loads.

To understand the crash behavior of complex structural systems such as the vehicle, one must first acquire the knowledge of crush characteristics of structural components. Most of the components designed for energy absorption are plate type columns with various shapes of sections and subjected to axial compression or bending. The elastic buckling and maximum strengths play an important role in characterizing the crush process of components. The problem of determining the theoretical load capacity of plate type composite components is mathematically complicated due to the local instabilities that precede structural collapse, and the understanding of material strength and failures.

In this study a simple analytical approach was developed and may serve as a design aid for plate type composite columns subjected to axial crush loads. The approach can be used to determine the elastic buckling and maximum strengths, which are related to section geometry, column length and the material properties. It is expected that the developed approach will prove capable of being implemented into a finite element beam-column code for the study of the crash behavior of structural systems.

LOCAL BUCKLING STRENGTH

When a plate type column is subjected to axial compressive load, its failure mode depends on the value of elastic buckling. If a plate has lower thickness to width ratio, the column would buckle locally. The local buckling of plate type columns is complicated and somewhat disordered subject. The critical load of a composite plate column has been addressed for some special cases either by exact or by approximate methods [1-3]. Earlier the author [1] obtained an approximate equation for the critical strength of a rectangular composite plate type column:

$$\sigma_{cr} = \phi_1 \pi^2 E^* (t/b)^2 \qquad (1)$$

where ϕ_1 is the elastic buckling coefficient of orthotropic column and is approximated as

$$\phi_1 = K_1 K_S \qquad (2)$$

in which K_1 is the elastic buckling coefficient of isotropic column [4] and is a function of the degree of restraints of unloaded longitudinal edges parallel to the compressive load. K_S is a coefficient depending on the material properties and layup configurations.

The above procedure has been implemented into a computer code, COMP-CRUSH, and further verified by comparing the local buckling loads with those obtained from PASCO [5,6]. The comparison is performed on the elastic buckling strengths of columns with rectangular sections. The lamina properties and layups are given in Table 1. The laminate has the stack sequence of (90/0/90/0/90/0/90) with total thickness equal to 0.07 inches. The comparison is provided in Table 2, from which it can be seen that the local buckling strengths from two codes are very closed to each other. However, COMP-CRUSH requires very little input and it can be used on a personal computer.

Table 1. The Mechanical Properties of Lamina

Property	Value	
E_1	4.5×10^6	psi
E_2	0.3×10^6	psi
E_{12}	0.8×10^6	psi
ν_{12}	0.3	
X_c	100000	psi
Y_c	20000	psi
S	10000	psi

Table 2   Critical Loads and Half-Wave Length of Columns with Rectangular Sections
----

d/b	λ_{cr} (inch)	P_{cr} (k) COMP_CRUSH	P_{cr} (k) PASCO	Error (%)
0.2	0.52	7337	7301	+0.5
0.4	0.55	7930	8405	-5.6
0.6	0.58	8839	8992	-1.7
0.8	0.61	9328	9270	+0.6
1.0	0.71	8877	8934	-0.6
1.2	0.74	7772	7753	+0.2
1.4	0.83	6579	6631	-0.8
1.6	0.93	5585	5709	-2.2
2.0	1.18	4196	4462	-5.9

----
b = 1.0 inch

ULTIMATE STRENGTH OF LAMINATE

In the design and analysis of automotive components for crush energy management it is important to predict the crush strength characteristics of various components. In metal type components the material strength can be easily represented by its stress-strain relationship. However, the ultimate strength of composite laminate can not be so easily characterized. In this study the lamination theory and various failure criteria are used to predict the ultimate strength of laminate.

The Lamination Theory

The lamination theory is based on the assumption of a linear strain field through the thickness which is compatible with the classical laminated plate theory. According to this theory, the laminate moduli can be calculated by

$$E_x = (A_{11}A_{22} - A_{12}A_{12})/A_{22} \qquad (3)$$

$$G_{xy} = A_{66} \qquad (4)$$

$$\nu_{xy} = E_x A_{12}/(A_{11}A_{22} - A_{12}A_{12}) \qquad (5)$$

where A_{ij} are the extensional stiffness of laminate. Refer to Figure 1. The laminate strains due to uniaxial compression are calculated knowing these laminate elastic properties,

$$\begin{Bmatrix} e_x \\ e_y \\ e_{xy} \end{Bmatrix} = \begin{Bmatrix} \sigma_x/E_x \\ -\nu_{xy} e_x \\ 0 \end{Bmatrix} \qquad (6)$$

These laminate strains are transformed to provide lamina strains in material directions,

$$\begin{aligned}
e_1 &= (\cos^2\theta - \nu_{xy}\sin^2\theta)e_x = B_1\sigma_x \\
e_2 &= (\sin^2\theta - \nu_{xy}\cos^2\theta)e_x = B_2\sigma_x \\
e_{12} &= -2\sin\theta\cos\theta(1+\nu_{xy})e_x = B_3\sigma_x
\end{aligned} \qquad (7)$$

From these lamina strains the lamina stresses in material directions are computed as

$$\begin{Bmatrix} \sigma_1 \\ \sigma_2 \\ \tau_{12} \end{Bmatrix} = \begin{bmatrix} Q_{11} & Q_{12} & 0 \\ Q_{12} & Q_{22} & 0 \\ 0 & 0 & Q_{33} \end{bmatrix} \begin{Bmatrix} e_1 \\ e_2 \\ e_{12} \end{Bmatrix} \qquad (8)$$

or

$$\begin{Bmatrix} \sigma_1 \\ \sigma_2 \\ \tau_{12} \end{Bmatrix} = \begin{Bmatrix} A_1 \\ A_2 \\ A_3 \end{Bmatrix} \sigma_x \qquad (9)$$

where

$$Q_{11} = E_1/(1-\nu_{12}\nu_{21}), \quad Q_{22} = E_2/(1-\nu_{12}\nu_{21})$$
$$\qquad (10)$$
$$Q_{12} = \nu_{21}E_1/(1-\nu_{12}\nu_{21}), \quad Q_{33} = E_{12}$$

E_1, E_2, ν_{12} and E_{12} are the four independent elastic constants of the lamina with respect to its material directions (1-2). Using these lamina strains or stresses the imminence of failure of each lamina can be evaluated by one of the following failure criteria.

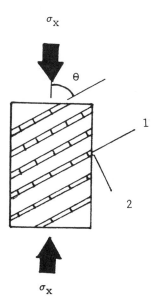

Fig. 1 A Laminate Subjected to Uniaxial
Compression

Lamina Failure Criteria

There are 50 failure criterion available
in the open literature, according to a recent
survey by Nahas [7]. The majority of these
criterion is based on the assumption of failure
at point. Specifically, whenever the stresses
or strain state at a point in the lamina
satisfies a certain critical condition, the
material at that point fails. When this
happens, the entire lamina fails. Four commonly
used failure criteria and their use in laminate
failure analysis are described in this paper.

1 - Maximum Strain Theory This theory [7]
is an extension of St. Venant's maximum
principal strain theory to anisotropic media.
For an orthotropic lamina, the strain
components must be referred to the material
axes. Thus three strain components can exist in
this criterion,

$$
\begin{array}{lll}
e_1 \leqslant e_{1c} & \text{or} & e_1 \geqslant e_{1t} \\
e_2 \leqslant e_{2c} & \text{or} & e_2 \geqslant e_{2t} \\
\gamma_{12} \leqslant \gamma_{12u} &
\end{array} \qquad (11)
$$

where e_{1c} (e_{1t}) and e_{2c} (e_{2t}) are the maximum
compressive (tensile) normal strains in the 1-
direction and 2-direction, respectively, γ_{12u}
is the maximum shear strain in the 1-2 plane.
In this theory, there is no interaction between
the strain components, but interaction exists
between the stress components caused by the
Poisson ratio effect.

2 - Maximum Stress Theory With this
theory [8], the stresses acting in a material
are transformed into 1 and 2 directions. The
stress state (σ_1, σ_2, τ_{12}) at a point will
cause failure when one or more of the
following conditions are satisfied:

$$
\begin{array}{lll}
\sigma_1 \leqslant X_c & \text{or} & \sigma_1 \geqslant X_t \\
\sigma_2 \leqslant Y_c & \text{or} & \sigma_2 \geqslant Y_t \\
\tau_{12} \leqslant S &
\end{array} \qquad (12)
$$

where X_c (X_t) and Y_c (Y_t) are the maximum
compressive (tensile) normal stress in 1-
direction and 2-direction respectively, S is
the maximum shear stress in 1-2 plane.

3 - Tsai-Hill Theory This failure theory
[9] predicts that failure would initiate when
the magnitudes of the stresses reach the
following condition,

$$
(\sigma_1/X)^2 - \sigma_1\sigma_2/X^2 + (\sigma_2/Y)^2 + (\tau_{12}/S)^2 = 1 \qquad (13)
$$

where X, Y and S are the uniaxial strengths
parallel and perpendicular to fibers, and in
shear. This theory does not consider different
strengths for tensile and compressive modes.

4 - Tsai-Wu Theory In an effort to more
adequately predict experimental results, Tsai
and Wu [10] proposed a lamina failure criterion
as follows,

$$
f(\sigma) = F_i\sigma_i + F_{ij}\sigma_i\sigma_j = 1, \quad i,j = 1,2,..6 \qquad (14)
$$

where F_i and F_{ij} are second and fourth order
lamina strength tensors. Under plane-stress
conditions, Tsai-Wu failure criterion becomes

$$
F_1\sigma_1 + F_2\sigma_2 + F_6\tau_{12} + F_{11}\sigma_1^2 + F_{22}\sigma_2^2 \\
+ 2F_{12}\sigma_1\sigma_2 + F_{66}\tau_{12}^2 = 1 \qquad (15)
$$

where

$$
\begin{array}{lll}
F_1 = 1/X_c + 1/X_t, & F_2 = 1/Y_c + 1/Y_t, & F_6 = 0, \\
F_{11} = -1/X_cX_t, & F_{22} = -1/Y_cY_t, & F_{66} = 1/S^2
\end{array} \qquad (16)
$$

For the determination of F_{12} biaxial tests are
required. In this study, F_{12} is considered
zero.

Ply by Ply Failure Analysis

In the prediction of the laminate strength
using the ply by ply failure analysis, the
laminate is considered to consist of bonded
layers. Each layer is considered to be
homogeneous and orthotropic. Lamination theory
is used to obtain the stresses and strains in
each layer as given by Eqs. 6 and 8. These
stresses and strains are transformed to the 1
and 2 axes before the failure criterion is
applied to each lamina. After the first lamina

fails, the laminate may or may not be able to
carry higher load. If the remaining laminae
cannot carry the redistributed load, the
laminate stress when the first laminae fail is
the predicted ultimate strength. If the
remaining laminae can carry the redistributed

Table 3 Compressive Strength of A-S/3501 Laminate $[(0/\pm45)_s]_{4T}$

Specimen No.	V_f (%)	Test[*]	Ultimate Compressive Strength (ksi) Prediction			
			Tsai-Wu	Max. Strain	Max. Stress	Tsai-Hill
CL-1	60.0	114.4	89.0	96.4	66.4	64.6
CL-2	60.0	110.4	-	-	-	-
CL-3	60.0	107.0	-	-	-	-
CL-4	60.0	109.7	-	-	-	-
CL-5	60.0	112.1	-	-	-	-
CL-7	60.0	98.9	-	-	-	-
CL-8	60.0	94.0	-	-	-	-
CL-9	60.0	95.2	-	-	-	-
CC-1	59.0	94.1	88.0	96.0	65.5	63.6
CC-2	59.0	92.5	-	-	-	-
CC-3	59.0	97.0	-	-	-	-
CC-4	59.0	96.1	-	-	-	-
CC-5	59.0	96.3	-	-	-	-
CC-6	59.0	95.6	-	-	-	-

* Test results from Reference [11]

Table 4 Compressive Strength of A-S/3501 Laminate $[\pm45_2/0_{12}/\pm45_2]_T$

Specimen No.	V_f (%)	Test[*]	Ultimate Compressive Strength (ksi) Prediction			
			Tsai-Wu	Max. Strain	Max. Stress	Tsai-Hill
1921	65.0	135.6	148.6	122.4	109.4	107.6
1922	65.0	127.9	-	-	-	-
1923	65.0	146.6	-	-	-	-

* Test results from Reference [11]

Table 5 Compressive Strength of T-300/5208 Laminates

Spec. No.	Laminate	V_f (%)	Test[*]	Ultimate Compressive Strength (ksi) Prediction			
				Tsai-Wu	Maximum Strain	Maximum Stress	Tsai-Hill
LFB-1-1-N	$[\pm45/0_4]_s$	73	133.9	141.2	137.6	168.5	136.8
LFB-1-2-N	,,	,,	126.1	,,	,,	,,	,,
4A-N	$[45/0_2/-45$	61.6	129.2	141.8	143.3	157.8	136.9
4B-N	$/0_4]_s$	,,	114.8	,,	,,	,,	,,
4C-N	,,	,,	129.0	,,	,,	,,	,,
4D-N	,,	,,	127.3	,,	,,	,,	,,
UFB-1-1-N	,,	69.0	129.6	155.4	154.3	176.1	150.4
UFB-1-2-N	,,	,,	124.0	,,	,,	,,	,,
UFB-2-1-N	,,	68.5	150.6	154.3	153.4	175.3	149.5
UFB-2-2-N	,,	,,	141.1	,,	,,	,,	,,
SIDE-3D-1-N	,,	68.6	131.8	154.5	153.6	175.6	149.7
SIDE-3D-2-N	,,	,,	134.2	,,	,,	,,	,,
POST-1B-1-N	$[45/0/-45$	71.3	132.8	144.5	141.6	170.4	140.3
POST-1B-2-N	$/0_2/45/0_2$	,,	142.0	,,	,,	,,	,,
	$/-45/0_3/0]_s$						

* Test results from Reference [12].

388

load, a yield strength has been predicted; the process can be successively repeated on the modified laminate to predict other yield strengths and/or the ultimate strength.

This procedure has been implemented into a program LSAP serving as a subroutine to COMP-CRUSH. To verify the procedure, the comparison has been made on the graphite laminates tested by Spier and Klouman [11,12]. The results are given in Tables 3, 4 and 5. From the comparison it can be seen that the ultimate strengths of A-S/3501 laminates have been successfully predicted by maximum strain theory and Tsai-Wu theory, while the ultimate strengths of T-300/5208 have been well estimated by Tsai-Hill theory and maximum strain theory. This means that the maximum strain theory has more generality than others, and probably this is why it is used more than any other failure theory. Maximum strain theory also determines the mode of failure of the failed lamina, and this facilitates the study of the behavior of the laminate after the first lamina failure.

MAXIMUM COLUMN STRENGTH

Maximum Crippling Strength

If a column experiences a local buckling its maximum crippling strength is generally a function of large deflections initiated at buckling and inelastic effects of the material. The theoretical post-buckling analysis of flat plates indicates a definite relationship among the maximum average strength, the elastic buckling stress and the ultimate stress. Mahmood, Zhou and Lee [1] obtained this relationship for plate type column in the following form:

$$\frac{\sigma_{max}}{\sigma_{ult}} = \left[\frac{\beta\,\sigma_{cr}}{\sigma_{ult}}\right]^{0.43} \qquad (17)$$

where $\sigma_{cr} \lesssim \sigma_{ult}$ and β is defined in reference [1]. The verification of Eq. 17 is provided in reference [1], which shows that the maximum strengths predicted by Eq. 17 is very closed to the test results for those columns exhibiting a local buckling prior to the maximum strength.

Maximum Crushing Strength

When a plate type column is subjected to axial crush load with local buckling strength σ_{cr} larger than laminate ultimate strength, the column will fractures progressively with a certain mean crush load or sever longitudinal or transverse crack and sudden lost of the crush load. To demonstrate this type of crush process, six square columns were designed and tested. The dimensions and layups of these columns are given in Table 6. The crush processes are shown in Figure 2.

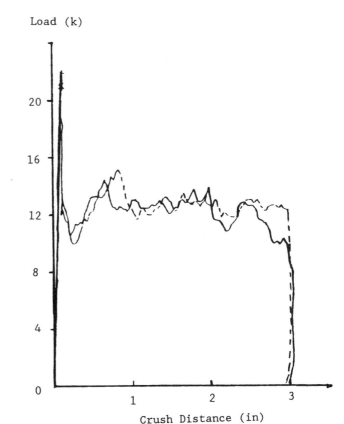

Load (k)

Crush Distance (in)

Fig. 2 Crushing Processes of Column 1A and 1B

For this type of crush mode the maximum crushing strength can be related to ultimate strength of laminated plate as follows,

$$\sigma_{max} = C_1\,\sigma_{ult} \qquad (18)$$

where C_1 is a strength reduction factor. In a very well controlled manufacturing process or idealized condition C_1 is equal to 1. However, from the test results of various components C_1 is estimated to be 0.6. The maximum crushing strengths of the tested columns are given in Table 7, from which it can be seen that Eq. 18 gives values very closed to the test results.

CONCLUSIONS

Crush characteristics of plate type composite columns is discussed. Two kinds of crushing processes are identified. The first starts with local buckling, crippling and then folding, and is characterized by elastic local buckling strength and the maximum crippling strength. The other is progressive material fracturing and characterized by the maximum crushing strength. For a thin walled column the first process is expected and the maximum crippling strength is a function of section geometry and material properties, while for a

389

thick walled column the material fracturing goes through crushing process and the maximum crushing strength is associated with the ultimate strength of laminate. The procedure of predicting the laminate ultimate strength is developed, which is based on lamination theory and certain failure criteria. Agreement with experimental data shows that the maximum strain failure criterion has more generality than any other failure criteria. Further tests of different geometry, layup configurtion and material property will be conducted to generalize the above models.

Table 6 Dimensions and Layups of Tested Columns

Specimen No.	Layup	Dimension	V_f^* (%)
1A	$(90/0/R_c)_4$	1.26x1.26x.18	28
1B	$(90/0/R_c)_4$	1.26x1.26x.18	28
2A	$(90/0/R_c)_2$	2x2x.18	14
2B	$(90/0/R_c)_2$	2x2x.18	14
3A	$(90/0/R_c)_4$	2x2x.18	28
3B	$(90/0/R_c)_4$	2x2x.18	28

*V_f - Fiber Volume Fraction
*R_c - Random Chopped Fiber

Table 7 Crush Characteristics of Tested Columns

Column No.	Predicted			Tested
	σ_{cr} (ksi)	σ_{ult} (ksi)	σ_{max} (ksi)	σ_{max} (ksi)
1A	138	35.6	21.4	23.4
1B	138	35.6	21.4	24.3
2A	37.9	28.2	16.9	18.3
2B	37.9	28.2	16.9	20.0
3A	54.9	35.6	21.4	25.6
3B	54.9	35.6	21.4	24.7

REFERENCES

1. H. F. Mahmood, J.H. Zhou, and M. S. Lee, "Axial Strength and Modes of Collapse of Composite Components'', Proceedings of the Seventh International Conference on Vehicle Structural Mechanics, SAE Paper No. 880891, April, 1988.

2. W. M. Banks and J. Rhodes, "The Postbuckling Behavior of Composite Box Sections'', Composite Structures, Edited by I. H. Marshall, Applied Science Publishers, London and New Jersey, 1981.

3. D. J. Lee, "The Local Buckling Coefficient for Orthotropic Structural Sections", Aeronautical Journal, Paper No. 575, July 1978, pp. 313-320.

4. H. F. Mahmood and A. Paluszny, "Design of Thin Walled Column for Crash Energy Management - Their Strength and Modes of Failure", Proceedings of the Fourth International Conference on Vehicle Structural Mechanics, SAE Paper No. 8-11302, Nov. 18, 1981.

5. W. Jefferson Stroud and M. S. Anderson, "PASCO: Structural Panel Analysis and Sizing Code. Capability and Analytical Foundations", NASA TM-80181, November 1981.

6. M. S. Anderson, W. J. Stroud, B. J. Durling, and K. W. Hennessy, "PASCO: Structural Panel Analysis and Sizing Code, User's Manual", NASA TM-80182, November 1981.

7. M. N. Nahas, "Survey of Failure and Post Failure Theories of Laminated Fiber-Reinforced Composites", Journal of Composite Technology and Research, Vol 8, 1986, pp. 138-153.

8. C. F. Jenkins, "Material of Construction Used in Aircraft and Aircraft Engines", Report to the Great Britain Aeronautical Research Committee, 1920.

9. R. Hill, "The Mathematical Theory of Plasticity", Oxford University Press, London, 1950.

10. S. W. Tsai and E. M. Wu, "A General Theory of Strength for Anisotropic Materials", Journal of Composite Materials, 5, pp. 58-80.

11. E. E. Spier and F. L. Klouman, "Ultimate Compressive Strength and Nonlinear Stress-Strain Curves of Graphite/Epoxy Laminates", Proceeding of 8th National SAMPE Conference. Bicentennial of Materials Progress - Part I", Seattle, WA, OCT. 1976.

12. E. E. Spier, "Stability of Graphite/Epoxy Structures with Arbitrary Symmetrical Laminates", Experimental Mechanics, Vol. 18, No. 11, Nov. 1978.

MECHANICAL FASTENERS FOR
THICK COMPOSITES

Chia-Chieh Chen
FMC Corporation
Central Engineering Laboratories
Santa Clara, California 95052   USA

ABSTRACT

Since composite structures cannot be welded, secondary attachments (brackets, hooks, etc.) have to be either bonded or mechanically fastened. For thick composite laminates, the use of blind fasteners can be an effective and economical method for small part attachment. In this study, the pull-out strength of a simple, threaded insert was determined for different combinations of insert and pilot hole sizes. A simple test fixture was used to perform both the 90° and 45° pull-tests. Microscopic examination of the sample after insert installation and testing showed both composite delamination and matrix cracking failure modes. A general linear trend was observed for the strength versus the amount of interference and the insert length. The interlaminar shear strength of the composite, the compressive strength of the matrix, the fastener size and the pilot hole size are all very important to the bolt installation torque and its pull-out strength.

INSIDE AN ALUMINUM personnel carrier, hundreds of studs, hooks and threaded holes were drilled or welded for the installation of boxes, brackets and other secondary structure components. If a thick composite is used to build the vehicle hull, a different method needs to be developed to install such secondary attachments.

Numerous types of fastener designs and sizes are commercially available and were evaluated for different composite applications[1,2,3]. Previously, several studies were conducted on fasteners in our laboratory to compare blind mechanical fasteners for thick composites. Of all the fasteners studied, a particular type of self-tapping insert (Figure 1) showed very high pull-out strength and was reasonably easy to install.

This type of fastener can either be an insert or a stud. They may be used for almost all necessary secondary attachments on the current aluminum vehicle hull at a reasonable cost.

The technical information received from different fastener suppliers was not consistent. Different sizes of pilot holes were recommended for the same size of insert. The pull-out strength of the insert after installation was often not reported for composites.

It was decided to evaluate this type of self-tapping, threaded insert to obtain the necessary information for design, so that a proper insert size and pilot hole size could be selected for any specific secondary attachment.

EXPERIMENTAL

To simulate an attachment to either an overhead or a sidewall location, both a 90° pull (the angle between the pull force direction and the surface of installation) and a 45° pull tests were conducted. The 45° test was chosen over a 0° test to avoid a pure shear test on the bolt. The test set-ups are shown in Figures 2 & 3. A bolt was threaded into the insert and used for the pull-tests.

COMPOSITE MATERIAL – Fiberite 7701 epoxy and $24oz/yard^2$, E glass woven roving prepreg was used to make 40 ply laminates. The ply orientation was $\{[(\pm45)/(0,90)]_{10}\}_s$.

After hand lay-up, the laminate was vacuum bagged and cured in an oven. A thermocouple was placed in the middle of the laminate to monitor the temperature during cure. The following cure cycle was used: Heat to 150°F and hold for one hour. Heat to 200°F and hold until the exotherm was completed. Heat to 250°F and hold for one hour.

THREADED INSERTS - The self-tapping inserts were purchased from two suppliers covering bolt sizes from 3/16 to 7/16. Some brass inserts as well as steel inserts were tested. These inserts are shown in Figure 1. These inserts have a tapered head for cutting. The diameter of the head (minimum) and the end (maximum) were measured for each insert.

SAMPLE FABRICATION - To fabricate the test samples, 2" x 2" square blocks of the composites were cut from the laminates. A pilot hole was drilled in the center of the block with a standard drill bit to give the desired interference between the insert and the composite.

The self-tapping insert was manually installed by using a bolt and a nut. After inserting the bolt and the insert, the nut was tightened against the insert. The insert was then carefully driven into the composite until it was flush with the surface. The maximum installation torque was recorded. The nut was broken loose while holding the bolt. The bolt and the nut were then backed out leaving the insert in place.

TESTING PROCEDURES - After the sample blocks were made, a bolt of proper size was threaded into the insert. The bolt was used to apply the pulling force. The crosshead speed was 0.1"/min. The load-deflection curve was recorded for each test. Three to five samples were tested for each size of insert and the average pull-out force was measured.

Two sets of tests were performed. For the first set of tests, the pilot holes were mostly drilled according to the size suggested by the manufacturer for thermosets. Only the 90° pull test was performed, but the results were not satisfactory.

Based on the results of the first test, the second test plan was prepared to study the effect of the pilot hole size to the torque and the resulting pull-out force. The tests covered 8 different inserts and 15 different pilot hole sizes. Both the 90° and the 45° pull tests were performed.

RESULTS AND DISCUSSION

The test results of the preliminary study are shown in Table I. Many of the inserts could not be tested because they were either too loose to test or they were too tight, causing bolt failure before testing.

Table II are the results of the second set of tests and very high pull-out forces were obtained when proper pilot hole sizes were selected.

The measured outer dimensions of each insert and the percent interference [(Insert O.D./hole dia.)-1] are listed in Table III. Since the strength is related to the extent of composite compression and the area of contact, the pull-out force is plotted against the fastener length times the interference. This is a greatly simplified view of the true mechanism because the effect of insert design, thread cutting, composite delamination, matrix cracking and other phenomena are all neglected.

Figure 4 is the 90° pull-out strength vs. the insert length times the interference. Except for the two data points of the 7/16 insert, a linear trend is observed. The same trend is seen for the 45° pull-out strength vs. the length times the interference (Figure 5). A better linear correlation appears to exist between the installation torque and the pull-out force (Figure 6). This fairly linear trend suggests that higher torque provides higher pull-out resistance. As long as the delamination is acceptable in installation, higher installation torque indicates higher pull-out force. This may provide some ideas for designing more effective inserts. Several other plots were made between the pull-out force and the insert dimension and similar trend was observed.

SAMPLE EXAMINATION - Figure 7 shows the sample blocks after the fastener installation and testing. For the 7/16 fastener, a slightly white area can be seen on the surface around the fastener even before testing. The composite delamination was caused by the installation.

Section views of the samples are shown in Figure 8. The threads created during insert installation were destroyed by pull-out tests. Compressive failure of the composite is observed for the sample after the 45° pull.

To study the mechanism, the sample (7/16, 0.590" hole) surface was polished smooth and dyed with black ink. The ink penetrated into the composite where surface cracks existed. The darker the area, the deeper the crack. The delamination can be clearly seen in Figures 9 and 10.

For the sample with a larger hole (0.605"), no delamination was observed and the threads were shallow as shown in Figure 11. The torque for installation is low and the pull-out strength is also low.

The effect of a 90° pull on the composite is shown in Figures 12 and 13. After the installation, the threads and the slight delamination are seen in Figure 12. The threads were destroyed and the delamination grew farther into the composite after testing (Figure 13). The interlaminar strength of the composite may be the determining factor to its pull-out strength.

When a 45° pull was applied, the composite showed not only delamination but also severe matrix cracking between plies (Figures 14, 15).

DISCUSSION - Proper selection of the pilot hole size is very important. A small change in the size could change the installation torque significantly.

The composite interlaminar strength appears to be the determining factor in the 90° pull test. Failure of the composite was characterized by the propagation of the delamination deeper into the composite. The compressive strength of the matrix could be the critical property in resisting the 45° pull where matrix cracking between plies was observed.

The delamination caused by installation could be observed on the surface by the color change around the hole. This indicates a very tight fit.

When the installation torque exceeded 100 in-lb, the threads of the brass insert started to strip and no test could be performed. The steel insert of the same size was able to withstand the torque and obtain high pull-out resistance.

A more quantitative analysis may be feasible to correlate the pull-out strength with the compressive properties of the matrix and the interlaminar shear properties of the composites.

If a high speed, automated installation machine is used, the viscoelastic response of the polymer and the increase in friction may cause higher installation torque and further composite delamination. Its effect on the pull-out strength needs to be investigated.

SUMMARY

When properly installed, the inserts can resist pull-out forces from a few hundred pounds to over 8000 pounds. The installation torque ranged from 28 to 330 in-lbs.

Steel inserts performed better than the brass inserts. The installation torque for steel insert was not much higher than that for the brass insert (see results for the 6-32 inserts), but the pull-out strength was much higher. Also, the brass inserts started to strip the threads once the torque reached a critical value.

The 45° pull test applied a combined tensile and compressive load on the composite. The resulting pull-out force is lower than that of the 90° pull. From the section view of the insert, we can see the composite delamination (due to tension) as well as matrix cracking (due to compression).

Table IV summarizes the insert size, the drill size for the pilot hole, the range of installation torque and the pull-out strength

from the test results. For any potential application of self-tapping inserts on the thick composite, a proper insert and a drill for the pilot hole may be selected from this table.

FUTURE STUDY

This study evaluated the strength of metal inserts under static load conditions. The fatigue properties and the long term creep effects need to be considered. A comparison with adhesive bonded structures or structures attached with both adhesive and insert will make the study more complete. The effect of the insert and the pilot hole on the integrity of the composite should also be evaluated.

REFERENCE

1. R.T. Cole, E.J. Bateh and J. Potter; Composites, (233-240) July 1982.
2. R.B. Starnes; Technical paper AD84-842, SME conference (1984).
3. J.W. Howard; Technical paper AD86-796, SME conference (1986).

TABLE I SCREENING TEST RESULTS

Fastener Size	Material	Maximum O.D.(in.)	Drill Size	Hole Size(in.)	Installation Torque(in-lb)	90° Pull-out Force(lb)
6-32	Brass	0.215	(1)	0.226	too loose	---
6-32	Steel	0.220	(1)	0.226	too loose	---
8-32	Brass	0.2558	(A)	0.236	100[1]	---
8-32	Brass	0.2558	(A)	0.236	100[1]	---
8-32	Brass	0.2558	(A)	0.236	70	473
8-32	Brass	0.2558	(1/4)	0.250	32	328
8-32	Brass	0.2558	(1/4)	0.250	30	338
10-32	Brass	0.3137	(L)	0.287	150[1]	---
10-32	Brass	0.3137	(L)	0.287	150[1]	---
10-24	Steel	0.2977	(L)	0.287	30	738
1/4-20	Steel	0.3761	(S)	0.343	120	2385
1/4-20	Steel	0.3761	(S)	0.343	120	2149
1/4-20	Steel	0.3761	(S)	0.343	90	1801
1/4/20	Steel	0.3761	(3/8)	0.375	14	62
1/4-20	Steel	0.3761	(3/8)	0.375	1[2]	---
1/4-20	Steel	0.3761	(3/8)	0.375	2[2]	---

(1) The bolt used for the installation was broken and the treads started to strip. No test could be performed.
(2) The insert was fit loosely in place and was pulled out without much force.

TABLE II AVERAGE PULL-OUT STRENGTH

Fastener Size	Material	Maximum O.D.(in.)	Drill Size	Hole Size (in.)	Installation Torque(in-lb)	Pull-out Strength(lb)	
						90ρ	45ρ
6-32	Brass	0.215	(6)	0.204	34	258	257
			(4)	0.209	30	302	243
6-32	Steel	0.220	(6)	0.204	30	684	514
			(4)	0.209	28	569	548
8-32	Brass	0.2558	(C)	0.242	100	441	435
10-32	Brass	0.3137	(M)	0.295	130	681	544
			(N)	0.302	100	763	596
10-24	Steel	0.2977	(J)	0.277	75	1207	1028
			(K)	0.281	50	1276	1147
1/4-20	Steel	0.3761	(11/32)	0.344	100	2056	1828
			(S)	0.348	100	2217	1639
5/16	Steel	0.4675	(27/64)	0.422	330	2637	1996
			(7/16)	0.438	300	2849	2199
7/16	Steel	0.6325		0.590	300	8976	5595
				0.605	190	6356	4599

TABLE III THE OUTSIDE DIMENSIONS, INTERFERENCE & PULL-OUT FORCES

Fastener & Pilot hold size (in.)	Outside dimension (in) & % Interference[1]		Installation Torque (in-lb)	Pull force (lbs)		Length x Interference (in)
	Maximum	Minimum		90°	45°	
6-32 (Brass)	0.215	0.212				
0.204	5.4%	3.7%	30	258	257	1.517
0.209	2.9%	1.2%	28	302	243	0.815
6-32 (Steel)	0.220	0.2175				
0.204	7.8%	6.6%	34	684	514	2.192
0.209	5.3%	4.1%	30	569	548	1.489
8-32 (Brass)	0.2558	0.2535				
0.242	5.7%	4.8%	100	441	435	1.870
10-32 (Brass)	0.3137	0.3112				
0.295	6.3%	5.5%	130	681	544	2.363
0.302	3.9%	3.2	100	763	596	1.463
10-24 (Steel)	0.2997	0.2915				
0.277	7.5%	5.2%	75	1207	1028	2.813
0.281	5.9%	3.7%	50	1276	1147	2.213
1/4-20 (Steel)	0.3761	0.3741				
0.344	9.4%	8.8%	100	2056	1828	4.550
0.348	8.1%	7.5%	100	2217	1639	3.920
5/16 (Steel)	0.4675	0.4646				
0.422	10.8%	10.1%	330	2637	1996	6.070
0.438	6.9%	6.2%	300	2849	2199	3.878
7/16 (Steel)	0.6325	0.6200				
0.590	7.2%	5.1%	300	8976	5595	5.623
0.605	4.6%	2.5%	190	6356	4599	3.593

(1) The percent interference was calculated as <u>outer diameter/hole diameter - 100%</u>. As can be seen in the table, a very small change in diameter can be very important.

TABLE IV INSERT SIZE, DRILL SIZE AND THE PULL-OUT FORCE[1]

	Insert size						
	6-32	8-32	10-32	10-24	1/4-20	5/16	7/16
Drill size	6,4	C	N	K	S,11/32	7/16	0.590
Pull-out Force(lbs)	514- 684	435- 441	544- 763	1028- 1276	1639- 2217	1996- 2849	4599- 8976

(1) In this study, the speed of installation was slow (manual). The effect of the high speed installation in a manufacturing environment should be studied. The installation torque will be increased and the interference may have to be reduced slightly to compensate the effect.

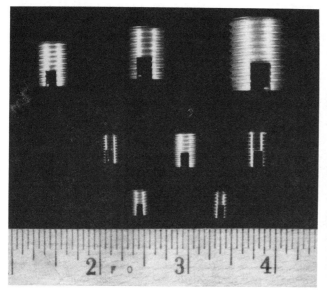

FIGURE 1   MAG = 1.2X

These are the fasteners used for the study.
the sizes of the fasteners are listed in
Table II.

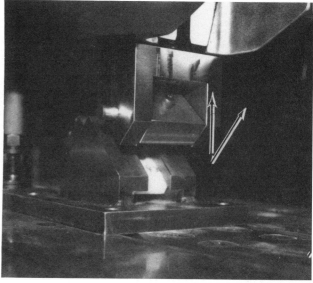

FIGURE 3   MAG = 0.2X

This is the test set-up for the 45ρ pull. A
fixture was designed to hold the sample block
at a 45° angle to the pulling direction.
The arrows show the angle.

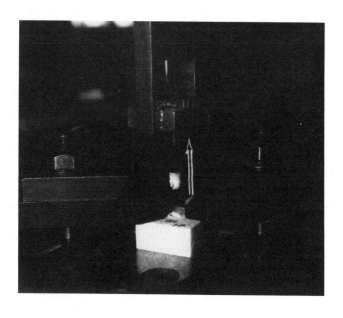

FIGURE 2   MAG = 0.3X

This is the test set-up for the 90ρ pull.
The sample block was fixed to a platform and
an extra long bolt was used to pull.

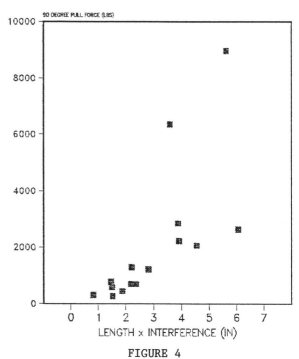

FIGURE 4

The 90° pull-out force is plotted against
the insert length times the interference.

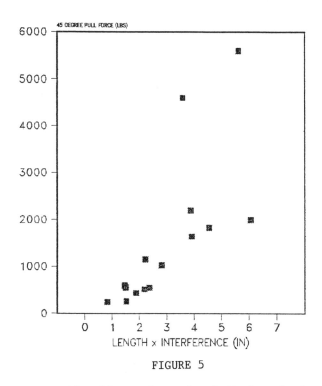

FIGURE 5

The 45° pull-out force is plotted against the insert length times the interference.

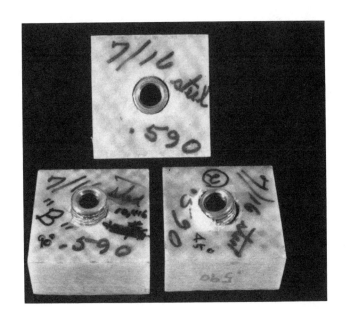

FIGURE 7   MAG = 0.75X

The 7/16 samples before and after testing, (0.590 hole). The sample on top was not tested.

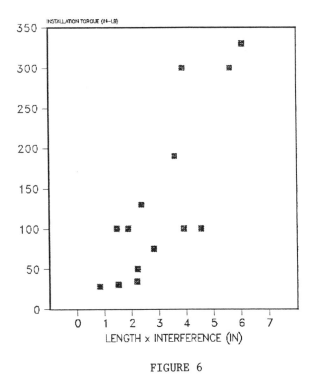

FIGURE 6

The installation torque is plotted against the insert length times the interference.

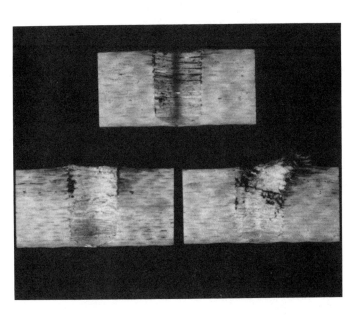

FIGURE 8   MAG = 1X

The section views of samples after insert installation, 90° test and 45° test.

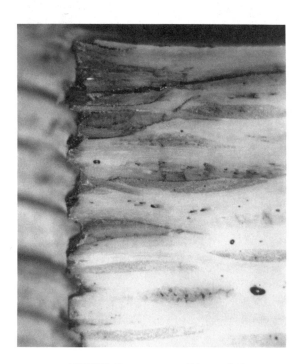

FIGURE 9   MAG = 9.7X

The delamination close to the surface of the
composite can be seen. The darker the color,
the deeper the crack.

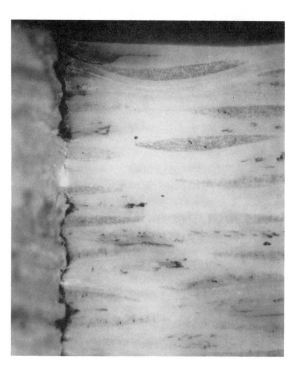

FIGURE 11   MAG = 9.7X

When the pilot hole is large (0.605"), the
delamination is seen only on the first layer
while the delamination in figure 10 is 3 to 4
layers deep.

FIGURE 10   MAG = 9.7X

The section of the insert (7/16, 0.590" hole)
is shown in the photo. The delamination was
caused by the installation of the insert.

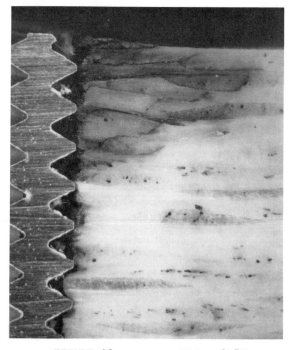

FIGURE 12   MAG = 9.7X

A smaller insert (1/4-20) is shown here after
installation. The delamination is again
about 3 to 4 layers deep.

398

FIGURE 13   MAG = 9.7X

After the 90° pull test, the delamination
propagated farther as well as deeper into the
composite.

FIGURE 15   MAG = 9.7X

The delamination and matrix cracking after
the 45° test are obvious with a large insert
(7/16).

FIGURE 14   MAG = 9.7X

After the 45° pull test, matrix cracking is
seen in addition to the composite
delamination.

METAL ALTERNATIVE DESIGN WITH MOLDABLE PHENOLIC COMPOSITES

John Arimond
Rogers Corporation
Rogers, Connecticut   USA

ABSTRACT

Moldable phenolic composites are being designed in to replace metals in an increasing number of automotive engine and transmission components. These components cannot be designed by simple material substitution, nor from the ASTM properties on suppliers' data sheets. Effective metal replacement requires a prototype design process closely integrated with material selection. Such a metal alternative design process is outlined and illustrated.

INTRODUCTION

Phenolic composites now find commercial applications in transmission stators, brake pistons, commutators, pulleys, sprockets, and other components demanding stiffness, strength, dimensional stability, chemical and creep resistance under high loads at elevated temperatures.

Prototype developments are underway in numerous intake manifolds, pump housings and transmission parts. A 4-cylinder, 2.3 liter engine unveiled this year by Polimotor Research, Inc. employs moldable phenolic composites in the engine block, head, oil pan and valve cover (Figure 1).

Composites are being designed in to replace metals for several reasons, but primarily to reduce costs. Cost reductions are possible through molding complex shapes which integrate several metal parts, thereby reducing downstream machining and assembly operations.

In the development of a phenolic transmission torque converter stator, for example, five metal parts were integrated into one phenolic molding. In addition, the density of molded stators was sufficiently uniform to eliminate the balancing operation required with cast aluminum stators[1].

Figure 1. The Polimotor Model 234 Engine.

Composite moldings require less machining than aluminum castings. The dimensional capability of phenolic molding is an order of magnitude more precise than sand casting, and significantly more precise than die casting.

Composite components offer weight
reductions. Reducing the weight of any
component provides some improvement in
fuel economy; reducing the weight of a
dynamic component can also offer
performance improvement through reduced
inertia and vibration.

Composite components can also offer
improved performance through smoother
molded surfaces, more uniform density
and higher mechanical loss tangent
(typically by two orders of magnitude).

METAL ALTERNATIVE DESIGN

The effective replacement of metal
components and assemblies requires more
than just the substitution of composite
materials into existing designs. When a
phenolic composite part is molded to
replace a cast or joined metal part, for
example, its load-bearing features may
need to be thicker or rounder than in
the original design to withstand the
same loads. This requires a metal
alternative design process.

Composite materials have a great variety
of properties and processing
requirements. Once a metal component or
assembly is targeted for composite
replacement, its prototype design
process must therefore be closely
integrated with material selection and
process design. An overall component
development process is illustrated in
Figure 2.

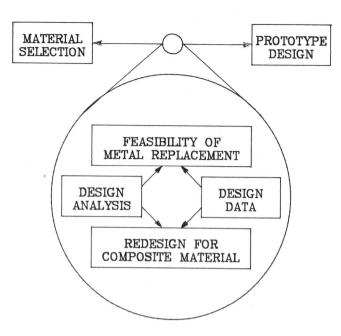

Figure 3. Metal Alternative Design Process

Metal alternative design integrates
prototype design with material
selection. Prototype design for metal
replacement involves two key questions
which depend on material selection:

(1) For which metal components and
    assemblies would composite
    replacements be feasible?

(2) Which walls, flanges, ribs and
    bosses in an existing component
    would need to be enlarged or
    rounded to redesign it in a
    composite?

Both of these questions require the same
tools: design analysis and design data.
Design analysis is needed to calculate
stresses for comparison with material
strength, and to calculate dimensional
changes in response to mechanical loads
and thermal excursions for comparison
with design requirements.

Design data are needed to perform and
use design analysis. The ASTM standard
property values provided on suppliers'
data sheets are useful for initial
material comparisons, but have limited
usefulness in part design. Composite
properties depend strongly on processing
conditions, so local properties within
parts usually vary significantly from
the ASTM values. Understanding the
effects of process-induced fiber
orientation and heat history on material
properties, prototype part design should
be integrated with process design and
with material selection[2].

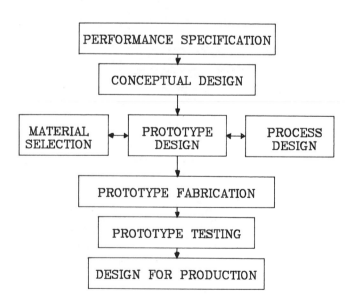

Figure 2. Component Development Process.

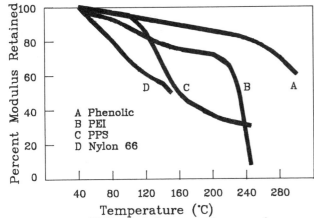

Figure 4. DMA Modulus vs. Temperature

The metal alternative design process shown in Figure 3 employs stress analysis and strength data to determine whether metal replacement is feasible, to determine whether part redesign is necessary, and if so to determine the extent of the required design changes.

MOLDABLE PHENOLIC COMPOSITES

Among the numerous types of composites available, phenolics will be the materials of choice in many components demanding excellent stiffness, strength, creep resistance, chemical resistance and dimensional stability under significant mechanical loads in the 250-400°F range[3].

Particularly in high temperature stiffness and creep resistance, phenolics outperform alternative composites. Modulus versus temperature curves and compressive creep data for phenolics and other moldable composites are shown in Figures 4 and 5.

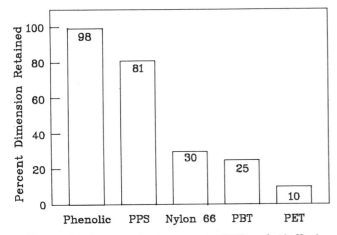

Figure 5. Compressive Creep at 150°C and 10 Kpsi.

The properties of phenolics can be tailored to the needs of particular applications. A moldable phenolic composite (MPC) is roughly one-third resin, two-thirds fibers, fillers and other additives. Properties such as stiffness, strength and thermal expansion can be controlled through fiber content; hardness, wear resistance and other properties can be controlled through intelligent choice of other fillers. The ranges of several mechanical properties which can be achieved in an MPC are summarized in Figure 6.

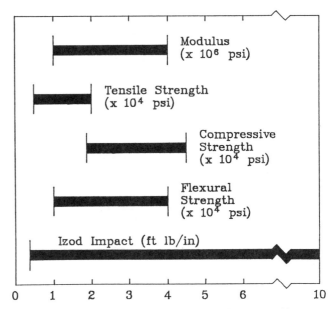

Figure 6. Range of Mechanical Properties Achievable with MPCs.

The properties of MPCs depend strongly on processing conditions. Part design, gate locations and molding conditions all affect the development of fiber orientation during molding, which in turn affects local part stiffness, strength and thermal expansion.

Near mold surfaces, shear stresses tend to align the fibers in the direction of flow, while fibers midway between mold surfaces tend to orient transverse to the flow direction. To illustrate the strong variation of properties with local fiber orientation, the tensile strength and coefficient of thermal expansion were measured on small specimens cut from the surface and the center of injection molded test bars of a Rogers phenolic composite[4]. These data are shown in Figure 7.

The degree of cure achieved during molding and postcuring affects an MPC part's final dimensions, creep resistance and retention of stiffness and strength at high temperatures[5]. These effects are illustrated in Figure 8, which shows the flexural strength of a Rogers phenolic composite measured as a function of temperature on both as-molded and postcured specimens.

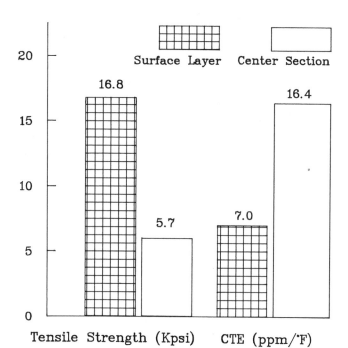

Figure 7. Properties of RX630 Injection Molded Bars.

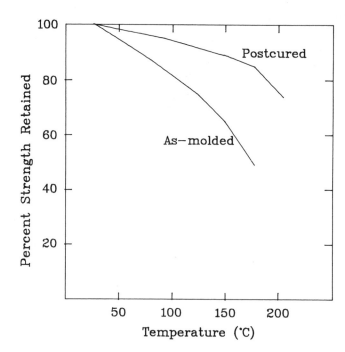

Figure 8. RX865 Flexural Strength vs. Temperature.

DESIGN EXAMPLE

To illustrate the metal alternative design process, consider its application to the development of phenolic automatic transmission torque converter turbines. A torque converter turbine transmits engine torque to a driveshaft via fluid coupling. This example will address two of the many mechanical requirements which a composite turbine must satisfy: it must withstand impinging transmission fluid without breaking, and it must deflect axially no more than a specified amount.

What material properties are needed to meet these requirements? What strength is needed to prevent part breakage, and what modulus is needed to meet the part stiffness requirement? The property requirements depend on prototype design, particularly on the thicknesses of various component features.

The effects of three design parameters have been investigated by finite element analysis. A turbine model, shown in Figure 9, was built from 322 3-D shell elements on a VAX minicomputer with ALGOR finite element software. The turbine was fixed at its inner edge, and a uniform axial load was distributed around its outer edge to model the impinging fluid load.

The turbine structure has three basic features: its blades, its shell (reinforced by the blades) and its hub section (inside the blades). Analyses were run in which the thickness of each of these features was varied independently.

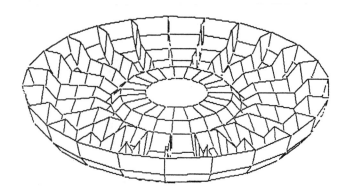

Figure 9. Finite Element Model of Phenolic Turbine.

The results of these analyses are summarized in Figure 10. The maximum principal stress was predicted to occur at the outer edge of the unreinforced hub section. Hub thickness was predicted to have the greatest effect on the maximum stress. Shell thickness was predicted to have the greatest effect on axial deflection. Blade thickness was predicted to have only a very small effect on these two mechanical requirements.

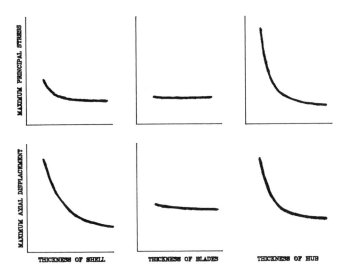

Figure 10. Design Analysis for Phenolic Turbine.

When a component or assembly is targeted for metal replacement, its performance specifications may be met by any of several materials. Mechanical requirements might be met either through material properties or through part design.

There is no such thing as an "optimal design" until a material is selected. Yet there can be no "best material" for an application until it is designed. The only solution to this dilemma is design integration. Metal alternative design offers the integration of prototype design with material selection to find the most cost-effective design/material combination.

CONCLUSIONS

Moldable phenolic composites (MPCs) are suitable materials for metal replacement in a variety of automotive engine and transmission components.

The effective use of MPCs in these applications requires metal alternative design, in which design analyses and design data are applied to assess the feasibility of metal replacement, to determine whether part redesign is necessary, and to redesign parts as necessary for successful manufacture in composite materials.

ACKNOWLEDGEMENTS

I would like to thank Bruce Fitts for his invaluable mentoring, and Peter Morse for preparing the figures for this paper.

REFERENCES

(1) D. Bradley, "Phenolic Transmission Reactor Replaces Metal, Cuts Cost," Plastics World, December 1981.

(2) J. Arimond and B.B. Fitts, "Design Data for Phenolic Engine Components," SAE Intl Congress and Expo, paper no. 880159, Detroit, March 1988.

(3) M.D. Bessette, B.B. Fitts and J. Arimond, "Properties, Performance and Applications of Phenolics," Proc SPI Phenolic Molding Division Tech Conf, Cincinnati, June 1987.

(4) B.B. Fitts, "Fiber Orientation of Glass Fiber-Reinforced Phenolics," Materials Engineering, November 1984.

(5) J. Arimond, B.B. Fitts and M.D. Bessette, "Computer Aided Design in Phenolics Technology," Proc SPI Phenolic Molding Division Tech Conf, Cincinnati, June 1987.

METAL INSERTS FOR EFFICIENT JOINING OF SHEET MOLDING COMPOUND STRUCTURES

S. V. Hoa, A. Di Maria
Concordia University
Montreal, Quebec Canada

INTRODUCTION

Composite materials introduce particular problems when components must be fastened together. A successful joint must be obtained in order to minimize component failure. The stainless steel inserts allow a convenient means of fastening Sheet Molding Compound (SMC). SMC is a brittle composite material which is made of polyester resin with glass-fiber reinforcement and calcium carbonate filler. It is widely used in the automotive industry for applications such as hoods, radiator supports, as well as subway seat benches.

Work done by previous researchers on the joining of SMC has placed emphasis on bonding and bolting [1]. It was found that epoxy adhesives give good bonding strength and that bolting gives stronger joints than bonding. Also, a previous design for the inserts has been investigated and reported in reference [2]. In that work, round inserts made of aluminum were used. Even though the strength of the inserted samples were comparable to the strength of the bolted samples (steel bolts), the aluminum inserts bend at large loads. In this paper, a modified design for the inserts is made. This design consists of a hexagonal head insert and the material is changed to stainless steel. These modifications produce significant improvements on the joint strength of the SMC. The advantages of no protrusions and ease in repeated assembly and disassembly is retained.

STAINLESS STEEL INSERTS

The molded-in inserts consist of a pair of interlocking disks which have a hexagonal head and a flaired body as shown in Figure 1. The flaired body protrudes from the large clamping head and allows the insert to be held rigidly within the SMC. Unlike the plain round insert, the hexagonal head prevents rotation of the insert when being torqued.[2] Interlocking is obtained through the existence of a shoulder male boss and a shoulder female counterbore. The two SMC components are joined by interlocking the mating inserts and using a special flat head cap screw to bolt them together. A flush assembly with no surface discontinuity is obtained.

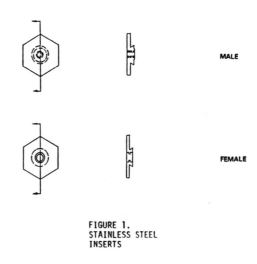

FIGURE 1.
STAINLESS STEEL
INSERTS

MALE

FEMALE

SPECIMEN PREPARATION

The Sheet Molding Compound used was Prepreg SMC R-30 and it was obtained from Jet Molding Compound Inc. (Ontario, Canada). The prepreg sheets were cut into 10" square sections. Holes of 0.350" diameter were punched into the prepreg and the inserts were placed in their appropriate positions. The sheets were then compression-molded under heat (350 F) and pressure (450 psi) for twenty minutes in a matched metal mold.

Once the cured plate was obtained, specimens were cut and shaped into four different configurations for five different tests. The first configuration consisted of two rectangular specimens, one containing a male insert and the other, a female insert. An effective hole diameter was established as 0.469" by calculating an average diameter for the entire insert. Ratios of edge distance (e) to effective hole diameter and width (w) to effective hole diameter were chosen as $e/d = 2$ and $w/d = 4$ respectively. See Figure 2. These ratios remained constant throughout the experiment. Both these specimens were 5 " long, 1.875 " wide and 0.095 " thick. SMC tabs were bonded to the ends of the specimens in order to be able to place them within the grips of the testing machine. Two-part epoxy (Ciba Geigy) was used to attach the tabs to the specimens. The samples were later assembled in a single lap joint configuration and pulled in tension. See Figure 2.

The second configuration consisted of a pair of specimens which were joined through a bolted connection. The individual specimens were 1 " wide, 5" long and 0.095" thick. The difference in width between the bolted and inserted samples is to ensure the same $w/d = 4$ ratio. Samples with width = 1.875" were also tested. SMC tabs were bonded to these samples as well. A hole (0.250" diameter) was drilled into each of the specimens at an equal distance of 0.5" from each edge. These specimens were assembled in a single lap joint configuration using a washer and a 0.250" diameter bolt. See Figure 2.

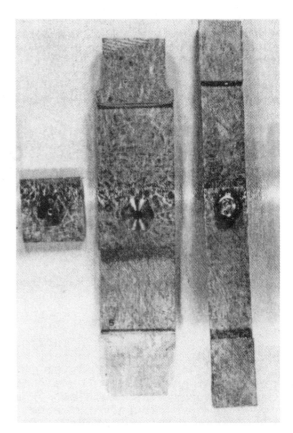

FIGURE 2. JOINTED SAMPLES

The third configuration consisted only of SMC material which was cut 1" wide, 10" long and 0.095 " thick. This was done in order to obtain the strength of the material to be used in calculating joint efficiencies. Joint efficiency is defined as the strength of the joint connection divided by the strength of the material.

The fourth configuration was to be used for a pull-out test as demonstrated in Figure 3. These specimens are 1" x 1" and contain an insert in the center. This was done in order to examine how rigid the insert was held within the SMC (with the aid of the flaired boss which resists the pull-out load). The same configuration of specimen was used in a torsion test in order to determine the insert's ability to resist torque.

Specimen dimensions are summarized in Table 1.

absorption, by weight, was found to be 1.3% for the insert connections, 2.65% for the bolted connections and 2.5% for the pull-out specimens. A graph describing the water absorption is shown in Figure 4.

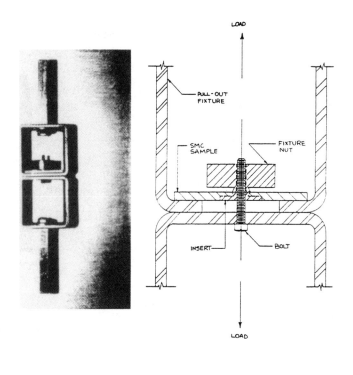

FIGURE 3. PULL-OUT TEST

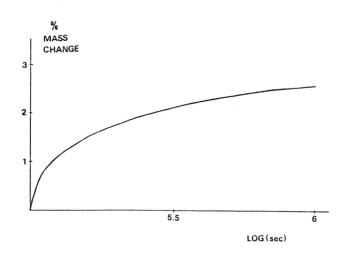

FIGURE 4. WATER UPTAKE OF SMC JOINT SAMPLES

TABLE 1. SPECIMEN DIMENSIONS

	WIDTH	LENGTH	THICKNESS
BOLTED			
(Smaller)	1.000	5	0.095
(Larger)	1.875	5	0.095
INSERT	1.875	5	0.095
CONTROL	1.000	10	0.095

NOTE: All dimensions are in inches

EXPERIMENTAL RESULTS

I. Water Absorption

Five samples of each configuration were placed in water at room temperature for 13 days. These water absorption specimens were weighed daily and the results documented. Water

Comparing the amount of water uptake to previous work [3] shows that the jointed samples absorbed more water than the regular material. It is also noted that the inserted samples absorbed less water than the bolted samples. This is due to the fact that edges were cut during hole drilling for the bolting. The insert connection samples did not have cut edges because the joint area was sealed during molding.

II. Mechanical Tests

Testing of all specimens was done on a MTS 810 Testing Machine of 100 KN capacity with crosshead speed of 5 mm/min. Load versus displacement plots were obtained for all tests and the results were tabulated in Tables 2 to 5. A torque of 60 in-lb was applied to all bolted connections prior to testing. All assemblies consisted of a single lap joint.

Tension tests were initially performed for the control specimens which were used as a basis to obtain the joint efficiencies. The loads sustained by these specimens are recorded in Table 4.

409

TABLE 2. STRENGTH OF INSERT CONNECTION

SPECIMEN #	LOAD (LB)	AVERAGE LOAD (LB)	TYPE OF FAILURE
1W	668.5		IV
2W	642.4		IV
3W	682.6	667.3	IV
4W	688.3		IV
5W	654.9		IV
1D	710.0		IV
2D	668.0		IV
3D	680.0	675.4	IV
4D	654.0		IV
5D	665.0		IV

D = DRY SPECIMEN
W = WET SPECIMEN

TABLE 3. STRENGTH OF BOLTED CONNECTION

SPECIMEN #	LOAD (LB)	AVERAGE LOAD (LB)	TYPE OF FAILURE
1W	584.3		I
2W	522.5		I
3W	505.6	547.2	I
4W	589.9		II
5W	533.7		I
1DS	674.2		I
2DS	589.9		III
3DS	662.9	625.0	II
4DS	573.0		III
1D	652.7		IV
2D	684.2		IV
3D	610.4	646.0	IV
4D	636.6		IV

DS = DRY SPECIMEN (WIDTH = 1 ")
 D = DRY SPECIMEN (WIDTH = 1.875 ")
 W = WET SPECIMEN

TABLE 4. STRENGTH OF CONTROL SPECIMENS

SPECIMEN #	LOAD (LB)	AVERAGE LOAD (LB)	STRESS (PSI)	AVERAGE STRESS (PSI)
1	674.2		7096.4	
2	719.1		7569.5	
3	662.9	730.3	6978.1	7687.8
4	865.2		9107.0	
5	730.3		7687.8	

An average stress of 7688 psi was obtained for the SMC control specimens. This is slightly lower than the results obtained from another batch of SMC (8400 psi) in a previous investigation [4]. Tension tests were also done for the insert and the bolted connection assemblies. Five sets of the water absorption specimens as well as five sets of the dry specimens were used for each type of connection. The plot of tensile load versus displacement for the insert and bolted joint is shown in Figure 5.

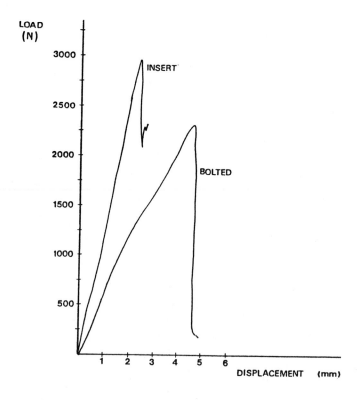

FIGURE 5. PLOT FOR INSERT AND BOLTED CONNECTION (WET)

The plot illustrates the more rigid joint of the insert connection. For the insert specimens, an average load of 675.4 lbs was obtained when dry, and 667.3 lbs when wet. The bolted specimens with the same width (1.875 ") sustained an average load of 646.0 lbs when dry. A load of 625.0 lbs for the dry and 547.2 lbs for the wet specimens of width = 1 " was obtained. A schematic diagram showing the types of failure which occured in the tests is shown in Figure 6. The type of fracture sustained by each of these specimens is indicated in their respective tables.

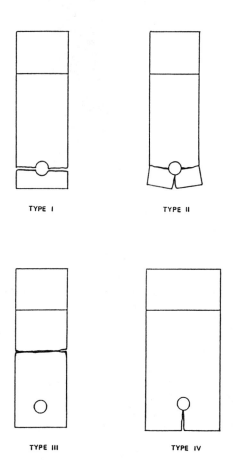

FIGURE 6. SCHEMATIC OF FRACTURED SPECIMENS

Pull-out tests were performed on five dry and five wet SMC specimens. They were tested in such a way that the inserts assumed the total applied load, which was perpendicular to the sample's surface as shown in Figure 3. The maximum load needed to pull out the insert, which was embedded into the specimen, is depicted on the plot of Figure 7 and the values recorded in Table 5. An average pull-load of 170.5 lbs was obtained for the dry specimens and 70.4 lbs for the wet specimens.

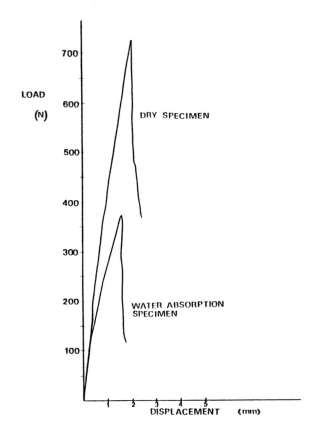

FIGURE 7. PLOT FOR PULL-OUT TEST

TABLE 5. STRENGTH OF INSERTS IN PULL-OUT MODE

SPECIMEN #	LOAD (LB)	AVERAGE LOAD (LB)
1W	59.6	
2W	67.4	
3W	69.7	70.4
4W	70.8	
5W	84.3	
1D	196.6	
2D	182.0	
3D	176.4	170.5
4D	168.5	
5D	129.2	

411

Torque retention was investigated by performing torsion tests. This was done by using a torque wrench and a 6-32 NC cap screw. The bolt was initially threaded into the insert finger-tight. Torque was then applied at a constant rate until the insert would begin to rotate within the SMC specimen. It was found that the Grade 8 cap screw with UTS = 180 000 psi broke without twisting the insert. Therefore, unlike the preliminary design of round inserts, the hexagonal head did not allow any rotation of the insert.

DISCUSSION

Due to the setup on the testing machine and the eccentricity of the applied load, bending of the single lap joint occured when pulled in tension. This bending at the joint consequently reduces the strength of the joint. Less relative movement was detected at the joint with the stainless steel inserts compared to the bolted connections. The preliminary design using aluminum inserts also showed a greater amount of rigidity in the joint, however, a slight bending of the insert flange was observed. The hexagonal head on the inserts was found to possess a greater advantage in torque retention when compared to the simple round inserts. None of the inserts were found to twist under the applied torque, instead the bolt yielded.

Different failure modes were obtained with the bolted connections of width = 1 ". However, the specimens with inserts and the bolted connections of width = 1.875 " failed with type IV mode only. Figure 8 shows the inserted specimens after the tensile test.

The specimens that absorbed water were found to be less efficient than the dry samples. A greater amount of noise, resulting from the cracking of fibers and matrix, was heard when the dry specimens were being loaded in tension compared to the wet specimens. Figure 7 illustrates a plot of a dry and wet specimen which were undergoing the pull-out test. Overall, the dry specimens were found to be 58.8 % stronger than their wet counterparts.

FIGURE 8. INSERTED SAMPLE AFTER TESTING

When tested in tension, little difference was obtained between the dry and wet specimens. The dry insert connections were found to be 1.2 % stronger than the wet specimens. However, a significant difference of 12.4 % can be observed between the dry and wet specimens for the bolted connections. This can be explained by the fact that the inserts seal the edges in the joint area which leads to less water absorption. The bolted connection, however, is exposed to the environment to a greater extent. The insert connections were found to be 49.3 % efficient and the bolted connections were found to be 47.2 % efficient in the tensile test.

A weight comparison for each type of connection was also obtained. The comparison was based on the pitch widths of the bolted and inserted connections. The pitch width is defined as the minimum width whereby failure occurs by shear. The pitch width for the bolted samples is 1.125" and that for the insert samples is 1.500". The weight for the bolted connection assembly was found to be 57.5 g and that for the inserted connection was found to be 70.0 g.

Due to the lower efficiency of bolts, a bolted structure would require more connections. As a result, the final structure would be heavier than a similar structure using the inserts.

CONCLUSION

The objective in using these stainless steel inserts is to increase fastening performance compared to the previous design of the aluminum inserts in terms of rigidity, tensile and pull-out strength as well as torque retention. Water absorbed directly from the environment decreases the joint efficiency. These inserts provide a more rigid joint compared to the bolted joint. A better wet performance of these molded-in inserts versus the bolted connection is obvious. Therefore, the stainless steel inserts provide a flush assembly as well as a convenient means of fastening and can be assembled repeatedly without damage to the SMC material.

REFERENCES

1. Hoa S.V. and Feldman D. " Joining Strength of Sheet Molding Components", Polymer Composites, Vol. 3, No. 1, Jan. 1982.

2. Hoa S.V., Lulham I. and Sankar T.S. "Aluminum Inserts for Fastening Sheet Molding Compounds ", Proceedings of the 3rd International Conference on Composite Structures, Paisley, Scotland, 1985.

3. Hoa S.V. and Ouellette P. " Liquid Absorption of a Sheet Molding Compound ", Polymer Composites, Vol. 2, No. 4, Oct. 1981.

4. Hoa S.V. " Notched Strength of Sheet Molding Compounds ", Polymer Composites, Vol. 2, No. 4, Oct. 1981.

FABRICATION STUDIES ON CARBON/THERMOPLASTIC —MATRIX COMPOSITES

Yasuhiro Yamaguchi, Yoshiaki Sakatani, Mikine Yoshida
Mitsubishi Heavy Industries, LTD.
Nagoya Aircraft Works
Nagoya, Japan

ABSTRACT

For aerospace structural applications, thermoplastic matrix composites have several advantages over conventional thermoset matrix composites including greater damage tolerance, easier processing and lower manufacturing costs.

This paper presents two experimental fabrication studies on continuous carbon fiber/thermoplastic -matrix composites, prepreg tape stacking-press molding and preconsolidated sheet-thermoforming processes.

In the press molding process, properties of two prepreg tapes AS4/PPS and IM7/PEEK were evaluated and processing parameters such as preheating time, pressure-temperature-time and cooling rate were optimized from the mechanical and thermal properties of these composites.

In the thermoforming process, the continuous bend forming of HM/PEEK composite beams with section rolls was studied for application in future large space structures. The process forms composite beams from thin flat sheet to hat-sections continuously with multi-step section rolls. Fabrication tests were carries out on experimental forming machine to optimize preliminary processing conditions. After the quality evaluations of the resultant composite beam specimens, the optimum forming roll configurations, temperature and roll feed rate were found.

PRESS MOLDING

The objective of this fabrication study was to determine the moldability of today's thermoplastic matrix prepreg tapes compared with the conventional thermosetting epoxy prepregs in aircraft structural parts. First, the properties of prepreg tapes were evaluated and press molding process variables such as preheating time, pressing time and cooling rate were optimized, using the quality and properties of the resultant molded laminates. Then angle shaped laminates were fabricated with the optimized press molding conditions to evaluate the processability of the prepreg tapes for complex shaped components. Toughness, one of the typical mechanical properties, was also evaluated, using the test method of compression strength after impact to compare the thermoplastic prepreg with today's epoxy prepregs.

MATERIALS—For the unidirectional prepreg tapes, the two materials were selected, AS4/PPS (Phillips) and IM7/PEEK (ICI) shown in Fig.1, because they were the only materials commercially available with properties which would potentially meet aerospace structure specifications. Table 1 shows the construction of these tapes provided by suppliers[1],[2] and Table 2 shows the properties obtained from as recieved prepreg tapes. Fig.2 shows the well impregnated cross section of the IM7/PEEK prepreg.

EXPERIMENTAL STUDY—Standard unidirectional laminates were press molded with the procedure provided by the prepreg suppliers, using the press molding systems shown in Fig.3. Processing parameters ; preheating time, pressure holding time and cooling rate were selected. Fig.4 shows the cross section of IM7/PEEK prepregs before and after molding. Both AS4/PPS and IM7/PEEK were well impregnated and consolidated after molding. The fracture surface of IM7/PEEK shown in Fig.5 indicates that adhesion between the fibers and the matrix was fairly good.

Table 3 shows the molding conditions and the resultant laminate properties. Preheating time and pressing time did not effect the laminate properties but cooling rate for PPS did have an effect on the Tg. A Slow rate raised the Tg to 400K causing much more crystallizations. PEEK is superior to PPS as a structural material because of its higher Tg and interlaminor shear strength but it requires higher molding temperature and pressure.

FABRICATION STUDY—To evaluate processability for complex shapes, angled components were press molded with IM7/PEEK prepreg tape according to the above optimized process. Fig.5 and Fig.6 show the fabrication sequence of the angled laminate. It's not as easy to lay-up and drape the prepreg tapes to the mold as it is with the thermosetting epoxy prepregs. The local heating and welding by soldering iron were found to be best for laying them up and draping them to the mold. However it's very easy to press them in the mold because of no outside resin flow and such a short pressing time, 10 min. . The resultant press molded laminates satisfied quality and dimension requirements.

TOUGHNESS EVALUATION—Thermoplastic matrix composites have many advantages over thermosetting epoxy composites, including greater toughness.

As the toughness evaluation, compression strength test after impact (CAI) often used for the aerospace structural composites was selected.

Thirty-two ply quasi-isotropic laminates of IM7/PEEK were press molded and impacted at three energy levels as shown in Fig.7 and 8. After NDI by ultra sonic to determine the damaged area, compression tests were carried out at room tempearature. Fig.9 and 10 show the resultant CAI and damaged area compared with today's carbon/epoxy composites including tough epoxy systems. Figures 11 to 14 show corss sections of the laminates just after impact and after compression tests. CAI of IM7/PEEK was superior to those of conventional epoxy and modified epoxy systems, but a little bit lower than that of tough epoxy 8551-7 (Hercules) and the damaged area was larger than that of epoxy systems. The reason(s) for these phenomena is not clear because the mechanism of the impact damage and compression failure is very complex and still under development. Further work will be required.

THERMO FORMING

The numbers of applications for carbon/resin composite which have light weight, high modulous and good dimensional stability have been increasing for aerospace structures. Studies on fabricating large structures in-space by using carbon/thermoplastic such as PS, PES have been carried out, mainly in the USA[3],[4]. However, forming techniques for carbon/thermoplastic composites are still under development[5]. Detailed techniques for high temperature resistant polymer matrix composite materials, such as PEEK, are not well known yet. Therefore basic studies of thermo-bendforming processes to make composite beam continuously were carried out. By this process, carbon /PEEK beams were continuously bend-formed from flat strip materials to hat-cross section beams through the multistep hot and cold section rolls as shown in Fig.15.

MATERIALS—Thin preconsolidated flat laminates of high modulous carbon fiber/PEEK composites were selected as the raw material strips. Table 4 and Fig.16 shows the construction, roperties and geometries of the laminate strips. For space applications high modulous carbon fibers were laminated in the longitudinal direction to obtain high modulous and good dimensional stability. And a glass scrim was laminated in the±45° direction in the center of the laminate to allow it to bend more easily.

EXPERIMENTAL FORMING MACHINE—For the forming studies, the basic experimental forming machine was designed and manufactured (Fig.18).
The machine was designed to form flat strip sheets into hat-shaped beams through the following steps.
Flat strip sheet was fed into the machine, softened at the intra-red heating unit, bend-formed into the desired shape step by step through hot forming rolls and is fully consolidated through cold rolls. The heating and cooling temperatures, forming speed and roll pressures are designed to be controlled by a controll unit.
The cross section of each roll was determined by a radius changing system in the design of a forming flower system as shown in Fig.17.
In this flower system, the bend radius of the first step roll was made larger. After that cross section changes during bend forming were gradually made. The forming line was located at the center.
For the continuous bend-forming of thermoplastic matrix composites by using multi-step section rolls without changing their qualities, it is necessary to heat-soften the materials which show elastic deformation at room temperature, and to bend gradually into the desired shape by the plastic deformation at high temperature. In this machine, 4 hot rolls were adapted to the shapes according to above forming flower system.
FORMING STUDY—After the evaluation of the raw material's thermal properties, the bendability by a static bending test at high temperature and the machine's heating capability, basic forming conditions such as forming temperature range and forming speed were selected. Then, under these settled conditions, experimental forming tests wer carried out.
From the results of these forming tests beams of good geometries and qualities (Fig.19) were obtained.

The conditions and properties shown in Table 5. The properties and qualities of the resultant formed beams were almost equal to those of the raw material laminates before forming. Through the forming tests it was found that the forming temperature has a great effect on the quality of the beams. At the proper temperature of 598K the material strip showed plastic deformation in the width direction with good quality. However, at low temperature, the strip showed local wrinkles and at high temperature delaminations.

CONCLUSION

Two experimental fabrication studies were carried out using continuous carbon fiber/thermoplastic-matrix composites which have potential of greater damage tolerance, easier processing and lower manufacturing cost.
In the prepreg tape lay up-press molding process primarily IM7/PEEK was evaluated and it showed good processability through angled shape laminate fabrication studies. And also press-molded IM7/PEEK laminate showed superior toughnes, CAI, to conventional epoxy composites.

In the continuous thermo bend-forming process using preconsolidated HM carbon/PEEK laminate, the optimum process conditions which form composite beam from thin flat sheet to hat-section with multi-step section rolls were found.
Both above process indicate the feasibility to be applied these thermoplastic composites to aerospace structural components with many advantages in fabrications.

ACKNOWLEDGMENTS

The authors wish to thank Phillips, ICI and Sumitomo Chemical Co. for providing materials.
The thermoforming process of this paper was performed under the management of the Research and Development Institute of Metals and Composites for Future Industries

as a part of the R & D project of Basic Technology for Future Industries sponsored by the Agency of Industrial Science and Technology, Ministry of International Trade and Industry, of Japan.

REFERENCES

1. Phillips Catalogue
2. ICI Catalogue
3. L.M. Povermo et al, Composite Beam Builder, SAMPE J., Jan./Feb., P7 (1981)
4. D.E. Beck, Continuous Graphite/Polysulfone RP Thermoforming for Large Space Structure Construction, Proceeding of 38th SPI, Session 20-C (1983)
5. M.T. Harvey, Thermoplastic Matrix Processing, Composite vol.1, ASM International, P544 (1987)

Fig 1.   Carbon/PPS · Carbon/PEEK Prepregs

Table 1. Construction of Material

Prepregs (Commercial Name)		Carbon/PPS (PPS AC40-60)	Carbon/PEEK (APC-2)
Supplier		Phillips	ICI
Carbon Fiber	Material	AS 4	IM7
	Specific Gravity	1.80	1.78
	Filament Counts	12,000	12,000
	Tensile Strength (GPa)	3.77	4.97
	Modulus (GPa)	233	281
	Strain (%)	1.53	1.70
Resin	Material	Polyphenylene Sulfide (PPS)	Polyetheretherketon (PEEK)
	Specific Gravity	1.34	1.30
	Tg (K)	360	416
	Tm (K)	560	607

Table 2.   Properties of Prepreg Materials

Prepregs	AS 4/PPS	IM7/PEEK
Resin Content (Wt%)	39.9	32.0
Vf (vol%)	53.0	61.0
Areal Fiber Weight (g/m²)	160	145
Laminate Ply Thickness(mm/ply)	0.150	0.125
(Thickness as Prepreg)	(0.21)	(0.17)
Range of Prepreg Thickness*(%)	+14.3	+ 5.9
	− 9.5	−11.8
Width of Prepreg (mm)	100	146

$$* \text{ Range (max)} = \frac{M_{max.} - M_{av.}}{M_{av.}} \times 100(\%)$$

$$\text{Range (min)} = \frac{M_{min.} - M_{av.}}{M_{av.}} \times 100(\%)$$

Fig 2. SEM of IM7/PEEK Prepreg

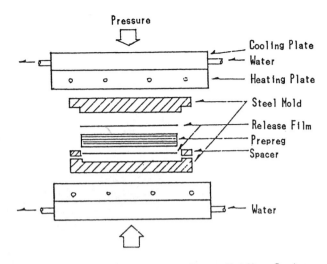

Fig 3. Fabrication Assembly of Press Molding System

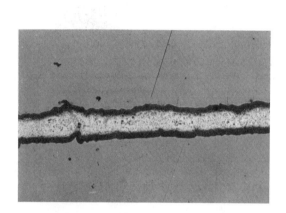

Prepreg

Molded Laminate   0.1mm

Fig 4. Cross-Section of IM7/PEEK Prepreg and Laminate

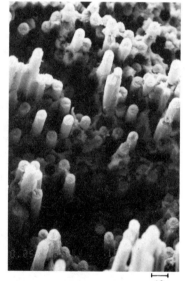

10 μm

Fig 5. SEM of Failed Surface

Table 3. Results of Molding Test of Uni-Directional Laminate

Material	Molding Conditions					Laminate Properties				
	Pre Heat Time (min)	Temperature (K)	Pressure (MPa)	Time (min)	Cool Down Rate (K/min)	Ply Thickness (mm/ply)	Vf (vol%)	Void Content (vol%)	Tg (K)	Short Beam Shear Strength (MPa)
AS4/PPS	17	588	9.8*	3	1	0.157	52.8	0	400	53
	17	588	9.8*	6	9	0.153	53.1	0	378	56
	3.7	588	9.8**	6	11	0.139	53.2	0	376	57
AS4/PEEK	25	653	13.7*	5	1	0.124	62.2	0	432	101
	5	653	13.7**	5	13	0.102	63.1	0	431	99
IM7/PEEK	10	653	13.7**	10	10	0.131	61.4	0	432	133

* Thickness Const
** Pressure Const

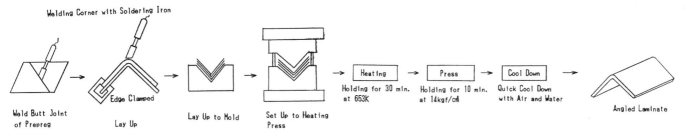

Welding Corner with Soldering Iron

Weld Butt Joint of Prepreg

Edge Clamped

Lay Up

Lay Up to Mold

Set Up to Heating Press

Heating — Holding for 30 min. at 653K

Press — Holding for 10 min. at 14kgf/cm²

Cool Down — Quick Cool Down with Air and Water

Angled Laminate

Fig 5. Fabrication Sequence of Angled Laminate

Weld Butt Joint of Prepreg

Lay Up

Lay Up to Mold

Set Up to Heating Press

Angled Laminate

Fig 6. Fabrication Sequence of Angled Laminate

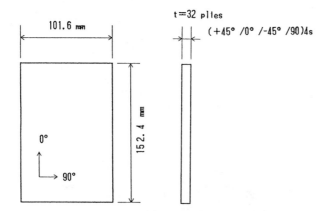

t=32 plies
(+45° /0° /-45° /90)4s

101.6 mm

152.4 mm

0°

→ 90°

Fig 7. Specimen

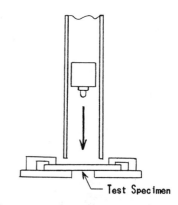

Test Specimen

Fig 8. Impact Test Method

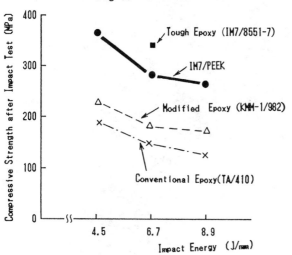

Tough Epoxy (IM7/8551-7)

IM7/PEEK

Modified Epoxy (KMM-1/982)

Conventional Epoxy(TA/410)

Fig 9. Results of Compression Test after Impact Test

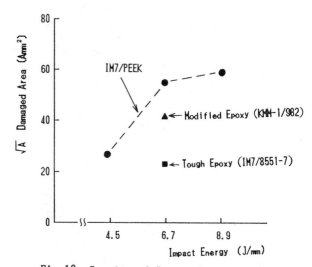

IM7/PEEK

Modified Epoxy (KMM-1/982)

Tough Epoxy (IM7/8551-7)

Fig 10. Results of Impact Test

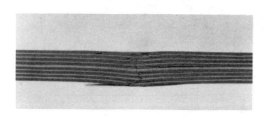

Fig 11. Modified Epoxy After Impact

Fig 13. Modified Epoxy After Compression Test

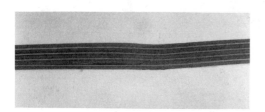

Fig 12. IM7/PEEK After Impact

Fig 14. IM7/PEEK After Compression Test

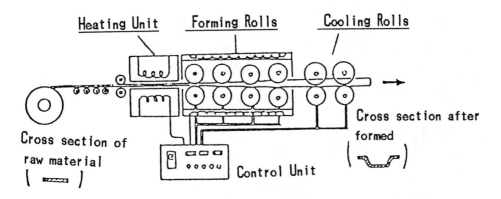

Fig 15. Basic Concept of Continuous Bend Forming Process

Table 4. Raw Material Sheet

Material (Supplier)	Material Constructions		Laminate Constructions	Properties of Carbon Fiber		Thermal Property	
	Reinforcement Fiber	Resin					
Carbon/PEEK Laminate (Sumitomo Chemical Co.)	High Modulus Carbon (HM-CF) Vf : 40~60%	Polyetheretherketon (PEEK)	(0°) with ±45° Glass Scrim	Filament Counts : 12000 Density : 1830kg/m³ Tensile Strength : 1910 MPa Tensile Modulus : 343 GPa		Tg 438K T_{HDT} 606K	

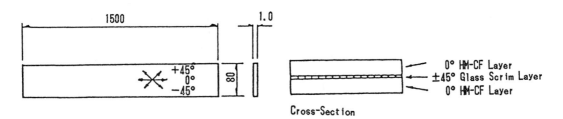

Fig 16. Test Specimen

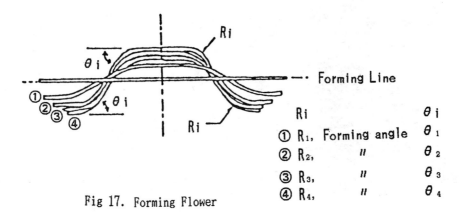

Fig 17. Forming Flower

	Ri		θi
①	R_1,	Forming angle	θ_1
②	R_2,	"	θ_2
③	R_3,	"	θ_3
④	R_4,	"	θ_4

Fig 18. Basic Experimental Forming Machine

Table 5. Results of Continuous Forming Tests

Forming conditions			Forming speed	Thickenss range ***	Void content	Tensile modulus (GPa)	Forming angle **
Forming temp. K	Forming roll angle * θ_A / θ_B	Cooling roll angle * θ_A / θ_B	m/h	%	vol%	(Data range%)	θ
598	80 / 70	69 / 64	18	+ 1.9 − 3.9	0.7 Max. 0.8	174 $\left(\begin{array}{c}+ 5.2 \\ - 4.0\end{array}\right)$	60~62°

*Roll angle   **Forming angle

$$*** \quad \text{Range (max)} = \frac{M_{max.} - M_{av.}}{M_{av.}} \times 100(\%)$$

$$\text{Range (min)} = \frac{M_{min.} - M_{av.}}{M_{av.}} \times 100(\%)$$

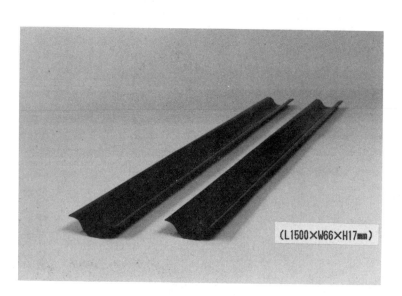

(L1500×W66×H17mm)

Fig 19. Beams After Formed

THERMOPLASTIC FOR
ADVANCED COMPOSITES

Steve Clemans, Tim Hartness
BASF Structural Materials, Inc.
Thermoplastic Composites
Charlotte, North Carolina   USA

Abstract

As high performance thermoplastic composites evolve as primary and secondary aircraft structural materials, there exists a need to offer to the community a material form that greatly broadens the selection of thermoplastic matrix materials. As one looks at the choices of thermoplastic prepregs available on the market today, an evaluation must be given to the pros and cons of that individual material. Today, hot-melt prepregs result in a stiff, boardy form with no tack. This form severely limits the handleability in lay up on complex contour tooling. If one attempts to offer tacky and drapeable prepreg, it is often at the expense of lengthy processing that allows solvent removal without void formation.

In the case of those materials that exhibit some reactivity, such as pseudothermoplastics, the incorporation of a solvent to this point has been unavoidable. The other major issue surrounding pseudothermoplastics often concerns the evolution of reaction by-products that must be managed usually by a time consuming and delicate process cycle.

The ability to offer thermoplastic prepregs that are drapeable and tacky, yet incorporate a wide range of thermoplastic and pseudothermoplastic matrix materials on a reinforcement of choice with no evolution of reaction by-products or solvents would greatly increase the choices of matrix material and at the same time offer improved processing in a cost effective manner. BASF has recently demonstrated this capability using a newly developed powder impregnation technology and is currently rapidly developing this technology as a commercial process.

WITH THE CONTINUING INTEREST and development of thermoplastic matrix composites, a decision was made to establish a separate division within BASF Structural Materials. This group called Thermoplastic Composites (TPC™) has the task of developing new and improved thermoplastic matrix composites. It is the purpose of this paper to report on the emergence of what is considered by some as a revolutionary development in the thermoplastic matrix composite community. The development consists of the ability to prepreg a wide range of thermoplastics that possess a wide range of melt viscosities without the use of solvents or hot-melt technology thus providing an increase in polymer selection, as well as improved processing, as demonstrated by improved handling and consolidation.

EXPERIMENTAL

POLYMER SELECTION - The current effort in polymer evaluation within TPC has been to evaluate those polymers that are considered aerospace grade. Most of these materials have been developed as matrix resins, but due to various reasons their potential has not been fully realized. When one evaluates the current prepregging techniques being utilized today, it becomes clear where potential problems exist. Prepregging techniques seem to be a function of the polymer type.   In the case of semicrystalline polymers, such as polyetheretherketone (PEEK) and polyphenylene sulfide (PPS), there are no solvents available in which to prepreg. The choice to date has been to melt impregnate. While this technique produces excellent wetting of the reinforcement, it also produces a prepreg that is stiff and boardy and lacks tack. In addition, when one must hot-melt impregnate high molecular weight polymers such as PEEK and PPS, one has obvious limitations on how high a melt viscosity one can handle in the prepregging operation without using diluents to lower the viscosity as well as the danger of thermally degrading the polymer. The next thermoplastic family that is currently being prepregged is the amorphous family. In this case, the choice has been to use solvents. The choice of which solvents is a function of the solubility of polymer. When high boiling polar solvents, such as NMP must be used, it is often difficult to remove these during processing. Although solvent impregnation often provides a drapeable and tacky product, this character can change as evaporation occurs. This amorphous family includes such polymers as the polsulfone family, polyetherimide, and some thermoplastic polyimides. Here again, consideration must be given to molecular weight as the solubility of a given polymer is a function of molecular weight. The last group of materials that are being considered as thermoplastic but more correctly called "pseudothermoplastic" are those that undergo some reaction chemistry during the processing cycle. These materials are more complex and should be understood. Presently, these materials must be handled as low molecular monomers or prepolymers. This allows the material in this form to be put into solution for prepregging. The solvent of choice has often been a high boiling solvent such as NMP or THF. These solvents historically have been very difficult to remove from the prepreg, yet are often necessary in order to give adequate wetting during the processing cycle. Great care has often been required to slowly diffuse out of the composite any residual solvent. Unsuccessful removal may result in blistering and thus rejection of that part. With some condensation polymers additional consideration must be given to condensation products that evolve during processing. This characteristic means that very careful processing must be followed in terms of heat up rates and the application of pressure. Pseudothermoplastics, such as polyamideimide and thermoplastic

polyimides, are considered in this class. Essentially, these materials process similar to thermosets, yet possess the quality of thermoplastics in terms of toughness. In the past, these materials could only be handled in the manner described. Recently, a thermoplastic polyimide (Larc-TPI)(1) was processed without going through the amic acid route. In this case, the partially imidized version was prepregged using a powder version. This concept may be applied to other pseudothermoplastics if certain technical issues are overcome. One of the main concerns centers around distributing a uniform amount of powder around the fiber and at the same time having adequate flow to generate a high quality composite. Table I shows a list of potential candidates that may be candidates for the powder technology. A brief description follows for each polymer.

POLYETHERETHERKETONE (PEEK) - There are three grades of neat resin available from ICI Americas. The 150P, 380P, and 450P grades are sold for different applications and are indicative of increasing molecular weights, as well as mechanical properties. It is believed that the prepreg version from ICI, called APC-2, falls in between 150P and 350P as far as molecular weight is concerned. As is observed, the Tg is the same for all three grades. It is expected that the higher grades should offer improvements in interlaminar fracture toughness. How much, has yet to be determined. Other issues, such as fiber resin interface development as a function of molecular weight need also to be considered. Preliminary mechanical property data has been generated on the three grades and is shown in Table II. The fiber used was Celion G30-500 unsized. As can be observed, excellent property translation is being generated. Additional data, such as Mode I and Mode II, will be necessary to answer the question of fracture toughness versus molecular weight.

POLYETHERKETONE (PEK), POLY (EKEKK) ULTRAPEK, AND POLY (EEKK) - The following polymers are available from BASF with some grinding trials indicating that PEK and PEEK are grindable into the correct form for prepregging. The Ultrapek has been less successful but efforts are still ongoing. No composite data has been generated to date, but if successful, would offer an increase in glass transition over the current PEEK product and possibly an improvement in mechanical properties.

POLYIMIDE (LARC-TPI) - This amorphous pseudothermoplastic has been under development for some time. This polymer was developed by NASA Langley, and has been licensed for manufacture by Mitsui Toatsui in Japan and Rogers Corporation. Larc-TPI has undergone some evaluation in the amic acid and powder form. The amic acid approach has been for the most part unsuccessful.(2) The powder approach has shown some promise due to its improved melt flow.(1) There has been some speculation that the LARC-TPI currently supplied by Mitsui Toatsui is inherently limited to a low molecular weight, thus resulting in less than optimum performance as a matrix resin. Recent work by Rogers Corporation to develop a version of LARC-TPI shows considerable improvement in neat resin properties (Table III) versus bars molded from Mitsui Toatsui powder. Table IV shows the mechanical properties generated from the Mitsui Toatsui LARC-TPI. No mechanical properties have been generated from the LARC-TPI developed by Rogers Corporation but the work is ongoing.

POLYETHERSULFONE (HTA) - This amorphous thermoplastic from ICI Americas is a high molecular weight form with an excellent glass transition. The structure is shown in figure I. The major concern with this polymer is its solvent resistance since it is an amorphous material. The material has been made into an acceptable form for prepregging and good progress is being made. POLYETHERIMIDE (ULTEM) - This amorphous thermoplastic from the General Electric Company has been available in several grades. The higher molecular weight with improved Tg and chemical resistance is of greatest interest. Efforts are ongoing to have useful polymer prepared for prepregging.

POLYIMIDE 2080 - This amorphous thermoplastic is a fully imidized and polymerized polyimide. Traditionally, the material has been put into solution in a high boiling solvent like NMP for prepregging. The material is now supplied from Lenzing in Austria. Positive attributes for this material are its exceptional high temperature properties, with a Tg of 310°C (589°F). Previous work indicates that molding pressures higher than standard autoclaves may be needed to mold higher quality laminates. Neat resin data indicates a strain-to-failure of 10% which indicates the polymers toughness.

POLYIMIDE-MATRAMIDE 9725 - This amorphous thermoplastic is also a fully imidized and polymerized polyimide. The material is supplied from Ciba-Geigy. The polymer offers outstanding thermal properties with a Tg in excess of 280°C (536°F). Neat resin stain-to-failure is reported to be 50%. The material possesses a unique characteristic in that it is soluble in a wide range of solvents yet becomes insoluble after apparently some crosslinking at 600°F for 16 Hrs. The material has the potential to offer exceptional properties if processing conditions can be developed.

PMR-15 - The inclusion of a thermoset with thermoplastics may seem inappropriate but a thermoplastic characteristic after imidization makes it an ideal candidate for powder technology. Traditionally this material has been supplied in an amic acid solution of alcohol with three monomers. This solution is then used to impregnate fibers which are then processed using conventional thermosetting procedures, usually in an autoclave. Recently Hysol Composites introduced a fully imidized PMR-15 powder to be used for molding or combining with reinforcements to make composites. The as received, fully imidized PMR-15 molding powder is in a temporary thermoplastic condition and can be taken advantage of if sufficiently low viscosity is reached around 220°C (430°F). This is all done without release of volatiles. The material is then taken to the 288-316°C (550-600°F) range where it is rapidly cured to a thermoset. It is also recommended that a post cure be completed in order to maximize thermo-oxidative stability. Currently, this is the most widely used high temperature polyimide. The ability to fabricate high quality laminates while eliminating many of the existing problems associated with the traditional material should make a significant impact on this high volume industry.

COMPOSITE PROPERTIES

To date limited composite properties have been generated from the matrix resins listed in Table I. Listed in Table II are typical composite properties generated on PEEK 150, 380, and 450. As is shown, excellent mechanical properties are being translated. Additional data is being generated, including fracture toughness in Mode I and Mode II. This may help decide the issue of increase in fracture toughness versus molecular weight. Typical photomicrographs and SEM's of the fracture surface are shown in figs 2 & 3. Date generated earlier on Mitsui Toatsui Larc-TPI is shown in Table IV. As was noted earlier, no composite mechanical properties have been generated on the Rogers Larc-TPI matrix resin. A SEM of the fracture surface is shown in fig 4. Initial flexural data has been generated on the polyethersulfone (HTA). A decision was made to evaluate this material on a higher modulus fiber. Initial work has been on Celion G40-700 unsized. Data generated to date is shown on Table V. One of the important issues concerning amorphous thermoplastics and intermediate modulus fibers, such as, G-40-700 is fiber/matrix interface development. The lack of good development is often the suspected cause of lower than expected compression properties. Examination of the fracture surface using SEM to examine the interface for resin adhesion is scheduled.

Data generated on PMR-15 powder impregnated Celion G30-500 is shown on Table VI. Photomicrographs indicate void free laminates with excellent translation of fiber properties (fig. 5) . Post cure conditions will remain the same as the solvent amic acid version of PMR-15. Table VII shows the current processing conditions for the powder version.

CONCLUSIONS

The ability to uniformly impregnate reinforcements that exhibit good tack and drape with a wide variety of thermoplastics and potentially thermosets as well, has been demonstrated. Composite properties generated to date show excellent property translation. The ability to offer composite materials with simple chemistry and ease of processing has many obvious advantages.

REFERENCES

1. J. T. Hartness, University of Dayton Research Institute, Preprint, 32nd International SAMPE Symposium and Exhibition, Vol. 32. P. 154-168 (1987).
2. N. J. Johnston and T. St. Clair, NASA Langley Research Center, Preprint, 18th National SAMPE Technical Conference (1987).

TABLE I
HIGH PERFORMANCE POLYMERS

POLYMER	FAMILY TREE	DRY Tg(°C)	Tm(C°)	PROCESSING TEMP.(°C)
Polyetherether Ketone .150P PEEK .380P ICI .450P	Semi-cry Thermoplastic	143	343	360-400
Polyether Ketone (PEK) BASF	Semi-cry Thermoplastic	165	365	400-450
Poly (EKEKK) BASF (Ultra Pek)	Semi-cry Thermoplastic	173	370	420-450
Polyimide (Larc-TPI) Mitsui Toatsui Rogers	Amorphous Pseudothermoplastic	250	325	~350
Polyether Sulfone (HTA) ICI	Amorphous Thermoplastic	260	-	400-450
Polyetherimide (Ultem 1000-6000) GE	Amorphous Thermoplastic	217	-	~350-400
Polyetherimide (P-IP) Mitsui Toatsui	Semi-cry Thermoplastic	270	380	380-420
Polyimide (PMR-15) Hysol	Amorphous Thermoset	320	-	315
Polyimide 2080 Lenzing	Amorphous Pseudothermoplastic	280	-	350
Polyimide (Matramide 9725) Ciba-Geigy	Amorphous Pseudothermoplastic	265	-	350

429

TABLE II
PEEK COMPOSITE
PHYSICAL TEST DATA FOR POWDER IMPREGNATION

TEST 0°-4 PT FLEX	150 PEEK/CELION G30-500	380 PEEK/CELION G30-500	450 PEEK/CELION G30-500
Strength (Ksi)	245 (323)*	258.4 (340)*	250.8 (330)
Std. Dev.	4	6	19
%CV	1.3	1.9	5.7
Modulus (Msi)	18.12 (19.7)*	17.9 (19.5)	19.2 (20.9)
Std. Dev/	0.53	0.47	0.36
%CV	2.7	2.4	1.7
90° Tensile			
Strength (Ksi)	9.2		10.1
Mod (Msi)	1.40		1.43
% Elong	0.70		0.71

*   ASTM D-790 4 Point Flex Test with Load Noise Modification

TABLE III
Larc-TPI Neat Resin
Mechanical Properties

Mitsui Toatsui

Tensile Strength (Ksi)	16.4
Tensile Modulus (Msi)	659
% Strain-to-failure	3

Rogers

Tensile Strength (Ksi)	22.2
Tensile Modules (Msi)	620
% Strain-to failure	8

TABLE IV

MECHANICAL PROPERTIES

0°

| Panel # | Temperature | | 3-Point Flex | | 4-Point Flex | | 4-Point Shear | SBS | | |
	°C	(°F)	Strength MPa (ksi)	Mod. GPa (msi)	Strength MPa (ksi)	Mod. GPa (msi)	MPa (ksi)	MPa (ksi)	FV %	Void %
3		RT	1946 (282.4)	121 (17.6					49.7	2.47
3	177	(350)	1288 (187.0)	126 (18.3)						
5	177	(350)	1609 (233.6)	126 (18.3)						
6		--	--	--			40 (5.8)			
9		RT	1631 (236.7)	141 (20.5)					57.10	2.00
13		RT	1617 (234.7)	116 (16.9)	983 (142.7)	92 (13.4)				
16		RT	1414 (205.3)	--						
17		RT	1962 (284.7)	113 (16.4)	951 (138.1)					
18		RT	--	--	--	--	54 (7.8)	63 (9.1)		
19		RT	2309 (335.1)	149 (21.6)	1753 (254.4)	112 (16.3)			58.05	4.11

Note: All specimens are an average of three.

3 Pt Flexural   Span-to-Depth Ratio 32-1, ASTM-D-790

4 Pt Flexural   Span-to-Depth Ratio 32-1, ASTM-D-790

4 Pt Shear   Span-to-Depth Ratio 16-1

SBS   Span-to-Depth Ratio 4-1

432

TABLE V
HTA/G40-600
0°-4-pT Flexural *

Strength (Ksi)   281

Modulus (Msi)   21.9

* ASTM D-790, 4 pT Flexural with load nose modification
(Span-to-depth Ratio 32:1)

TABLE VI
HYCOMP M-100 CELION G30-500

TEST	STRENGTH	MODULUS	CONDITION
RT-0°-4pT. Flex	284 Ksi	17.5 Msi	Mod. D790
	243 Ksi	16.0 Msi	No postcure Std. D790 No postcure
	295 Ksi	19.2 Msi	Mod. D790 Postcure*
RT-90°-4pT Flex	6.87 Ksi	1.38 Msi	No postcure
	6.23 Ksi	1.02 Msi	Postcure
500°F--90°-4pT.	8.63 Ksi	1.42 Msi	Postcure
RT-0°-Tensile	254.1 Ksi	19.3 Msi	Postcure
RT-90°-Tensile	4.42 Ksi	1.05 Msi	No postcure
RT-90°-Tensile	4.37 Ksi	0.98 Msi	Postcure
RT-SBS	10.9 Ksi		No postcure
RT-SBS	16.7 Ksi		Postcure

*Postcure: 13 Hr. 600°F cir air oven

TABLE VII
PMR-15 HY COMP M-100
PROCESSING CONDITIONS

- PREDRY STACK 1 HOUR @ 250 F IN VACUUM OVEN.

- PREHEAT PRESS TO 430 F.

- INSERT MOLD IN PRESS AND APPLY 200 PSI.

- HEAT RT-----> 430 F. AS FAST AS POSSIBLE.

- HOLD 430 F FOR 5 MINUTES.

- HEAT 430 F-----> 560 F @ 5 F /MINUTES

- HOLD 1 HOUR @ 560 F.

- COOL UNDER PRESSURE @ 5 F/ MIN. TO 200 F OR LOWER

- REMOVE

- POSTCURE: STANDARD OR AS FUNCTION OF USE TEMP.

FATIGUE PERFORMANCE OF GLASS REINFORCED THERMOPLASTICS

David W. Adkins, Ronald G. Kander
E. I. du Pont de Nemours & Company
Wilmington, Delaware   USA

ABSTRACT

New data show that fatigue strength for glass reinforced thermoplastics declines linearly with the logarithm of load cycles out to about one million cycles. This behavior is similar to that of glass filled composites made from a variety of other resins and fiber arrangements described in the literature. The rates of decline are within a fairly narrow range, allowing good fatigue strength estimates from a simple rule. Beyond a million cycles the relationship between fatigue strength and load cycles appears to be more complicated. Testing and damage mechanics analysis are underway to define and explain long term performance.

BACKGROUND AND INTRODUCTION

Composites for automotive applications benefit distinctly from thermoplastic matrix resins. In addition to their inherent mechanical properties, such as toughness, thermoplastic composites are amenable to fast reproducible molding and can be reclaimed. Thermoplastics are solvent free, non-toxic and easy to handle safely. Automotive composites are reinforced almost exclusively with glass because it is inexpensive, strong and relatively stiff.

Understanding long term mechanical behavior is important for automotive structural applications. This paper presents fatigue data for several new composite systems, as shown in Table I. The resins include nylon, polycarbonate and several modifications of thermoplastic polyester. The polyesters are modified with conventional additives, and are identified as "toughened", "crystallized", and "stabilized". The fiber arrangements include continuous unidirectional, random chopped strands, and injection molded short fibers. The unidirectional and chopped strands were coated with 50 volume percent resin by a proprietary DuPont process.

TESTING PROCEDURE

As shown in Table I, continuous fiber systems were loaded in three point bending. The test geometry followed ASTM D790 with span to depth ratio of 16. The discontinuous fiber systems were loaded in tension. The random chopped strand composite was tested as ASTM D3039 coupons because of longer reinforcement. Short fiber reinforced composites were tested as D638 Type 1 coupons. All materials were tested at room temperature. The random chopped strand composites were also tested at 120C.

All samples were fatigued by cycling the load sinusoidally from a selected level to one tenth of that level (R = 0.1). Minimum and maximum values of both load and deflection were recorded during flexural tests. The load frequency was fixed for each coupon at a rate low enough to avoid hysteresis heating. All tensile tests were performed at five Hertz and flexural tests were performed at one to three Hertz depending on stress level. Enough samples and stress levels were tested for most systems to provide design allowable data according to ASTM E739.

This typically requires at least fifteen total tests at five different stress levels.

Tensile fatigue tests were run to catastrophic coupon failure. Failure in flexural fatigue tests was determined by monitoring changes in the coupon apparent bending stiffness. This stiffness is defined as the cyclic load range divided by the cyclic deflection range for a given cycle. This value is equivalent to the dynamic spring constant of the coupon beam.

Figure 1 shows apparent bending stiffness versus flexural fatigue cycles for a typical coupon. Stiffness remained essentially constant for most of the test and then dropped rapidly. We defined flexural fatigue life as the cycle count corresponding to a ten percent stiffness decrease. Complete failure always occurred shortly after this point. This definition provided unambiguous endpoints that consistently corresponded with coupon failure. All visible damage of the unidirectional flexural coupons occurred on the tensile side.

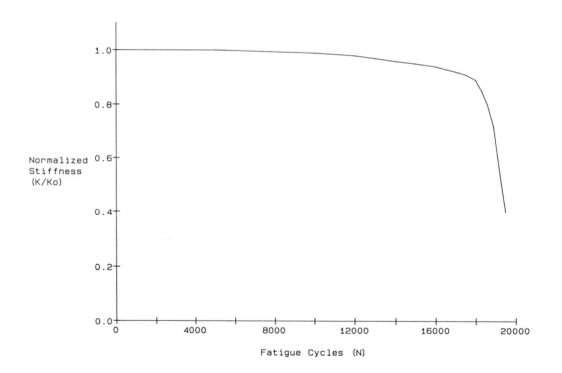

Figure 1   Flexural coupon apparent bending stiffness versus fatigue cycles

Table I
MATERIALS AND TEST CONFIGURATIONS

Matrix Resins	Fiber Arrangement	Test Configuration	Temp(C)
Unmodified PET	Continuous Uni.	3-Pt Bend D790	RT
Toughened PET	Continuous Uni.	3-Pt Bend D790	RT
Crystallized PET	Continuous Uni.	3-Pt Bend D790	RT
Stabilized PET	Continuous Uni.	3-Pt Bend D790	RT
Nylon 66	Continuous Uni.	3-Pt Bend D790	RT
Polycarbonate	Continuous Uni.	3-Pt Bend D790	RT
PET	Inj. Molded Rynite* 545	Tension D638	RT
Nylon 66	Inj. Molded "GRZ" 033	Tension D638	RT
Unmodified PET	Random Chopped Strand	Tension D3039	RT
Toughened PET	Random Chopped Strand	Tension D3039	RT
Toughened PET	Random Chopped Strand	Tension D3039	120
Crystallized PET	Random Chopped Strand	Tension D3039	RT
Crystallized PET	Random Chopped Strand	Tension D3039	120
Stabilized PET	Random Chopped Strand	Tension D3039	RT
Stabilized PET	Random Chopped Strand	Tension D3039	120

* DuPont's registered trademark for polyester
   engineering thermoplastic resin

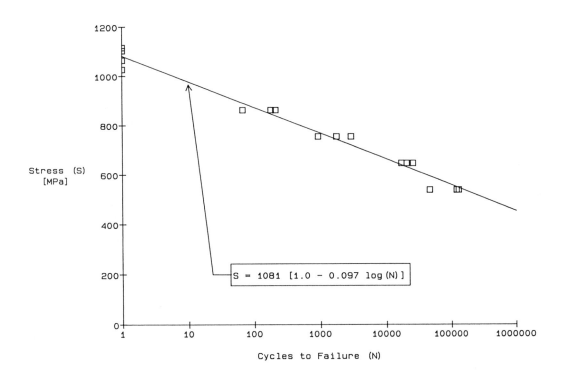

$$S = 1081 \left[1.0 - 0.097 \log (N) \right]$$

Figure 2   Flexural stress versus fatigue cycles to failure for
            unidirectional E-glass in "crystallized" PET

439

DATA ANALYSIS

Figure 2 shows flexural fatigue data for one of the continuous glass reinforced polyesters. For less than one million cycles the fatigue stress decreases linearly with the logarithm of cycles to failure. This data can be represented by an equation of the form [6]

$$S = Su [1 - b \log(N)]$$

where S is the applied fatigue stress, Su is the ultimate (static) stress, and N is the cycle count. The variable b is the slope of the S-N curve and in this case equals 0.097 for the best data fit.

This equation suggests normalizing the data by the ultimate static stress:

$$S/Su = 1 - b \log(N)$$

Normalized flexural fatigue data for the six unidirectional composites are shown in Figure 3. Fatigue data for random chopped strand and injection molded composites are shown in Figures 4 and 5.

Interestingly, all fifteen sets of data follow a common S-N curve with a slope, b, of approximately 0.1. Table II lists the ultimate stresses, best fit slopes and number of data points for the individual data sets. Note the wide variety of resins, fiber arrangements, and test conditions represented by these data.

Literature data for a wide range of glass reinforced systems also follow this common S-N curve. Table III includes representative ultimate stresses and S-N slopes for several systems reported in the literature, along with references.

The mean slope of all the systems presented in Tables II and III is 0.104 with a standard deviation of 0.012. More than 90% of the data can be captured by the relation

$$S/Su = 1 - 0.1 \log(N)$$

with +/- 10% variation of Su, as shown by the dotted lines in Figures 3, 4, and 5.

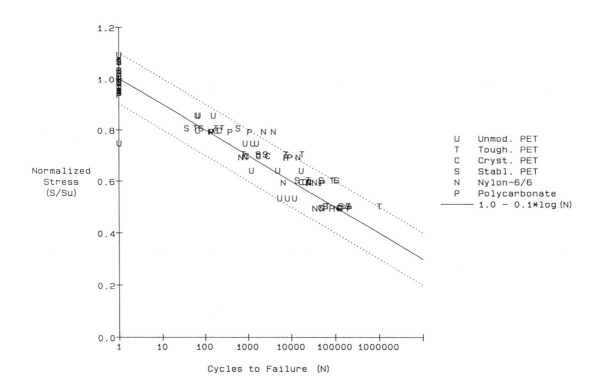

Figure 3   Normalized flexural stress versus fatigue cycles to failure
for six continuous unidirectional thermoplastic composites

440

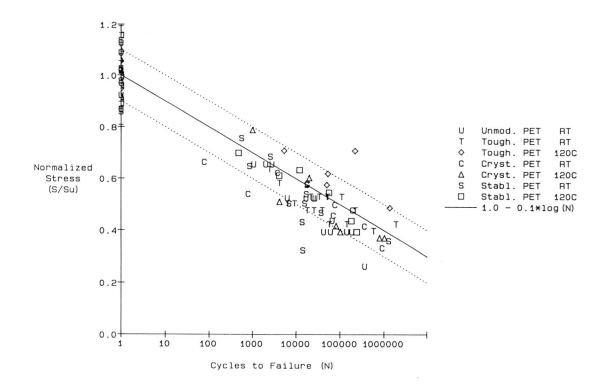

Figure 4   Normalized tensile stress versus fatigue cycles to failure
for four random chopped strand thermoplastic composites at
two temperatures

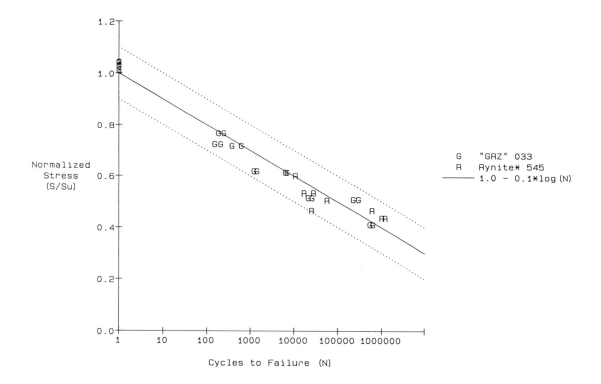

Figure 5   Normalized tensile stress versus fatigue cycles to failure
for two injection molded thermoplastic composites

441

Table II

FATIGUE DATA SUMMARY

Matrix Resins	System Description		Su (MPa)	Slope (b)	Data Pnts
Unmod. PET	Continuous Uni. Flex.	(RT)	1020	0.099	17
Tough. PET	Continuous Uni. Flex.	(RT)	871	0.088	16
Cryst. PET	Continuous Uni. Flex.	(RT)	1081	0.097	16
Stabl. PET	Continuous Uni. Flex.	(RT)	1029	0.091	16
N-6/6	Continuous Uni. Flex.	(RT)	915	0.094	15
PC	Continuous Uni. Flex.	(RT)	865	0.089	18
Unmod. PET	Random Chopped Tension	(RT)	160	0.121	17
Tough. PET	Random Chopped Tension	(RT)	130	0.106	28
Tough. PET	Random Chopped Tension	(120C)	78	0.077	7
Cryst. PET	Random Chopped Tension	(RT)	163	0.104	14
Cryst. PET	Random Chopped Tension	(120C)	149	0.108	14
Stabl. PET	Random Chopped Tension	(RT)	194	0.121	18
Stabl. PET	Random Chopped Tension	(120C)	154	0.105	10
PET	Rynite * 545 Tension Injection Molded	(RT)	103	0.100	9
N-6/6	"GRZ" 033 Tension Injection Molded	(RT)	163	0.104	20

Table III

SELECTED LITERATURE DATA
(Room Temperature Tensile Fatigue)

Matrix Resins	System Description	Su (MPa)	Slope (b)	Data Pnts	Ref.
PP	"Azdel" Swirl Mat	96	0.118	--	10
N-6/6	Injection Molded	181	0.108	20	5
PC	Injection Molded	161	0.113	24	5
PPS	Injection Molded	180	0.109	31	5
PAI	Injection Molded	202	0.105	18	5
Epoxy	(0/90) Laminate	418	0.102	47	4
Epoxy	Impregnated Strand	1160	0.093	18	6
Vinyl Ester	Chopped Strands (Theroset RTM)	238	0.129	10	1
Poly-ester	Thermoset (SMC50)	208	0.113	26	6

SUMMARY AND CONCLUSIONS

Design allowables data presented here for new glass reinforced thermoplastic composites show that their fatigue performance is similar to many other composite systems. Fatigue strengths decrease linearly with the logarithm of cycles for a wide variety of materials and fiber arrangements up to about one million cycles. Although specific material data are best, the slopes of these curves are similar enough to provide a useful fatigue life approximation. This simple rule is adequate for preliminary design and material selection.

In contrast, this simple rule does not adequately explain some aspects of fatigue behavior. For example, many materials deviate from linear S-N behavior at low stress levels. Preliminary random chopped strand fatigue data shown in Figure 6 suggest that our composites also deviate from linear S-N behavior at high cycles. In future work we will continue testing our composites at low stresses to better define this behavior.

Studying mechanisms of fatigue behavior over a wide range of stress and strain levels is important for applications and essential for fundamental understanding [9,11]. Future work will also include detailed observation of fatigue damage accumulation [3]. Acoustic emission monitoring, microscopy and residual property measurements will show the mechanisms, location and rates of damage [2]. This information will be used to develop damage mechanics models for predicting fatigue behavior [3,7,8,11].

ACKNOWLEDGEMENTS

Shoibal Banerjee and Don Huang provided the fatigue data for injection molded "GRZ" 033. Bob Boyle, Don Brill, Drew Graham, Bruce Moran, and Joe Scheese measured the remaining static and fatigue data.

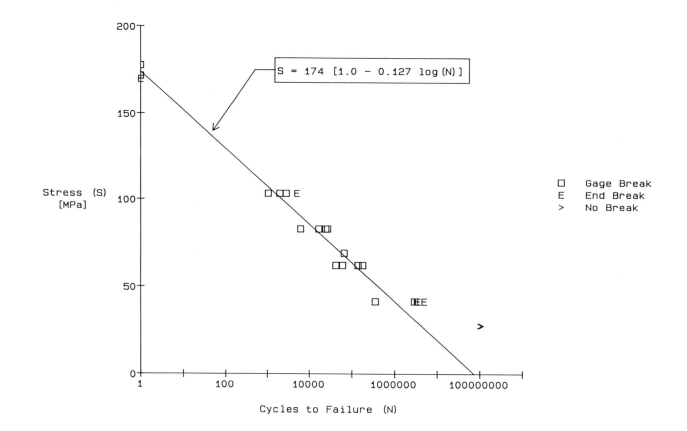

Figure 6   Tensile stress versus fatigue cycles to failure for
random chopped E-glass in "unmodified" PET

REFERENCE

[1]   Babbington, D. A., J. Enos,
J.   M.   Cox   and   J.
Barron,"Fast-Cure   Vinyl   Ester
Meets   Detroit's   Structural
Demands",Modern Plastics, August,
1987.

[2]   Chang,   F. H., et al., "A
Study   of   Fatigue   Damage   in
Composites   by   Nondestructive
Testing   Techniques",   in:
Fatigue of Filamentary Composite
Materials , ASTM STP 636, K.
L. Reifsnider and K. N. Lauraitis
(eds.),   American   Society
for   Testing   and   Materials,
Philadelphia, 1977.

[3]   Lorenzo,   L.   and H. T. Hahn,
"Fatigue   Failure   Mechanisms   in
Unidirectional   Composites",   in:
Composite   Materials: Fatigue and
Fracture , ASTM STP 907, H. T.
Hahn (ed.),   American Society
for   Testing   and   Materials,
Philadelphia,   1986.

[4]   Mandell, J. F., "Fatigue Crack
Growth   in   Fiber   Reinforced
Plastics", Polymer Composites, 2
(1), 22 (1981).

[5]   Mandell, J. F., D. D. Huang and F.
J.   McGarry,   "Fatigue   of
Glass   and   Carbon   Fiber
Reinforced   Engineering
Thermoplastics",   Polymer
Composites, 2 (3), 137 (1981).

[6]   Mandell, J. F., "Fatigue Behavior
of   Short   Fiber   Reinforced
Composite Materials", in: The
Fatigue   Behavior   of   Composite
Materials , K. L. Reifsnider
(ed.), Elsevier Applied Science
Publishers Ltd., New York, 1987.

[7]   Manson, S. S. and U. Muralidharan,
"Fatigue   Life   Prediction   in
Bending   from   Axial   Fatigue
Information", Fatigue & Fracture
of Eng. Mat. & Structures, 9 (5),
357 (1986).

[8]   Srivastava, V. K. and R. Prakash, "Fatigue Life Prediction of Glass Fibre Reinforced Plastics Using the Acousto-Ultrasonic Technique", Int. J. Fatigue, 9 (3), 175 (1987).

[9]   Talreja, R., _Fatigue of Composite Materials_, Technomic Publishing Co., Lancaster, Pennsylvania, 1987.

[10]   Tomkinson-Walles, G. D., "Performance of Random Glass Mat Reinforced Thermoplastics", J. Thermoplastic Composite Materials, 1, 94 (1988).

[11]   Volume 4 of the proceedings from _The Sixth International Conference on Composite Materials and Second European Conference on Composite Materials_, Matthews, F. L., et al. (ed.), Elsevier Applied Science Publishers Ltd., New York, 1987.

445

CONTINUOUS SILICON CARBIDE FIBER REINFORCED METAL MATRIX COMPOSITES

Melvin A. Mittnick, John McElman*
Textron Specialty Materials
Lowell, Massachusetts 01851 USA

Abstract

Continuous silicon carbide (SiC) fiber reinforced metals (FRM) have been successfully applied on numerous aerospace development programs fulfilling primary design objectives of high specific strength over baseline monolithic materials. This presentation will review the current state-of-the-art in silicon carbide fiber reinforced metals through discussion of their application. Discussion will include a review of mission requirements, program objectives and accomplishments to date employing these FRMs.

THE TERM COMPOSITE USUALLY signifies a combination of two or more constituent elements (in this case a high-performance, low-weight filler embedded in a metal) to form a bonded quasi-homogeneous structure that produces synergistic mechanical and physical property advantages over that of the base elements. Theoretically, there are three types of composites: (1) a particulate based material formed by the addition of small granular fillers into a binder that generally derives an increase in stiffness but not strength; (2) a whisker/flake filler that realizes a greater proportion of the filler strength due to its higher aspect ratio and hence a greater ability to transfer load; and (3) a continuous fiber system (i.e., Fiber Reinforced Metal, RFM) that, due to fiber continuity, derives the full properties of the high-performance fiber (strength and stiffness).

There are various metal matrix systems commercially available today that conform to the three types of composites discussed above, the mechanical properties of these various systems vary significantly. The major import is the clear distinction between continuous fiber systems and the discontinuous

(particulate and whisker/flake) systems. In the latter, only greater stiffness is generally realized, whereas in the former, both the modulus and the strength of the continuous reinforcing fiber are fully translated into the composite. In practice, if all of the filler strength is to be translated into the composite, then (a) the filler must have sufficient aspect ratio to transfer the load from "filler" to "filler" through the matrix, (b) the filler particles must be in close proximity to each other to avoid large matrix distortions (high volume loading), and (c) a large number of "filler termination sites" placed in close proximity must be avoided to preclude stress concentrations that overload the filler and matrix. In practice it is very difficult to satisfy all of the requirements, and hence achieving the theoretical potential of the filler is close to impossible.

Although the mechanical properties of the continuous fiber composite systems (FRM) are superior to those of the discontinuous systems, the ease of providing material isotrophy is a significant advantage, for there are many stiffness-controlled applications that benefit from discontinuous reinforced material. These applications generally involve complex geometries where it is difficult to position continuous fibers during the fabrication process. However, where there is sufficient volume of material to allow for orientation of fibers, the FRM systems will usually provide the most efficiency (weight savings).

SILICON CARBIDE FIBER

Since the advent of high-strength, high-modulus, low-density boron fiber, the role of fibers produced by chemical vapor deposition (CVD) in the field of high-performance composites has been well established. Although best known for its use as a reinforce-

*deceased

ment is resin-matrix composites,[1,2] boron
fiber has also received considerable attention in the field of metal matrix composites.
[3,4,5] Boron/aluminum was employed for tube-
shaped truss members to reinforce the Space
Shuttle orbiter structure, and has been investigated as a fan blade material for turbo-
fan jet engines. There are drawbacks, however,
in the use of boron in a metal matrix. The
rapid reaction of boron fiber with molten
aluminum[6] and long-term degradation of the
mechanical properties of diffusion-bonded
boron/aluminum at temperatures greater than
480°C (900°F) preclude its use both for high-
temperature applications and for potentially
more economically feasible fabrication methods
such as casting or low-pressure, high-
temperature pressing. These drawbacks have
led to the development of the silicone carbide
(SiC) fiber.

SILICON CARBIDE FIBER PRODUCTION
PROCESS - Continuous SiC filament is produced
in a tubular glass reactor by CVD. The process
occurs in two steps on a carbon monofilament
substrate which is resistively heated. During
the first step, pyrolytic graphite (PG)
approximately 1 μm thick is deposited to
smooth the substrate and enhance electrical
conductivity. In the second step, the PG
coated substrate is exposed to silane and
hydrogen gases. The former decomposes to form
beta silicon carbide (βSiC) continuously on
the substrate. The mechanical and physical
properties of the SiC filament are:

Tensile Strength = 3400 MPa
(500 ksi)
Tensile Modulus = 400 GPa
(60 msi)
Density = 3.045 g/cm^3
(0.11 lb/in.^{3})
Coefficient of Thermal Expansion
CTE = 1.5 x 10^{-6}/°C
(2.7 x 10^{-6}/°F)
Diameter = 140 mm (0.0056 in.)

Various grades of fiber are produced,
all of which are based on the standard βSiC
deposition process described above where a crystalline structure is grown onto the carbon substrate. The βSiC is present as such across
all of the fiber cross-section except for the
last few microns at the surface. Here, by
altering the gas flows in the bottom of the
tubular reactor, the surface composition and
structure of the fiber are modified by, first,
an addition of amorphous carbon that heals
the crystalline surface for improved surface
strength, followed by a modification of the
silicon-to-carbon ratio to provide improved
bonding with the metal.

PROCESSING CONSIDERATIONS - As in any vapor
deposition or vapor transport process, temperature control is of utmost importance in
producing CVD SiC fiber. The Textron process
calls for a peak deposition temperature of

about 1300°C (2370°C).

Temperatures significantly above this temperature cause rapid deposition and subsequent
grain growth, resulting in a weakening of tensile strength. Temperatures significantly below
the optimum cause high internal stresses in the
fiber, resulting in a degradation of metal
matrix composite properties upon machining
transverse to the fiber.[7]

Substrate quality is also an important
consideration in SiC fiber quality. The carbon
monolifament substrate, which is melt-spun coal
tar pitch, has a very smooth surface with occa-
sional surface anomalies. If severe enough, the
surface anomaly can result in a localized area
of irregular deposition of PG and SiC which
is a stress-raising region and a strength-
limiting flaw in the fiber. The carbon monofilament spinning process is controlled to
minimize these local anomalies sufficiently
to guarantee routine production of high-strength
>3450 MPa (>500 ksi) SiC fiber.

Another strength-limiting flaw which can
result from an insufficiently controlled CVD
process is the PG flaw.[8] This flaw results
from irregularities in the PG deposition. Two
causes of PG flaws are: (1) disruption of the
PG layer due to an anomaly in the carbon
substrate surface and (2) mechanical damage to
the PG layer prior to the SiC deposition.
PG flaws often cause a localized irregularity
in the SiC deposition, resulting in a bump on
the surface. Poor alignment of the reactor
glass can result in mechanical damage to the
PG layer by abrasion. A series of PG flaws
results in what is called a "string of beads"
phenomenon at the surface of the fiber. The
mechanical properties of such fiber are severely degraded. These flaws are minimized by
careful control of the PG deposition parameters,
proper reactor alignment, and the minimization
of substrate surface anomalies.

The surface region of Textron's SiC fibers
is tyically carbon rich. This region is
important in protecting the fiber from surface
damage and subsequent strength degradation.
An improper surface treatment or mishandling
of the fiber (e.g., abrasion) can result in
strength-limiting flaws at the surface. Surface flaws can be identified by an optical
examination of the fiber fracture face. These
flaws are minimized by proper process control
and handling of the fiber (minimizing surface
abrasion).

Typical mechanical properties for the
Avco CVD SiC fiber consist of average tensile
strength of 3790 to 4140 MPa (550 to 600 ksi)
and elastic moduli of 400 to 415 GPa (58 to
60 msi). A typical tensile strength histogram shows an average tensile strength of
4000 MPa (580 ksi) with a coefficient of
variation of 15%.

FIBER VARIATIONS - The surface region of
the SiC fiber must be tailored to the matrix.

SCS-2 has a 1 μm carbon-rich coating that increases in silicon content as the outer surface is approached. This fiber has been used to a large extent to reinforce aluminum. SCS-6 has a thicker (3 μm) carbon-rich coating in which the silicon content exhibits maxima at the outer surface and 1.5 μm from the outer surface. SCS-6 is primarily used to reinforce titanium.

SCS-8 has been developed as an improvement over SCS-2 to give better mechanical properties in aluminum composites transverse to the fiber direction. The SCS-8 fiber consists of 6 μm of very fine-grained SiC, a carbon-rich region of about 0.5 μm, and a less carbon-rich region of 0.5 μm.

COST FACTORS - From an economic standpoint, SiC is potentially less costly than boron for three reasons: (1) the carbon substrate used for SiC is lower cost than the tungsten used for the boron; (2) raw materials for SiC (chlorosilanes) are less expensive than boron trichloride, the raw material for boron; and (3) deposition rates for SiC are higher than those for boron, hence more product can be made per unit.

COMPOSITE PROCESSING

The ability to readily produce acceptable SiC fiber reinforced metals is attributed directly to the ability of the SiC fiber to (a) readily bond to the respective metals and (b) resist degradation of strength while being subjected to high-temperature processing. In the past, boron and Borsic^{TM} fibers have been evaluated in various aluminum alloys and, unless complex solid-state (low-temperature, high-pressure) diffusion binding procedures were adopted, severe degradation of fiber strength has been observed. Likewise in titanium, unless fabrication times are severely curtailed, fiber/matrix interactions produce brittle intermetallic compounds that again drastically reduce composite strength.

In contrast, the SCS grade of fibers has surfaces that readily bond to the respective metals without the destructive reactions occurring. The results is the ability to consolidate the aluminum composites using less-complicated high-temperature casting and low-pressure (hot) molding. Also for titanium composites, the SCS-6 filament has the ability to withstand long exposure at diffusion bonding temperatures without fiber degradation. As a result, complex shapes with selective composite reinforcement can be fabricated by the innovative superplastic forming/diffusion bonding (SPF/DB) and hot isostatic pressing (HIP) processes.

In the following discussion, further details of fabrication techniques will be discussed; however, first the production of intermediary products such as preforms and fabrics used in the component fabrication are described. These are required to simplify the loading of fibers into a mold and to provide correct alignment and spacing of the fibers.

COMPOSITE PREFORMS AND FABRICS - "Green tape" is an old system consisting of a single layer of fibers that are collimated/spaced side by side across a layer, held together by a resin binder, and supported by a metal foil. This layer constitutes a prepred (in organic composite terms) that can be sequentially "laid up" into the mold or tool in required orientations to fabricate laminates. The laminate processing cycle is then controlled so as to remove the resin (by vacuum) as volatilization occurs. The method normally used to wind the fibers onto a foil-covered rotating drum, overspraying the fibers with the resin, followed by cutting the layer from the drum to provide a flat sheet of "prepreg".

"Plasma-sprayed aluminum tape" is a more advanced "prepreg" similar to "green tape" but replaces the resin binder with a plasma-sprayed matrix of aluminum. The advantages of this material are (a) the lack of possible contamination for resin residue and (b) faster material processing times because of the hold time required to ensure volatilization and removal of the resin binder is not required. As with the green tape system, the plasma-sprayed preforms are laid sequentially into the mold as required and pressed to the final shape.

"Woven fabric" is perhaps the most interesting of the preforms being produced since it is a universal preform concept that is suitable for a number of fabrication processes. The fabric is a uniweave system in which the relatively large-diameter SiC monofilaments are held straight and parallel, collimated at 100 to 140 filaments per inch and held together by a cross-weave of a low-density yarn or metallic ribbon. There are now two types of looms that can be specially modified to produce the uniweave fabric. The first is a single-arm Rapier-type loom capable of producing continuous 60 in. wide fabric with the SiC filament oriented in the "fill" (60 in. width) direction. The other is a shuttle-type loom in which the SiC monofilaments are oriented in the continuous direction with the light-weight yarn a metal ribbon in the "fill" axis. The shuttle loom can weave fabric up to 6 in. wide. Various types of cross-weave materials have been used, such as titanium, aluminum, and ceramic yarns.

PROCESSING METHODS - "Investment casting" is a fabrication technique that has been used for many years but is still universally accepted as a very cost effective method for producing complex shapes.

The aerospace business has for some time rejected aluminum castings due to the low strengths that are typically achieved; however, with a material that is now fiber dependent and not predominantly matrix controlled, significant structural improvements have been derived so as to revive the interest in this low-cost procedure. The investment casting technique, sometimes called the "Lost Wax" process, utilizes a wax replicate of the intended shape to form a porous ceramic shell mold where, upon removal of the wax (by steam heat) from the interior, a cavity for the aluminum is provided. The mold includes a funnel for gravity pouring, with risers and gates to control the flow of the aluminum into the gage section. A seal is positioned around the neck of the funnel, allowing the body of the mold to be suspended into a vacuum chamber. By a combination of gravity and vacuum (imposed through the porous walls of the shell mold), the total cavity is filled with aluminum.

The SiC fibers are installed in mold using the fabric described above by either first placing the fabric into the wax replica or simply splitting open the mold and inserting the fabric into the cavity after the wax has been removed. At present, the latter approach is usually used due to contamination and oxidation of the fibers during wax burnout. At some future date, the necessary techniques for including the fiber in the wax (thereby reducing the processing costs) will probably be developed.

"Hot molding" is a term coined by Textron to describe a low-pressure hot pressing process that is designed to fabricate shaped SiC-aluminum parts at significantly lower cost than the typically diffusion bonding, solid-state process. As stated previously, the SCS-2 fibers can withstand molten aluminum for long periods; therefore, the molding temperature can now be raised into the liquid-plus-solid region of the alloy to ensure aluminum flow and consolidation at low pressure, thereby negating the requirement for high-pressure die molding equipment.

The best way of describing the hot molding process is to draw an analogy to the autoclave molding of graphite epoxy where components are molded in an open-faced tool. The mold in this case is a self-heated, slip-cast ceramic tool embodying the profile of the finished part. A plasma sprayed aluminum preform is laid into the mold, heated to a near molten aluminum temperature, and pressure consolidated in an autoclave by a "metallic" vacuum bag. The mold can be profiled as required to produce near net shape parts including tapered thicknesses and section geometry variations.

"Diffusion bonding of SiC/titanium" is accomplished by hot pressing (diffusion bonding) technology, using fiber preforms (fabric) that are stacked together between titanium foils for consolidation. Two methods are being developed by aircraft and engine manufacturers to manufacturer complex shapes. One method is based on the HIP technology, and uses a steel pressure membrane to consolidate components directly from the fiber/metal preform layer. The other method requires the use of previously hot pressed SCS/titanium laminates that are then diffusion bonded to a titanium substructure during subsequent super-plastic forming operations.

This is typical of the first fabrication procedure noted above. The fiber preform is placed onto a titanium foil. This is then spirally wrapped, inserted, and diffusion bonded onto the inner surface of a steel tube using a steel pressure membrane. The steel is subsequently thinned down and machined to form the "spline attachment" at each end. Shafts are also being fabricated for other engine fabricators without the steel sheath.

The concept developed for superplastic forming of hollow engine compressor blades. Here the SCS/titanium laminates are first diffusion bonded in a press. These are then diffusion bonded to form monolithic titanium sheets, with "stop-off" compounds selectively positioned to preclude bonding in desired areas. Subsequently, the "stack-up" is sealed into a female die. By pressurizing the interior of the "stack-up", the material is "blown" into the female die to form the desired shape, stretching the monolithic titanium to form the internal corrugations.

These processes typically require long times at high temperature. In the past, all of the materials used have developed serious matrix-to-fiber interactions that seriously degrade composite strength. SCS-6, however, due to its unique surface characteristics, delays intermetallic diffusion and retains its strength up to 7 hours in contact with titanium at 925°C (1700°F).

COMPOSITE PROPERTIES

Since continuous SiC reinforced metals have been in existence for a relatively short period of time, the property data base has been developed sporadically over this period depending on funded applications.

SiC/ALUMINUM - The most mature of the SiC reinforced aluminum (SiC/Al) consolidation approaches is hot molding, and therefore the greatest mechanical property data base has been developed using this material. The design data base for hot molded SCS-2/6061 aluminum includes static tension and compression properties, in-plane and interlaminar shear strengths, tension-tension fatigue strengths (SN curves), flexure strength, notched tension data, and fracture toughness data. Most of the data have been developed over a temperature range of -55°C to 75°C (-65°F to 165°F) with static tension test results up to 480°C (900°F). As can be seen from these data, the inclusion of a high performance, continuous SiC fiber in 6061 aluminum yields a very high strength 1378 MPa (+200 ksi) high-modulus 207 GPa (30 msi) anisotropic composite material having a density just slightly greater, 2.85 g/cm^3 (0.103 lb/in.^{3}) than baseline aluminum. As in organic matrix composites, cross or angle plying produces a range of properties useful to the designer.

The property data developed to date for investment cast SCS/aluminum have been limited to static tension and compression. Fiber volume fractions are lower (40% maximum) than the hot molded laminates (47% typical) due to volumetric constraints in dry loading the shell molds; however, good rule-of-mixture (R.O.M.) tensile strengths and excellent compression strengths (twice the tensile strength) are being achieved.

The use of 6061 aluminum as the matrix material and the capability of the SiC fiber to withstand molten aluminum has made conventional fusion melding a viable joining technique. Although welded joints would not have continuous fiber across the joint to maintain the very high strengths of the composite, baseline aluminum weld strengths can be obtained. In addition to fusion welding, traditional molten salt bath dip brazing has been demonstrated as an alternative joining method.

An important consideration for emerging materials is corrosion resistance. Testing has been performed on SCS-2/6061 hot molded material at the David W. Taylor Naval Ship R&D Center[9] under marine atmosphere, ocean splash/spray, alternate tidal immersion, and filtered seawater immersion conditions for periods of 60 to 365 days. The SCS/aluminum material performed well in all tests, exhibiting no more than pitting damage comparable to the baseline 6061 aluminum alloy.

SiC/TITANIUM - SCS-6/Ti 6-4 composites were originally developed at high temperature. There has been a successful program to reinforce the beta titanium alloy 15-3-3-3 with SCS fiber and superior composite properties have been achieved at 1585 to 1930 MPa (230 to 280 ksi) tensile strengths[10]. Fabrication of titanium parts has been accomplished by diffusion bonding and HIP. The HIP technique has been particularly successful in the forming of shaped reinforced parts (e.g., tubes) by the use of woven SiC fabric as a preform. The high-strength, high-modulus properties of SCS-6/Ti represent a major improvement over B$_4$C-B/Ti composites in which the modulus of composite is increased relative to the matrix, but the tensile strength is not as high as would be predicted by the rule of mixture.

SiC/MAGNESIUM AND SiC/COPPER - SCS-2 has been successfully cast in magnesium.[11] Under a recent Naval Surface Weapons Center (NSWC) program,[12] development of SiC-reinforced copper has been initiated. At present, about 85% of R.O.M. strengths have been achieved at a volume fraction of 20 to 33%.

APPLICATIONS

The very high specific mechanical properties of SiC reinforced metal matrix composites have generated significant interest within the aerospace industry, and as a result many research and development programs are now in progress. The principal area of interest is for high-performance structures such as aircraft, missiles, and engines. However, as more and more systems are developing sensitivities to "performance" and "transportation weight", other and less sophisticated applications for these newer materials are being considered. The following paragraphs describe a few of these applications.

"SiC/aluminum wind structural elements" are currently being developed. Ten foot long "Zee" shaped stiffeners are to be hot molded and then subsequently riveted to wing planks for full-scale static and fatigue testing. Experimental results obtained to date have verified material performance and the design procedures utilized.

"SiC/aluminum bridging elements" are being developed for the Army to be used for the lower chord and the king post of a 52m assault bridge. Future plans call for development of the top compression tubes of the new Tri-Arch bridge being developed by Fort Belvoir.

"SiC/aluminum internally stiffened cylinders" are being developed using the previously discussed investment casting process. A wax replica is first fabricated that incorporates the total shape of the shell including internal ring stiffeners and the end fittings. The fabric containing the SiC fibers is then wound onto the inner shell mold, the two halves of the shell are remated and sealed, and infiltration of the aluminum is then accomplished.

"SiC/aluminum fins" for high-velocity projectiles are in the process of evaluation.

"SiC/aluminum missile body casings" have been fabricated utilizing a unique variation of filament winding. An aluminum motor case is first produced in the conventional manner; this time, however, with significantly less wall thickness than normally required. The casing is then over-wrapped with layers of SiC fibers, where each layer is sprayed with a plasma of aluminum to build up the matrix thickness. No final consolidation of the 90% dense system is required, for the hydrostatic internal pressure on the circular body imposes no (or very minimal) shear stresses on the matrix. It is hoped that further development of this technique will permit full consolidation of the matrix by vacuum bagging the total section and hot isostatic pressing.

"SiC-titanium drive shafts" are being developed and fabricated by the hot isostatic pressing process described previously. These are generally for the core of an engine, requiring increased specified stiffness to reduce unsupported length between bearings and also to increase critical vibratory speed ranges. SiC-Ti tubes up to 5 ft. in length have been fabricated and have incorporated into their ends a monolithic load transfer section for ease of welding to the splined or flanged connections.

"SiC discs for turbine engines" are currently under development. Initially discs were made by winding SiC-Ti monolayer over a mandrel followed by hydrostatic consolidation (hot isostatic pressing). The concept now being developed utilizes a "doily" approach where single fibers are hoop wound between titanium metal ribbons to be subsequently pressed together in the axial direction, reducing the breakage of fibers and simplifying the production of tapered cross-sections.

"Selectively reinforced SiC-titanium hollow fan blades" are being developed.

"SiC/copper materials" have been fabricated and tested for high-temperature missile applications. Also, SiC/bronze propellers have been case for potential Navy applications where more efficient/quiet propellers are required.

FUTURE TRENDS

The SiC fiber is qualified for use in aluminum, magnesium, and titanium. Copper matrix systems are under development and reasonably good results have been obtained using the higher temperature titanium aluminides as matrix materials. The SCS-6 fiber demonstrates high mechanical properties to above $1400°C$ ($2550°F$). It is natural, then, to project systems such as SiC-nickel aluminides/ iron aluminide/superalloys, etc., all of which, on an R.O.M. basis at least, project very useful properties for "engine" and hypersonic vehicle" applications. Work required in this area includes diffusion barrier coatings and matrix alloy modifications to facilitate high-temperature fabrication processes. Also required is the detailed investigation of any detrimental thermal/mechanical cycling effects that may occur as a result of the mismatch in thermal expansion coefficients between matrix and fiber.

REFERENCES

1. DeBolt, H., "Boron and Other Reinforcing Agents," in Lubin, G., ed., Handbook of Composties, Van Nostrand Reinhold Co., New York, 1982, Chapter 10.

2. Krukonis, V. J., "Boron Filaments," in Milweski, J.V., and Katz, H.S., ed., Handbook of Fillers and Reinforcements for Plastics, Van Nostrand Reinhold Co., New York, 1977, Chapter 28.

3. McDaniels, D.L., and Ravenhall, R., "Analysis of High-Velocity Ballistic Impact Response of Boron/Aluminum Fan Blades," NASA TM-83498, 1983.

4. Salemme, C. T., and Yokel, S. A. "Design of Impact-Resistant Boron/Aluminum Large Fan Blades," NASA CR-135417, 1978.

5. Brantley, J. W., and Stabrylla, R. G., "Fabrication of J79 Boron/Aluminum Compressor Blades," NASA CR-159566, 1979.

6. Wolff, E., "Boron Filament, Metal Matrix Composite Materials," AF33 (615)3164.

7. Suplinskas, R. J., "High Strength Boron". NAS3-22187, 1984.

8. Aylor, D. M., "Assessing the Corrosion Resistance of Metal Mtrix Composite Materials in Marine Environments," DTNSRDC/ SMME-83/45, 1983.

10. Kumnick, A. J., Suplinskas, R. J., Grant, W. F., and Cornie, J. A., "Filament Modification to Provide Extended High Temperature Consolidation and Fabrication Capability and to Explore Alternative Consolidation Techniques," N00019-82-C-0282, 1983.

11. Cornie, J.A., and Murty, Y., "Evaluation of Silicon Carbide/Magnesium Reinforced Castings," DAAG46-80-C-0076, 1983.

12. Marzik, J. V., and Kumnick, A. J., "The Development of SCS/Copper Composite Material," N60921-83-C-0183, 1984.

MORE COST EFFECTIVE
MOLDING WITH HYBRIDS

H. R. "Dan" Edwards
Amoco Chemical Company
Naperville, Illinois   USA

ABSTRACT

For sensitive applications demanding multiple
and uniform properties, higher costs are often
expected. However, research has led to the
development of materials that can offer
economies through rapid molding cycles, and
cured resin resilience such that high
strengths can be achieved with less use of
expensive mat reinforcement.

These new materials, hybrid resins, are
two-component liquids that are custom
formulated to achieve specific processing and
performance requirements. This paper will
document the broad ranges of properties
available that make hybrids appeal to many
processing techniques. It will contrast the
chemistry of hybrids with conventional resins
to explain some of the advantages. Included
are viscosities, reaction times, emissions and
cured part toughness.

Advantages in productivity and material
selections often combine to make hybrids more
economical. Examples will be given.
Suggestions for the optimal use of and
applications for hybrid resins will include a
method to achieve excellent surface appearance
with low temperature molding.

COST EFFECTIVE MOLDING of high performance
thermosetting plastics can lead to an expanded
marketplace and replacement of many
traditional materials of construction. A new
material has recently been introduced by Amoco
Chemical Company. Xycon™ hybrid resins,
custom-formulated, two-component liquid
systems, are generally priced between low-cost
unsaturated polyesters and higher-priced
epoxies, urethanes and vinylesters. That
hybrids offer unique processing advantages and
unique combinations of cured resin mechanical
properties has been well-established in recent
years. Now that hybrids are commercial
entities, these demonstrated advantages can be
translated into economic savings for molders
and fabricators. The principles to achieve
this are detailed in this paper. Among the
contributors are a means to achieve high
quality surfaces in low temperature molding of
glass reinforced articles.

EXPERIMENTAL

While hybrids are most efficiently delivered
via machinery with two-component, internal mix
capabilities, the systems used in developing
the data for this paper were batch-mixed and
poured over glass fibers which were placed in
a mold cavity. The press was then closed on
stops to assure the desired thickness and thus
achieve the prescribed filler, glass and resin
ratios. Room-temperature gel times of ten
minutes were used in the unheated mold. A
post-cure of two hours at 105°C followed. The
filler was calcium carbonate and was premixed

into the "B Side" (polyol) prior to introduction of the "A Side" (isocyanate). The reinforcement was 0.75 ounce per square foot continuous glass roving.

Physical properties, as shown in the various attachments, were run in accordance with the appropriate ASTM test methods.

DISCUSSION

Using more expensive resin to make a better part is understandable, but to make a less expensive part, too...? It is not only theoretically possible, but has been put into practice. Here's how:

Principle 1: Hybrids can <u>gel faster</u>. Due to their urethane heritage, hybrids can be catalyzed to gel in just seconds even at room temperature.

Principle 2: Hybrids <u>cure thoroughly faster</u>. Hybrids have considerably less styrene monomer than conventional polyesters. This contributes to fewer styrene emissions. Less obvious though is that lower levels of styrene monomer react faster* since a greater percentage of it has available its favorite co-reactant, the

maleate/fumarate unsaturation in the polyester backbone. A faster gel time, resulting in more concentrated exotherm, enhances the thoroughness of the reaction.

Principle 3: Hybrids are <u>more stress-resistant</u> due to their very high linear molecular weight development. Most unsaturated polyesters are dependent on glass reinforcement to defray stresses that otherwise cause them to fail catastrophically, and prematurely. With approximately 25% glass, tensile and flexural, strengths are not so much enhanced as they are made more uniform. Hybrids benefit from lower levels of glass, especially in notched Izod impact strengths, but seldom require it for unnotched Izod impact, flexural and tensile strengths.

To be sure, glass increases moduli, but fillers and foam do also at much less cost.

Principle 4: Hybrids are <u>inherently thinner</u> at room temperature even though they contain less reactive diluent. They can, therefore, be pumped faster and filled higher.

Principle 5: Hybrids <u>can foam</u>. This trait

*Amoco Chemical bulletin IP-70, pp. 23-24

enhances stiffness and reduces the density of systems highly filled to achieve fire retardancy. It can also lead to improved surfaces by providing the positive pressure in a mold at room temperature that thermoplastic additives contribute to SMC (sheet molding compound) at high molding temperatures. RIM (reaction injection molding) urethanes use a similar approach to achieve "Class A" surfaces when they are frothed. Of course, frothed resins do not penetrate glass fibers as efficiently as resins that generate foam chemically after injection into the mold.

Now, by employing these Principles, let's see how production economies can be achieved with hybrids that will more than offset their higher initial raw material cost (RMC). Two processes, spray-up and resin transfer molding (RTM), will be assessed. Hybrids will be contrasted to conventional unsaturated polyesters, although one could as easily "plug in" values for more expensive urethanes, epoxies and vinylesters. In the latter cases, the advantage for hybrids would be even more dramatic.

SPRAY-UP - Table I shows an economic

scenario wherein hybrids use:

1. less glass (less necessary for equal strengths), but more inexpensive filler (achievable by the lower viscosity);
2. shorter fibers, which combined with item 1, can eliminate the need for rolling;
3. a faster gel time (no roll-out step makes it possible to take advantage of systems gelling in 10-20 seconds at room temperature); and
4. the associated faster demold time of a low styrene level system.

For the purposes of this calculation, some costs and cycle times were estimated. They may not exactly reflect everyone's experiences, but they are meant to be "in the ballpark" for the size part and application (fire retardancy required) represented. Try your own values as a check on the validity of this exercise.

Some elaboration on the scenario may be in order. Firstly, a conventional UV stable gel coat was assumed. In the hybrid case, less drying time was needed since subsequent glass was not to be compacted (rolled). UV stable hybrid gel coats, while more expensive, could significantly reduce the time demanded in this phase of production. Secondly, because less of the denser filler and glass were used in the hybrid case, equal volumes were achieved at a 6% weight reduction in the part. Thirdly, the higher production rates for hybrids allow unit overhead costs to be spread over more parts, thus reducing its impact.

Furthermore, fewer molds would be required, and where styrene emissions limit production, the lower styrene levels of hybrids would allow better plant utilization by increasing these photochemical limited production levels.

RESIN TRANSFER MOLDING - An economic scenario for RTM is shown in Table II. As for spray-up, RMCs and times are typical, albeit perhaps not exact for everyone. For RTM, one can take advantage of hybrids in these ways:

1. A lower viscosity allows more filler, faster fill cycles and thus shorter gel and cure times are permitted.

2. Higher inherent strengths allows less use of glass (a cost savings in itself) and thus less cutting time.

We could, of course, show a gel coat step here as well. Gel coats seem to be much less a way of life in RTM than in spray up, however. Furthermore, uses to date show less need for gel coats in hybrid RTM, perhaps due to lower styrene levels and accordingly less shrinkage and glass prominence.

The scenario shown here involves a relatively large part. Fill times, while shorter than for conventional resins, did not allow hybrids to fully exploit their gel/cure time advantages. With smaller parts, the economics would favor hybrids more dramatically as the cycle time differential increases.

The use of a 90% volume figure for hybrid costs is predicated on their ability to achieve equal load bearing properties with thinner sections. The justification for this follows in section C1.

Structural RIM economics parallel the approach for RTM except cycle times are shorter and thus favor hybrids even more.

OTHER CONSIDERATIONS - Making more parts faster and with less labor is only part of the hybrid story, as is using less glass and more filler. Thinner parts that provide equal load bearing properties help reduce material consumption and weight. A related approach is to reduce weight and material use through foam. In this case, thicker parts meet stiffness requirements, but still reduce weight. Here are some examples:

Reduced Thickness - Table III shows some actual values for commercial hand lay-up, conventional resin FRP. It also shows the property values necessary to produce equivalent stiffness and load bearing properties at two reduced thicknesses. These are determined by placing the actual measured loads and slopes (from tensile and flexural tests) into theoretical calculations for the hypothetical, reduced thicknesses.

Composites of several hybrid formulations were then molded and tested for physical properties. These were done at two thicknesses for each system, but when glass and filler levels were consistent, mechanical properties within a system were also, even though thicknesses changed. Remember, loads and slopes, however, were greatly affected by thickness. Table IV shows the physical properties of three hybrid systems.

A quick examination of these properties compared with those in Table III suggests that hybrids can be formulated to deliver equivalent moduli and strengths at a 5 to 10% reduction in part thickness while at the same time replacing half the glass with low cost fillers. Low aspect ratio fillers improve stiffness at the expense of strength, especially impact strength. One might conclude that hybrid composite impact strength would benefit from more glass and less filler. Such a system is reported in Table V.

Foam - Hybrid foam provides an excellent route to rigidifying a plastic or composite while reducing cost and material consumption. The most common outlet for this approach also calls for fire retardancy. Thus the hybrid foam shown in Table VI is filled 40% by weight with hydrated alumina. On a volume basis, adjusted to yield equal rigidity, the hybrid foam is a 10%

more economical stiffener than conventional FRP.

Class A Appearance Via Foam - The hand labor involved in finishing an FRP part to meet automotive surface standards has always been a liability. SMC, molded at high temperatures (usually about 150°C) and employing thermoplastic "low profile additives" presented a quantum improvement over contact molded, non-gel coated parts. Unfortunately, lower temperature, lower pressure molding methods cannot achieve comparable performance from the low profile additives. They do, however, achieve more uniformity in physical properties.

It is interesting that RIM urethanes can achieve Class A surfaces under relatively low temperature (about 65°C) condition, by frothing the resin about 5% prior to injection. Here heated nitrogen provides the same positive pressure in the mold that thermoplastics (greatly expanding upon heating) provide in SMC.

Hybrids can be formulated to
chemically produce the same effect.
Although to date we have only visual
appraisals of the surface which
indicate extraordinary improvements,
physical properties have been run
and are shown in Table VII. The
slightly expanded (low specific
gravity) version has the expected

higher modulus, but also
demonstrates comparable strengths.
This innovation could lead to the
production of Class A, non-gel
coated parts from low pressure
injection, low temperature methods
using soft tooling.

CONCLUSION

The unique processing and mechanical
properties of Xycon™ hybrid resins can be
employed to lower weight, increase production,
reduce labor and improve surface finish while
delivering equal or better load bearing
properties. Incredibly, unit costs are
reduced at the same time. This novel
technology can help the FRP industry grow by
becoming more competitive with alternative
materials of construction.

TABLE I

ECONOMICS OF SPRAY-UP

		Conventional	Hybrid
A.	Gel Coat		
	Workers	One	One
	Material	7% (80¢/lb)	7% (80¢/lb)
	Time	15'sp + 45'dry	15'sp + 15'dry
B.	Structural		
	Workers	Two	Two
	Material		
	Resin	41% (60¢/lb)	50% (110¢/lb)
	Filler	27% (15¢/lb)	33% (15¢/lb)
	Glass	25% (80¢/lb)	10% (80¢/lb)
	Time	15'	15'
C.	Roll-Out		
	Workers	Three	0
	Time	30'	0
D.	Demold Time	60'	15'

	Conventional	Hybrid
Total Time	165'	60'
Man Hours	135'	45'
Material Invested, ¢/lb	54.25	73.55

Assume 50 lb part (or equivalent volume)*
Assume wages/benefits = $7/hr
Assume overhead = $3/hr

	Conventional	Hybrid
RMC per part	$27.13	$34.57
Labor per part	15.75	5.25
Overhead per part	8.25	3.00
Cost per part	$51.13	$42.82

*Sp Gr of Hybrid is 6% less

TABLE II

Economic Analysis
of RTM Hybrids

	Conventional	Hybrid
Continuous Strand Mat		
% of composite	30	18
¢/lb	150	150
Filler		
% of composite	35	42
¢/lb	12	12
Resin		
% of composite	35	40
¢/lb	81	120
Specific Gravity	2.11	2.07
RMC, ¢/lb	77.55	80.04
RMC (assume 50 lb part, and equal volume) dollars/part	38.78	39.26
RMC at 90% of conventional volume, $/part	---	35.33
Differential Labor (cutting and placing mat)		
man hours	0.5	0.3
cost, $ ($7/hr)	3.50	2.10
Productivity		
fill time, hours	0.125	0.083
cure cycle, hours	0.375	0.25
* Parts per hour Includes consideration of cutting and placing time.	1.0	1.6
Overhead (assume $3/hr) $	3.00	.87
Normalized cost/part, $	45.28	43.23
at 90% volume, $	---	39.30

Table III

Properties of Actual Conventional FRP
and Theoretical Equivalents for Thinner Parts

	Actual	For Equivalent Stiffness and Load Bearing Properties	
Part Thickness, mm	3.8	3.6	3.4
Flexural Strength, psi	18,100	20,800	22,300
Flexural Modulus, psi	530,000	652,000	727,000
Tensile Strength, psi	13,000	13,900	16,300
Tensile Modulus, psi	790,000	846,000	987,000
Tensile Elongation, %	1.9	---	---
Izod Impact Strength, ft lbs/inch			
Notched	8.1	8.6	10.3
Unnotched	11.4	12.0	14.4
Glass Content, wt%	30	---	---
Filler Content, wt%	0	---	---

Table IV

Properties of Hybrid Composites

Resin System	A	B	C
Part Thickness, mm	*	*	*
Flexural Strength, psi	17,100	20,100	21,500
Flexural Modulus, psi	510,000	780,000	960,000
Tensile Strength, psi	13,300	13,700	9,900
Tensile Modulus, psi	870,000	1,000,000	970,000
Tensile Elongation, %	2.4	1.7	1.1
Izod Impact Strength, ft lbs/inch			
Notched	4.4	3.6	4.6
Unnotched	9.2	6.8	6.1
Glass Content, wt%	15	15	15
Filler Content, wt%	15	15	15

*Varied from 2.5 to 3.6, but physical properties remained the same within a
resin system when glass and filler levels were consistent.

Table V

The Effect of Higher Glass Levels

Hybrid System	C	C
Flexural Strength, psi	21,500	25,800
Flexural Modulus, psi	960,000	980,000
Tensile Strength, psi	9,900	12,700
Tensile Modulus, psi	970,000	1,090,000
Tensile Elongation, %	1.1	1.1
Izod Impact Strength, ft lbs/inch		
Notched	4.6	5.7
Unnotched	6.1	8.7
Glass Content, wt%	15	20
Filler Content, wt%	15	15

Table VI

Foam for Rigidity

	Hybrid Foam	Equivalent Properties in Conventional FRP
Thickness, mm	7.1	*3.2
Flexural Strength, psi	2,800	14,000
Flexural Modulus, psi	130,000	*1,460,000
Specific Gravity	0.64	1.86
Filler, wt%	40	25
Glass Fiber, wt%	zero	25

*If the conventional FRP has a flexural modulus of 800,000 psi, it would require a thickness of 3.9 mm to match the stiffness of the hybrid foam.

RMC, ¢/lb		
Resin	120	60
Glass	---	80
Filler	18	18
Cost[1]($/cubic foot)	$4.207	$8.414
Cost(adjusted for thickness)	7.659	---

[1](Cost Per Pound) x (Spec. Grav.) x (8.3)

Table VII

The Effect of Low Level Foaming

Hybrid System	D	D
Flexural Strength, psi	24,200	18,800
Flexural Modulus, psi	840,000	860,000
Tensile Strength, psi	10,700	11,400
Tensile Modulus, psi	930,000	1,030,000
Tensile Elongation, %	1.3	1.4
Izod Impact Strength, ft lbs/inch		
Notched	5.0	6.1
Unnotched	7.2	8.1
Glass Content	15	15
Filler Content	15	15
Specific Gravity	1.419	1.368
Relative Hybrid System Expansion, %	---	+6.7
Relative Part Expansion, %	---	+3.7

EFFECTIVE USE OF CARBON FIBER
IN THE AUTOMOTIVE INDUSTRY

Scott Beck
Quality & Productivity Center
GM Truck & Bus Group
Flint, Michigan   USA

INTRODUCTION

Advanced composites can reduce system-costs if used in applications which can benefit from the unique properties they exhibit. The application engineer must now understand almost all of the aspects of advanced composites in order to make a wise decision which will have a positive impact on cost, quality and performance.

On a straight dollars-per-pound basis composite materials are hard pressed to compete with steel or aluminum, but a closer look is justified. In certain performance categories advanced composites can offer cost benefits if an appropriate application philosophy is used.

The cost of the high performance reinforcements is indeed high but some indications point to a downward trend. We must be positioned to use such materials as the costs come in line -- with lower costs comes increased opportunity. This paper will deal with the use of carbon fiber for applications which may currently be cost effective or have excellent potential with further fiber cost reductions. The information contained herein represents averages of currently available material properties from various sources and is not intended to be used as a design guide in any way.

BACKGROUND

Three dominant fiber reinforcements have emerged as candidates for use in the automotive industry; glass fiber, aramid fiber and carbon fiber. Each fiber has its own set of unique properties and outstanding features such as the low cost of glass, the high impact strength of aramid or the high modulus of carbon fiber. The properties of different types of composites can be summed up and compared to steel and aluminum on a stress-strain diagram (See Figure 1.).

It is apparent that unidirectional carbon fiber composites have the stiffness of steel with superior strength, that Kevlar and S-glass have stiffnesses close to that of aluminum with superior strength and that E-glass has tensile strengths superior to aluminum with about half the stiffness. Despite the excellent properties displayed by some of these advanced composite materials they haven't seen widespread application in the automotive industry because of their high costs (See Figure 2.). The negative cost stigma is particularly apparent in the case of carbon fiber. As carbon fiber came out of the aerospace industry and became available for commercial use its cost

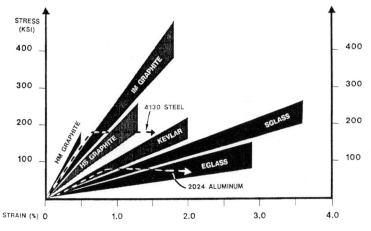

Figure 1.   TENSILE STRESS/STRAIN COMPARISON

was still high. There were a few engineers who wanted to exploit the properties of this unique material and thus made some prototype parts (even vehicles) from carbon fiber composites. But it boiled down to needing a lower cost fiber to make it economical to productionize some of these applications. The fiber companies showed charts that future fiber costs would fall below $10/lb. with volume usage. But a commitment to use enough fiber to bring the cost down to an economical level was not forthcoming from the automakers.

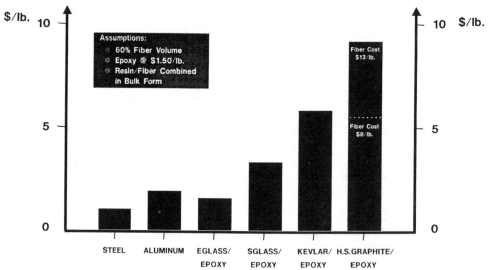

Figure 2.   COST COMPARISON - 1987

So who was going to go first? Would the fiber companies lower the
price in hopes of volume usage catching up in the future or would the
automakers take the first step and start using the high cost fiber,
building the volume with time and bringing the cost down over a period
of several years? No one was willing to take the first step and incur
any potential losses to further the usage of carbon fiber.

Nearly a decade later the same question arises. Who will take the
next step? A great number of things have changed since then that have
a tremendous impact on how the problem may be solved. Our
understanding of the material and its unique properties has grown
tremendously. New developments and refinements have occured at the
fiber end of the business. New and improved composites manufacturing
processes are available for our use. Some parts using carbon fiber
have already made it into production and blazed the trail for new
applications. Our "micro" view of how a section of a component can be
switched to carbon fiber composites and made to work has given way to
the "macro" approach where the impact of carbon fiber use can be
determined in an overall "system approach". Better analysis tools, a
larger selection of resins, a greater data base to work from, and
increasing public awareness all contribute to an increasing propensity
to use carbon fiber composites where they make sense. Carbon fibers
are principally used in the aerospace industry followed by the recreational products market and finally automotive use (See Figure 3.).

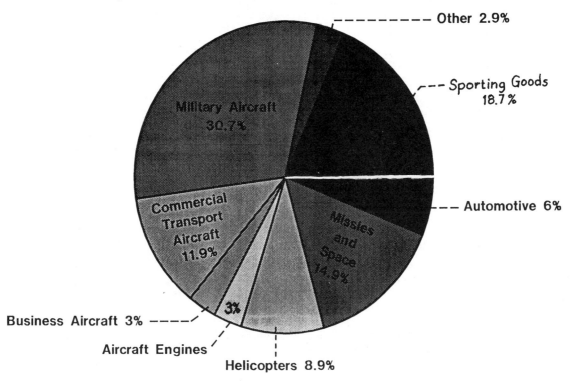

Figure 3.   CARBON FIBER MARKET SEGMENTATION

Some suppliers of carbon fibers are firmly entrenched in the aerospace market and are not making an effort to produce the low cost carbon fiber needed by the automotive and recreational products markets. It is also apparent that the automotive industry is not ready and willing to commit to any wholesale changes in its products. These two phenomena are most likely a carryover reaction from the work which was pursued in the mid to late 1970's when talk was heard and no action was taken, resulting in some disillusionment on both sides.

With the agressive growth of composites usage predicted in the aerospace industry (See Figure 4) it is unlikely to see the automotive market segment substantially change size in the very near future even with agressive applications. Most individual component applications in the automotive arena would average carbon fiber consumption on the order of 100,000 pounds per year, which is only a small percentage of production and becoming smaller as aerospace makes the market even larger.

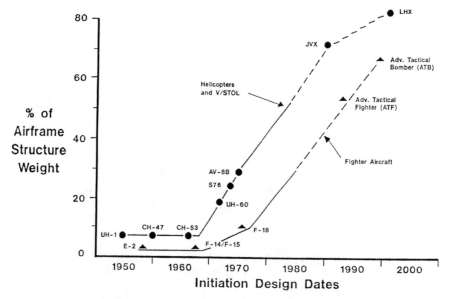

Figure 4.   PREDICTED AEROSPACE INTEGRATION OF COMPOSITES TO AIRCRAFT STRUCTURES. (COURTESY OF GRUMMAN AEROSPACE CO.)

Growth in the automotive market depends on three major factors:

o   Lower carbon fiber costs.
o   Smart application to product.
o   Elimination of opinioneering.

Lower fiber costs naturally expand the cost effective opportunity base in areas where it's currently used, and create opportunities in new areas where its never been cost effective before. Smart applications are made by engineers who understand the multiple disciplines involved with composites;

- Material properties.
- Unique characteristic exploitation.
- Manufacturing techniques and limitations.
- Tooling requirements.
- Costing factors (Production speed, burdens, capital expenditures, comparison techniques etc.).

And finally, "opinioneering" (as opposed to engineering) can be reduced if the engineer is able to properly report the benefits of the correct application of composites.

New technologies and refinement of existing methods are bringing carbon fiber costs down to the point where automotive use becomes practical.

Several companies are hard at work trying to make their carbon fibers "fit" the automotive market. Automotive applications of carbon fiber generally do not require aerospace quality and inspection, and sometimes don't require the degree of physical properties. Automotive grades of carbon fiber that would fit into a specific category (such as stiffness critical or conductive as described later in the report) could help the manufacturer to lower costs by eliminating the steps that are needed in making the fiber better than it has to be.

PAN (Poly Acrylo Nitrile) - precursor in large textile grade bundle sizes, (40K-50K Filaments per bundle) is available for making a low cost fiber product. But there are limitations in what you can manufacture with that size tow such as in fabrics and filament winding (depending on needs). But the use of textile grade large tow size material is very attractive to the automotive industry because of potential price breaks. Indications from key suppliers is that carbon fiber prices can fall below the $10/lb. range if the right conditions exist.

Also, a new method of producing carbon fiber by pyrolysis of vapor phase hydrocarbons (natural gas, benzene) is heralded as a means of achieving tremendous cost reduction. The vapor grown fibers exhibit many of the characteristics of PAN based fibers and with a subsequent heat treatment process exhibit electronic, magnetic and thermal transport characteristics of single crystal graphite. It is conceivable that the costs of vapor grown fiber could fall below $5/lb. making it economical for many automotive applications. A major limitation of this fiber type however is the format: Non-continuous fibers are produced similar to "chopped" fiber formats. Unless an orientation/pre-preg device is used, the material cannot be used in applications which require unidirectional continuous reinforcement.

Carbon fiber usage is growing beyond actual automotive componentry into areas of engineering support, and manufacturing assembly. The broadening of scope comes not only with low cost but with an understanding of the material and its unique properties. Carbon fiber composites can offer unique solutions to some tough manufacturing and engineering problems, and this sort of penetration should manifest itself as a linear increase in the use of carbon fiber in our industry. This is opposed to the jagged "stepped" increases due to usage in a particular component of a vehicle line or vehicle family such as a driveshaft (See Figure 5.). In summary, a slow acceleration in automotive usage of carbon fiber is the consensus of opinion today, slow steady growth will be augmented by large volume components which are converted to carbon fiber composites.

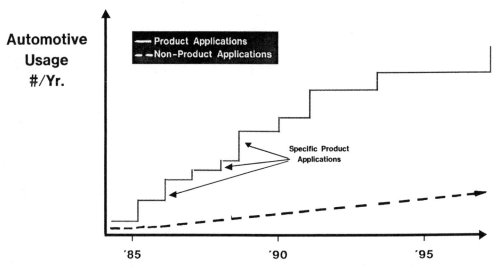

Figure 5.   DEPICTION OF GROWTH BY APPLICATION USE AND OTHER USE

APPLICATION PHILOSOPHY

There exists numerous metallic components which can be readily converted to composites. But some of these would require large amounts of time and money to develop and some would never be cost effective. A strategy ought to be used in determining which items should be made from advanced composites.

Listed below are several areas of opportunity to look for when judging the potential for composites application:

1. Part consolidation
2. Part elimination
3. Existing component is hard to manufacture (Deficient quality)
4. Current use of premium material
5. Measureable performance improvements to be gained

Part consolidation is one of the most common and important opportunities to be applied. The consolidation of several parts and elimination of their subseqent assembly can go a long way towards offsetting the premium cost of the raw material. An example is a composite one-piece floor pan which replaces a multi-piece steel assembly (See Figure 6.). Here not only the assembly costs are eliminated but also a sealing operation and the joints which reduce structural stiffness. Also, the advantage in the number of tools required is substantial. Multiple forming dies are used to make the steel pieces and only one tool is used for the composite. The savings in tooling costs are only apparent in low volume; because of cycle time limitations molding composites in large volume could require greater tooling investment than with steel.

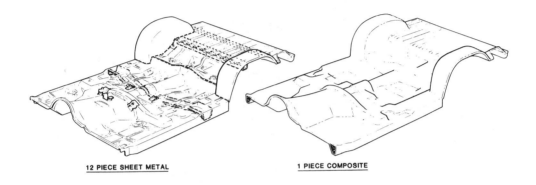

12 PIECE SHEET METAL   **1 PIECE COMPOSITE**

Figure 6.   FLOOR PAN COMPARISION – SHEET METAL vs. COMPOSITE

The flexibility of the molding processes should be used to their utmost to achieve maximum part consolidation, but not overused, which could result in a component which cannot be manufactured.

Part elimination has the same net effect as consolidation but by a different means. If the part function and/or performance is improved substantially other parts may not become necessary. A well known example is the one-piece composite driveshaft which replaces a two-piece steel assembly (See Figure 7.). Because of the composites high specific modulus the entire center bearing is eliminated and in some cases the support member for the center bearing also. Also the dynamic dampening properties of the composite may allow the elimination of add-on noise and vibration dampeners, all told a tremendous parts elimination potential.

Sometimes an assembly can be particularly difficult to fabricate resulting in a lack of consistency, high equipment cost, worker fatigue and the like. If a composite assembly can be made to eliminate the deficiency this can be a possible selling point of the product. For example the floor pan that was discussed earlier and shown in Figure 6. Spot welding the front section to the rear section requires the use of a specially made, extremely large spot weld gun which is very heavy and unwieldy. Because of the equipment, human operator and other factors, consistency may be lacking.

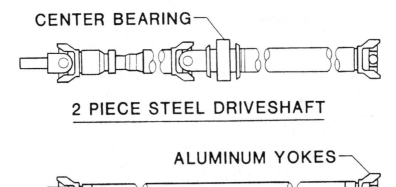

CENTER BEARING

2 PIECE STEEL DRIVESHAFT

ALUMINUM YOKES

1 PIECE COMPOSITE DRIVESHAFT

Figure 7.   ONE-PIECE COMPOSITE DRIVESHAFT/TWO-PIECE STEEL DRIVESHAFT

The one-piece composite floor pan in Figure 6 would eliminate the need for the welding operation and thereby eliminate related inconsistencies. There is also the opportunity to make improvements to the welding gun to overcome its size and weight deficiencies by the use of composites.

Current use of premium materials dictates the need for materials with unique characteristics. It may be the case that an aerospace grade of aluminum is being forged into a suspension component. Perhaps it is very desirable to reduce the weight in this dynamic component in which case some composites can offer an improved reduction over that of the premium material. This type of case is extremely difficult to find and may even be difficult to implement based on the situation and requirements of the component.

If higher performance is a net result of weight reductions, an opportunity exists for advanced composites. Reciprocating or moving components such as connecting rods, push rods, turbocharger turbines etc. can greatly benefit from the unique properties of composites and weight reduction potential. Highlighting the unique property characteristics of carbon fiber composites is its superior strength and stiffness-to-weight ratios (See Figure 8.)

If applications can be located that exploit these unique properties then almost surely a performance improvement will result. A strategy must be set-up to stay away from head-on competition with steel, to exploit the unique properties of the composite materials and finally to do things with the composite which cannot be done with steel or

aluminum.   A slow acceleration is taking place and care must be taken
to   stay   in   a   position   where   future   breakthroughs   in   cost   and
properties   will   be   able   to   be translated to economic advantages in
future products.

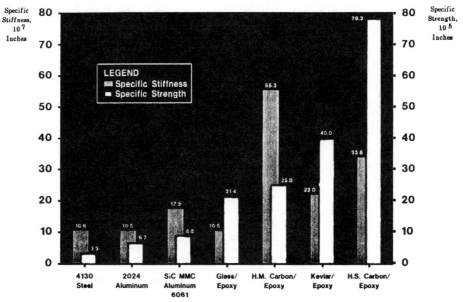

FIGURE 8.   SPECIFIC STRENGTH AND STIFFNESS COMPARISON

APPLICATIONS

Effective   application of carbon fiber composites range from structural
uses   to   in-plant   uses.   Once   again   the   applications   that   are
successful   exploit   the   unique properties available to do what cannot
be   done   with   other materials.   Not all of the applications mentioned
here   represent   a   cost   savings,   but these items do represent niches
where   the use of carbon fiber makes sense because they're doing unique
things   for strength and stiffness, shielding, quality, timing, inertia
reduction, etc.

Stiffness   critical applications include such things as driveshafts and
vehicle   structures.   Driveshafts   benefit   greatly   from   the   high
specific   stiffness of carbon fiber composites because of their dynamic
nature.   Greatly   simplified, the equation which determines one-piece
driveshaft   "critical speed" or first bending mode show us why specific
stiffness counts in this situation;

$$R.P.M. = 33 \; \frac{d}{L^2} \; \sqrt{\frac{E}{\rho}}$$

Vehicle   packaging   will dictate the maximum allowable diameter (d) and
the   length   (**L**)   of   the shaft (from universal joint centerline at the
transmission   to universal joint centerline at the pinion flange).   The
vehicle   driveline   combinations   will   dictate   a   target   value   for
critical   speed,   leaving   only   the   specific   stiffness   to determine
the suitability of a material to meet the bending mode criteria.

The   majority   of   the   market   for   composite   driveshafts   is seen in
current   two-piece   steel   assemblies.   In GM Truck products, well over

471

1/2 million two-piece shafts are used. When these shafts are designed as composites an average usage of carbon fiber can be determined. For these light truck applications the average usage is 0.9 pounds of carbon fiber per shaft. The cost of carbon fiber has a direct impact on the penetration of composite driveshafts into this potential market as shown in Figure 9.

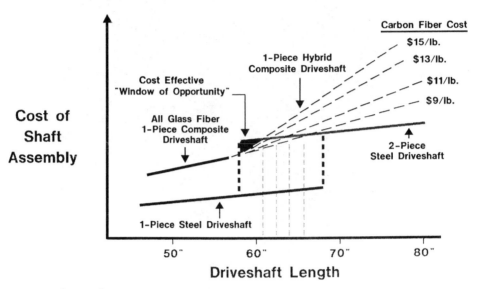

Figure 9.   DRIVESHAFT PENETRATION VERSUS CARBON FIBER COST

The cost of steel driveshafts is shown versus length as a very low sloped line which jumps at a point where it becomes necessary to use a two-piece assembly. The jump represents the added cost of the center bearing and related components. On the other hand, the cost of a composite shaft is non-linear with length because of the increased usage of carbon fiber with increased length. An "area of opportunity" is defined by the composite shaft curve where it falls below that of the two-piece steel line. This "area of opportunity" is expanded by the downward price shifts of carbon fiber denoted by the dashed lines.

Another stiffness critical application is in vehicle structures. All three major automakers have made composite vehicle structures and have learned that if glass fibers are used exclusively there will be a slight weight penalty versus steel. This is because of the low specific stiffness of the glass fiber composites, particularly random glass composites. On the other hand, using an appropriate process a graphite structure could save as much as 70% of the weight of a steel structure. Configurations can range from body-frame-integral, plat- form frames (Figure 10), body-on-frame or spaceframe type of construction, the merits of each will not be discussed. It is sufficient to say that weight savings is best achieved by using a higher modulus material such as carbon fiber composites.

It is also important to note that the most effective way to use carbon fiber composites in this situation is to hybridize the composite. By placing unidirectional carbon fiber in areas of high

Figure 10.   PLATFORM FRAME CONCEPT

flexural stress such as a rocker top and bottom (farthest from neutral axis), a "sandwich" type of laminate is formed which yields the best utilization of the carbon fiber (See Figure 11.). Composite structures benefit from joint continuity which can yield up to 30% higher stiffness than a spot-welded joint and from increases in structure section modulus in some areas due to increased wall thickness and incorporation of interior trim pieces, all of which may yield a more optimized vehicle structure design.

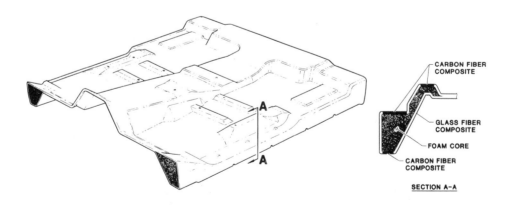

Figure 11.   OPTIMUM USAGE OF CARBON FIBER - ROCKER SECTION

Strength critical applications tend to be the hardest in which to justify the use of carbon fiber composites. One such application is an upper control arm where tensile and compressive strength drive the design of the part. The control arm is a suspension component which attaches the steering knuckle to the frame of the vehicle and allows travel of the wheel relative to the frame (See Figure 12.).

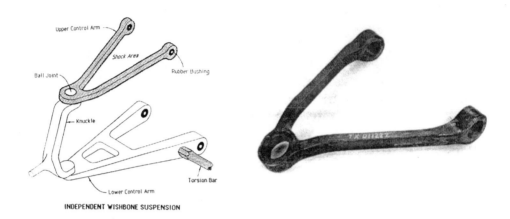

Figure 12.   COMPOSITE UPPER CONTROL ARM

The loading condition and intensity determined the use of uni-directional carbon fibers in epoxy to carry the tensile/compressive loads seen in the arms. Although an 80% weight savings was achieved versus steel and 25% versus aluminum, a cost savings was not achieved in this part but with lower fiber prices this part could become cost effective. Depending on the application, the decreased mass may be beneficial to future "active ride" suspension systems where inertia needs to be minimized. In this application as much as 0.5 pounds of carbon fiber would be used in each arm.

A logical extension of the work which has been done in driveshafts is the composite torsion bar spring. This part is not stiffness critical as much as it is shear strength critical. In order to achieve the appropriate spring rates and angular deflection maximum while staying within a small enough "package" carbon fiber must be used. A well defined load path again determines the $\pm 45°$ orientation of the unidirectional carbon fiber composite. A 70% weight reduction was achieved but again, a cost penalty was incurred on this part when compared with steel. A part of this nature may contain as much as 2 pounds of carbon fiber (See Figure 13.).

Carbon fibers have another unique property in that they are good electrical conductors. By incorporating a nonwoven carbon fiber paper into a glass fiber composite panel EMI/RFI shielding is achieved in the otherwise "transparent" panel. This "shielding" characteristic leads the way to applications in composite hoods where adequate RFI shielding is not possible with the current glass fiber composite designs (See Figure 14.). The use of an integral distribution of

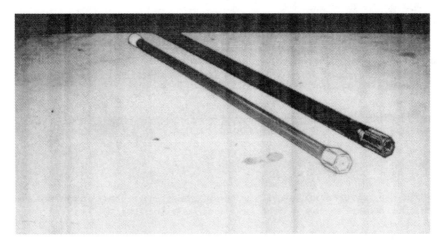

Figure 13.   STEEL AND CARBON FIBER TORSION BARS

carbon fiber is a much more cost effective way of providing shielding than incorporating a foil or metallic piece into the hood assembly as separate pieces. As much as 2.5 square yards of carbon fiber non-wovens could be used in each hood assembly depending upon manufacturing process and shielding requirements.

Also, if a composite hood were put on a steel bodied truck in production it would need to be either off-line painted or coated with a conductive primer for subsequent electrostatic paint application.

CARBON FIBER PAPER
SWIRL MAT GLASS
GLASS WEAVE

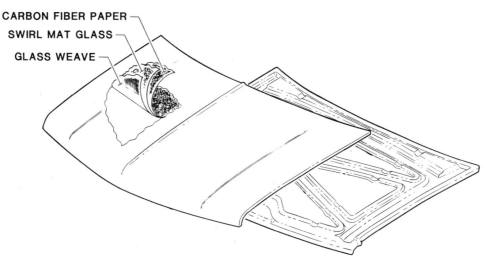

Figure 14.   COMPOSITE HOOD W/CARBON FIBER FOR SHIELDING

Assuming this paper was uniformly distributed over the entire part and was very close to the surface it could potentially serve a dual purpose: Besides the EMI/RFI shielding the layer could serve as the conductor necessary for electrostatic paint operations. This could potentially save a tremendous amount of capital investment and/or piece cost.

Another product application of carbon fiber that has been reported by Ashland Petroleum Company is use in composite friction materials, i.e. brake pads. It was stated that carbon fibers;

- Enhanced friction coefficients
- Showed longer wear characteristics
- Uniformly distributed heat across friction surface
- Could replace asbestos fibers

This is another application where penetrations into the automotive market is dominated mostly by price. In a component of such high annual volume the cost of constituent materials is extremely important.

There are several areas outside of the actual product offerings where the automotive industry may potentially use significant amounts of carbon fiber. These areas do not represent high volume of any one part and therefore do not totally rely upon the need for low cost carbon fiber. But these uses constitute a real and growing utilization of advanced materials in the automotive industry.

Engineering mock-ups are increasingly being changed to carbon fiber composites from vacuum formed thermoplastic sheeting (such as polycarbonate or acrylic). Quality of the mock-up is the main reason for switching materials. Real part thicknesses (that of sheet metal) can be achieved while maintaining a reasonable amount of structure. Making thermoplastic sheet to sheet metal thicknesses results in a very flimsy unrepresentative part, but with better than a magnitude improvement in modulus the carbon fiber composite makes a stiff, yet light part. Another quality advantage is in shrinkage and variation. Closer tolerances can be held because the composites vacuum bagging process is done at room temperature, while vacuum forming is done at high temperatures. This coupled with the fact that the thermoplastics have a relatively high coefficient of thermal expansion while the composites have relatively low coefficient prove that the dimensional accuracy of the mock-up can be greatly improved by switching materials. Pending some ongoing trials a more aggressive move toward changing materials may occur.

The use of carbon fiber in lightweight gages and fixtures is also expanding rapidly in the automotive industry. Due to the carbon fiber's extreme light weight, low coefficient of thermal expansion and high strength and stiffness it is a natural for this application (See Figure 15.). Quality is the biggest factor in switching materials from steel, aluminum or glass fiber composites to carbon fiber composites. The dimensional stability of the gages or fixtures with respect to temperature variation is extremely important in the

476

assembly plant.   For reasons of set-up, mobility, and ease of use, reduced weight is important to the assembly plant. Another important advantage is the speed at which these gages and fixtures can be fabricated and changed compared to steel and aluminum. This affords more flexibility and shorter lead times for manufacturing engineers.

Figure 15.   TUBULAR CONSTRUCTION GAGE AND FIXTURE

Finally, the spot weld guns of the future are being designed today from composite materials, namely carbon fiber composites (See Figure 16.).   If manufacturing costs can be kept low and optimum use of carbon fiber is employed a cost effective welding gun can be made with the following advantages;

    o   Reduced mass
          - Smaller and faster robots used in automatic applications
          - Less worker fatigue in manual applications

    o   Smaller profile, greater work envelope

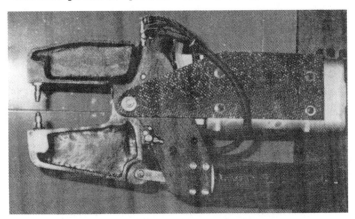

Figure 16.   COMPOSITE WELD GUN

Relative to the current copper structures on most weld guns today, a carbon fiber composite can offer tremendous increases in specific stiffness and strength which translate into a much smaller, lighter package. With further development this application could see increasing production usage with possible proliferation in the 1990's.

All told, there are many potential applications of carbon fiber composites not all of which are cost effective. Good judgement by composites application engineers will allow successes which could possibly drive the cost of carbon fiber composites down. Along with the advent of new technologies which can further reduce fiber costs its apparent that a future strategy must be determined and some work be done today on finding those good applications. A slow but steady growth in the automotive use of carbon fiber should be supplemented by strong spurts of growth where product applications such as driveshafts add large amounts of annual usage. Overall usage may not be as high as some might have predicted, but fiber costs are not as low as some have predicted either. The keys to greater penetration and usage are the cost of carbon fiber and the knowledge of the engineers applying it.

REFERENCES:

"Growth Patterns in Carbon-Fiber Markets," Advanced Composites, May/June 1986.

Segal, Charles L., "An Analysis of Current and Future Markets for High Performance Carbon, Organic, and Glass Fibers," OMNIA Study, February 1987.

Beetz, C.P., G.G. Tibbets, M. Endo, "Carbon Fibers Grown from the Vapor Phase; A Novel Material," SAMPE Journal September/October 1986, pp. 30-34.

Composite Materials Workshop, Proceedings of February 1985 Meeting, Section K-5B.

Krock, Richard P., Don Carlos, D. Chris Boyer, "Versatility of Short Pitch-Based Carbon Fibers in Cost Efficient Composites," 42nd Annual SPI Composites Institute Conference, February 2-6, 1987.

Beck, S.A., "Design, Manufacture and Dynamic Characteristics of Composite Driveshaft," General Motors Institute Thesis, May 15, 1985.

Knollmeyer, Paul A., "Technical Fiber Nonwovens for Composite Applications," 42nd Annual SPI Composites Institute Conference, February 2-6, 1987.

Beck, S.A., "Concepts for High Speed Manufacture of Quality Structural Composites," SME #EM87-453, AUTOCOM '87, June 1-4, 1987.

MECHANICAL PROPERTIES AND MICROSTRUCTURE OF SHORT ALUMINA FIBER REINFORCED MAGNESIUM ALLOY

Mamoru Sayashi, Harumichi Hino, Mikiya Komatsu
Department No. 1
Materials Research Laboratory
Central Engineering Laboratories
Nissan Motor Co., Ltd.
Yokosuka, Japan

Masato Sasaki
Material Research Section
Research and Development
Department
Atsugi Motor Parts Co., Ltd.
Atsugi, Japan

Abstract

An investigation was made of squeeze-cast short alumina fiber reinforced AZ91 in order to examine the possibilities of using this lightweight material to reduce the weight of automotive parts and thereby achieve higher vehicle performance. Problems related to material fabrication were resolved, fiber preforms were improved and the properties of the material were clarified. It was found that AZ91 could be reinforced using short alumina fibers by optimizing the fiber preforms and fabrication conditions. Young's modulus of the material, containing 9.5 % fiber by volume and without being heat treated, was smaller than that of a T6 heat treated 336 aluminum alloy. Although its tensile strength and fatigue strength were inferior at room temperature, they were superior at 573 K to those of the T6 heat treated 336 alloy. The material showed excellent specific strength at the higher temperature, but its creep strength was inferior to that of the 336 alloy. While tensile strength and fatigue strength were still unsatisfactory at near room temperature, it is thought that those problems can be solved by improving the material and production technologies. It is expected that the application of fiber reinforced magnesium alloys to automotive parts in the future will achieve significant weight reductions which will translate into higher levels of vehicle performance.

REDUCING THE WEIGHT of components through the application of lightweight materials is one important way of achieving higher vehicle performance and lower energy consumption. High tensile strength steels, aluminum alloys, magnesium alloys, plastics and fiber reinforced plastics are some of the lightweight materials that are now being used to make automotive parts. In addition to these materials, the authors have been examining the possible application of fiber reinforced metals[1]. This paper presents the results of research done on a short alumina fiber reinforced magnesium alloy, about which little is mentioned in the literature.

AZ91, the most commonly used structural magnesium alloy, was employed as the matrix alloy. Short alumina fibers were used as reinforcement because of their excellent mechanical properties, chemical stability, isotropy of properties and formability into complex shapes, although boron, alumina or graphite continuous fibers had been used in other investigations[2-4]. Squeeze casting, in which the molten alloy was infiltrated into the fiber preform under pressure, was used to fabricate composites consisting of AZ91 and short alumina fibers, referred to as SAFR AZ91 in this report. Through this work, techniques were established for squeeze casting sound composites, fabricating sound fiber preforms and estimating material properties. Various problems were identified and approaches to resolving them were clarified.

EXPERIMENTAL PROCEDURE

AZ91 was selected as the matrix alloy because the initial aim of this work was to gain a good understanding of the material properties, identify potential problem areas and analyze various phenomena in connection with the fabrication and use of the material. Because of its suitability for melting and casting and also low price, AZ91 is the most widely used structural magnesium alloy. The chemical composition of the magnesium matrix alloy is given in Table 1.

Although continuous [2-4] boron, alumina or graphite fibers had been used in other investigations, short fibers were employed in

Table 1 – Chemical composition of matrix alloy

AZ91	mass %
Al	9.8
Zn	0.7
Si	0.58
Fe	0.005
Mn	0.28
Cu	0.002
Ni	0.000
Be	0.001
Mg	bal.

Table 2 – Outline of alumina fiber

Chemical composition, mass %	
Al_2O_3	96 – 97
SiO_2	3 – 4
Mean fiber diameter, μm	3
Mean fiber length, μm	150
Tensile strength, GPa	2
Young's modulus, GPa	300
Crystal phase	mainly delta alumina

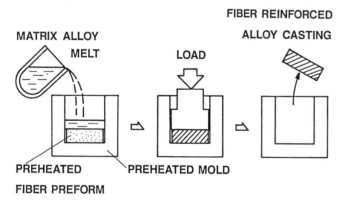

Fig. 1 – Schematic illustration of fabrication process for fiber reinforced alloy casting

this work because it was assumed that the material would be applied to automotive parts having complex shapes. Alumina was chosen as

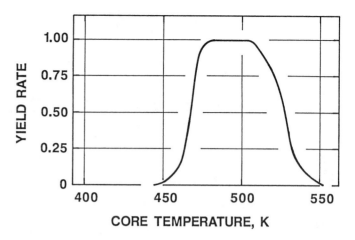

Fig. 2 – Yield rate in production of cup-like SAFR AZ91 test piece castings (80 mm dia. x H 80 mm x T 3–20 mm)

the fiber material because of its superior tensile strength, high Young's modulus and excellent thermal and chemical stability. It was also thought that few reaction-related problems would occur with alumina during fabrication and use. An outline of the alumina fiber reinforcement is shown in Table 2.

The squeeze casting method used to fabricate the composites is illustrated schematically in Fig. 1. The molten matrix alloy was infiltrated under pressure into a preheated fiber preform set in a mold. This process can produce complex shaped castings of fiber reinforced alloys.

Tensile strength at room and elevated temperatures, fatigue strength, creep strength, wear resistance and thermal dimensional stability of SAFR AZ91 were evaluated. The microstructure was also analyzed using an optical microscope and a scanning electron microscope.

RESULTS AND DISCUSSIONS

SQUEEZE CASTING – In previous work the authors had established a squeeze casting technique for aluminum alloy matrix composites. At first, an attempt was made to manufacture SAFR AZ91 under optimized fabrication conditions for the aluminum alloy matrix composites. It was found, however, that sound magnesium alloy matrix composites could not be obtained under those conditions because of the formation of cold shut and misruns and deformation of the fiber preform.

The yield rate is given in Fig. 2 for the production of cup-like trial castings, having a diameter of 80 mm, height of 80 mm and thickness of 3 – 20 mm. The thermal properties of AZ91 and the 336 aluminum alloy are given in Table 3. The high liquidus temperature, small specific heat and small latent heat of AZ91 increase the

Table 3 - Thermal properties of AZ91 and 336
aluminum alloy

	AZ91	336
Solidification temperature range, K	869 –741	881 – 839
Specific heat, MJ/m^3	1.8	2.6
Latent heat, GJ/m^3	0.67	1.03

tendency for solidification to occur during squeeze casting. Consequently, in order to obtain sound SAFR AZ91 castings, the mold temperature, preheat temperature of the fiber preforms and pouring temperature had to be set higher than the optimized conditions for 336 aluminum alloy matrix composite castings. The actual optimum squeeze casting conditions tended to vary considerably depending on the casting machines used and the shapes of the castings. The suitable conditions for casting cup-like test pieces of SAFR AZ91 using a casting machine like that shown in Fig. 1 were found to be: a mold temperature of 723 K, core temperature of 473 – 523 K, preheat temperature of the fiber preform of 723 K, pouring temperature of 1000 K, moving velocity of the pressurizing punch of 7 mm/s and maximum pressure of 70 MPa. Under these conditions, the squeeze casting technique yielded sound SAFR AZ91 castings.

FIBER PREFORM - Tensile strength data and photomicrographs of the fracture surface of the material, when it was squeeze cast under the optimum conditions mentioned above using a fiber preform from an earlier stage of development, are shown in Fig. 3. The strain rate during testing was 8.3×10^{-4} s^{-1}. The predicted tensile strength of SAFR AZ91 at 473 K was about 200 MPa. That value was derived from the tensile strength of the matrix AZ91(F) and alumina fiber at 473 K, the volume fraction of the fiber, Vf, and the fiber orientation. The tensile strength shown in Fig. 3 was less than one-half of the predicted value. It was found that there were some fiber free zones which were about 500 μm in diameter, as seen in the microstructure in Fig. 3. This indicated that the fiber preform was not uniform and was constructed of fiber balls having a diameter of about 1 mm. The SEM micrograph in Fig. 3

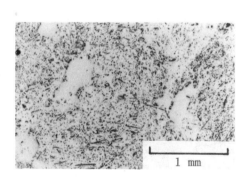

a) Optical micrograph

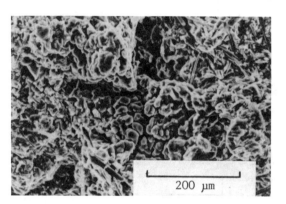

b) SEM micrograph of fracture surface

Tensile Strength at 473K: 91 MPa

Fig. 3 - Photomicrographs and tensile strength of SAFR AZ91 made with fiber preform containing voids

Table 4 - Outline of fabrication techniques for fiber preform

	Vacuum process	Pressing process	Casting process
Dimensional accuracy, mm	1	0.1	0.1
Accuracy of volume fraction of fiber, %	20	5	5
Range of volume fraction of fiber, %	5 – 10	7 – 20	7 – 30
Density distribution	poor	fair	good
Fiber dispersivity	good	poor	good
Fiber orientation	good	poor	good
Formability into complex shape	poor	good	good

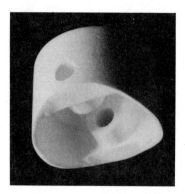

Fig. 4 – Example of complex shaped fiber
preform

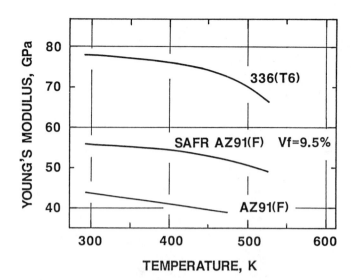

Fig. 6 – Young's modulus of SAFR AZ91 (F,
Vf: 9.5%), AZ91(F) and 336
aluminum alloy(T6) as a function
of temperature

shows that there were micro-shrinkage cavities in the fiber free zones and that the cavity surface was covered with dendrites. The low tensile strength of the material was thought to result from the presence of the fiber free zones containing micro-defects.

Improvements were then made to the technique for fabricating the fiber preforms so as to prevent the formation of the fiber free zones. An outline of different fabrication techniques is given in Table 4. The defective fiber preform mentioned above was produced using the vacuum process shown in the figure. Recently it has become possible to obtain complex shaped fiber preforms with a homogeneous fiber distribution using the casting process. An example of such complex preborms is shown in Fig. 4. Tensile strength data and photomicrographs of the microstructure and the fracture surface of the material, when it was squeeze cast using a fiber preform made with the casting process, are shown in Fig. 5. Fiber distribution was improved and

tensile strength of about 180 MPa was obtained, which was near the predicted value.

MATERIAL PROPERTIES – The material properties of SAFR AZ91, which was not heat treated and had a Vf of 9.5 %, are shown in Figs. 6 through 12. Test pieces were cut from cast disks which were 80 mm in diameter and 10 – 40 mm thick.

Young's modulus of the material was 49 – 56 GPa, which was smaller at all temperatures from room temperature to 573 K than that of the 336 aluminum alloy and almost equal to the predicted value (Fig. 6). Although the effect of the fiber reinforcement was clearly evident in the tensile strength, the value at room

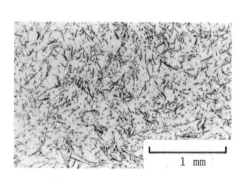

a) Optical micrograph

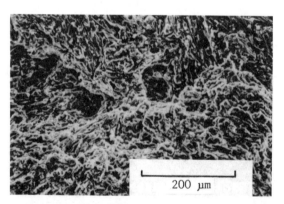

b) SEM micrograph of
fracrure surface

Tensile strength at 473 K: 180 MPa

Fig. 5 – Photomicrographs and tensile strength of SAFR AZ91 made with sound
fiber preform

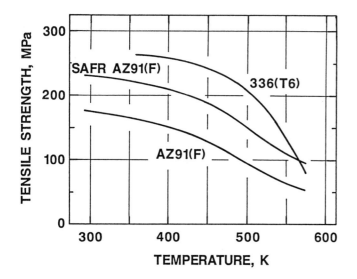

Fig. 7 – Tensile strength of SAFR AZ91 (F,
Vf: 9.5%), AZ91(F) and 336
aluminum alloy(T6) as a function
of temperature

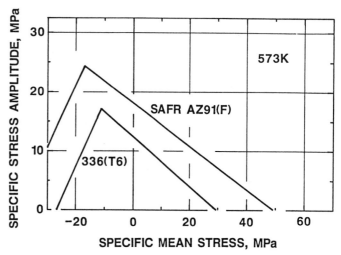

Fig. 9 – Fatigue limit diagrams at 573 K for
SAFR AZ91(F, Vf: 9.5%) and 336
aluminum alloy(T6) in terms of
specific strength

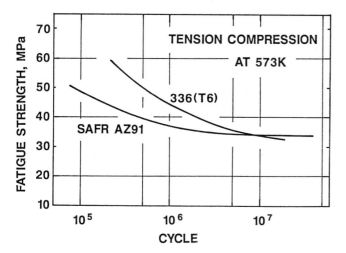

Fig. 8 – Fatigue strength of SAFR AZ91 (F,
Vf: 9.5%), and 336 aluminum alloy
(T6) at 573 K

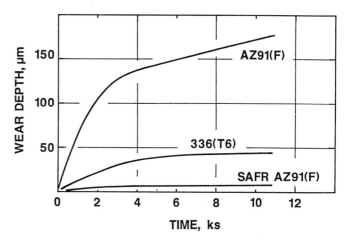

Fig. 10 – Wear resistance of SAFR AZ91 (F,
Vf: 9.5%), AZ91(F) and 336
aluminum alloy(T6)
Test pieces (20 mm dia. x 10
mm thick) came in contact with a
rotating gray cast iron disk (50 mm
dia. x 24 mm thick) in 353 K motor
oil.
sliding velocity: 0.3 m/s
load: 980 N

temperature was between those of the 336
aluminum alloy (T6) and AZ91(F) and smaller than
that of the 336 alloy (Fig. 7). However, the
temperature dependence of the tensile strength
was smaller than that of the 336 aluminum alloy.
The tensile strength at 573 K was superior to
that of the 336 alloy. The fatigue strength
characteristics of the material were the same as
those of the tensile strength. Fatigue strength
was inferior at temperatures from room
temperature to 523 K, but superior at 573 K to
that of the 336 aluminum alloy (T6) (Fig. 8).
The experimental results obtained at 573 K were
summarized in a fatigue limit diagram in terms
of specific strength (Fig. 9). Although it was

thought in the early stage of the development
work that the wear resistance of the material
was poor, it was found to be significantly
improved as a result of adding the fiber
reinforcement and was superior to that of the
336 aluminum alloy (T6) (Fig. 10).

The fiber reinforcement also had some effect
on creep strength, but the creep strength level
was still lower than that of the 336 aluminum
alloy (Fig. 11). Dimensional change, similar to

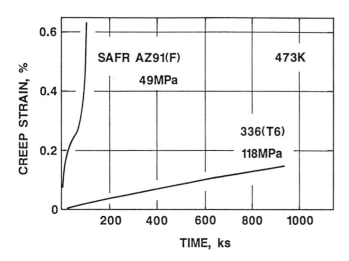

Fig. 11 – Creep strain curves of SAFR AZ91 (F, Vf: 9.5%) and 336 aluminum alloy(T6)

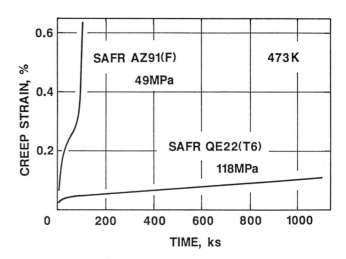

Fig. 13 – Creep strain curves of SAFR AZ91 (F, Vf: 9.5%) and SAFR QE22(T6, Vf: 9.5%)

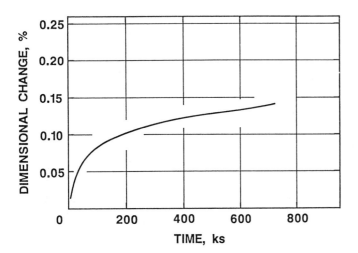

Fig. 12 – Expansion of SAFR AZ91 (F, Vf: 9.5%) in direction of fibers when held at 573 K

that seen in aluminum matrix composites having an oriented fiber distribution[5], was also evident in SAFR AZ91 when it was subjected to sustained elevated temperature (Fig. 12). This change was thought to result from induction of thermal stress during cooling because of the difference in the thermal expansion coefficients of the matrix alloy and the fiber reinforcement, and relaxation of the stress at higher temperature. Improvement of the creep strength and dimensional stability of SAFR AZ91 was thought to be necessary in order to apply the material to parts in which deformation and dimensional change at high temperature would not be permitted.

FUTURE ISSUES - As mentioned above, Young's modulus, tensile strength, fatigue strength,

creep strength and wear resistance of AZ91 were improved by reinforcing the material with short alumina fibers. It was particularly significant that the tensile strength and fatigue strength at high temperature and wear resistance of SAFR AZ91 were superior to those of the T6 heat treated 336 aluminum alloy.

It was mentioned earlier that AZ91 was selected as the matrix alloy because the aim at the first stage of the development work was to identify the general material properties and problems areas of the SAFR magnesium alloy. Optimization of the matrix alloy to allow application to a wide variety of parts was an issue that was left for future work. The data obtained on the properties of SAFR AZ91 indicated that tensile strength and fatigue strength would have to be improved over a temperature range from room temperature to 523 K and creep strength would also have to be increased, in order for the material to play a major role in producing lightweight automotive parts in the future. These problems are thought to be solvable by developing a more suitable matrix alloy, and investigations are now under way to find a matrix alloy that will provide higher strength, and heat resistance. Maintaining dimensional stability at high temperature is another important issue that must be addressed if the material is to be applied to parts that are used under high temperatures. This issue can be resolved by optimizing the matrix alloy and heat treatment procedure.

An example of the results obtained in development work now under way is shown in Fig. 13. Creep strength has been improved by modifying the matrix alloy. A prototype short alumina fiber reinforced magnesium alloy piston is shown in Fig. 14 as an example of the application of the material to engine parts.

Fig. 14 – Prototype short alumina fiber
reinforced magnesium alloy
piston (82 mm dia.)

SUMMARY AND CONCLUSIONS

The results of this investigation into the alumina fiber reinforced magnesium alloy, which was squeeze cast and not heat treated, are summarized below.

(1) In order to obtain a sound material, the preheat temperature of the fiber preform, core and mold temperatures and pouring temperature had to be raised to higher levels than the corresponding temperatures in the fabrication of aluminum alloy matrix fiber reinforced materials. The reason is that the magnesium alloy has a higher liquidus temperature, smaller latent heat and smaller specific heat.

(2) The properties of the material obtained with a suitable squeeze casting process were strongly affected by the quality of the fiber preforms. In order to reinforce AZ91, it was necessary to establish a fabrication technique that would yield homogeneous preforms.

(3) Young's modulus of the material was about 50 GPa, which was between those of AZ91 and the 336 aluminum alloy and smaller than that of the latter. The tensile strength and the fatigue strength were inferior to those of the T6 heat treated 336 aluminum alloy over a temperature range from room temperature to 523 K and superior at 573 K. Wear resistance and creep strength were inferior to those of the (T6) 336 alloy.

(4) Improvement of the tensile strength and fatigue strength in a temperature range from room temperature to 523 K, enhancement of creep strength and improvement of dimensional stability at high temperature are issues that must be addressed in future work, as seen from the results of the present investigation. These problems are thought to be solvable by developing and selecting an improved materials system and by optimizing the heat treatment process.

The results of this work indicate that short alumina fiber reinforced AZ91 shows a number of excellent material properties without being heat treated. Some problems relating to material properties remain to be resolved, however, before the material can be applied to a wide variety of automotive parts. Since there is still sufficient latitude for further development and selection of optimum alumina short fiber preforms and magnesium matrix alloys, these problems should be resolvable in future work. Consequently, it is expected that short alumina fiber reinforced magnesium alloys will come to play an important role in fabricating lightweight automotive parts in the future.

REFFERENCES

1. Komatsu, M. and Hino, H, J. Soc. Automotive Engineers of Japan. Manuscript submitted April, 1988
2. Stroganova, V.F, Composite Materials, 238-50(1979)
3. Hack, J.E, Page, R.E. and Levereant, G.R, Metallurgical Transactions A, 15, 1389-96(1984)
4. Kural, M.H. and Min, B.K, J. Composite Materials, 18, 519-35
5. Yanagisawa, T, Yano, T. and Sagahara, T, J. Japan Institute of Metals, 49, 569-76(1985)

PLASTIC WASTES AND THEIR IMPACT ON THE ENVIRONMENT

Susan E. Selke
School of Packaging
Michigan State University
East Lansing, Michigan   USA

ABSTRACT

A growing fraction of municipal solid waste is plastic materials of various types. As landfill disposal capacity is strained, alternative disposal methods such as incineration and recycling increase in importance. Composite plastic materials pose some special problems for these alternative disposal technologies. As the use of plastic composites grows, so does the negative attention they draw in solid waste management planning. The perceptions as well as realities of the impact of plastic materials, including composites, on solid waste disposal alternatives are discussed.

266 BILLION POUNDS of municipal solid waste were produced in the United States in 1984, of which 19 billion pounds, 7.2 %, was plastics [1]. As landfill capacity decreases, concerns about groundwater contamination grow, and disposal costs escalate, the search for alternatives to landfill for disposal of solid waste is becoming more pressing. Incineration and especially recycling are among the most promising and popular options.

Disposal of plastics has become a focus of special public concern. Because multicomponent materials may pose special processing difficulties, especially for recycling, plastic-containing composite structures are beginning to draw negative attention from environmental groups and legislators. As the use of plastic-containing composite materials grows, it is likely that negative attention directed toward them will grow as well.

The purpose of this paper is to discuss both the public perception and the reality of the impact of plastic wastes on solid waste management, with special attention to packaging materials and to the role of composite structures containing plastic as a component.

SOLID WASTE DISPOSAL

LANDFILLS - Currently in the United States about 85% of municipal solid waste is landfilled, 5% is incinerated, and the remaining 10% is diverted from the disposal stream by recycling [1]. However, plentiful and inexpensive landfill space has become a thing of the past in large segments of the U.S. It is estimated that 25% of major U.S. cities will be out of landfill capacity by 1991 [2].

Landfill capacity has decreased significantly for two reasons - old landfills simply filling up, and landfills becoming identified as sources of groundwater pollution and therefore being closed. Over 200 sites on the Superfund list were municipal landfills. Unlined landfills on the banks of streams or in proximity to underground aquifers can cause serious water pollution problems.

Both siting and construction requirements for municipal landfills have been tightened considerably. A large number of existing landfills have been closed in the last ten years, and few new ones have been constructed. Even when geologically sound sites have been identified, public opposition to new landfills is often intense.

As a result, costs have skyrocketed. In the 1970's, tipping fees of under $10 per ton were commonplace. Such costs are now unheard of in much of the United States, especially the East Coast. Fees in New Jersey are now in excess of $100 per ton. Some communities on Long Island, New York, were paying $150 per ton for waste disposal in 1987, up from $5 per ton in 1984 [2].

INCINERATION - Incineration is an option which is growing rapidly, usually incorporating energy recovery as a way to recoup value from the waste and thus partially offset the very high cost of modern incineration facilities incorporating state-of-the-art emission controls. Such controls on air emissions are essential to public confidence in the safety of these facilities.

There is a widespread public perception that incineration facilities produce large quantities of highly toxic substances, and that plastics in particular are the source of these dangerous pollutants, especially dioxin. Though recent studies have shown no significant relationship betwen the amount of polyvinyl chloride in the feedstream and dioxin emissions, provided proper operating conditions (especially temperature) are maintained [3], past poor performance by municipal solid waste incinerators fuels public concern [4,5].

Disposal of ash from incinerators is another growing concern. Top ash in particular has been demonstrated to sometimes contain relatively high concentrations of heavy metals, leading to pressure for requirements that it be disposed of in special hazardous waste landfills rather than ordinary municipal landfills. The official Environmental Protection Agency policy on ash disposal has not yet been finalized [6].

Plastic materials are major contributors to the energy production of municipal solid waste incinerators. Plastics also tend to increase the operating temperatures of these facilities, which is now recognized to be desirable in emission control, especially to limit dioxin production [3]. It is, of course, necessary that the incinerator be designed to accomodate these higher temperatures and the tendency of plastics to melt. Composite materials containing only plastics or only plastics and paper pose no special problems in incineration beyond those of single-resin plastic structures. Aluminum-containing structures will produce aluminum oxide, which may be a pollution concern.

SITING - Both siting of new landfills and siting of incineration facilities has become very difficult. In addition to meeting the requirements of the governing political agencies, the builder of the facility must usually face intense public opposition, which may even culminate in a prolonged court battle.

RECYCLING

In contrast to siting of new landfills or incineration facilities, recycling tends to be a politically popular alternative, for the most part, especially if proposed as a voluntary rather than mandatory program. Recycling of metals, glass and paper has a long history. Recycling of plastics is much newer. At the industrial scrap level, recycling of plastics grew rapidly after the increase in oil prices of the mid 1970's, and is now commonplace. However recycling of post-consumer plastic wastes is still in its infancy.

PET - The post-consumer plastic material most frequently recycled is the polyethylene terephthalate (PET) beverage bottle. Because of deposit legislation in nine states (Connecticut, Delaware, Iowa, Maine, Massachusetts, Michigan, New York, Oregon, and Vermont), large supplies of returned bottles are collected and are therefore available for recycling. Since recycled PET has useful properties, this collected material has an economic value sufficient to support its processing for reuse. It is estimated that 20% of PET beverage bottle production is currently being recycled [7]. Major markets include fiberfill, strapping, and carpet backing. The high density polyethylene (HDPE) base cups which are present on most bottles are also being recycled after separation from the PET. Some of this material is used in molding new base cups.

OTHER PLASTICS - Recycling of other types of plastic materials is much less common. Some collection of HDPE milk bottles is done, as is some collection of low density polyethylene (LDPE) stretch film from pallet wrapping. Telephones and battery cases have also been recycled. Overall, however, it is estimated that only about 1% of post-consumer plastic waste is diverted from disposal by recycling [7].

COMMINGLED PLASTICS - One of the difficulties in establishing recycling programs for plastics is the necessity of obtaining homogeneous materials in order to secure reasonably good mechanical properties. As is well documented, plastic resins of various types tend to be incompatible with each other and form homogeneous domains within a structure when they are mixed, contributing to easy failure of the item when exposed to stress. In addition, widely differing melting temperatures and thermal stabilities can cause some materials in a plastic mixture to be unmelted while others suffer degradation from excessive heating, thus contributing to poor mixing and inferior properties.

Two approaches to this problem have been taken. One is to concentrate on acquiring relatively homogeneous feedstocks by relying on separation of resins by generic type (PET, HDPE, LDPE, etc.). This approach poses obvious difficulties for composite materials, where separation is likely to be virtually impossible.

The other approach is to develop equipment designed specifically for handling mixed plastic streams (commingled plastics) and produce items in which the relatively poor mechanical properties obtained are sufficient for the application. Those composite structures which contain only plastics are obvious candidates for this approach to plastics recycling.

The ET-1 extruder produced by Advanced Recycling Technology, Ltd. is probably the most popular equipment for processing commingled plastics at present, though other machines also exist. Products are generally large cross-section items falling in the general category of wood or concrete substitutes. Landscape timbers, parking stops for automobiles, and fencing are typical applications.

One very interesting aspect of equipment designed for processing commingled plastics is that it can commonly accomodate significant fractions of non-plastic materials as well. The ET-1, for instance, can utilize substantial fractions of fillers such as paper, wood chips, soft metals, etc. [8]. Thus these types of equipment may prove to be suitable for recycling composite structures which contain paper, foil, or other materials as components along with plastic, as well as those containing only plastics.

COMPOSITES IN PACKAGING

The trend in selection of packaging materials in the last several years has been towards the use of lighter-weight materials and the replacement of rigid containers with flexible or semi-rigid structures. Composite materials of various types have played a primary role in these changes.

Composite packaging materials fall into two categories. Structures combining paper or aluminum foil with plastic are designed to obtain from this combination of materials, properties either impossible or uneconomical to obtain with one material alone. While not really new, these materials have proliferated in recent years. "Brick-pack" structures used in aseptic packaging, for instance, have replaced tin-plated steel cans with laminated structures containing paper, aluminum foil, and low density polyethylene. Metallized films in bag-in-box structures have replaced cans for bulk food distribution, and have replaced glass for wine.

All-plastic composite materials are a newer development. Many of these materials were made possible by the development of coextrusion technology combined with production of new polymers with improved barrier to oxygen permeation. Plastic barrier bottles have replaced glass for ketchup, mayonnaise, etc. Plastic cans are appearing as replacements for metal.

Coating technology is also used, as in the plastic soft drink can. Blends of plastics are increasing in use as well. Some of these blends arise from attempts to reuse manufacturing scrap from coextrusions. Others are deliberately formulated to achieve desirable properties.

LEGISLATIVE ACTIVITY

In response to the concerns about solid waste disposal discussed above, many states and communities are considering legislation to affect municipal solid waste production. Because about a third of municipal solid waste is packaging materials, packaging is a focus of many of the attempts at legislative remedies. The other prime target of legislation is plastics.

The targeting of plastics appears to due to public perception that plastics are more of a problem in waste disposal than other materials. Their nondegradability is often cited. The public sees plastics as a prime contributor to emission problems from municipal solid waste incineration. Plastics are perceived to be nonrecyclable. They are also perceived, with some legitimacy, as prime contributors towards the "throw-away" mentality common in the United States. Another valid consideration is that plastics, though accounting for only about 7% of municipal solid waste by weight, account for a considerably larger fraction by volume.

Legislation currently being proposed tends to have as a focus either waste reduction, encouragement of recyclable materials over non-recyclables, or encouragement of the use of degradable over nondegradable materials. Types of legislation include taxes, bans, mandatory recycling, and deposits.

The deposit laws for beverage containers which are currently the law in nine states, as discussed above, were originally formulated as anti-litter measures, rather than solid waste control. Thus though they provide a strong incentive for collection of containers, they commonly do not address the disposal of the collected material. Much of this material is in fact recycled, but this is due to its economic value, not to legislative mandate. Laws now in effect in approximately eleven states require that plastic bundling rings for beverage six-packs be degradable. These were also anti-litter measures.

In 1984 New Jersey gained widespread attention from the plastics and packaging industries when a bill was introduced which would have placed deposits on a wide variety of plastic containers, banned the "brick-pack" laminated containers discussed above, and probably banned polyvinyl chloride (PVC) and polyvinylidene chloride (PVDC) containers. Though this bill was substantially modified before its mandatory recycling components

eventually passed, it serves as a signal of public concerns about plastics and especially plastic packaging. It should also be noted that the reason for proposing the ban on brick-pack containers was their non-recyclability.

In current legislative sessions, at least nine states (Hawaii, Iowa, Maine, Massachusetts, Minnesota, New York, Vermont, Washington and West Virginia) are proposing taxes and twelve (California, Florida, Georgia, Hawaii, Maine, Minnesota, Missouri, New York, Rhode Island, Vermont, Washington and West Virginia) are proposing bans on various types of packaging. Some states are including bans on certain non-packaging "throw-away" items as well, the most common targets being disposable razors and disposable diapers. Similar legislation is also being introduced at county and municipal levels, especially on the East Coast. Though most of the legislation being proposed will not pass, the fact that so much of it is being introduced is indicative of the seriousness of the problem and the need for industry responsiveness to these very real public concerns.

Both Minnesota and Iowa have enacted legislation which provides for a packaging review mechanism. If a package is determined after review to cause an environmental problem, that package can be banned. Composite packages with the difficulty they pose for recycling operations are an obvious target for such reviews. At present no packaging reviews have been carried out in either of these states.

It should also be noted that the ability of a state to ban a certain type of package has been upheld by the U.S. Supreme Court. Minnesota banned plastic milk bottles in 1977. An injunction delayed enforcement of the law during the appeal process. After the law was upheld in 1981, it was immediately repealed, so never was implemented. Nonetheless, the legal precedent was established.

SUMMARY AND CONCLUSIONS

Concerns about municipal solid waste disposal are real and growing. They are based on real problems of lack of landfill capacity and escalating costs. The search for alternatives is currently concentrating on incineration and recycling.

Composite materials containing plastic and paper alone do not pose any special concerns for incineration beyond those posed by single resin plastics. Aluminum-containing composites may be a pollution concern.

In recycling, however, single-component plastics can be reprocessed into single resin streams which have more desirable properties than mixed-plastic streams, and hence are of higher value. Composite materials by their nature will remain commingled even if they contain only plastics. The addition of paper or aluminum renders these materials even less desirable, though it may be possible to use them as a component of filled commingled plastics in undemanding end uses.

Unless the problems plastics and especially composites cause in solid waste disposal are addressed responsibly by the industries involved, there is a very real possibility of legislation being passed which will either place special taxes on these materials or ban them completely.

REFERENCES

1. Franklin Associates, Ltd., "Characterization of Municipal Solid Waste in the United States, 1960 to 2000", Prairie Village, Kansas (1986)

2. Business Week, May 25, p. 150 (1987)

3. Midwest Research Institute, "Results of the Combusion and Emissions Research Project at the Vicon Incinerator Facility in Pittsfield, Massachusetts", New York State Energy Research and Development Authority, Albany, NY (1987)

4. Rice, F., Fortune, April 11, p. 96 (1988)

5. Wasson, J. M. and S. Pollack, Science for the People 19(6), 3-14 (1987)

6. Porter, J. W., "Proceedings, Fourth Annual Conference on Solid Waste Management and Materials Policy", New York City, Jan. 29, 1988, New York State Legislative Commission on Solid Waste Management, New York, NY (in press)

7. Weis, R. S., "Proceedings, Recycled Plastics: Applications and Developments", C. Lai, K. Yam, S. Selke, eds., School of Packaging, Michigan State University, E. Lansing, MI, pp. 31-35 (1987)

8. Maczko, J. and R. Kukla, "Proceedings, Recycled Plastics: Applications and Developments", C. Lai, K. Yam, S. Selke, eds., School of Packaging, Michigan State University, E. Lansing, MI, pp. 61-69 (1987)

PLASTIC RECYCLING—
A STRATEGIC VISION

Wayne Pearson
Plastic Recycling Foundation, Inc.
Washington, D. C.   USA

Introduction

Virtually overnight municipal solid waste officials have discovered that they are rapidly running out of viable landfill space, and further, that waste-to-energy incineration capacity is not being installed at a sufficient rate to handle the increasing quantity of municipal garbage. Moreover, there is a growing belief among legislators that materials that are perceived to be non-recyclable should not be permitted to grow in the market place. Increasingly, plastics packaging is coming under attack as a material that is not considered to be recyclable.

The main purpose of my talk will be to describe the strategic vision that the Plastics Recycling Foundation has developed for recycling of post-consumer plastics packaging. I will, also, highlight the research we are sponsoring at the Center for Plastics Recycling Research at Rutgers and various other universities to validate this vision.

The trend of the data developed so far clearly indicates that plastic containers hold the potential for being recycled on a nationwide scale. Indeed, plastics may ultimately become the most recyclable of packaging materials due to the broad property ranges, the durability and the versatility of these valuable materials.

Municipal Solid Waste Management Plan

The primary means of disposal of municipal trash in the United States today is by landfill.

MUNICIPAL SOLID WASTE MANAGEMENT PLAN

MEANS OF DISPOSAL	CURRENT (%)
Landfill	85
Incineration	5
Recycle	10
Total	100

SOURCE: Franklin Associates

With diminishing landfill space, many states are now developing municipal solid waste management plans with goals to significantly increase the quantity of materials recovered. New Jersey, for example, has established a recycle goal of 25%. Other states and the EPA are discussing similar goals. The remainder of household trash that could not be economically recycled would be disposed of in waste-to-energy incinerators. Landfills would continue to be required long term to dispose of the residual ash from the incinerators plus large items, like household appliances, which could not be processed through an incinerator.

MUNICIPAL SOLID WASTE MANAGEMENT PLAN

MEANS OF DISPOSAL	CURRENT (%)	GOAL
Landfill	85	10
Incineration	5	65
Recycle	10	25
Total	100	100

SOURCE: New Jersey Office of Recycling

MUNICIPAL SOLID WASTE COMPOSITION

On a national basis, municipal solid waste is being generated at a 150 million ton annual rate. The approximate breakdown of this trash by material is as follows:

MUNICIPAL SOLID WASTE COMPOSITION

MATERIAL	PERCENTAGE
Paper	42
Yard Waste	16
Glass	9
Steel	8
Plastic	7
Aluminum	1
Miscellaneous	17
Total	100

SOURCE: Franklin Associates

Paper products are the largest component at 42 weight percent of the municipal solid waste stream. Yard wastes are second at 16 percent. Plastic materials make up about 7 percent.

CURRENT RECYCLING OF MUNICIPAL SOLID WASTES

This table shows what materials are currently being recovered from the municipal waste stream.

MUNICIPAL SOLID WASTE-CURRENT RECYCLING

MATERIAL	PERCENT IN MSW (%)	CURRENT RECOVERY RATE (%)	PERCENT OF MSW RECYCLED (%)
Paper	42	21	8.6
Glass	9	7	0.7
Steel	8	3	0.2
Plastics	7	<1	0.1
Aluminum	1	27	0.4
Miscellaneous	33		0.2
Total	100		10.2

SOURCE: Franklin Associates

Paper products make up about 85% of the recycled material. Paper products with the highest recovery rates include corrugated, newspapers, and office paper. To increase the recycle rate from 10% to 25% as is being proposed by several states, a tremendously expanded collection infrastructure must be set up.

PLASTICS RECYCLING FOUNDATION

As most of you know, the Plastics Recycling Foundation was formed in 1985 by concerned members of the Plastics Packaging Industry who felt that research was needed to make recycling of post-consumer plastics packaging a viable enterprise on a national scale.

The Foundation established the Center for Plastics Recycling Research at Rutgers University to conduct research to improve the operating efficiency of recycling systems and the quality of recycled products.

Based on research during the past three years, we have developed a strategic vision of how to expand recycling of post consumer plastics packaging and current programs are aimed at confirming the key parts of this vision.

For any recycling system to be successful there are four
components which must be in place. First, the recyclable
material must be collected. Next, the material must be sorted
into generic type, if the collection system involves mixed
products. Then the quality of the recovered material must be
enhanced through reclamation. And finally, the recycled material
must be sold into adequate end-use markets.

RECYCLING SYSTEM

- o   Collection
- o   Sorting
- o   Reclamation
- o   End-Use Markets

The United States has been recycling newspaper, aluminum, and
glass for over forty years. By contrast, plastics is a relative
newcomer. Most of the plastics packaging we have grown to enjoy
because of its unique and versatile characteristics is less than
ten years old and most is a lot less than that.

Nevertheless, with only a fraction of the collection
infrastructure that has been established for newspaper, aluminum,
and glass, about 1% of post-consumer plastics packaging is
already being recycled, and the percentage is growing.

You can think of this separation activity as "Mining" the waste
stream. It has been happening with PET soft drink bottles and
HDPE milk jugs and is expanding to other containers.

CURBSIDE COLLECTION SYSTEM - STRATEGIC VISION

This brings us to a detailed examination of our strategic vision
for the collection, separation, reclamation and re-use of post
consumer plastics packaging.

We envision a collection system where these materials will be
picked up at the curbside.

POTENTIAL CURBSIDE RECYCLABLES

- o   Newspapers
- o   Aluminum and Steel Cans
- o   Glass Bottles and Jars (Clear, Green, and Amber)
- o   Plastic Beverage Containers

  -    PET Soft Drink Bottles
  -    HDPE Milk Jugs
  -    Other Beverage Bottles

Beverage containers typically consist of the containers shown here.

NEW JERSEY MANDATORY SOURCE SEPARATION AND RECYCLING ACT

The New Jersey Mandatory Source Separation and Recycling Act which requires curbside collection of recyclables has given us a model upon which to build.

COLLECTION/SORTING SYSTEM - STRATEGIC VISION

```
┌─────────────────────┐
│  CURBSIDE           │
│  RECYCLABLES        │
│  CONTAINERS         │
└─────────────────────┘
          ↘
┌─────────────────────┐
│  COLLECTION         │
│  VEHICLES           │
└─────────────────────┘
          ↘
┌─────────────────────┐
│  MATERIALS          │
│  RECOVERY           │
│  FACILITIES         │
└─────────────────────┘
```

SOURCE: PLASTICS RECYCLING FOUNDATION

Under this new system, mixed recyclables, including beverage containers and newspapers, would be placed at the curbside by the homeowner for collection by the local municipality. These recyclables would, then, be hauled to a central sorting location, probably at the County level. At these sorting centers, which are called Materials Recovery Facilities or MRF's.

At the MRF, newspaper, aluminum, glass, and plastics would be separated into components for subsequent marketing. The glass would be manually sorted into clear, green and amber to imporve the sales value of the cullet. Plastics would also be separated into PET soft drink bottles and HDPE milk, water and fruit juice jugs and other containers as desired. Finally, plastics that cannot be separted into generic-type would be dealt with in a commingled mixture.

Research being conducted in a pilot plant at Rutgers has confirmed that it is possible to get good separation of the

497

plastic containers as described above.

It is interesting to note that plastics are the second most valuable recyclable in the municipal waste stream currently selling for about 6 cents/lb. In fact, plastic bottles will contribute as much to collection and sorting revenues as will newspapers.

CONTRIBUTION TO COLLECTION AND SORTING REVENUES

COMPONENT	PERCENT IN MSW (%)	SELLING PRICE (C/lb.)	REVENUE CONTRIBUTION (%)
Newspaper	8	1.5	13
Glass Bottles	9	2.0	20
Aluminum Cans	1	50.0	54
Plastic Bottles	2	6.0	13

SOURCE: Plastics Recycling Foundation

During 1986, 13 billion pounds of plastics were used in packaging in the United States. A little over half was consumed in making rigid containers and the remainder in flexible packaging.

From these data it can be seen that about three billion pounds of the rigid containers were bottles and about half of the bottles were beverage containers. PET from the soft drink bottle and HDPE from the milk jug were the primary polymers consumed in the "Beverage Containers" category. We believe that the proposed curbside collection system for plastics should initially focus on bottles and more specifically on beverage containers.

Once the infrastructure is set-up to collect and sort plastic beverage containers, we believe that it will be easy to add collection of all other bottles and potentially most rigid containers in the future.

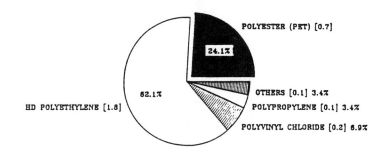

PLASTICS IN BOTTLES - 1986
2.9 BILLION POUNDS

POLYESTER (PET) [0.7]

24.1%

OTHERS [0.1] 3.4%
POLYPROPYLENE [0.1] 3.4%
POLYVINYL CHLORIDE [0.2] 6.9%

62.1%

HD POLYETHYLENE [1.8]

[BILLION POUNDS]

SOURCE: MODERN PLASTICS

A closer inspection of the "Other Bottles" category reveals that over half the polymer is polyethylene. This means that containers such as detergent and bleach bottles have the potential to be collected and separated into generic HDPE polymer in the future. Experience is showing that these containers are already being channeled back to the generic stream.

Because of aluminum's high value, it theoretically will be the dominant contributor to revenues. However, its existing buy-back infrastructure may reduce the amount that will be available at the curbside. For perspective, if no aluminum cans were collected at the curbside, the contribution of plastics would increase from 13 to almost 30 percent.

PET SOFT DRINK BOTTLE RECLAMATION

PET soft drink bottles are currently being reclaimed and sold into non-food markets because the economics of recovered PET favor it versus virgin resin for many applications.

Materials in Typical 2-Liter PET Beverage Bottles

Aluminum Cap - 1g
PET - 52g
Label & Adhesive - 4g
HDPE Base Cup & Adhesive - 18g

About 20 percent of these containers are being recovered today. Since 1979, the rate of recycling has increased dramatically as the infrastructure for collecting PET soft drink bottles has expanded.

PET BEVERAGE CONTAINER RECYCLING
(MILLIONS OF POUNDS)

1979	1982	1984	1986
8	40	110	130

SOURCE: Center for Plastics Recycling Research

This has happened with only a fraction of the potential collection systems in place. Prospective buyers and users of reclaimed PET are anxious to buy the material that is collected. We believe they will be willing to invest to develop the markets once an assured supply is available from the new curbside collection system.

The PET soft drink bottle enjoys the high degree of recyclability that it does because collection, sorting, reclamation and re-use exist in harmony.

The PET soft drink bottle is being collected from programs that run the gamut from deposit legislation to voluntary drop-off to curbside collection. The technology for separating and reclaiming the polymers in this complex package is well established, and we like to think that the research sponsored by

the Plastics Recycling Foundation at the Center for Plastics
Recycling Research at Rutgers University has helped expand the
technology for processing the soft drink bottle. Rutgers is
presently licensing the PET reclamation technology it has
developed to entrepreneurs on a non-exclusive basis, for a
nominal fee of $3,000 to encourage nationwide commercialization.
More than 100 firms from around the world have expressed interest
in acquiring the technology. At least three companies have
already signed licensing agreements.

The Rutgers process is capable of recovering high quality generic
PET which is readily finding new uses.

RECYCLED PET FROM POST-CONSUMER SOFT DRINK BOTTLES

	PERCENT
PET	99.9
PE	0.03
ALUMINUM	0.01
OTHER	0.06

SOURCE: Center for Plastics Recycling Research

The economics of PET reclamation are favorable. Flake material
can be sold profitably for about 25 cents per pound, f.o.b.
plant.

ECONOMICS
PET RECLAMATION

CAPACITY	20 MILLION POUNDS/YEAR
INVESTMENT	$2.5 MILLION
PURITY	99.9% PET
PRICE	25 CENTS/POUND F.O.B. PLANT (FLAKE)
REVENUE	$5 MILLION/YEAR
PROFIT	$1 MILLION/YEAR
NET RETURN	20%

SOURCE: Center for Plastics Recycling Research

The flake can be converted to pellet for about 5 cents/lb. giving
a 30 cent/lb. price, roughly half the price of virgin PET. Our

market research studies indicate that the demand for PET, recovered from containers, has the potential to grow from 130 mm lbs. today to 600 mm pounds per year by the mid 1990's; assuming the infrastructure is set up for collection.

A potential list of end-uses is shown here:

        Fiber Fill
        Strapping
        Engineering Plastic
        Geotextile
        Carpeting

HDPE MILK JUG RECLAMATION

The economics for reclaiming HDPE bottles are also attractive.

                      ECONOMICS
              HDPE RECLAMATION FROM MILK JUGS

            PURITY   99.9% HDPE
            PRICE   17 cents/lb.
                            f.o.b. plant-flake

SOURCE: Center for Plastics Recycling Research

It is much simpler to reclaim this generic material because the packages are less complex than the soft drink bottle. This product can be sold for about 17 cents per pound versus virgin resin which is approaching 40 cents per pound.

Collection systems are minimal and consist almost entirely of buy back programs.

The typical end uses for HDPE are listed here:

        Base Cups
        Pipes
        Septic Tanks
        Flower Pots
        Non-Food Containers

Our market research indicates that requirements for reclaimed HDPE have the potential to increase from 50 mm pounds per year today to 600 mm pounds per year by the mid 1990's.

The potential exists for separating other generics from the wate stream. For example, we believe polystyrene from fast food waste can be collected and technology developed to separate and recover it for re-use in generic polystyrene markets.

Finally, commingled plastics, materials that can't be economically separated and restored to their generic form, can be processed into useful product.

We have installed a pilot facility for making these products at Rutgers. This is a specialized extruder molder for making profiles that resemble lumber.

Some typical end-uses are as shown:

COMMINGLED PLASTICS

- Plastic Lumber
- Boat Docks
- Landscaping Ties
- Outdoor Furniture
- Car Stoppers
- Farm Structures
- Railroad Ties

SOURCE: Center for Plastics Recycling Research

Conclusions

In conclusion, the markets for recycled plastics are developing rapidly and in direct proportion to the collection of these materials from the post-consumer waste stream. The encouraging news is that the consumer will separate recyclables from household garbage in sufficient quantity to make a meaningful impact on reducing the municipal solid waste that must be sent to landfill. New technologies for the collection and separation of the recyclables into their components is being put in place in a number of communities across the country. Once the recyclables are separated into their generic components, the markets will develop. This has long been the experience with newspaper, glass and aluminum, and is now becoming the case for plastics such as PET and HDPE.

Perhaps the most exciting news is that markets are developing and additional ones can be envisioned for commingled plastics, which are those materials that cannot be economically separated into their generic forms. If this technology can be commercialize on a substantial scale, it could well turn plastics into the most versatile of packaging materials with respect to recyclability, and provide the plastics engineer with substantial volumes of a new source of valuable polymeric materials.

THE RECYCLING OF POST CONSUMER PLASTIC PACKAGING MATERIALS

D.R. Morrow
The Center for Plastics Recycling Research
Rutgers, The State University of New Jersey
Piscataway, New Jersey 08855   USA

THE   RECYCLING   OF   POST
CONSUMER PLASTIC PACKAGING
        MATERIALS

   Darrell R. Morrow, Ph.D.
          Director
Center for Plastics Recycling
          Research
Rutgers, The State University
        of New Jersey

        The effect implementation
of a large-scale post-consumer
plastics   recycling   program
requires   a   sound   technical
base   and   the   creation   of   a
well organized infrastructure.
The   Center   for   Plastics
Recycling Research through its
national research and process
development program is making
significant   progress   in
generating   technical
information and guidelines for
the   four-component   recycling
system   (collection,
s o r t i n g / p r e p a r a t i o n ,
reclamation,   end-use
manufacturing/marketing).   An
overview will be provided of
the activities and strategic
visions of the Center.

I) Introduction

Increasingly, plastics are the materials of choice for today's new packaging products. Their light weight, shatter resistance, durability and unparalleled performance have made them the preferred packaging materials for innumerable beverages, foods and other consumer products.

But this growth has gone unnoticed by a public increasingly concerned about protecting the environment. Both consumers and legislators are raising questions about the fate of many products-including packaging - once their usefulness is over.

Landfill shortages, skyrocketing waste disposal costs and other problems such as litter are becoming rallying cries for action. Environmentalists, public officials, solid waste professionals and, ultimately regulators and legislators, are pushing for answers now. And recycling undoubtedly will have an important role to play in a multi-faceted solution.

Unlike paper, glass and metals, which have been around for many years, plastics are relative newcomers to the packaging field. Similarly, unlike the recycling business that have been built up around these "traditional" materials, plastics recycling is a young and fledgling industry.

The manufacturers of plastics products and resins, as well as soft drink companies and users of plastics packaging, are committed to seeing plastics recycling grow from its current status - an estimated 20 percent of all plastic soda bottles and a smaller amount of plastic milk bottles are recycled annually - into a full-scale national reality. This they are doing not only by improving recycling technical, but also by developing collections systems, markets and new reclamation technologies.

The principal way that these industries are accomplishing these goals is through the research and process development activities of the Center for Plastics Recycling Research. In 1985, the Plastics Recycling Foundation established the Center for Plastics Recycling Research (CPRR) on the campus of Rutgers - The State University of New Jersey. Initially named the Plastics Recycling Institute, the Center is an Industry/University Cooperative Research Center of the National Science Foundation and as an Advanced Technology Center of the New Jersey Commission on Science and Technology. The CPRR has a successful pilot plant operating for resin recovery as part of its Process Development Division, along with a (plastic) materials processing facility (Pilot Plant #3) and a commingled plastics processing facility (Pilot Plant #2) and serves as the focal point of projects funded by the Foundation, the NJCST and the NSF. The CPRR operates the pilot plastics recycling plant on the Rutgers campus where high technology and automation have been applied to the technology of plastics recycling. To date, the plant's operations have focused on short-lived rigid plastic packaging, specifically PET (polyethlylene terephalate) containers for soft rinks and high density polyethylene (HDPE) milk bottles. The CPRR also has established as its third operating division the Information Services Division. It is the responsibility of the Information Services Division to accumulate, process and disseminate technical/economic data and information on plastics recycling. A computerized data base has been established by the Information Services Division of the Center for Plastics Recycling Research, which provides a wealth of information about numerous aspects plastics recycling. Included in the data base are technical reports, news articles and editorials, historical information and other data dealing with collection and separation methods, quality improvement, market development, and business developments in the recycler community. The Center for Plastics Recycling Research is a very actively engaged in research and process (systems) development in all four (4) areas of the requisite infra-structure for plastics recycling: namely; collection, sorting/pre-paration, reclamation and re-use. The overall objectives of the CPRR are listed below.

II) OBJECTIVES OF THE CENTER FOR PLASTICS RECYCLING RESEARCH

GENERAL OBJECTIVE

Develop and demonstrate the technical and economic viability of the recycling and re-use of plastic materials with a particular emphasis on plastic packaging materials, separately or in combined with other packaging materials.

SPECIFIC OBJECTIVES

- Develop recycling technology for all plastics. The CPRR will focus on efforts aimed at advancing the science and technology of <u>plastics collection, sorting/preparation, reclamation (including separation and cleansing) and re-use/re-manufacture of a wide variety of products (end-use markets)</u>.
- Address the social and economic aspects of recycling. Motivation and cost effectiveness studies will be included.
- Disseminate technical and socio-economic information about plastics recycling. The CPRR will gather recycling information to form a publicity available database. Information is to be disseminated widely by the CPRR through newsletters, conferences, seminars and the publication of research results in appropriate journals.
- Transfer of technology. Mechanisms for the transfer of technology in all areas are being developed , with the

intent of stimulating the formation of new business and creating new jobs.

- Aid in the solution of environmental problems of concern to New Jersey, the Nation and the World. These efforts can be expected to result in the increased use of recycled plastics and decreased use of land fills.

- Determine the extent to which recovery of post-consumer plastics of any given species or category should/can be recovered in the form of a resin, a component in a commingled mixture (of plastics), or a distribution between resin and commingled component feedstocks. Any such determination must be based on a combination of technical feasibility and economic justifications.

III) ORGANIZATION AND MANAGEMENT STRUCTURE

The organization and management structure has been designed to involve each of the Program's participants in as much of the decision making process as possible. The decisions reached must, of course, be consistent with university policy and the spirit of cooperative research. Within these constraints, the participants are able to aid in the determination of the research activities, policies, and future directions of the CPRR. The CPRR is comprised of three divisions; namely, the research division, the process development division and the information services division, as seem in Figure 1. The main

activity of the Process Development Division is the operation of a pilot plant facility to demonstrate the technical and commercial feasibility of the recycling of plastic materials. There is also an equipment evaluation laboratory in which new processes and their requisite properties can be assessed, along with the development of appropriate Q/C methodology.

An advanced system designed for the recycling of mixed plastics wastes has been installed in a new pilot plant operation, Pilot Plant #2, on the Kilmer campus of Rutgers, The State University of New Jersey. It is employed primarily for the processing of mixed thermoplastic waste materials, such as the beverage, bleach, detergent, food and shampoo containers, typically consigned to landfills. The heart of Pilot Plant #2 is an ET/1 extrusion-molding machine developed by the Advanced Recycling Technology of Brankel, Belgium.

A key feature of the system is its ability to accommodate mixtures of plastics previously considered incompatible. Equally important, it will tolerate high percentages of such contaminants as paper, dirt, and thermoset plastics, which are incorporated in the finished products as fillers.

The end product, which is similar to structural lumber both in applications and handling, can be planed, sawed, drilled or nailed and will take standard wooden screws.

508

A third facility associated with the Process Development Division is the sorting and material preparation unit-Pilot Plant #3. The original concepts of the Process Development Division called for the sorting and preparation stages of the recycling process to be part of the collection efforts. Experience has shown that the previous model of collection, reclamation and end-use markets requires modification to include a sorting/preparation step. Hence, the need to carry out research efforts in process development involving all three pilot facilities.

The Information Services Division is responsible for the collection, organization and dissemination of technical/marketing information associated with plastics recycling. Included in the activities of this division are a newsletter, technology transfer and training programs, conferences, literature identification and accumulation. It is the goal of the Information Services Division to evolve into a national/international information clearing house on plastics recycling. A major accomplishment of the CPRR & the ISD has been the preparation of a "how to" manual on a resin recovery processing system. The manual has been made available through low cost licensing program in which 10 companies (world-wide) have enrolled to date. A second "how to" manual dealing in this care with collection and sorting is scheduled for release early in

October, 1988.

The Research Division of the CPRR is responsible for soliciting, organizing and carrying out research projects relating to all phases of plastics recycling, including collection, processing, manufacturing and end-use markets.

Academic institutions outside of New Jersey, such as Michigan State University, the University of Toledo Rensselear Polytechnic Institute, NY and Case Western Reserve University Cleveland are, used as resources for the Center.

- Faculty of other New Jersey institutions and of institutions in other states are invited during each research funding cycle to propose research responsive to the CPRR's objectives: thus, there is a national scope to the resources utilized by the Center.

Research activities of the CPRR both for last year (1987-88) and the current funding cycle (1988/89) are listed below.

- 1987-88

- Recycling of High Density Polyethylene Milk Bottles Susan E. Selke, Project Manager; Christopher C. Lai, Joseph Miltz, Eric A. Grulke, and David I. Johnson,

Coinvestigators-MSU

- Operating Parameters, Process Improvements, Product Purity, and Economics in a Plastics Recycling Pilot Plant Frank W. Dittman, Project Manager-Rutgers

- Quality Assurance for Plastic Recycling Edward M. Phillips, Darrell R. Morrow, and Susan L. Albin, Project Managers; Rutgers

- Utilization of Commingled Plastics Diverted from Municipal Solid Wastes; Sidney Rankin and Darrell R. Morrow, Project Managers; Rutgers

- Collection Systems for Plastics in Municipal Solid Waste; C. Neale Merriam and Darrell R. Morrow, Project Managers-Rutgers

- Market Research on the Plastics Recycling Industry Robert A. Bennett, Project Manager; Toledo

- Experimental and Engineering Comparisons of Four Methods of Separating Aluminum Chips from PET Chips Frank W. Dittman, Project Manager; Benjamin R. Hannigan Jr. and Jose R.

Fernandes, Co-investigators; Rutgers

- 1988-89

- New Product Applications of Database on the Plastics Recycling Industry Robert A. Bennet, Project Manager; Toledo

- Development of Quality Evaluation and Statistical Quality Control Test Methods for Generic and Commingled Plastics Materials; Edward M. Phillips and Henry Frankel, Project Managers: Susan A. Albin

- Evaluation of Collection Systems for Plastics in Solid Waste; C. Neale Merriam and Darrell R. Morrow, Project Manager; Rutgers

- Research and Development for Sortation and Handling-Pilot Plant #3; Frank W. Dittman and Henry Frankel, Project Managers:Jose R. Fernandes and Benjamin R. Hannigan Jr., Coinvestigators

- Cryogenic Granulation of Recycled Beverage Bottles; Frank W. Dittman, Project Manager; Jose R. Fernandes, Benjamin R. Hannigan Jr., Edward M. Phillips, and David Stein,

Coinvestigators

Utilization of Commingled Plastics Diverted from Municipal Solid Wastes and Experiments to Design Products and Optimize the Processing; Susan L. Albin, Darrell R. Morrow, and Thomas J. Nosker, Project Managers

- Application of Compositional Quenching Technology to the Recycling of Polymers from Mixed Waste Streams; E. Bruce Nauman, Rensselaer; Polytechnic Institute, Project Manager

- 1988-89

- Investigation of the Recyclability and the Recycling of Polystyrene; Darrell R. Morrow, Project Manager;Rutgers

- Cryogenic Granulation Processing for Recycled Plastic Packaging Materials; Darrell R. Morrow and Jose R. Fernandes, Project Managers; Rutgers

- Processing/Struction/Property Relations in Commingled Plastics Anne Hiltner and Darrell R. Morrow; Project Manager-Case Western Reserve University and Rutgers

IV) Strategic Vision

Curbside Collection Strategic Vision

- The CPRR/PRF envision a collection system where at least four materials will be picked up at the curbside. These materials should be glass, aluminum/steel cases, newspapers and mixed bottles & jars plastic containers. To maximize the quantity of materials that can be obtained from each curbside collection point, pick up should be in a single container for all recyclables, unsorted.

- Sorting System-Strategic Vision

- For the sorting phase of the CPRR/PRF strategic vision, mixed recyclables will be transported to centrally located MRF, probably in each county. These facilities can be highly mechanized operations using minimal labor, or they can be fairly manual operations, creating employment opportunities for the local community.

At the MRF, newspaper, aluminum, glass, and plastics would be separated into components for subsequent marketing. The glass would be manually sorted into clear, green and amber to improve the sales value of the cullet. Plastics would, also, be separated into PET soft drink bottles and HDPE milk, water and fruit juice jugs and other containers as desired. Finally, plastics that cannot be separated into generic-type would be dealt with in a commingled mixture.

- <u>Reclamation System-Strategic Vision</u>

- Resin Recovery Systems for reclamation of purified generic materials for use as a feed stock for subsequent plastics molding operation

- Commingled plastics processing via various molding technologies to produce useful products from mixed plastics

- <u>Revise Markets-Strategic Vision</u>

- Use of generic materials in existing and new markets.

- Use of commingled structures for birds variety of s t r u c t u r a l applications

V) <u>Conclusions</u>

From its inception in 1985, the Center for Plastics Recycling Research (CPRR) has focused on the problems presented by the disposition of billions of pounds of rigid plastics, most of which presently find their way into landfills throughout the United State. Through research, development and engineering, the Center seeks answers to the questions of how such plastics can be collected, processed, marketed and utilized in a manner that takes full advantage of their unique properties. In doing so, a major step is being taken toward the protection of our environment and the conservation of a valuable and finite resource.

In 1988 the United States is projected to produce nearly 55 billion pounds of plastic materials most of which will be landfilled if a solution to the problem of disposing of solid waste is not found. Current solid waste disposal sites in this country are at or near capacity and plastic waste is continually being generated, requiring both an immediate and long-term solution. Reduction of solid waste can result in a reduction of tipping fees (which are increasing at an astronomical pace) paid by the municipalities to landfills, thereby having significant positive impact on the economies of the communities throughout the Nation. Since the plastics industry has not yet developed its recycling technology sufficiently to allow for recycling of post-consumer plastics, this has become the mission of the CPRR

Plastics are a non-renewable resource. Thus a technically and economically sound recycling system is a far better means of handling the materials after their first use than is incineration to produce energy. Through its research program, process development efforts, and information services programs, the Center for Plastics Recycling Research is currently working on the development technologically feasible, as well as economically viable recycling solutions. The recycling infrastructure for all recyclables, including a broad spectrum of plastic materials, must, can and will be established, to the benefit of the environment, the public and industry.

Figure 1

CENTER FOR PLASTICS RECYCLING RESEARCH

Organization Chart- Adminstration Division

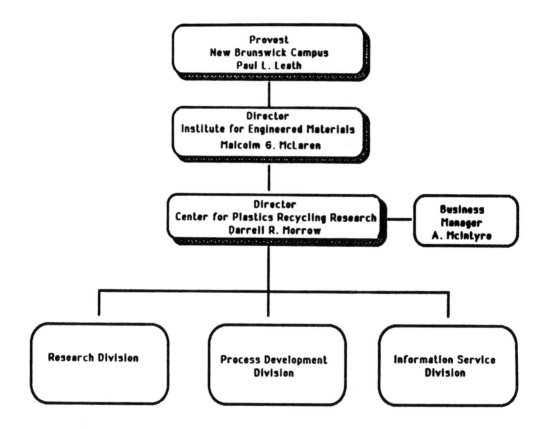

ADVISORS TO THE IEM DIRECTOR AND THE CPRR DIRECTOR

* PRF INDUSTRIAL ADVISORY COUNCIL

* NJCST EXECUTIVE DIRECTOR, EDWARD COHEN

* NSF OBSERVER, ALEX SCHWARZKOPF

* CPRR PROGRAM EVALUATOR, S. GEORGE WALTERS

* ACADEMIC ADVISORY COUNCIL

RECYCLED PLASTICS—
NEW PRODUCT APPLICATIONS

Robert A. Bennet
College of Engineering
The University of Toledo
Toledo, Ohio 43606   USA

ABSTRACT

Traditional uses and markets for recycled
plastics are reviewed. New markets and products using recycled plastics are presented
followed by a discussion of future opportunities in some potentially high volume product
areas. Physical testing, which is used to
determine recycled plastic product applications, will be described.
Results show that present markets for recycled
PET are fiberfill, unsaturated polyester,
polyols for rigid urethane foam, strapping,
engineering plastics and extruded products.
New applications, such as thermoformed products
and textiles/geotextiles offer additional
opportunities to utilize post consumer plastic
scrap. Markets for HDPE are soft drink basecups, plastic lumber, containers, drum pails
and various types of pipes. Some products
manufactured from mixed plastics include
parking lot car stoppers, plastic lumber,
sheeting and landscape timber. Slides showing
products made from recycled plastics will be
presented at the conference.
The presentation will also describe an
electronic database which was developed to
keep track of plastics recycling activity
nationally. This database utilizes a desktop
personal computer and a commercially available
software package. Recycling information in
the database is disseminated through the
Plastics Recycling Foundation and the Center
for Plastics Recycling Research.

GOALS

THIS RESEARCH PROJECT WAS DIRECTED AT gaining
an improved understanding of new product
opportunities, markets and the economics
associated with the emerging plastics recycling
business. Goals include investigating potential
new products and markets for recycled plastics
and developing an electronic database on the
plastics recycling industry.

Research was performed in conjunction with
other plastics recycling research projects
funded through The Plastics Recycling Foundation/
Center for Plastics Recycling Research. As
part of this research, companies have been
contacted to determine interest in utilizing
recycled plastics in manufacturing products.
Questionnaires have been sent to recyclers to
determine quantity, products and types of
plastics being recycled. Information is entered
into a personal computer database which allows
efficient and convenient access. Physical
testing is being performed to better understand
performance of plastic lumber applications.

BACKGROUND

In 1987, over 57 billion pounds of plastics
were sold in the U.S. (Exhibit 1). These sales
volumes were obtained from Modern Plastics,
Jan. 1988 and the Textile Economics Bureau, Inc.,
Roseland, NJ. Polyethylene (high and low
density) was the dominant resin with a total
of over 17 billion pounds or about 30% of the
total plastics sales. Polyester resin for
nontextile products accounted for 1.8 billion
pounds or 3.15% of the total. Polyester used
in textiles accounted for another 3.5 billion
pounds or 6.1% of total plastic sales. Polyethylene and polyester are the dominate plastics
currently used in recycling post consumer
plastics. These billions of pounds of thermoplastics offer recycling opportunities. Reuse
of thermoplastics will reduce raw material
costs to manufacturers and reduce the burden
caused by plastics on the solid waste stream.
Recent price increases for plastic resin will
provide further incentives for recycling
plastics.

Exhibit 1 - U.S. PLASTIC SALES 1987*

Material	Million lb.
ABS	1194
Acrylic	665
Alkyd	305
Cellulosics	88
Epoxy	404
Nylon	471
Phenollic	2764
Polyacetal	122
Polycarbonate	387
Polyester, (PBT,PET)	1800
Polyester, unsaturated	1315
Polyethylene, high density	7824
Polyethylene, low density	9499
Polyphenylene-based alloys	175
Polypropylene and copolymers	6472
Polystyrene	4857
Other styrenics	1250
Polyurethane	2681
Polyvinyl chloride and copolymer	8055
Other vinyls	1050
Styrene acrylonitrile	102
Thermoplastic elastomers	441
Urea and melamine	1565
Others	260
Sub Total	**53,746**
PET Textile	
** Polyester filament yarn	1180
Polyester Staple and tow	2369
Sub Total	**3,549**
U.S. SALES	**57,295**

* Source of data: <u>Modern Plastics</u>, Jan., 1988
* * Source of data: Textile Economics Bureau, Inc.,
   Roseland, NJ

HISTORY OF RECYCLING POST CONSUMER
PLASTICS SCRAP

In the beverage industry aluminum containers were quickly recognized as being a
valuable material which should not be
discarded. About 20 years ago the aluminum
companies initiated recycling programs.
Aluminum cans are well-known as being
recyclable. Glass bottles have been mass
produced since the turn of the century and are
being recycled into new containers. The
plastic beverage container, made of polyester
(PET) and polyethylene, was introduced
nationally in 1978. Recycling began almost
immediately through the efforts of small
entrepreneur recyclers. These early plastic
recyclers recognized the intrinsic value of
this hi-tech polymer. In 1979, only one year
after the plastic PET bottle's introduction,
eight million pounds of bottles were recycled.

This poundage of recycled beverage containers
grew to 40 million pounds in 1982 and by 1985,
an estimated 110 million pounds of bottles
were being recycled. In 1987, this research
estimated that 145 million pounds of PET beverage
bottles were recycled. Deposit legislation on
soft drink containers in 9 states influenced
this rapid growth. Recycling legislation
aimed at reducing solid waste is expected to
continue the growth of plastics recycling along
with other materials. The following (Exhibit 2),
shows the rapid increase in recycling plastic
beverage containers.

Exhibit 2
PET BEVERAGE CONTAINER RECYCLING

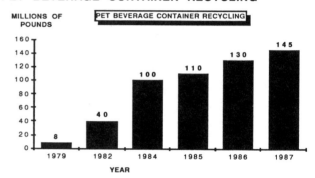

Polyester (PET) is the plastic being
recycled most as post consumer scrap in the
United States. Wellman, Inc. was identified
as being the major recycler in this area.
Results show that present markets for recycled
PET are fiberfill, unsaturated polyester,
polyols for rigid urethane foam, strapping,
engineering plastics and extruded products.
New applications, such as thermoformed products
and textiles/geotextiles offer additional
opportunities to utilize post consumer plastic
scrap. See Exhibit 3 for current PET markets.

In 1987, there was 740 million pounds of
PET used in manufacturing soft drink containers.
Estimates showed that there exists a potential
market for 500 million pounds of this material
in non food applications. Since only 150
million pounds are currently being recycled,
this market is far from being saturated. A
10% penetration into the PET textile would
alone represent approximately a 350 million
pound market since filament yarn, staple and
tow made from virgin PET markets is over 3.5
billion pounds.

Exhibit 3
MARKETS FOR RECYCLED PET PRODUCTS

> **CIVIL ENGINEERING**
> Geotextile
> Urethane Foam
> **RECREATIONAL**
> Skis
> Surfboards
> Sailboat Hulls
> **INDUSTRIAL**
> Carpet Backing
> Fence Posts
> Fibrefill
> Fuel Pellets
> Industrial Paints
> Strapping
> Unsaturated Polyester
> Paint Brushes
>
> Total Volume
> 150 Million lbs.*

HDPE RECYCLING

Total sales of HDPE in 1987 was in excess of 7.8 billion pounds. Market research shows that a potential market of an estimated 440 million pounds could be developed to utilize recycled HDPE. Major potential markets for HDPE are soft drink basecups, plastics lumber, containers, drum pails and various types of pipes. Currently only about 52 to 58 million pounds per year of HDPE have been identified as being recycled. Markets identified for recycled HDPE are listed in Exhibit 4.

Exhibit 4
MARKETS FOR RECYCLED HDPE PRODUCTS

> **AGRICULTURE**   **RECREATIONAL**
> Drain Pipes   Toys
> Pig and Calf Pens
>
> **MARINE ENGINEERING**   **INDUSTRIAL**
> Boat Piers (lumber)   Drums/Pails
>   Kitchen Drain Boards
> **CIVIL ENGINEERING**   Matting
> Building Products   Milk Bottle Carriers
> Curb Stops   Pallets
> Pipe   Soft Drink Basecups
> Signs   Trash Cans
> Traffic-Barrier Cones
>
> **GARDENING**
> Flower Pots
> Garden Furniture
> Golf Bag Liners
> Lumber
>
> Total Volume
> 2.3 Million lbs. recycled in 1986

MIXED PLASTICS

Plastic scrap often is collected as a mixture of many types of plastics. Separation of this mixture into its various plastic components would be costly. It is possible to process the mixed plastics into noncritical product applications. This type of mixed plastics recycling has already begun in Europe and Japan. Products with high volume which could be manufactured from mixed materials are being identified. Lumber substitutes for miscellaneous outdoor furniture, posts, and farm structures are ideal markets. Many items have been mentioned by ART/Europe (Advanced Recycling Technology S.A. Ltd.) and are listed in Exhibit 5.

Exhibit 5
RECYCLED MIXED PLASTICS PRODUCTS

> **AGRICULTURE**   **RECREATIONAL**
> Barrier Retainers   Flower Pots
> Duck Boards   Flower and Tree Boxes
> Electric Fences   Golf Course Walkways
> Erosion Control Timber   Park Benches
> Fruit Tree Supports   Picnic Tables
> Gates   Playground Equipment
> Horse Stalls   Sand Box Kits
> Markers   Stadium Seating
> Pig and Calf Pens   Storage Bins
> Poultry Construction
> Vine Stakes   **INDUSTRIAL**
> Ranch Fences   Car Stops
> Tree-guards   Fencing
>   Flooring
>   Footings, Posts, and
> **MARINE ENGINEERING**   Sill Plates
> Beach Erosion Control   Highway Construction
> Boardwalks   Loading Dock Rails
> Boat Docks   Markers
> Coast Erosion Protectors No-load Grating
> Dock Side Fenders   Pallets
> Fishing Boat Wear Plates Pipe Racks
> Lobster Traps   Planks
> Pier Impact Protectors   Sign Posts
> Rub Rails   Slab Separators
> Sea Walls   Stair Treads
> Trawler Net Rollers   Traffic Barriers
>   Truck Flooring
> **CIVIL ENGINEERING**   Wire Racks
> Barriers
> Bearing Pads   **GARDENING**
> Fences   Compost Enclosures
> Road Delineators   Fences, Gates, Enclosures
> Traffic Direction Posts   Garden Boundary Retainers
>   Landscaping Timbers
>   Retainer Walls

MARKET ESTIMATES FOR LANDSCAPE TIMBER

From the calculation below, approximately 1/2 billion pounds of mixed plastics could be utilized in relatively noncritical landscape applications. Landscape timbers made of recycled plastics possess the advantage over wood by being rot resistant, splinterless and not requiring pressure treating.

```
Sales Mid 1980's  $60-70 million
@ $4-7 each
$60 million/$5 per pc. = 12 million pcs.
4 x 3 1/2 x 8 ft (treated vol 0.78 ft³)
12*0.78 ft³ = 9.36 million ft³
9.36 million ft³*(56-60 lbs/ft³) =
(533-561) million lbs.
```

PRODUCT TESTING

Testing is being performed to better understand the limitations associated with using recycled plastic resins. Tests have been made on samples of 2x4's manufactured from mixed plastics. A series of nail and screw pull-out experiments shows plastic lumber to behave quite differently than wooden lumber. Nail pull-out tests were performed on wood and recycled plastic specimens to compare their nail holding strength. The specimens were cut out of 8 feet (2x4) construction members. Actual dimensions of the specimens were: 1 1/2" x 3 7/16" x 8'. Nails used have a diameter = 0.15". Nail penetration was 1 1/2" throughout the tests. Results of these tests show that mixed plastics hold nails approximately 40% better than typical wood when nails are perpendicular to the grain. When nails are driven parallel to the grain, wood significantly loses by approximately 50% the ability to hold nails while plastics maintain relatively the same retention capability. However, this advantage of recycled plastic lumber is rapidly lost at elevated temperature of approximately 149°F (65°C).

The following Exhibits show the performance of wood versus mixed recycled plastics.

Exhibit 6.1
Nail Pullout Test - Perpendicular to Grain (face)

Wood Pullout Force (lbs)	Recycled Plastic Pullout Force (lbs)	% Difference
112.33	157	41%

Nails were driven perpendicular to grain for the wood specimen on the 3 7/16" side, and on the same side for the recycled plastic specimen. Conclusion - Recycled plastic has 41% more strength to hold the nails than wood, for nails driven as described previously.

Exhibit 6.2
Nail Pullout Test - Perpendicular to Grain (side)

Wood Pullout Force (lbs)	Recycled. Plastic Pullout Force (lbs)	% Difference
111	155	40.0%

Nails were driven perpendicular to grain for the wood specimen on the 1 1/2" side, and on the same side for the recycled plastic specimen. Conclusion - Recycled plastic has 40.0% more strength to hold the nails than wood, for nails driven as described.

Exhibit 6.3
Nail Pullout Test - Parallel to Grain

Wood Pullout Force (lbs)	Recycled Plastic Pullout Force (lbs)	% Difference
46	166	261%

Nails were driven parallel to grain for the wood sample on the 1 1/2" side, and on the same side for the recycled plastic sample. Conclusion - Recycled plastic has 261% more strength to hold the nails than wood, for nails driven as described.

Exhibit 6.4
Nail Pullout Test - Heated Samples

Wood Pullout Force (lbs)	Recycled Plastic Max Pullout Force (lbs)	% Difference
116	88.3	-24%

In this experiment, both wood and recycled plastic samples were soaked in water at the same time with the nails driven 1 1/2" perpendicular to grain. Then the water was heated until the water temperature reached 149 °F [65 °C]. After that the heating stopped, and the samples were kept in water for 10 minutes and then were taken immediately for nail pullout testing. The first and third tests were done starting with the recycled plastic sample. The second test was done starting with the wood sample, no effect was found on the experiment due to to the order change. The results are tabulated as follows: Conclusion - Recycled plastic has 24% less strength to hold the nails than wood for the set up described previously.

Exhibit 6.5
Percentage loss in strength due to heating samples (Comparison between Exhibits 6.1& 6.4)

	Room Temp	Heated 65 °C Bath	% Loss
Wood Maximum Pullout force-lbs.	112	116	3.2%
Plastic Maximum Pullout force-lbs.	157	88	-43.7%

Conclusion - Wood retained much of its strength while being heated and soaked. Recycled plastic lost 43.7% of its nail holding strength under the same conditions.

Exhibit 6.6
Screw Pullout Test

Wood Pullout Force (lbs)	Recycled Plastic Max Pullout Force (lbs)	% Difference
713	353	-50.5%

(#8 x 2") wood screws were hand driven into the wood and recycled plastic samples, and then pulled out for comparison. Twenty threads of the screws were driven in the samples. The results are tabulated as follows. Conclusion - Recycled plastic has 50.5% less strength to hold the hand driven screws than wood. In this experiment, when the screws were pulled out from the recycled plastic sample, the threads spaces were filled with plastic material due to the hand driving of the screws.

CONCLUSION FROM TESTING

Wood retained much of it's strength while being heated in a water bath. Recycled plastic lost 43.7% of its nail holding strength due to the same conditions. Consequently, utilization of lumber made from recycled plastics must be carefully evaluated regarding the environment in which it will be used in order to avoid inappropriate applications.

PLASTICS RECYCLING FUTURE

Reduction of the solid waste stream has become a national issue. The ability of many plastics to be easily recycled is now being recognized as one way to reduce solid waste economically. Material costs savings obtained by using recycled plastics will also be a driving force in accelerating the growth of the recycling industry. It is necessary that informational networking be increased to assist recyclers in locating business opportunities. The electronic database developed in this research address that need. The database contains the names of contacts in the plastics industry. This database will prove to be a valuable asset for future research work and understanding plastics recycling activities nationally.

ACKNOWLEDGEMENTS

Research funded by: The Plastics Recycling Foundation/Center for Plastics Recycling Research and Ohio Department of Natural Resources, Division of Litter Prevention & Recycling

PLASTICS DISPOSAL—
A NATIONAL ISSUE,
A TECHNICAL OPPORTUNITY

Douglas R. Barr
Mobay Corporation
Pittsbugh, Pennsylvania   USA

"Abstract"
The continuing growth of plastics and composites in manufactured products focuses attention on the ultimate disposal of these products at the end of their life-cycle. Plastics and composites are not amenable to current means of disposal, i.e. for wood and metal products. This is an increasing consumer and industrial problem that must be addressed today. The alternatives of landfill, material recycling, chemical recycling and waste-to-energy reclamation will be explored; stressing that the ultimate concern for the chemical industry continues to be "to produce quality materials for quality products while protecting our environment and remaining cost competitive". Disposal is a technical challenge to us all.

PLASTICS HAVE GIVEN THE AUTOMOTIVE INDUSTRY what they want and need: materials that are high strength, light weight, cost competitive, and non-corrosive.

Their combination of properties have fueled the desires and dreams of design engineers.

However, this ever increasing use of plastics and composites in automobiles presents chemical suppliers, molders, assemblers, and consumers with a critical opportunity. The opportunity - What to do with plastics after the completion of their initial use?

Although plastics in automobiles consume only about 5% of the annual plastic usage, the centralized disposal of cars and trucks, Federal and State legislation and our own concern as responsible manufactures and consumers urge us to address the issue of disposal today. The alternatives to this opportunity are simple: 1) landfill,

2) material and chemical recycling, and
3) waste-to-energy reclamation. As responsible manufacturers which do we pursue? The answer is not simple and, as do many of our decisions, must be based on economics.

The US is currently sitting on a national disaster of waste, trash, and garbage. New York City alone generates over 25,000 tons of municipal waste a day. Annually the US generates over 130,000 MM tons a year. Landfills are full, contaminated, and decreasing in numbers. Barges are loaded with garbage with nowhere to go. History is teaching us, the members of the plastics industry, a valuable lesson today: "You can't pick up waste or scrap if you don't have a place to put it down."

Although municipal garbage is not directly related to the automobile industry, it continues to use up our remaining landfill capacity, thereby undermining the economic alternatives available to the plastic industry today. We as a group, however, need to focus our attention on the 5% or approximately 2.5 billion pounds of plastic consumed in the automobile industry yearly. This will increase to over 3.0 billion pounds in the early 1990's. The issues of waste management and how it relates to the automotive industry are certainly difficult. Let us first state certain factors that surround this critical opportunity:

FACTS:
1)   There is 2.5 billion pounds of potentially usable material.
2)   Plastics are not compatible to traditional, municipal, and incineration methods.
3)   Plastics have a poor public image regarding disposal, reusability, and the environment.

4) Importation of automobiles increases the amount of North American automotive plastic waste.

5) Automotive manufacturers demand quality parts and materials that meet ever stringent consumer and federal requirements, new and improved technology and products, increase productivity, and reduction in overall costs.

6) Petroleum prices continue to be stable.

7) There exists a diversity of plastic materials for disposal.

8) Economics of plastic reuse versus landfill continues to be a business obstacle.

9) The recycling of aluminum, glass, paper and some plastic PET products indicates the potential success of such methods.

10) The need to establish a consistent consumer philosophy emphasizing the reduction of waste, a clean environment, and economic opportunity and incentives.

11) The need for Government support (not legislation) and industry unification on a major world-wide issue.

With this information in hand, let us explore the possibilities.

LANDFILLS:

There are two types of waste sites, hazardous and non-hazardous. Our conversation focuses on non-hazardous (municipal) sites, since materials in question are reacted and do not pose a toxic danger. All sites hazardous and municipal are decreasing in number. Costs are continuing to rise and their long term viability is questionable. Therefore, landfill as an option is not deemed feasible (long-term). However, landfills can not be eliminated. Whether we use material recycling, chemical recycling, or waste-to-energy reclamation residues will still be generated. We must reduce the volumes of "untreated" wastes going to our landfills, thereby conserving volume for wastes (or residues) generated from recycling, reclamation, or incineration. Giving economic considerations to residue from such options as recycling, reclamation or incineration could focus greater attention and corporate manpower

and finances to this end. The SPI has stated in it's Policy Statement on Municipal Solid Waste Disposal that the presence of plastics in a landfill is environmentally safe and is therefore viable if recycling and/or incineration are inappropriate or ineffective.

The issue of landfilling for plastics is an economic one. As the cost of landfill approaches $200/ton, segregation of plastic materials for reuse becomes more viable. "A Recycler" is always caught in a precarious situation. He has no one to pass his costs onto; therefore his very existence in such a market must be questioned. However, his importance should not be overlooked.

MATERIAL RECYCLING:

Much has been said about material recycling. It has merit, but is peculiar to applications, processing, and markets. Thermosets, unlike some thermoplastics, do not readily lend themselves to total material recycling. The automotive industry has the unique opportunity of the recycling of plastic material after the useful life of the car/truck is complete. Recycling of the plastic material at a molder seems viable if paint, attached hardware, and subsequent process heat degradation still produce a recyclable material meeting specifications. (Let us not forget the lesson the Japanese have taught us regarding the need for quality in materials as well as in automobiles). The real issue is the approximately 2.5 billion pounds of plastic waste that will end up in our junk yards. This waste is composed of many different types of plastics shown below:

AUTOMOTIVE PLASTICS:
1) ABS
2) Acrylic
3) Epoxy
4) Nylon
5) Phenolics
6) Polycarbonate
7) Polyester, Thermoplastics
8) Polyester Unsaturated
9) Polyethylene, HD, LD
10) Polypropylene
11) Polystyrene
12) Polyurethanes
13) Polyvinyl chloride and copolymers
14) Other Vinyls
15) Thermoplastic Elastomers
16) Others

It would certainly take a trained staff to separate parts by materials, then grind, package, and ship them to the appropriate source to be reused. In addition, the feasibility of such recycling as presented ignores such issues as; adhesion retention, contamination, age degradation of the different materials, and subsequent chemical costs necessary to purify such streams to produce quality materials and parts.

The PUR flexible foam market has economically used foam scraps to produce rebond for carpet underlayment. In fact, some scrap foam has been imported for such use. Volumes in excess of 175 MM/lbs annually of scrap foam (with ≈ 20-30MM/lbs of PU binder) are consumed each year. This is an example of a particular market with economic advantages. The PET recycle market (primarily from beverage bottles) is also a prime example. Of the 875 MM/lbs of virgin PET used each year, more than 150 MM/lbs is recovered and reused. This material, however, is reused for other than bottles, since federal legislation prohibits it's reuse for food grade containers. The success of the six major recyclers of PET rests with their ability to produce value added upgraded products made from recovered resin that sell for a price/lb well in excess of what raw recyclable PET can sell for. This is not unlike the above mentioned PUR rebond foam market, and of course such markets that must be established by any potential plastic recycler. Certainly, another problem surrounding the viability of material recycling.

Mr. Morrow, the Director of the Center for Plastic Recycling Research (CCPRR) at Rutgers University in New Jersey, states that the biggest challenges for plastic recyclers is the reuse of mixed plastic (i.e. from municipal streams). Today, little economic incentive exists for recyclers to process mixed plastics. A nickel or dime per pound does not justify the plastic waste being sorted, packaged, and purchased. Some investigations have centered around the use of such mixed plastics that evolved from our cars and trucks for use in construction applications where an extrusion of a mixed plastic is utilized to produced materials to replace wood and/or virgin ABS.

Comments from many experts in the field of recycling, again emphasize that the final question is economics. In some instances, they may exist in the future for automotive plastic parts. Any portion that can be recycled will of course reduce the load on our landfills and ultimately help our national problem.

CHEMICAL RECYCLING:

For thermosets, chemical recycling is technically achievable. Companies like Bayer AG, Texaco, Chardonol Corporation, Freeman Chemical, and Foam Systems hold patents professing the recycling of polyurethane (PUR) materials recovered from other plastic waste streams. To this end, Bayer/Mobay is investigating chemical means to reclaim processed PUR scrap for reuse in RIM molding operations. These processes are designed to be useable by the molders, thereby eliminating transportation costs, eliminating outsource processing costs, and ultimately reducing molding costs. In this manner, the molder can be insured of his recycle quality as well as final customer specifications. In general, more complicated chemical recycling involves three major chemical processes:
1) Pyrolysis - thermocracking of polymers into hydorcarbon fractions.
2) Hydro-cracking - thermoprocessing under extreme pressure in the presence of hydrogen to yield different and more functional hydrocarbon fraction.
3) Hydrolysis - processing of polycondesation polymers into starting or similar materials for reuse in polymer production.

PET bottle scrap is currently hydrolyzed creating products for reuse (approximately 20-30 MM/lbs annually) for later manufacture into rigid polyurethane foams.

Collectively these processes are commercially limited, due to the cost advantage of the petroleum based virgin material. However, future market situations may merit these processes more economically feasible.

WASTE-TO-ENERGY RECLAMATION:

Particularly for municipal wastes, waste-to-energy facilities are needed to meet waste disposal needs first and create energy second. These facilities are strategic to our economic and environmental survival. Therefore, additional costs needed to maintain or improve the environment must not be overlooked or ignored. The alternative of waste-to-energy through reclamation has been a world-wide focal point for decades of study and research. The focus has been to develop new recycling and incineration processes for hazardous and non-hazardous wastes as an alternative to landfill. This recovery technique offers many potential benefits: 1) development of long-term controllable and reliable means of meeting waste-disposal needs, 2) implementation of waste disposal systems reducing liability for landfilling hazardous waste, 3) reduction of landfill volume leaving capacity for residues from higher priority waste management options, and 4) supplementation of our domestic energy supply, therefore reducing usage and demand for petroleum energy sources and subsequently energy cost savings.

The incineration of polyurethanes and related waste materials has been prominent outside of North America:

1) Waste treatment facilities at VEB-Synthesis plant in Schwarzheide, East Germany (a major producer of PUR and agrochemicals) is incinerating waste efficiently meeting existing legal pollution limits.

2) Ebara Corporation, Tokyo, Japan has created a two bed pyrolysis system effectively gasifying municipal plastic waste into high calorie fuel gas (Stardust-'80 recovery system).

3) Dyckerhoff AG of West Germany utilizes plastic scrap as an energy source in the manufacture of cement; plastic generated energy provides ≈ 5% of the energy with no environmental problems.

4) VKE (in conjunction with SPI and leading European plastics suppliers, i.e. Bayer AG) in Munich, West Germany has established a pilot plant to investigate the potential of pyrolysis of mixed plastic waste. Results are expected to be published later this year.

Studies by SPI and CSI Resources Inc. shows incineration of PE, PS, PUR, and PVC to be environmentally acceptable, and they provide significant operating and energy savings.

Under the US Department of Energy, the Office of Industrial Development is looking to establish a waste-to-energy facility to burn collective automotive plastic wastes. They expect 100 M cars/yr to produce 25 MW of power.

These studies and the experiences from the facilities discussed above indicate that: 1) a continuous waste-to-energy process is possible, 2) plastic previously cumbersome in traditional municipal and incineration waste streams can be handled and produce usable fuel gas, 3) processes have been shown to be environmentally acceptable, and 4) the economic balance can be favorable.

CONCLUSIONS:

"We will bury you", was popular phrase voiced in the 1950's by the Russian's. They won't have to! Unless we, as suppliers, manufacturers and consumers address the opportunities concerning waste, we will bury ourselves.

Municipal waste management has inadequately addressed its problem and is now presented with a potential disaster. We in the plastic community have seen what has happened and must address our opportunity "today". The solution is not a simple one, but one that will involve recovering resources from waste in every form: material recycling, chemical recycling, and waste-to-energy reclamation. Our opportunities will help this pending world disaster we call waste. As plastic suppliers and users, we are not isolated from these global waste issues.

Your business, our business involves, waste and/or scrap, so we are in the middle of it. Solid waste disposal problems are real; they will not go away. We need to do everything economically feasible to create opportunities to use our scrap products. Our economic analysis of such opportunities must not be analyzed as other investment opportunities. These programs may save our very business.

We at Bayer/Mobay are committed to turn these problems into opportunities. Working with organizations such as the ASM International, SPI, Federal and State agencies and legislators, and members of the plastic community, we must focus on these opportunities. We, at Bayer/Mobay, are committed to supply our customers quality products. This quality must not be sacrificed by our activities to resolve our waste management opportunities and for this reason, waste-to-energy reclamation seems to be the most prudent approach.

Our livelihood and our environment rests on our decisions.

REFERENCES

1. Smoluk, George R., PET Reclaim Business
   Picks Up New Momentum, Modern Plastics,
   February 1988, 87-91

2. Boyd, Gordon M., Lessons of the Barge, On
   the Mark (an ESD publication).,
   vol. 1, no. 1, March 1988, 6-7

3. Ishii, Iskii, Hiroyama, Ito, Disposal
   of Municipal Refuse and RDF in Japan
   by a Two-Bid Pyrolysis System; National
   Institute for Environmental Studies,
   1987.

4. Basta, N., Plastics Recycling Grows Up,
   Chemical Engineering, 11/23/87, 22-27.

5. Leaversuch, R.D., Practicality is the
   Key in New Strategies for Recycling,
   Modern Plastics 8/87, 66-67.

6. Leaversuch, R.D., Industry Begins to
   Face Up to the Crisis of Recycling,
   Modern Plastics, 8/87, 44-47.

7. Volland, Dr. R.P., Disposal of Plastics
   - A Critical Opportunity,
   1988 SAE Presentation, 3/3/88.

THE RECYCLING OF COMPOSITE PLASTIC STRUCTURES

Brian Harper
Polymer Products
RR 3, Box 182
Iowa Falls, Iowa 50126   USA

Abstract
Plastic waste is seen as the major
problem in dealing with solid waste.
The development of processes and
machines to process mixed plastic waste
is described. Composite structures
used in packaging are processed as a
special case of mixed waste.
The type of product and the problems
encountered are described and the
effects of fibre reinforced resin
products, both thermoset and thermoplastic, are covered.
Potential developments in machine
systems are discussed with particular
reference to high output machine
systems.

THE PURPOSE OF THIS PAPER is to outline the current situation in relation
to the recycling of composite plastic
structures. In this paper the term
composite will usually apply to multilayer thermoplastic structures but the
processing of fibre reinforced plastics
will also be covered.

BACKGROUND - In dealing with solid
waste, although plastics represent about
8% by weight in volume terms, they are
about 30% of the stream. Consequently,
plastics are perceived to be the major
problem in the disposal of solid waste.
In landfills plastics do not biodegrade
and in incineration they can produce
air pollution problems and these facts
add to the perception of plastic as the
major problem in solid waste disposal.

Since the late sixties, development work has been carried out to devise
systems which would allow mixed plastic
waste, either industrial or post-
consumer, to be recycled into a useful,
beneficial form. The earliest patented
machine system was introduced by
Mitsibushi with their Reverzer. This
was a high output machine and was not a
commercial success. However, this
machine showed that mixed plastic waste
could be processed into useful objects.

The next machine and process was
patented in Europe. This machine was
of lower output than the Reverzer but
it has not been commercially exploited.
It is in production processing waste on
two sides in Europe and appears to be
commercially viable.

Development of mixed plastic waste
recycling in the United States started
late but the last machine and process
patented was in the U.S. and it is the
experience gained with this machine
which forms the basis for this paper.

NARRATIVE - In general terms the
original concept of mixed plastic waste
related to post-consumer waste, which
was chiefly polyolefinic. When the
first machines were produced, the systems
for the collection of such waste did not
exist and are not yet fully developed.
Consequently most mixed waste has been
obtained from industrial sources. At
this time the producers of industrial
waste are best organized to make their
waste available. The concept of mixed
polyolefinic waste still holds with the
largest producers being the plastic
packaging industry. The Reverzer was
conceived around the reprocessing of
polyolefines and polyvinylchloride
which were the common packaging plastics.
With such a feed-stock, large moldings
were produced with no problems with the

decomposition of the polyvinylchloride.

The next machine was conceived around the need to recycle polyolefine fibre waste and whilst it could handle other plastics, restrictions did exist on the proportions of other plastics which could be admixed.

When the process and machine for the American development were devised, the concept was to chiefly use polyolefines which were seen to be the majority resins. However, between concept and realization there was a major change in packaging concepts with the development of multilayer structures. These materials range from simple bilayer structures usually of plastics of widely divergent properties e.g. polyester and polyethylene to structures consisting of five layers organized to give optimum barrier properties. Current developments use seven layers with the middle layer being used to carry recycled waste from the system.

Our experience is chiefly with bilayer structures and we have processed laminates of polyester/polyethylene (50/50 by weight), cellulose acetate/ polyethylene (70/30 by weight). In the processing of these laminates, we do add polyolefines to assist in flow into the molds but we maintain a high proportion of polyester.

THE PROCESS - In the process the received waste, which can be in the form of edge-trim, sheets or reels, is comminuted to pass a 3/8" screen. These raw materials are then blended to control the properties of the finished product within the limits available using waste. From the mixer the feedstock passes into our patented machine where it is plasticized and transferred into molds which pass into a cooling tank. A novel feature is that molding takes place at relatively low pressure allowing inexpensive molds. After cooling, the moldings are removed from the molds and then allowed to cool in air until they reach room temperature. Once they are thermally stabilized, finishing operations, such as drilling, can be performed.

THE PRODUCTS - The usual product from a mixed plastic recycling operation is conceived as plastic lumber. As a market, this is very restrictive as processing costs ensure that the plastic lumber will be more expensive than real lumber. Our approach has been to consider the mixed waste as producing a material with specific properties and then to look for markets where these properties are advantageous. We have also looked for markets where the economies of the process in low mold costs give an advantage.

Products where the properties provide the market base are curb stops, speed bumps, and landscape ties. These are our proprietary products and we also produce custom moldings. Areas where inexpensive molds give an advantage are generally short run work where injection mold costs would be prohibitive. It is also possible to produce moldings with metal inserts.

THE PROBLEMS - In processing mixed plastic waste there are more problems than with virgin plastic. The first major problem is non-plastic contamination i.e. paper, cardboard, and metal. Whilst paper and cardboard can go through the process with little problem, tramp metal will cause damage to all parts of the equipment. Magnetic traps will remove some metal but non-magnetic metals and alloys can only be caught by vigilance by operators.

The second major problem is the variability which is associated with waste. No two batches are the same and considerable blending has to be done to ensure that any changes will be slow and machine conditions or the blend of plastics can be changed to compensate for the changes.

A third problem is the presence of thermally unstable polymers, such as polyvinylideneclichloride. These degrade during plasticizing, producing hydrogen chloride and leaving a char which does not mold.

These problems apply to all types of waste but if post-consumer waste is considered, then a multitude of problems can occur. Oil containers which sill contain oil, food packaging which still contain food, and containers which contain tramp metal are examples of the problems which can occur.

It is the context of the domestic waste stream that we have processed fibre reinforced plastic. Fibre-glass serving trays etc. have been processed by us as part of the stream. The fibre reinforced plastic, processed through comminution, flowed through the process and had no effect on the moldings. There is no reason to believe that significant quantities of fibre reinforced plastics, both thermoplastic and thermoset, could not be processed in admixture with conventional thermoplastics through machine systems which have been designed to process mixed plastic waste.

THE FUTURE - Current machines are of relatively low output in relation to the size of the problem. On average, U.S. consumption of plastic is 4 pounds/head/annum the majority of which goes to landfill eventually.

The problem in increasing machine output is removing the heat contained in the molding. At low outputs, the machine system for cooling can be fairly simple and contained in a reasonable machine size. Heat loss from the moldings is controlled by the heat conductivity of the plastic, and this is low.

Future machines will process large quantities of plastic and more complex mold handling systems will be needed to give adequate cooling time. It is also likely that machines will be dedicated to handling single source waste.